ENCYCLOPEDIA OF PHYSICS

CHIEF EDITOR

S. FLÜGGE

VOLUME XXV/2 a

LIGHT AND MATTER Ia

EDITOR

L. GENZEL

WITH 164 FIGURES

SPRINGER-VERLAG
BERLIN · HEIDELBERG · NEW YORK
1967

HANDBUCH DER PHYSIK

HERAUSGEGEBEN VON

S. FLÜGGE

BAND XXV/2a

LICHT UND MATERIE Ia

BANDHERAUSGEBER

L. GENZEL

MIT 164 FIGUREN

SPRINGER-VERLAG
BERLIN · HEIDELBERG · NEW YORK
1967

ISBN-13: 978-3-642-46076-0 e-ISBN-13: 978-3-642-46074-6
DOI: 10.1007/978-3-642-46074-6

Softcover reprint of the hardcover 1st edition 1967

Library of Congress Catalog Card Number A56-2942

Title-Nr. 5764

Contents.

Optical Constants and their Measurement.

By

E. E. Bell.

With 20 Figures.

A. Introduction.

1. The optical constants of a material are numbers which describe the manner in which a plane electromagnetic wave progresses through the material. The constants, a pair for every frequency, measure the speed and the attenuation of the wave. These constants have value not only for describing the wave progress but also for their intimate relation to the fundamental constitution of the material. The frequency dependence of the optical constants gives a large amount of information about the physical nature of the material. Because of the electric nature of the particles which make up the material, the electromagnetic wave is a natural handle with which to shake the material and thereby learn about its constitution; to find its resonance frequencies; to test its uniformity; etc.

This chapter will seek to explain the relations between the wave and the optical constants and to suggest useful methods of measuring these constants. The presentation is not intended to be an encyclopedia of all possible measuring techniques, but rather to be an introduction to experimental methods of general applicability to a wide range of materials. The ideas presented are limited to those directly applicable to isotropic materials and to methods which may be used in the infrared range of the spectrum, where the observations are made by a single detector measuring the power from a test apparatus. These methods may be in great contrast to the methods which depend upon the subjective observation by a human observer in order to recognize a pattern and to adjust an instrument. The chapter will not contain a discussion of the problems associated with anisotropic materials, nor will be the ideas necessary for an extension to X-rays, ultraviolet, or waves in a wave guide be indicated. The chapter will be especially devoted to problems pertinent to the infrared region of the spectrum and to measurements on solids, particularly. It will be evident that many of the methods of measurement and the descriptions of the phenomena will be general enough to be applicable in other regions of the spectrum.

One should expect at the outset that the method which is most useful in determining the optical constants of a particular material will be strongly dependent upon the accuracy with which the measurement needs to be made; upon the magnitude of the constants themselves; whether the material is transparent or opaque; and upon the specimen of the material available for measurement — its size, shape, and surface conditions. A number of methods for measuring optical constants are necessary in order to meet the various circumstances. It would be unusual, for example, to have a large, polished prism of a material that has not already been measured. One must frequently be content with a crude specimen and the challenge of obtaining as much information as possible. It is for the purpose of presenting an outline of elementary theory and practice in these situations that this chapter is devoted. It will be, of necessity, a discussion of idealized situations. The deviation in practice from the ideal must be the concern of each experimenter in each particular measurement if the optical constants derived from the measurements are to have their maximum utility. Compromises with the ideal are inevitable not only in the experimental measurements themselves, but also in the precision of the theory which is applied to the interpretation of the experimental data. A firm background knowledge of the nature of the wave, its relation to the optical constants, and the influence of the geometry of the sample is needed in order to evaluate the degree of compromise which can be tolerated in the measurements. For this reason the chapter includes a short introduction to the nature of the plane electromagnetic wave in isotropic materials and the dependence of the reflected and transmitted waves upon the sample's geometry and optical constants. This introductory material will also serve to define the symbols and conventions which will be used to describe the measuring techniques.

The sophistication of optical measurements in the infrared has been increasing with the development of better commercial spectrometers, better detectors, better gratings, more window materials, grid polarizers, and the increased use of interferometric techniques. The improved theoretical bases for understanding the interaction of the electromagnetic wave with solid material has increased the desire for experimental verification and the study of new effects. New devices — transistors, lasers, etc. — have made the measurements of more than academic use. The development of high speed computers has made Fourier transform spectroscopy practical and makes possible new techniques for optical constant measurements. The computer can be used to calculate the optical properties of a specific sample from the optical constants and can be programmed to do the inverse calculation in many situations. The experimenter will have to be wary that the computer is not fed poor data and expected to produce good results; to this end the experimenter must understand the limitations of the experimental techniques which he employs and the limitations of the computations which he uses. Because computers will be able to extract the two optical constants from almost any pair of independent measurements the chief burden will be the determination of possible inaccuracies in the experimental measurements and the influence of these inaccuracies upon the values of the optical constants derived.

B. Optical properties of an absorbing medium.

I. Maxwell's equations.

2. In this section the equations for a plane wave progressing through an absorbing medium will be developed from Maxwell's equations and the macroscopic constitutive relations. The general nature of the plane wave will be investigated without reference to the microscopic nature of the material through which it is travelling. The relationships between the microscopic parameters and the macroscopic parameters will be touched upon in a later section only to show that there are some necessary interrelationships between the optical constants and to typify certain general frequency relationships of the optical constants. The development here is helpful in establishing the symbolic notation for the discussion about measurements in later sections.

The problem of notation in the description of the optical problems is somewhat troublesome and there will be no attempt in this presentation to simplify the situation. The notation and the units which will be used those which were used by BORN and WOLF [1] in the book "Principles of Optics" because this book is well known and widely studied by workers in this field. The deviations from their notation will be few and will be obvious.

MAXWELL'S equations for the electromagnetic fields in a material may be expressed in the following form:

$$\operatorname{div} \boldsymbol{D} = 4\pi \varrho, \tag{2.1}$$

$$\operatorname{div} \boldsymbol{B} = 0, \tag{2.2}$$

$$\operatorname{curl} \boldsymbol{H} = c^{-1} \partial \boldsymbol{D}/\partial t + 4\pi c^{-1} \boldsymbol{j}, \tag{2.3}$$

$$\operatorname{curl} \boldsymbol{E} = -c^{-1} \partial \boldsymbol{B}/\partial t \tag{2.4}$$

in which c is a constant equal to the speed of light in vacuum and the symbols for the electromagnetic quantities have their universal significance. In these equations the units for $\boldsymbol{E}$, $\boldsymbol{D}$, and $\boldsymbol{j}$ and ϱ are in electrostatic cgs units and $\boldsymbol{B}$ and $\boldsymbol{H}$ are in electromagnetic cgs units. In a homogeneous, isotropic medium the constitutive relations are

$$\boldsymbol{D} = \varepsilon \boldsymbol{E}, \quad \boldsymbol{B} = \mu \boldsymbol{H}, \quad \text{and} \quad \boldsymbol{j} = \sigma \boldsymbol{E}.$$

The constant ε is the permittivity (or dielectric constant in these units), μ is the permeability, and σ is the conductivity. The constant μ will be presumed to be unity for all of the descriptions which follows but will be left in the literal form in the equations in part B only as an aid to those who will wish to change the formulae to other units. MAXWELL'S equations will prescribe the progress of an electromagnetic wave in the medium once the wave has been launched. The values of ε and σ will be constant only insofar as the frequency of the wave is a constant. The values are dependent upon the frequency, but in this section we will presume that they have the specific values required to keep the equations valid at the frequency of the wave proposed. It is through the measurements of the optical constants as a function of frequency that the dependence of the constitutive constants ε and σ will be determined, and thus the microscopic nature of the material will be probed.

II. Characteristics of plane electromagnetic waves.

3. We shall focus our attention on the electric field in the description of the electromagnetic wave because of its strong coupling with the charges in the material through which the wave passes. The magnetic field is not so effective in coupling with the material in usual situations. We shall study the plane wave travelling according to the representation

$$\hat{\boldsymbol{E}} = \hat{\boldsymbol{E}}_0 \exp i(\hat{\boldsymbol{K}} \cdot \boldsymbol{r} - \omega t), \tag{3.1}$$

where the bold letters indicate the vectorial nature of the quantity and the circumflex indicates that the quantity is represented by a complex value with real and imaginary components. It is only the real portion of the complex expression for the total wave which represents the wave. The imaginary portion is useful in manipulations which will serve to indicate the phase relationships between various real waves. The real and imaginary parts of some complex quantities will be indicated by primes and double primes; viz. $\hat{\boldsymbol{K}} = \boldsymbol{K}' + i\boldsymbol{K}''$. The ω in the equation is the angular frequency in radians per second, t is the time in seconds, and $\boldsymbol{r}$ is the vector displacement in space.

The quantity $\hat{\boldsymbol{K}}$ in the wave equation is called the complex propagation constant or wave vector. By writing the wave equation in the form

$$\hat{\boldsymbol{E}}=\hat{\boldsymbol{E}}_0\exp(-\boldsymbol{K}''\cdot\boldsymbol{r})\exp i(\boldsymbol{K}'\cdot\boldsymbol{r}-\omega t) \tag{3.2}$$

it is easy to see that the vector $\boldsymbol{K}'$ is normal to the surfaces of constant phase and that $\boldsymbol{K}''$ is normal to the surface of constant amplitude. $\boldsymbol{K}''$ measures the space rate of decay of the amplitude in cm^{-1}; and $\boldsymbol{K}'$ measures the spatial sinusoidal fluctuation rate in radians per centimeter. If $\boldsymbol{K}'$ and $\boldsymbol{K}''$ have the same direction in space, then the wave will be called homogeneous, otherwise the wave will be called inhomogeneous.

In the medium there will be an associated magnetic field wave represented by

$$\hat{\boldsymbol{H}}=\hat{\boldsymbol{H}}_0\exp i(\hat{\boldsymbol{K}}\cdot\boldsymbol{r}-\omega t) \tag{3.3}$$

with a corresponding interpretation for the symbols. It will be true that the constitutive relations will need to indicate the possible phase relations in the following manner: $\hat{\boldsymbol{j}}=\hat{\sigma}_1\hat{\boldsymbol{E}}$, $\hat{\boldsymbol{D}}=\hat{\varepsilon}_1\hat{\boldsymbol{E}}$, and $\hat{\boldsymbol{B}}=\mu\hat{\boldsymbol{H}}$. The complex values $\hat{\sigma}_1$ and $\hat{\varepsilon}_1$ are still scalars for the isotropic medium. The subscripts on $\hat{\sigma}_1$ and $\hat{\varepsilon}_1$, are to identify these quantities as belonging to this development. In part *C* the two quantities will be combined into a single complex dielectric constant $\hat{\varepsilon}$ used without the subscript.

Our concern will be only with the electric and magnetic fields of plane waves, therefore the differential operators in Maxwell's equations will operate only on plane wave fields and the equations may be recast in simpler algebraic form. The differential operators operating on the electric wave field give the following simple results:

$$\partial\hat{\boldsymbol{E}}/\partial t=-i\omega\hat{\boldsymbol{E}}, \tag{3.4}$$

$$\operatorname{div}\hat{\boldsymbol{E}}=i\hat{\boldsymbol{K}}\cdot\hat{\boldsymbol{E}}, \tag{3.5}$$

$$\operatorname{curl}\hat{\boldsymbol{E}}=i\hat{\boldsymbol{K}}\times\hat{\boldsymbol{E}}. \tag{3.6}$$

Corresponding to the usual development of the differential wave equation from Maxwell's equations, we find that

$$\left.\begin{aligned}\operatorname{curl}\operatorname{curl}\hat{\boldsymbol{E}}&=-\mu c^{-1}\operatorname{curl}\partial\hat{\boldsymbol{H}}/\partial t\\&=-\mu\hat{\varepsilon}_1 c^{-2}\partial^2\hat{\boldsymbol{E}}/\partial t^2-4\pi\mu\hat{\sigma}_1 c^{-2}\partial\hat{\boldsymbol{E}}/\partial t\end{aligned}\right\} \tag{3.7}$$

becomes

$$i\hat{\boldsymbol{K}}\times i(\hat{\boldsymbol{K}}\times\hat{\boldsymbol{E}})=\mu\hat{\varepsilon}_1\omega^2c^{-2}\hat{\boldsymbol{E}}+i4\pi\mu\hat{\sigma}_1\omega c^{-2}\hat{\boldsymbol{E}}. \tag{3.8}$$

This relation is

$$-(\hat{\boldsymbol{K}}\cdot\hat{\boldsymbol{E}})\hat{\boldsymbol{K}}+(\hat{\boldsymbol{K}}\cdot\hat{\boldsymbol{K}})\hat{\boldsymbol{E}}=\omega^2c^{-2}[\mu(\hat{\varepsilon}_1+i4\pi\hat{\sigma}_1\omega^{-1})]\hat{\boldsymbol{E}}. \tag{3.9}$$

For later convenience, we shall define $\hat{n}$ so that

$$\mu(\hat{\varepsilon}_1+i4\pi\hat{\sigma}_1\omega^{-1})=\hat{n}^2 \tag{3.10}$$

and $\hat{n}$ will be the usual complex refractive index for the medium.

By taking the scalar product of $\hat{\boldsymbol{K}}$ with the left and right member of Eq. (3.9) it is found that

$$(\omega^2c^{-2}\hat{n}^2)(\hat{\boldsymbol{K}}\cdot\hat{\boldsymbol{E}})=0 \tag{3.11}$$

so that either $\hat{n}=0$ or $\hat{\boldsymbol{K}}\cdot\hat{\boldsymbol{E}}=0$. Ordinarily $\hat{n}$ is not zero and it follows that $\hat{\boldsymbol{K}}\cdot\hat{\boldsymbol{E}}$ must be zero. This is the condition that the wave be transverse. The relation may be reinterpreted as div $\hat{\boldsymbol{E}}=0$, which corresponds to the fact that there is no free charge density associated with this plane wave, that is div $\hat{\boldsymbol{D}}=4\pi\hat{\varrho}=0$. It should be noted that $\hat{n}$ may be nearly zero in some situations so that the influence of the effects of oscillating charge distributions on the boundaries of the medium may produce a wave for which $\hat{\boldsymbol{K}}\cdot\hat{\boldsymbol{E}}$ is not zero. Such a wave would have a longitudinal component. In all that follows it will be assumed that the wave is transverse and that $\hat{\boldsymbol{K}}\cdot\hat{\boldsymbol{E}}=0$.

Associated with the electric field wave is a magnetic field wave which may be obtained from the electric field by MAXWELL's equations as

$$i\hat{\boldsymbol{K}}\times\hat{\boldsymbol{E}}=-\mu\, c^{-1}\partial\hat{\boldsymbol{H}}/\partial t. \tag{3.12}$$

For the $\boldsymbol{E}$ wave and the $\boldsymbol{H}$ wave to satisfy this relation at all times, the $\boldsymbol{H}$ wave must have the same exp $i(-\omega\, t)$ time dependence as the $\boldsymbol{E}$ wave. Also, then,

$$\hat{\boldsymbol{K}}\times\hat{\boldsymbol{E}}=\mu\,\omega\, c^{-1}\hat{\boldsymbol{H}} \tag{3.13}$$

and this can be true at all positions only if the $\boldsymbol{H}$ wave has the same exp $i(\hat{\boldsymbol{K}}\cdot\boldsymbol{r})$ space dependence as the $\boldsymbol{E}$ wave. The equation

$$\hat{\boldsymbol{H}}=\hat{\boldsymbol{H}}_0 \exp i(\hat{\boldsymbol{K}}\cdot\boldsymbol{r}-\omega\, t) \tag{3.14}$$

can represent the $\boldsymbol{H}$ wave associated with the $\boldsymbol{E}$ wave if

$$\hat{\boldsymbol{K}}\times\hat{\boldsymbol{E}}_0=\mu\,\omega\, c^{-1}\hat{\boldsymbol{H}}_0 \tag{3.15}$$

and this allows the $\hat{\boldsymbol{H}}_0$ to be calculated from $\hat{\boldsymbol{E}}_0$, $\hat{\boldsymbol{K}}$, and ω. The calculation of $\hat{\boldsymbol{E}}_0$ from $\hat{\boldsymbol{H}}_0, \hat{\boldsymbol{K}}$, and ω can be carried through with the relation, which is easily derived from the above,

$$\left.\begin{aligned}\hat{\boldsymbol{E}}_0&=-\mu\,\omega\, c^{-1}(\hat{\boldsymbol{K}}\times\hat{\boldsymbol{H}}_0)(\hat{\boldsymbol{K}}\cdot\hat{\boldsymbol{K}})^{-1}\\&=-\mu\, c\,\omega^{-1}\hat{n}^{-2}(\hat{\boldsymbol{K}}\times\hat{\boldsymbol{H}}_0).\end{aligned}\right\} \tag{3.16}$$

It should also be noted that, since $\hat{\boldsymbol{K}}\cdot\hat{\boldsymbol{K}}\times\hat{\boldsymbol{E}}=0$, it must follow that

$$\hat{\boldsymbol{K}}\cdot\hat{\boldsymbol{H}}=0 \tag{3.17}$$

which is the transversality condition for the magnetic field wave.

Because of the transversality condition $\hat{\boldsymbol{K}}\cdot\hat{\boldsymbol{E}}=0$, Eq. (3.9) shows that $\hat{\boldsymbol{K}}\cdot\hat{\boldsymbol{K}}=\omega^2\, c^{-2}\,\hat{n}^2$ and

$$\left.\begin{aligned}(c\,\omega^{-1}\hat{\boldsymbol{K}})\cdot(c\,\omega^{-1}\hat{\boldsymbol{K}})&=\hat{n}^2=(n+i\,k)^2\\&=\mu\,(\hat{\varepsilon}_1+i\,4\pi\,\hat{\sigma}_1\,\omega^{-1}).\end{aligned}\right\} \tag{3.18}$$

For $\hat{\varepsilon}_1$ and $\hat{\sigma}_1$ real and positive, n and k would be real and positive. The real part of the complex refractive index $\hat{n}$ will be called the index of refraction n, and the imaginary part will be called the extinction coefficient k.

The frequency dependence of the optical constants, n and k, follows immediately from any theoretical model of the medium which gives the frequency dependence of the constitutive parameters $\hat{\varepsilon}_1$ and $\hat{\sigma}_1$. Microscopic models of the media have been very fruitful in this respect and, conversely, have allowed the measurements of the optical constants to give measured values for the parameters in the microscopic model. H. A. LORENTZ [2] in his book "Theory of Electrons", first published in 1909, gathered a great many phenomena together with a unifying microscopic model of material media. The successful application of such models makes it possible to extend the knowledge gained from a few optical constant measurements, at a finite number of frequencies, for example, to other spectral regions and to other phenomena. A comprehensive history of electromagnetic theories and optics has been written by E. T. WHITTAKER [3] and should be consulted for further references.

III. Classification of wave types and modes.

4. The relation (3.18) shows that a knowledge of the nature of the medium, and thus the optical constants n and k, does not determine completely the complex wave vector $\hat{\boldsymbol{K}}$. There are many possible values of $\boldsymbol{K}'$ and $\boldsymbol{K}''$ which can satisfy the relation (3.18) for any particular medium and frequency. These various kinds of waves are determined by the processes through which the wave is launched into the medium. It is possible, for example, to launch homogeneous waves, $\boldsymbol{K}'$ parallel to $\boldsymbol{K}''$, or inhomogeneous waves, $\boldsymbol{K}'$ not parallel to $\boldsymbol{K}''$, in the same medium. In order to understand the optical constants and also to understand the various techniques for measuring the optical constants, it is necessary to be acquainted with the types of waves which may be propagated in the medium. These will be discussed in the following paragraphs.

a) Homogeneous and inhomogeneous waves.

5. Consider, for its simplicity, the situation in which $\hat{n}$ is real and positive — a nonabsorbing dielectric — and notice that the relation (3.18) with $\hat{\boldsymbol{K}}=\boldsymbol{K}'+i\boldsymbol{K}''$ takes the form

$$\boldsymbol{K}'\cdot\boldsymbol{K}'-\boldsymbol{K}''\cdot\boldsymbol{K}''+i\,2\boldsymbol{K}'\cdot\boldsymbol{K}''=\omega^2\,n^2\,c^{-2}. \tag{5.1}$$

It must be, therefore, since $\boldsymbol{K}''\cdot\boldsymbol{K}'=0$, that $\boldsymbol{K}''$ is perpendicular to $\boldsymbol{K}'$ or that $\boldsymbol{K}''$ is zero. Thus the inhomogeneous wave is possible in the nonabsorbing dielectric. The wave neighboring the external surface of a dielectric in the case of total internal reflection is such an inhomogeneous wave having $\boldsymbol{K}''$ perpendicular to the surface and $\boldsymbol{K}'$ parallel to the surface. This is a special situation, however, and in the bulk of a nonabsorbing material the wave must have $\boldsymbol{K}''=0$ and behave like a homogeneous wave.

Consider, next, the situation for an absorbing medium in which the physical angle between the directions of the $\boldsymbol{K}''$ and the $\boldsymbol{K}'$ wave vectors is Θ. In this situation

$$\left.\begin{aligned}\hat{\boldsymbol{K}}\cdot\hat{\boldsymbol{K}}&=K'^2-K''^2+i\,2K'K''\cos\Theta,\\&=\omega^2\,c^{-2}\left((n^2-k^2)+i\,2nk\right)\end{aligned}\right\} \tag{5.2}$$

where $K'=|\boldsymbol{K}'|$ and $K''=|\boldsymbol{K}''|$. From the known values of n, k, and Θ it is possible to find the values of K' and K''. These are:

$$K'=\omega\,c^{-1}\,2^{-\frac{1}{2}}\left[\left((n^2-k^2)^2+4n^2\,k^2/\cos^2\Theta\right)^{\frac{1}{2}}+(n^2-k^2)\right]^{\frac{1}{2}} \tag{5.3}$$

$$K''=\omega\,c^{-1}\,2^{-\frac{1}{2}}\left[\left((n^2-k^2)^2+4n^2\,k^2/\cos^2\Theta\right)^{\frac{1}{2}}-(n^2-k^2)\right]^{\frac{1}{2}}. \tag{5.4}$$

We note that K' is always greater than K''.

For a homogeneous wave, $\Theta=0$, $K''=\omega\,c^{-1}\,k$, and $K'=\omega\,c^{-1}\,n$. This simplicity makes for a ready interpretation of the optical constants in terms of the wave vectors and of the wave vectors in terms of the optical constants. The optical constants give an immediate picture of the homogeneous wave which could progress through the medium.

The inhomogeneous wave may have properties approximately like those of the homogeneous wave in certain limiting situations. If the inhomogeneous wave has an angle Θ between the wave vectors $\boldsymbol{K}''$ and $\boldsymbol{K}'$ which is small enough, or if the value of k is small enough so that n is very large compared to $k/\cos\Theta$, then it will be approximately true that $K''=\omega\, c^{-1} k/\cos\Theta$ and $K'=\omega\, c^{-1} n$. These are useful relations which hold in many situations for which the radiation is propagated through a material with so little attenuation that it is possible to measure the transmitted radiation.

In general in an inhomogeneous wave $\omega\, c^{-1}\hat{n} \neq K' + iK''$, therefore $\hat{K}$ will not be used as a symbol for $\omega\, c^{-1}\hat{n}$ as it would not follow the normal nomenclature for its real and imaginary parts, that is, $\hat{K} \neq K' + iK''$, but we keep $K' = |\boldsymbol{K}'|$ and $K'' = |\boldsymbol{K}''|$.

The characteristics of the wave motion in an absorbing isotropic medium are described by $\boldsymbol{K}'$ and $\boldsymbol{K}''$ and we shall say that two waves have the same "character" if they both have the same magnitudes of $\boldsymbol{K}'$ and $\boldsymbol{K}''$ and have the same angle between the directions of $\boldsymbol{K}'$ and $\boldsymbol{K}''$. Two waves of the same character may have different directions of propagation or may have different modes — transverse electric or transverse magnetic.

b) Transverse electric and transverse magnetic waves.

6. Two common inhomogeneous modes of plane wave propagation are the transverse electric, TE, and the transverse magnetic, TM, modes. The transverse electric mode has $\boldsymbol{E}'$ and $\boldsymbol{E}''$ parallel and both perpendicular to the plane of $\boldsymbol{K}'$ and $\boldsymbol{K}''$. The complex amplitude vector for such a wave can be written as $\boldsymbol{E}_0 \exp i\Phi_E$. By a shift of the time origin in the wave equation the real amplitude $\boldsymbol{E}_0$ will suffice for this TE wave. The associated magnetic wave has an amplitude vector given by Eq. (3.15), and therefore $\boldsymbol{H}'$ and $\boldsymbol{H}''$ are in the plane of $\boldsymbol{K}'$ and $\boldsymbol{K}''$ and are not parallel.

The transverse magnetic wave mode has $\boldsymbol{H}'$ and $\boldsymbol{H}''$ parallel and both perpendicular to the plane containing $\boldsymbol{K}'$ and $\boldsymbol{K}''$. The complex amplitude $\hat{\boldsymbol{H}}_0$ can be replaced by $\boldsymbol{H}_0 \exp i\Phi_H$, or with a shift in the time origin, by simply $\boldsymbol{H}_0$. The associated electric wave has the amplitude vector given by Eq. (3.16), and therefore $\boldsymbol{E}'$ and $\boldsymbol{E}''$ are in the plane of $\boldsymbol{K}'$ and $\boldsymbol{K}''$ and are not parallel.

Any inhomogeneous transverse plane wave can be represented by a combination of a TE and a TM wave. This resolution is helpful in the solution of complicated wave problems.

IV. Intensity of an electromagnetic wave.

7. The intensity of a plane wave is measured by the power flow through a unit area. This quantity is conveniently represented by the Poynting vector

$$\boldsymbol{S} = (4\pi)^{-1} c\,(\boldsymbol{E}\times\boldsymbol{H}) \tag{7.1}$$

which gives the instantaneous power flow per unit area both in magnitude and in direction. It is instructive to evaluate the Poynting vector for the plane electromagnetic waves in order to gain insight into some of the peculiarities of the inhomogeneous wave. By expressing the $\boldsymbol{E}$ and the $\boldsymbol{H}$ for the plane wave as the real parts of the $\hat{\boldsymbol{E}}$ and $\hat{\boldsymbol{H}}$, forming the vector product $\boldsymbol{E}\times\boldsymbol{H}$, and obtaining the time average for a full cycle, it is found that

$$\boldsymbol{S}_{\text{ave}} = (8\pi)^{-1} c\,(\boldsymbol{E}_0'\times\boldsymbol{H}_0' + \boldsymbol{E}_0''\times\boldsymbol{H}_0'')\exp(-2\boldsymbol{K}''\cdot\boldsymbol{r}). \tag{7.2}$$

It is evident that the intensity of the wave depends upon the position vector $\boldsymbol{r}$ and that this intensity decreases exponentially with $\boldsymbol{r}$ in the direction of the $\boldsymbol{K}''$

vector. This is consistent with the interpretation of the $\boldsymbol{K}''$ vector as being normal to the surfaces of constant amplitude. It is also evident that the direction of the flow of power, $\boldsymbol{S}_{\text{ave}}$, is given by the direction of $(\boldsymbol{E}_0' \times \boldsymbol{H}_0' + \boldsymbol{E}_0'' \times \boldsymbol{H}_0'')$.

For a transverse electric wave, TE mode, $\hat{\boldsymbol{E}}_0 = \boldsymbol{E}_0$ and

$$\boldsymbol{E}_0' \times \boldsymbol{H}_0' = \boldsymbol{E}_0 \times (\boldsymbol{K}' \times \boldsymbol{E}_0)\,\mu^{-1}\,\omega^{-1}\,c = E_0^2\,\boldsymbol{K}'\,\mu^{-1}\,\omega^{-1}\,c\,. \tag{7.3}$$

Thus

$$\boldsymbol{S}_{\text{ave}} = (8\pi\,\mu\,\omega)^{-1}\,c^2 E_0^2\,\boldsymbol{K}' \exp(-2\boldsymbol{K}''\cdot\boldsymbol{r}) \tag{7.4}$$

and the flow of power in the TE mode wave is in the direction of $\boldsymbol{K}'$, that is, the flow is normal to the surfaces of constant phase.

For a transverse magnetic wave, TM mode, the magnetic field amplitude $\hat{\boldsymbol{H}}_0$ is equal to $\boldsymbol{H}_0$ and

$$\hat{\boldsymbol{E}}_0 = -\mu\,\omega^{-1} c\,(n^2 + k^2)^{-2}\left[(n^2 - k^2) - i\,2n k\right](\hat{\boldsymbol{K}} \times \boldsymbol{H}_0)\,. \tag{7.5}$$

From this it can be found that the power flow in the TM wave is represented by

$$\boldsymbol{S}_{\text{ave}} = \frac{\mu\,c^2}{8\pi\,\omega}\left\{\frac{(n^2 - k^2)\,\boldsymbol{K}' + 2n k\,\boldsymbol{K}''}{(n^2 + k^2)^2}\right\} H_0^2 \exp(-2\boldsymbol{K}''\cdot\boldsymbol{r})\,. \tag{7.6}$$

In this TM wave, as in the TE wave, the intensity of the wave decreases exponentially in the $\boldsymbol{K}''$ direction. The power flow in the TM mode, however, is not directed entirely along the $\boldsymbol{K}'$ direction nor entirely along the $\boldsymbol{K}''$ direction. It flows at an angle to the $\boldsymbol{K}'$ direction, an angle which increases with the increase in the absorption of the material. In the usual weak absorber, for which n is much larger than $k/\cos\Theta$, Θ being the physical angle between $\boldsymbol{K}'$ and $\boldsymbol{K}''$, the $\boldsymbol{K}''$ component of power flow is much smaller than the $\boldsymbol{K}'$ component.

For homogeneous waves the distinction between TE and TM modes is lost. The power flow is in the direction of the $\boldsymbol{K}''$ and the $\boldsymbol{K}'$ vectors because they are parallel and the intensity decreases exponentially along this same direction. The intensity decays with distance $|\boldsymbol{r}|$ in the direction of propagation according to $\exp(-2\omega\,c^{-1}\,k|\boldsymbol{r}|)$ and so $2\omega\,c^{-1}k = \alpha$ is the absorption coefficient.

The electric and magnetic fields in any inhomogeneous transverse wave can be resolved into a TE and a TM mode. The power flow in such a resolved wave is not the power flow of the TE component wave plus the power flow of the TM component wave. Cross product terms between the fields of the two modes occur in the POYNTING vector in this case. One would expect, however, that weakly absorbing materials would still have the major power flow along the $\boldsymbol{K}'$ direction. It would still be true that the wave intensity would decrease exponentially with the distance in the $\boldsymbol{K}''$ direction, regardless of the complexity of the mode of the wave.

The interest in the POYNTING vector and the power flow in the medium is only to understand the influence of the optical constants on the wave mode and the transport of energy in the medium. In the measurement of the optical constants of a material the radiation detector usually will not be submerged in the material but will be outside in a standard medium. The detector will be used to measure reflectances or transmittances, that is, it will be used to measure power ratios. In some cases only the sample position or the sample geometry for a maximum or minimum power to the detector will be measured. Thus it will be generally sufficient to calculate the square of the amplitude of the field vector $\hat{\boldsymbol{E}}_0$ and to consider that the power is proportional to this square without concerning ourselves about the proportionally constant.

V. Reflection and transmission at a plane surface.

8. The power of a wave incident upon a surface between two materials with different optical properties will be divided between a reflected wave and a transmitted wave. The measurement of the optical properties of a solid material depends upon the measurement of the characteristics of these waves. Thus the most important relations of use in optical constant measurements are those which relate the reflectance and the transmittance of this boundary to the optical properties of the bounding media. More complicated situations which involve several surfaces can be analyzed by the use of these simpler one boundary relations together with the superposition principle which allows the total wave to be constructed from a series of partial waves. The single boundary will be analyzed first.

A particularly simple and useful relation gives the description of the wave field at one space position in terms of the wave field at another position in the same medium and the vector displacement between the positions. This relationship is

$$\hat{\boldsymbol{E}}\,(\text{at }\boldsymbol{r}_2)=\hat{\boldsymbol{E}}\,(\text{at }\boldsymbol{r}_1)\exp i\,\hat{\boldsymbol{K}}\cdot(\boldsymbol{r}_2-\boldsymbol{r}_1)=\hat{\boldsymbol{E}}\,(\text{at }\boldsymbol{r}_1)\hat{a}\,. \tag{8.1}$$

The complex quantity $\hat{a}$ indicates the manner in which the amplitude and phase of the wave at $\boldsymbol{r}_2$ are altered from the values at $\boldsymbol{r}_1$.

A set of conditions on the character of the waves reflected and transmitted at a boundary are given by the geometry of the situation. The amplitudes of the waves are determined by the electromagnetic boundary conditions. We will impose the electromagnetic conditions on the waves after the character of the waves has been determined from the geometrical considerations.

a) Character of the reflected and transmitted waves.

9. Let us consider a plane surface between two media with a unit vector $\boldsymbol{u}$ parallel to the surface and a coordinate frame with an origin in the surface. Let the incident wave be $\hat{\boldsymbol{E}}_1$, the transmitted wave be $\hat{\boldsymbol{E}}_2$, and the reflected wave be $\hat{\boldsymbol{E}}_3$, each in their respective media. These waves may be represented by the equations:

$$\hat{\boldsymbol{E}}_1=\hat{\boldsymbol{E}}_{01}\exp(-i\,\omega_1\,t)\exp(\hat{\boldsymbol{K}}_1\cdot\boldsymbol{r})\qquad\text{incident wave}, \tag{9.1}$$

$$\hat{\boldsymbol{E}}_2=\hat{\boldsymbol{E}}_{02}\exp(-i\,\omega_2\,t)\exp(\hat{\boldsymbol{K}}_2\cdot\boldsymbol{r})\qquad\text{transmitted wave}, \tag{9.2}$$

$$\hat{\boldsymbol{E}}_3=\hat{\boldsymbol{E}}_{03}\exp(-i\,\omega_3\,t)\exp(\hat{\boldsymbol{K}}_3\cdot\boldsymbol{r})\qquad\text{reflected wave}. \tag{9.3}$$

If these are the only waves in the media then it is necessary that the waves match properly at all positions $\boldsymbol{r}=r\boldsymbol{u}$ on the surface and at all times. These matching conditions are those that will assure that one complex number $\hat{t}$ will exist such that $\hat{\boldsymbol{E}}_2=\hat{t}\,\hat{\boldsymbol{E}}_1$ at every point on the surface and at all times and that also one complex number $\hat{r}$ should exist such that $\hat{\boldsymbol{E}}_3=\hat{r}\,\hat{\boldsymbol{E}}_1$ everywhere on the surface and at all times. These complex numbers $\hat{t}$ and $\hat{r}$ are the amplitude transmittance and reflectance coefficients for this incident wave. The waves can match only if they all have the same frequency so that the subscripts on the frequency may be omitted. The matching at all points on the boundary can be satisfied only if

$$\boldsymbol{K}_1'\cdot\boldsymbol{u}=\boldsymbol{K}_2'\cdot\boldsymbol{u}=\boldsymbol{K}_3'\cdot\boldsymbol{u}, \tag{9.4}$$

and if

$$\boldsymbol{K}_1''\cdot\boldsymbol{u}=\boldsymbol{K}_2''\cdot\boldsymbol{u}=\boldsymbol{K}_3''\cdot\boldsymbol{u} \tag{9.5}$$

for all directions $\boldsymbol{u}$ parallel to the surface. Thus it is seen that $\boldsymbol{K}_1'$, $\boldsymbol{K}_2'$, and $\boldsymbol{K}_3'$ are coplanar and also that $\boldsymbol{K}_1''$, $\boldsymbol{K}_2''$ and $\boldsymbol{K}_3''$ are coplanar. These matching conditions relate the directions and magnitudes of the wave vectors in the two media and correspond to the law of reflection — angle of incidence equals the angle of reflection — and to SNELL's law of refraction.

These matching relations will now be used to find the character of the waves which pass through a set of parallel lamellae of absorbing materials. The incident wave will be presumed to be a homogeneous wave propagating through a non-absorbing medium. This is to model the usual situation with a source at some distance from a parallel lamella of the material which is to be investigated

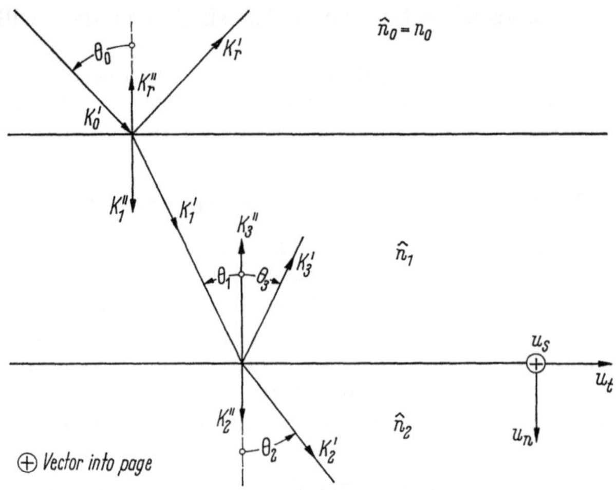

Fig. 1. The wave vectors in a system of three media with two parallel interfaces. These waves arise from an incident homogeneous wave with $\hat{\boldsymbol{K}}_0 = \boldsymbol{K}_0'$ in the non-absorbing dielectric, n_0. $\boldsymbol{K}_r''$ is shown to be zero in the text discussion. The unit vector $\boldsymbol{u}_s$ is parallel to the interface and directed into the page.

Fig. 1 illustrates the geometrical relationships. The plane of the figure is the plane of incidence containing $\boldsymbol{u}_n$ and $\boldsymbol{K}_1'$. The unit vectors are chosen so that $\boldsymbol{u}_t$ is parallel to each boundary surface and is in the plane of incidence, which is the plane containing the normal to the surface and the normal to the constant phase surface of the incident wave. The unit vector $\boldsymbol{u}_n$ is normal to each boundary surface and in the general direction of the progress of the transmitted power. The unit vector $\boldsymbol{u}_s$ is parallel to the boundary surface in the righthanded direction determined by $\boldsymbol{u}_t \times \boldsymbol{u}_n = \boldsymbol{u}_s$. The angles Θ_1, Θ_2, and Θ_3 are the usual angles of incidence, refraction, and reflection that would result with nonabsorbing media. The angle Θ_0 is the angle of incidence on the first boundary from a medium with $\hat{n} = n_0$. The total wave in each medium will be the sum of many partial waves arising from the reflections and transmissions through many surfaces. For the moment we wish to treat only the partial waves shown in the figure.

At the first surface the matching conditions are $\boldsymbol{K}_r'' \cdot \boldsymbol{u}_t = \boldsymbol{K}_1'' \cdot \boldsymbol{u}_t = 0$ since $\boldsymbol{K}_0'' = 0$. Thus $\boldsymbol{K}_1''$ and $\boldsymbol{K}_r''$ are normal to the surface or they are zero, and the following argument will show that $\boldsymbol{K}_r''$ is zero. In the incident medium, $\boldsymbol{K}_r''$ must be zero or it must be perpendicular to $\boldsymbol{K}_r'$ because the medium is nonabsorbing. Since $K_0'^2 = \omega^2 c^{-2} n_0^2 = K_r'^2 - K_r''^2$ from Eq. (5.2) it must be that $K_r'^2 \geqq K_0'^2$. However from $\boldsymbol{K}_0' \cdot \boldsymbol{u}_t = \boldsymbol{K}_r' \cdot \boldsymbol{u}_t$ we see that if $\boldsymbol{K}_r'$ were parallel to the surface then $K_r' < K_0'$. This can not be, so $K_r'' = 0$ and $K_r' = K_0'$, and the angle of reflection must be equal the angle of incidence.

The transmitted wave at the first boundary must obey the relation $\boldsymbol{K}_0' \cdot \boldsymbol{u}_t = \boldsymbol{K}_1' \cdot \boldsymbol{u}_t$ which is the law of refraction, $K_0' \sin \Theta_0 = K_1' \sin \Theta_1$. If this first layer can absorb radiation, then $\boldsymbol{K}_1''$ has a value and it must be perpendicular to the surface. The wave in the absorbing medium is an inhomogeneous wave and Θ_1 is the angle between $\boldsymbol{K}_1'$ and $\boldsymbol{K}_1''$. The relations (5.3) and (5.4) could be used to obtain the values of K_1' and K_1'' from n_1, k_1, and Θ_1 if it were desirable to do so. There are simpler formulations of the results, however, which will be shown later.

The condition at the second surface in Fig. 1 is complicated by the fact that the incident wave is inhomogeneous and the fact that we wish to find the reflected and transmitted waves in absorbing materials. The simplifying fact is that the $\boldsymbol{K}''$ vectors in all of the media are normal to the boundaries. From $\boldsymbol{K}_1'' \cdot \boldsymbol{u}_t = 0$ at the first surface we see that $\boldsymbol{K}_1'' \cdot \boldsymbol{u}_t = 0$ at the second surface of this parallel lamella. The matching relations then show that $\boldsymbol{K}_2'' \cdot \boldsymbol{u}_t = 0$ and $\boldsymbol{K}_3'' \cdot \boldsymbol{u}_t = 0$, so that $\boldsymbol{K}_2''$ and $\boldsymbol{K}_3''$ are normal to the boundary surface. This results in all $\boldsymbol{K}''$ vectors being normal to the boundaries throughout this laminated system. The matching conditions also give

$$K_1' \sin \Theta_1 = K_2' \sin \Theta_2 = K_3' \sin \Theta_3 \tag{9.6}$$

and these are equal to $K_0' \sin \Theta_0$. These relations are SNELL's law applied to the absorbing media and show that the law applies for the normal to the constant phase surfaces.

The geometry of the situation in Fig. 1 shows that

$$\hat{\boldsymbol{K}}_2 = (K_2' \sin \Theta_2)\, \boldsymbol{u}_t + (K_2' \cos \Theta_2 + i K_2'')\, \boldsymbol{u}_n. \tag{9.7}$$

This can be written as

$$\hat{\boldsymbol{K}}_2 = \omega\, c^{-1} \hat{n}_2 [\boldsymbol{u}_t \sin \hat{\Theta}_2 + \boldsymbol{u}_n \cos \hat{\Theta}_2] \tag{9.8}$$

in which the complex angle is defined to make

$$\omega\, c^{-1} \hat{n}_2 \sin \hat{\Theta}_2 = K_2' \sin \Theta_2 \tag{9.9}$$

and

$$\omega\, c^{-1} \hat{n}_2 \cos \hat{\Theta}_2 = K_2' \cos \Theta_2 + i\, K_2'' \tag{9.10}$$

which together with

$$\sin^2 \hat{\Theta}_2 + \cos^2 \hat{\Theta}_2 = 1 \tag{9.11}$$

gives

$$\omega\, c^{-1} \hat{n}_2 \cos \hat{\Theta}_2 = [\omega^2 c^{-2} \hat{n}_2^2 - K_2'^2 \sin^2 \Theta_2]^{\frac{1}{2}}. \tag{9.12}$$

The relations

$$\omega\, c^{-1} \hat{n}_2 \cos \hat{\Theta}_2 = K_2' \cos \Theta_2 + i\, K_2'' = [\omega^2 c^{-2} \hat{n}^2 - K_0'^2 \sin^2 \Theta_0]^{\frac{1}{2}}, \tag{9.13}$$

$$\omega\, c^{-1} \hat{n}_2 \sin \hat{\Theta}_2 = K_0' \sin \Theta_0 = \omega\, c^{-1} n_0 \sin \Theta_0 \tag{9.14}$$

then serve as a basis for the calculation of the transmitted wave vectors in the third medium from the characteristics of the wave in the second or the first medium.

For the reflected wave a similar set of relations hold. These are

$$\hat{\boldsymbol{K}}_3 = (K_3' \sin \Theta_3)\, \boldsymbol{u}_t + (K_3' \cos \Theta_3 + i\, K_3'')\, (-\boldsymbol{u}_n) \tag{9.15}$$

and

$$\hat{\boldsymbol{K}}_3 = \omega\, c^{-1} \hat{n}_1 [\boldsymbol{u}_t \sin \hat{\Theta}_3 - \boldsymbol{u}_n \cos \hat{\Theta}_3] \tag{9.16}$$

where the negative sign on the $\boldsymbol{u}_n \cos\hat{\Theta}_3$ term is used to keep the complex angle $\hat{\Theta}_3$ easily reducible to the ordinary angle of reflection. From these, the relations

$$\omega\, c^{-1} \hat{n}_1 \sin\hat{\Theta}_3 = K_3' \sin\Theta_3 = \omega\, c^{-1} n_0 \sin\Theta_0 = K_1' \sin\Theta_1, \tag{9.17}$$

$$\left.\begin{aligned}\omega\, c^{-1} \hat{n}_1 \cos\hat{\Theta}_3 &= K_3' \cos\Theta_3 + i\, K_3'' \\ &= \omega\, c^{-1} [\hat{n}_1^2 - n_0^2 \sin^2\Theta_0]^{\frac{1}{2}} = K_1' \cos\Theta_1 + i\, K_1''\end{aligned}\right\} \tag{9.18}$$

follow. Thus it is seen that $K_3'' = K_1''$, and, from (5.3) and (5.4), there must be only one value of K' and Θ in this medium for this value of K''; therefore $K_3' = K_1'$, $\Theta_3 = \Theta_1$, and $\hat{\Theta}_3 = \hat{\Theta}_1$. The reflected wave is just like the incident wave except for the direction of propagation and, as will be seen later, the intensity of its TE and TM component modes.

For this parallel layered system with a homogeneous source wave indicent upon the system in a medium with a real index of refraction, it is seen that the partial waves in all of the layers behave according to the relations

$$\hat{\boldsymbol{K}} = (K' \sin\Theta)\, \boldsymbol{u}_t + (K' \cos\Theta + i\, K'')\, (\pm \boldsymbol{u}_n), \tag{9.19}$$

$$\hat{\boldsymbol{K}} = \omega\, c^{-1} \hat{n}\, (\boldsymbol{u}_t \sin\hat{\Theta} \pm \boldsymbol{u}_n \cos\hat{\Theta}), \tag{9.20}$$

$$\hat{n} \sin\hat{\Theta} = c\, \omega^{-1} K' \sin\Theta = n_0 \sin\Theta_0, \tag{9.21}$$

$$\hat{n} \cos\hat{\Theta} = c\, \omega^{-1} (K' \cos\Theta + i\, K'') = (\hat{n}^2 - n_0^2 \sin^2\Theta_0)^{\frac{1}{2}}, \tag{9.22}$$

The upper and lower signs on $\boldsymbol{u}_n$ are determined by the direction of the wave expressed. These relations completely determine the character and direction of all of the waves in this parallel laminated system. The use of the complex angles in these particular formulas will not apply if the $\boldsymbol{K}''$ vectors are not normal to the boundary surfaces. The special angle $\Theta_0 = 0$ makes $\hat{\Theta} = 0$ for every layer.

b) Amplitude of the reflected and transmitted waves.

10. The amplitudes of these transmitted and reflected waves will be determined by the electromagnetic boundary conditions on the tangential components of the electric and magnetic fields. The total wave in each layer will be the sum of all of the partial waves in that layer. These partial waves arise from the many reflections in the system. The total wave in any layer will consist of a wave progressing in the $+\boldsymbol{u}_n$ direction and a wave progressing in the $-\boldsymbol{u}_n$ direction. Each of the total waves will have the character of any of the partial waves from which it is combined because all of the partial wave components will have the same character according to the relations (9.19) to (9.22).

The total wave in any layer can be found by adjusting the amplitude of each of the total waves in the system so that the electromagnetic boundary conditions are fulfilled simultaneously at all of the boundaries. It is also possible to find all of the partial waves resulting from the multitude of internal reflections and transmissions with each set of incident, reflected, and transmitted waves separately satisfying the electromagnetic boundary conditions; the sum of the partial waves will be the total wave. The partial wave analysis is particularly simple for the systems of interest here, and it will be used in the following analyses. The partial wave technique has the advantage of suggesting certain simplifications and approximations from a physical rather than a mathematical point of view. The advantage of the physical picture is important to the experimenter who must devise the approximate measurement method for the particular sample which is to be investigated.

The more complicated situation which arises when an inhomogeneous wave is incident upon a surface with $\boldsymbol{K}''$ not perpendicular to the surface will not be considered here. It is evident that waves in weakly absorbing materials will have an energy transport mainly along the $\boldsymbol{K}'$ direction and that, in many situations, it will be reasonable to calculate the absorption as though it produced an exponentially decreasing intensity along the direction of the energy transportation. Thus the directions of the $\boldsymbol{K}'$ vectors may be found by assuming absorption to be zero and then introducing the absorption as though it does not alter the direction of propagation.

The character and direction of the waves reflected and transmitted at a plane boundary were determined by the condition (9.4) and (9.5)

$$\hat{\boldsymbol{K}}_1 \cdot \boldsymbol{u} = \hat{\boldsymbol{K}}_2 \cdot \boldsymbol{u} = \hat{\boldsymbol{K}}_3 \cdot \boldsymbol{u} \tag{10.1}$$

and the values of the complex refractive indices $\hat{n}_1$ and $\hat{n}_2$. The amplitudes of these waves are determined by the two electromagnetic boundary conditions that there is no free charge on the boundary and no surface current from the motion of free charges. These boundary conditions give rise to the conditions: the tangential component of the electric field $\boldsymbol{E}$ must be the same on both sides of the boundary; and the tangential component of the magnetic field $\boldsymbol{H}$ must be the same on both sides of the boundary. In order to see how these conditions determine

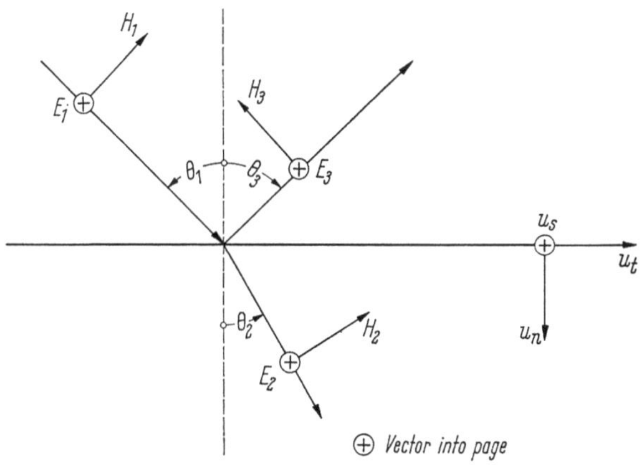

Fig. 2. The positive direction convention for the electric and magnetic field components of the incident, transmitted, and reflected waves at the interface between two media. The waves are s-polarized.

the wave amplitudes, consider the situation shown in Fig. 2. The incident wave has its electric field directed normal to the plane of incidence, that is $\boldsymbol{E}_1 = E_1 \boldsymbol{u}_s$. A wave with this plane of polarization will be called s-polarized. The reflection and transmitted waves will also be s-polarized, TE, electric waves with their positive values in the $+\boldsymbol{u}_s$ direction. The boundary condition on the electric field gives

$$\hat{\boldsymbol{E}}_1 + \hat{\boldsymbol{E}}_3 = \hat{\boldsymbol{E}}_2, \tag{10.2}$$

and the magnetic field boundary condition gives

$$\hat{\boldsymbol{H}}_1 \cdot \boldsymbol{u}_t + \hat{\boldsymbol{H}}_3 \cdot \boldsymbol{u}_t = \hat{\boldsymbol{H}}_2 \cdot \boldsymbol{u}_t. \tag{10.3}$$

From (3.13) and (9.19) we find

$$\left.\begin{aligned} \boldsymbol{u}_t\cdot\hat{\boldsymbol{H}} &= \mu^{-1}\,\omega^{-1}\,c\,\boldsymbol{u}_t\cdot\hat{\boldsymbol{K}}\times\hat{\boldsymbol{E}} = \mu^{-1}\,\hat{n}\,[\boldsymbol{u}_t\times(\boldsymbol{u}_t\sin\hat{\Theta}\pm\boldsymbol{u}_n\cos\hat{\Theta})]\cdot\hat{\boldsymbol{E}} \\ &= \mu^{-1}\,\hat{n}\,\boldsymbol{u}_s\cdot\hat{\boldsymbol{E}}\,(\pm\cos\hat{\Theta}) = \mu^{-1}\,\hat{n}\,\hat{E}\,(\pm\cos\hat{\Theta}) \end{aligned}\right\} \tag{10.4}$$

with the minus sign for the reflected wave. From Eq. (10.3) and (10.4) it is found that

$$\hat{E}_1 - \hat{E}_3 = \hat{\chi}_s\,\hat{E}_2 \tag{10.5}$$

with the characteristic parameter $\hat{\chi}_s$ defined by

$$\hat{\chi}_s = (\hat{n}_2\cos\hat{\Theta}_2)\,(\hat{n}_1\cos\hat{\Theta}_1)^{-1} \tag{10.6}$$

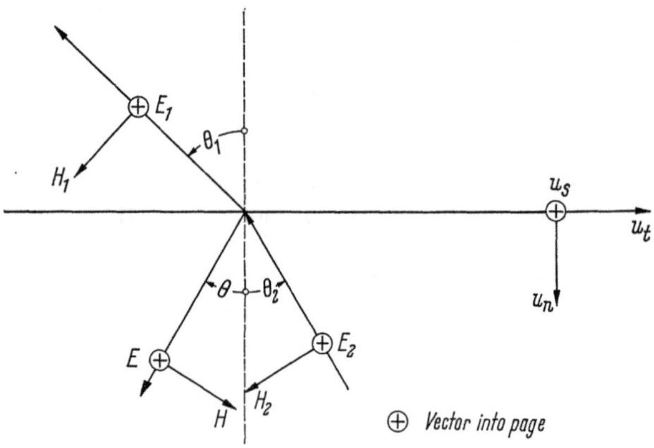

Fig. 3. The positive direction convention for the situation arising from the reversal of the direction of the transmitted wave of Fig. 2.

the relations (10.5) and (10.2) give the ratio $\hat{E}_3/\hat{E}_1$ for the amplitude coefficient of reflectance as

$$\hat{r}_s = (1-\hat{\chi}_s)\,(1+\hat{\chi}_s)^{-1} \tag{10.7}$$

and the corresponding amplitude coefficient of transmittance $\hat{E}_2/\hat{E}_1$ as

$$\hat{t}_s = 2\,(1+\hat{\chi}_s)^{-1} \tag{10.8}$$

for a TE wave polarized with the E-vector perpendicular to the plane of incidence. The subscript s on the quantities is to indicate this "senkrecht" polarization. The well known relations for real indices of refraction and real angles are seen to be special cases of the above relations.

A second important situation is indicated in Fig. 3. In this situation the transmitted wave of the previous situation is considered to be impinging as an incident wave on the same boundary as though it were reversed in its direction of propagation. The angle of incidence of this wave is the same as the angle of refraction of the transmitted wave in the previous paragraph. The same kind of analysis as given in the previous paragraph applied to this situation gives an amplitude coefficient of reflectance

$$\hat{r}'_s = -(1-\hat{\chi}_s)\,(1+\hat{\chi}_s)^{-1} = -\hat{r}_s \tag{10.9}$$

in which the prime indicates the reversed direction. The amplitude coefficient of transmittance is

$$\hat{t}'_s = 2\hat{\chi}_s\,(1+\hat{\chi}_s)^{-1} = \hat{\chi}_s\,\hat{t}_s\,. \tag{10.10}$$

Another useful form of the relation is

$$\hat{t}_s \hat{t}'_s - \hat{r}_s \hat{r}'_s = 1. \qquad (10.11)$$

These relations may also be derived by choosing a new $\hat{\chi}_s$ which is the reciprocal of the $\hat{\chi}_s$ given in (10.6) and using the relations given previously. This is equivalent to exchanging the subscripts No. 1 and 2 in the equations. It is important to note that (10.9) and (10.10) are the complex form of the STOKES' relations and that they will be used frequently in the application of the partial wave technique to multilayer problems.

The quantity $\hat{\chi}_s$ characterizes the boundary and may be written as

$$\hat{\chi}_s = (\hat{n}_2 \cos \hat{\Theta}_2)\,(\hat{n}_1 \cos \hat{\Theta}_1)^{-1} = \hat{n}_{2s}/\hat{n}_{1s} \qquad (10.12)$$

where $\hat{n}_{2s}$ and $\hat{n}_{1s}$ represent the "effective" indices of refraction of the media or the TE wave polarized perpendicular to the plane of incidence.

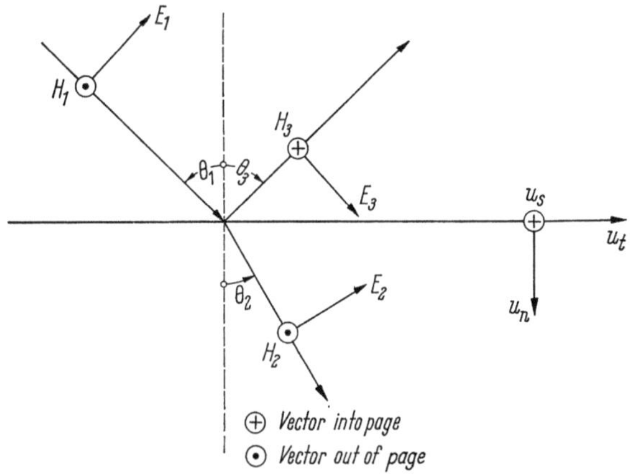

Fig. 4. The positive sign convention for the electric and magnetic field components of p-polarized waves. At normal incidence, the $\boldsymbol{E}$ vectors will all have the same positive sense and will correspond to the convention for s-polarization in this situation.

A similar analysis for the TM wave with the $\boldsymbol{E}$ vector components in the plane of incidence and therefore with the $\boldsymbol{H}$ perpendicular to the plane of incidence gives the following results:

$$\hat{r}_p = (1 - \hat{\chi}_p)\,(1 + \hat{\chi}_p)^{-1}, \qquad (10.13)$$

$$\hat{t}_p = 2\,(1 + \hat{\chi}_p)^{-1} (\cos \hat{\Theta}_1)\,(\cos \hat{\Theta}_2)^{-1}, \qquad (10.14)$$

$$\hat{\chi}_p = (\hat{n}_2/\cos \hat{\Theta}_2)\,(\hat{n}_1/\cos \hat{\Theta}_1)^{-1} = \hat{n}_{2p}/\hat{n}_{1p}, \qquad (10.15)$$

$$\hat{r}'_p = -\hat{r}_p, \qquad (10.16)$$

$$\hat{t}'_p = \hat{\chi}_s \hat{t}_p \qquad (10.17)$$

$$\hat{t}_p \hat{t}'_p - \hat{r}_p \hat{r}'_p = 1. \qquad (10.18)$$

The geometric orientation of the positively directed fields for this p-polarized wave is shown in Fig. 4. These directions were chosen so that the relations (10.13) to (10.15) would be identical to those for s-polarization at normal incidence. Because normal incidence and unpolarized radiation is frequently chosen in experiments and in calculations, it is convenient to have this situation represented by a single set of relations and conventions that is the limiting case for either polarization. This is not the convention used by BORN and WOLF [1] and many others. The results for the TM wave are similar to those for the TE wave except that the effective indices $\hat{n}_{2p}$ and $\hat{n}_{1p}$, are different than for the TE wave. The subscript p indicates that the $\boldsymbol{E}$ vector is parallel to the plane of incidence.

It is useful in optical constant measurements to be able to transform from values of $\hat{r}$ to values of $\hat{\chi}$. The relation reciprocal to (10.7) and (10.13) with suppression of the polarization designation, is

$$\hat{\chi}=(1-\hat{r})(1+\hat{r})^{-1}. \tag{10.19}$$

From this it is readily found that

$$\left.\begin{aligned}\chi'&=(1-r^2)(1+r^2+2r\cos\Phi_r)^{-1},\\ \chi''&=(-2r\sin\Phi_r)(1+r^2+2r\cos\Phi_r)^{-1}\end{aligned}\right\} \tag{10.20}$$

with $\hat{r}=r\exp i\,\Phi_r$ defining the phase angle Φ_r.

The relations for $\hat{r}_s$, $\hat{r}_p$, $\hat{t}_s$, and $\hat{t}_p$ in terms of the complex refractive indices and the angles of incidence and refraction are commonly called FRESNEL'S relations. Since these are frequently the starting point for further developments it is well to have them in a convenient form. The following forms for these relations are often used:

$$\hat{r}_s=\frac{\hat{n}_1\cos\hat{\Theta}_1-\hat{n}_2\cos\hat{\Theta}_2}{\hat{n}_1\cos\hat{\Theta}_1+\hat{n}_2\cos\hat{\Theta}_2}=\frac{\sin\hat{\Theta}_2\cos\hat{\Theta}_1-\sin\hat{\Theta}_1\cos\hat{\Theta}_2}{\sin\hat{\Theta}_2\cos\hat{\Theta}_1+\sin\hat{\Theta}_1\cos\hat{\Theta}_2}=\frac{\sin(\hat{\Theta}_2-\hat{\Theta}_1)}{\sin(\hat{\Theta}_2+\hat{\Theta}_1)}, \tag{10.21}$$

$$\hat{r}_p=\frac{\hat{n}_1\cos\hat{\Theta}_2-\hat{n}_2\cos\hat{\Theta}_1}{\hat{n}_1\cos\hat{\Theta}_2+\hat{n}_2\cos\hat{\Theta}_1}=\frac{\sin\hat{\Theta}_2\cos\hat{\Theta}_2-\sin\hat{\Theta}_1\cos\hat{\Theta}_1}{\sin\hat{\Theta}_2\cos\hat{\Theta}_2+\sin\hat{\Theta}_1\cos\hat{\Theta}_1}=\frac{\tan(\hat{\Theta}_2-\hat{\Theta}_1)}{\tan(\hat{\Theta}_2+\hat{\Theta}_1)}, \tag{10.22}$$

$$\hat{t}_s=\frac{2\hat{n}_1\cos\hat{\Theta}_1}{\hat{n}_1\cos\hat{\Theta}_1+\hat{n}_2\cos\hat{\Theta}_2}=\frac{2\sin\hat{\Theta}_2\cos\hat{\Theta}_1}{\sin\hat{\Theta}_2\cos\hat{\Theta}_1+\sin\hat{\Theta}_1\cos\hat{\Theta}_2}=\frac{2\sin\hat{\Theta}_2\cos\hat{\Theta}_1}{\sin(\hat{\Theta}_2+\hat{\Theta}_1)}, \tag{10.23}$$

$$\left.\begin{aligned}\hat{t}_p&=\frac{2\hat{n}_1\cos\hat{\Theta}_1}{\hat{n}_1\cos\hat{\Theta}_2+\hat{n}_2\cos\hat{\Theta}_1}=\frac{2\sin\hat{\Theta}_2\cos\hat{\Theta}_1}{\sin\hat{\Theta}_2\cos\hat{\Theta}_2+\sin\hat{\Theta}_1\cos\hat{\Theta}_1}\\ &=\frac{2\sin\hat{\Theta}_2\cos\hat{\Theta}_1}{\sin(\hat{\Theta}_2+\hat{\Theta}_1)\cos(\hat{\Theta}_2-\hat{\Theta}_1)}\end{aligned}\right\} \tag{10.24}$$

The ratio between the two polarized components are also useful; these are

$$\frac{\hat{r}_p}{\hat{r}_s}=\frac{\hat{n}_2\cos\hat{\Theta}_2\cos\hat{\Theta}_1+\hat{n}_1\sin^2\hat{\Theta}_1}{\hat{n}_2\cos\hat{\Theta}_2\cos\hat{\Theta}_1-\hat{n}_1\sin^2\hat{\Theta}_1}=\frac{\cos\hat{\Theta}_2\cos\hat{\Theta}_1+\sin\hat{\Theta}_2\sin\hat{\Theta}_1}{\cos\hat{\Theta}_2\cos\hat{\Theta}_1-\sin\hat{\Theta}_2\sin\hat{\Theta}_1}=\frac{\cos(\hat{\Theta}_2+\hat{\Theta}_1)}{\cos(\hat{\Theta}_2-\hat{\Theta}_1)}, \tag{10.25}$$

$$\frac{\hat{t}_p}{\hat{t}_s}=\frac{\hat{n}_1\cos\hat{\Theta}_1+\hat{n}_2\cos\hat{\Theta}_2}{\hat{n}_1\cos\hat{\Theta}_2+\hat{n}_2\cos\hat{\Theta}_1}=\frac{1}{\cos\hat{\Theta}_2\cos\hat{\Theta}_1+\sin\hat{\Theta}_2\sin\hat{\Theta}_1}=\frac{1}{\cos(\hat{\Theta}_2-\hat{\Theta}_1)}. \tag{10.26}$$

From FRESNEL'S relations in complex form it is easy to find the corresponding special form for non-absorbing dielectrics and to find the forms for the power reflectances: $\varrho_s=r_s^2$, $\varrho_p=r_p^2$. For such materials the angles reduce to the usual real angles of incidence and refraction.

The simple relation between the power reflectance and the E vector amplitude reflectance is the result of the fact that the incident and the reflected waves have the same character and are in the same medium. The relations $\varrho_s=r_s^2$ and $\varrho_p=r_p^2$ are true even for absorbing materials and would be the ratio of the reflected to the incident intensitics. These would also be the ratio of the reflected power per unit of interface area to the incident power per unit interface area. The power transmittance is not so simply related to the E vector amplitude reflectance, however, because the direction, the character, and the medium for the incident and transmitted powers are different. The ratio of the transmitted power per unit interface area to the incident power per unit interface area can be obtained in simple compact forms. From (10.7), (10.8), (10.9) and (10.10) we find that the power transmittances, in this special sense, are the following:

$$1-r_s^2=\tfrac{1}{2}(\hat{\chi}_s+\hat{\chi}_s^*)\,t_s^2, \tag{10.27}$$

$$1-r_p^2=\tfrac{1}{2}\left|\cos\hat{\Theta}_2/\cos\hat{\Theta}_1\right|^2(\hat{\chi}_p+\hat{\chi}_p^*)\,t_p^2 \tag{10.28}$$

with the superscript * indicating the complex conjugate quantitity. These are not the ratios of the intensities of the transmitted and incident powers.

c) Special angles for reflection.

11. There are several angles of incidence for which the reflected radiation has special characteristics that are useful for optical constant measurements. These are: the critical angle for total internal reflection; BREWSTER's angle for complete polarization; and the characteristic angles for producing circular polarization of the reflected radiation.

12. Critical angle. It is evident from Fig. 2 and SNELL's law that the angle of refraction for radiation incident upon a dielectric at high angles of incidence can not exceed the value arcsin (n_1/n_2) that occurs for an angle of incidence of $\pi/2$. For angles of refraction exceeding this, the reversed wave would have a reflectance r', see Fig. 3, that has a magnitude unity as there can be no corresponding transmitted wave. Thus the radiation in a high index material incident upon the interface with a low index material will be totally reflected if the angle of incidence exceeds a critical value.

The magnitude of the reflectance for radiation from a material with a complex refractive index $\hat{n}_1$ onto a material with a complex refractive index $\hat{n}_2$ is determined by the characteristic parameter $\hat{\chi}$ given in Eqs. (10.7) and (10.13). For $\hat{\chi}$ a pure imaginary number, $\hat{r}=(1-\hat{\chi})(1+\hat{\chi})^{-1}$ will have unity magnitude since $1-\hat{\chi}$ and $1+\hat{\chi}$ will both have the same magnitude. From (10.6) and (10.12), together with (9.22), etc., it is found that the characteristic parameters of s- and p-polarizations are

$$\hat{\chi}_s=(\hat{n}_2^2-n_1^2\sin^2\Theta_1)^{\frac{1}{2}}(n_1\sin\Theta_1)^{-1}, \tag{12.1}$$

$$\hat{\chi}_p=(\hat{n}_2^2-n_1^2\sin^2\Theta_1)^{-\frac{1}{2}}\,n_1^{-1}(\hat{n}_2^2\cos\Theta_1) \tag{12.2}$$

for the situation that $\hat{n}_1$ is real. If also $\hat{n}_2$ be real ,then $\hat{\chi}_s$ and $\hat{\chi}_p$ will both be pure imaginary for $n_2<n_1\sin\Theta_1$. The angles Θ_1 greater than the critical angle Θ_c, given by

$$\Theta_c=\arcsin(n_2/n_1)\,. \tag{12.3}$$

will satisfy this criterion for total reflection.

The character of the wave in the second medium is determined by Eqs. (9.19) to (9.22). The angle Θ_2 must be $\pi/2$; K' must be $\omega\,c^{-1}\,n_1\sin\Theta_1$; K'' must be $\omega\,c^{-1}(n_2^2-n_1^2\sin^2\Theta_1)^{\frac{1}{2}}$; $\boldsymbol{K}'$ must be parallel to the surface; $\boldsymbol{K}''$ must be perpendicular to the surface.

It is important to note that a small of absorption in the second medium, so that $k_2>0$, will result in a reduction of the reflectance and r will not be unity at angles of incidence greater than Θ_c. In this case the $\boldsymbol{K}'$ vector will not be parallel to the surface; the s- and p-polarizations will not have the same values of reflectance; and the radiation carries power only a short distance into the second medium — a skin depth of only a fraction of a wavelength.

13. BREWSTER's angle. The radiation reflected from a dielectric is completely s-polarized for the angle of incidence Θ_B called BREWSTER's angle. At this angle the reflectance for p-polarized radiation is zero. For the reflectance to be zero it must be that $1-\hat{\chi}$ must be zero according to (10.7) and (10.13); thus $\hat{\chi}$ must be real and equal to unity. We will consider only the case for $\hat{n}_1$ real.

For $\hat{\chi}_s=1$ it must be that $\hat{n}_2/n_1=1$ according to (12.1). This is the trivial case of no interest except to inform us that there is no angle for zero reflectance of s-polarized radiation in any other situation.

For $\hat{\chi}_p=1$, it must be that $\hat{n}_2/n_1=\tan\Theta_B$ according to (12.2). Thus $\hat{n}_2$ must be real and Θ_B is given by

$$\Theta_B=\operatorname{arc}\tan(n_2/n_1). \tag{13.1}$$

Unpolarized radiation incident upon a non-absorbing dielectric at the angle Θ_B will have only its s-polarized component reflected and will therefore become completely s-polarized. It is evident that weak absorbers, $k_2 \ll n_2$, also will not have an appreciable component of p-polarized reflectance.

14. Characteristic angles. Plane polarized radiation with s- and p-polarization components given by E_s and E_p has an azimuth angle defined by $\tan\Psi_P=E_s/E_p$. Such radiation will become elliptically polarized upon reflection. For a special angle of incidence Θ_P and a special azimuth angle Ψ_P, the reflected radiation will be circularly polarized, that is, the reflected s- and p-polarized components will be equal and the phase between the components will be $\pi/2$. This special pair of angles, the principal angle of incidence Θ_P and the principal azimuth Ψ_P, are called the characteristic angles.

For the relative phase between the s- and p-polarized components to be $\pi/2$, it is necessary that the ratio $\hat{r}_p/\hat{r}_s$, see Eq. (10.25), be a pure imaginary number. The magnitude of this ratio determines the ratio of the amplitudes of the E_p and E_s reflected waves and thus the azimuth angle must be given by

$$\hat{r}_p/\hat{r}_s=i\tan\Psi_P \tag{14.1}$$

for equal reflected wave components.

By the use of (10.25) with (14.1) for the special case with $n_1=1$ it is easily found that

$$\left.\begin{aligned}(\hat{n}_2^2-\sin^2\Theta_P)^{\frac{1}{2}}(\sin\Theta_P\tan\Theta_P)^{-1}&=(1+i\tan\Psi_P)(1-i\tan\Psi_P)^{-1}\\&=\cos 2\Psi_P+i\sin 2\Psi_P,\end{aligned}\right\} \tag{14.2}$$

and

$$\hat{n}_2=\{(\cos 4\Psi_P+i\sin 4\Psi_P)(\sin\Theta_P\tan\Theta_P)^2+\sin^2\Theta_P\}^{\frac{1}{2}}. \tag{14.3}$$

The usefulness of these principal angles is seen to be the fact that their measurement determines $\hat{n}_2$ completely.

For many situations of interest the complex refractive index is large so that $|\hat{n}_2| \gg \sin\Theta_p$ and (14.3) can be simplified to be

$$\hat{n}_2=(\cos 2\Psi_P+i\sin 2\Psi_P)\sin\Theta_P\tan\Theta_P. \tag{14.4}$$

Thus the optical constants may be calculated from the two equations

$$n_2 = \sin\Theta_p \tan\Theta_p \cos 2\Psi_p, \tag{14.5}$$

$$k_2 = \sin\Theta_p \tan\Theta_p \sin 2\Psi_p, \tag{14.6}$$

with the interesting relation $k_2\, n_2^{-1} = \tan 2\Psi_p$. In the other limit, for $k_2 = 0$, $\hat{n}_2$ is real and therefore $\Psi_P = 0$ from (14.2). As a result (14.3) becomes $n_2 = \tan\Theta_p$, and therefore the principal angle of incidence approaches BREWSTER's angle as k_2 approaches zero.

VI. Reflection and transmission of a plane, parallel faced plate.

15. A most useful configuration for a measurement sample is that of a plate with plane parallel surfaces. The size of the sample is to be large enough so that the edges play no part in the measurement of the optical properties of the sample. Such a sample will be called a lamella in order to emphasize that the surfaces are plane and parallel.

The total reflectance and transmittance of a lamella can be obtained very simply in terms of the reflectance and transmittance coefficients of the boundary surfaces. In addition to the reflectance and transmittance coefficients for the surfaces, the attenuation of the partial wave in its traversal through the lamella must be evaluated. We shall presume that the thickness of the sample is b and use the attenuation factor for a single partial wave component as $\hat{a} = \exp(i\hat{\boldsymbol{K}}_2 \cdot b\boldsymbol{u}_n) = a \exp i\,\Phi_a$ following Eq. (8.1). This attenuation factor $\hat{a}$ evaluates the amplitude and phase of the partial wave inside the material at a point on the exit face of the lamella in terms of the wave inside the lamella at a point directly opposite on the entrance face. We use these related points even though the incident beam may not be normal to the surface, and the constant phase surfaces may not be parallel to the sample surfaces. Because $\hat{\boldsymbol{K}}_2 \cdot b\,\boldsymbol{u}_n = \omega\, c^{-1}\, \hat{n}_2 \cos\hat{\Theta}_2$, the attenuation factor $\hat{a}$ becomes

$$\hat{a} = \exp i(\omega\, c^{-1}\, \hat{n}_2\, b \cos\hat{\Theta}_2) \tag{15.1}$$

and this factor does not depend upon the polarization of the incident wave.

If an incident wave of unit $\boldsymbol{E}$-vector amplitude is incident upon the lamella, then the total reflected wave be given by the sum of its partial wave components as

$$\hat{R} = \hat{r} + \hat{t}\,\hat{t}'\,\hat{r}'\,\hat{a}^2(1 + \hat{a}^2\,\hat{r}'^2 + \hat{a}^4\,\hat{r}'^4 + \cdots). \tag{15.2}$$

$\hat{R}$ is the amplitude reflectance of the lamella. The values of $\hat{r}, \hat{t}, \hat{r}', \hat{t}'$, and $\hat{a}$ may be evaluated by the formulae of the previous sections for any values of the complex indices of refraction, incident angle, or polarization. The series is readily simplified by the STOKE's relations (10.9), (10.11), (10.16), and (10.18), and thus the amplitude reflectance may be put in the following forms:

$$\left.\begin{aligned}\hat{R} = R \exp i\,\Phi_R &= \hat{r}(1-\hat{a}^2)(1+\hat{a}^2\,\hat{r}^2+\hat{a}^4\,\hat{r}^4+\cdots)\\ &= (\hat{a}^{-1}-\hat{a})\left((\hat{a}\,\hat{r})^{-1}-(\hat{a}\,\hat{r})\right)^{-1}.\end{aligned}\right\} \tag{15.3}$$

The series form is convenient for approximations if $|\hat{a}\,\hat{r}| \ll 1$, so that only a few terms are sufficient for practical use. Such approximations are of particular value in obtaining an appreciation of the dependence of the reflectivity upon the various parameters.

A similar summation of the transmitted partial waves gives the total amplitude transmittance as:

$$\left.\begin{aligned}\hat{T}&=T\exp i\,\Phi_T=(1-\hat{r}^2)\hat{a}\,(1+\hat{a}^2\,\hat{r}^2+\hat{a}^4\,\hat{r}^4+\cdots)\\&=(\hat{r}^{-1}-\hat{r})\left((\hat{a}\hat{r})^{-1}-(\hat{a}\hat{r})\right)^{-1}.\end{aligned}\right\}\tag{15.4}$$

$\hat{T}$ is the amplitude transmittance of the lamella.

The usual application of these relations will be for a plate immersed in a dielectric medium having no absorption so that the incident reflected and transmitted power can be measured at a distance from the sample lamella. The measured beams will all be measured in the same medium and therefore the power in the beams will be proportional to the square of the E-vector amplitude, and the coefficients of power reflectance and power transmittance will be $\varrho=\hat{R}\hat{R}^*=R^2$ and $\tau=\hat{T}\hat{T}^*=T^2$. These power reflectance and power transmittance coefficients are the following:

$$\varrho=R^2=\frac{a^{-2}+a^2-2\cos 2\Phi_a}{(a r)^{-2}+(a r)^2-2\cos(2\Phi_r+2\Phi_a)}\,,\tag{15.5}$$

$$\tau=T^2=\frac{r^{-2}+r^2-2\cos 2\Phi_r}{(a r)^{-2}+(a r)^2-2\cos(2\Phi_r+2\Phi_a)}\tag{15.6}$$

in which the angles correspond to the notation $\hat{r}=r\exp i\,\Phi_r$ and $\hat{a}=a\exp i\,\Phi_a$. These equations were put in the following very useful form by BARNES and CZERNY [4]:

$$\varrho=r^2\left\{\frac{(1-a^2)+4\,a^2\sin^2\Phi_a}{(1-a^2 r^2)^2+4\,a^2 r^2\sin^2(\Phi_a+\Phi_r)}\right\},\tag{15.7}$$

$$\tau=a^2\left\{\frac{(1-r^2)^2+4\,r^2\sin^2\Phi_r}{(1-a^2 r^2)^2+4\,a^2 r^2\sin^2(\Phi_a+\Phi_r)}\right\}.\tag{15.8}$$

The values of the parameters in the reflectance and transmittance expressions are simplified for the situation in which the lamella is embedded in a non-absorbing dielectric of index of refraction n_0 and the angle of incidence is Θ_0. The amplitude attenuation factor becomes

$$a\exp i\,\Phi_a=\exp i\left[\omega\,c^{-1}\,b\,(\hat{n}^2-n_0^2\sin^2\Theta_0)^{\frac{1}{2}}\right]\tag{15.9}$$

and the amplitude reflectance factor becomes

$$r_s\exp i\,\Phi_{rs}=\frac{n_0\cos\Theta_0-(\hat{n}^2-n_0^2\sin^2\Theta_0)^{\frac{1}{2}}}{n_0\cos\Theta_0+(\hat{n}^2-n_0^2\sin^2\Theta_0)^{\frac{1}{2}}}\tag{15.10}$$

for s-polarization, and

$$r_p\exp i\,\Phi_{rp}=\frac{n_0(\hat{n}^2-n_0^2\sin^2\Theta_0)^{\frac{1}{2}}-\hat{n}^2\cos\Theta_0}{n_0(\hat{n}^2-n_0^2\sin^2\Theta_0)^{\frac{1}{2}}+\hat{n}^2\cos\Theta_0}\tag{15.11}$$

for p-polarization. These last two formulae come from FRESNEL'S relations (10.21) and (10.22).

A further simplification can be made for weak absorbers, that is for those materials which have an appreciable transmission through a sample greater than a few wavelengths in thickness. Such materials have optical constants such that $n\gg k/\cos\Theta$. According to the logic which follows Eqs. (5.3) and (5.4) that $K''=\omega\,c^{-1}\,k/\cos\Theta$ and $K'=\omega\,c^{-1}\,n$, we find that (9.19) becomes

$$\hat{\boldsymbol{K}}=\boldsymbol{u}_t\,\omega\,c^{-1}\,n\sin\Theta\pm\boldsymbol{u}_n(\omega\,c^{-1}\,n\cos\Theta+i\,\omega\,c^{-1}\,k/\cos\Theta)\tag{15.12}$$

and

$$\hat{n}\sin\hat{\Theta}=n\sin\Theta=n_0\sin\Theta_0\tag{15.13}$$

and

$$\hat{n}\cos\hat{\Theta} = n\cos\Theta + i\,k/\cos\Theta. \tag{15.14}$$

It should be noted that with this approximation $\hat{n}^2\cos^2\hat{\Theta} + \hat{n}^2\sin^2\hat{\Theta}$ is only approximately equal to $\hat{n}^2$.

A further important simplification is for the incident medium to have an index of refraction equal to one, $n_0 = 1$, and to have the angle of incidence equal to zero, $\Theta_0 = 0$. The equations for the parameters in the reflectance and transmittance relations then become

$$a\exp i\,\Phi_a = \exp(-b\,\omega\,c^{-1}\,k)\exp i(b\,\omega\,c^{-1}\,n), \tag{15.15}$$

$$r_s\exp i\,\Phi_{rs} = r_p\exp i\,\Phi_{rp} = (1-(n+i\,k))\,(1+(n+i\,k))^{-1} \tag{15.16}$$

giving

$$r_s^2 = r_p^2 = ((1-n)^2+k^2)\,((1+n)^2+k^2)^{-1} \tag{15.17}$$

and

$$\tan\Phi_{rs} = \tan\Phi_{rp} = 2k\,(n^2+k^2-1)^{-1}. \tag{15.18}$$

The simplicity of the expression $a^2 = \exp(-\alpha\,b)$ for the attenuation of the power in passing from the entrance face to the exit face inside of the lamella, gives rise to the name absorption coefficient for the quantity α and to the relation

$$\alpha = 4\pi\,\nu\,k \tag{15.19}$$

in which ν measures the "frequency" in reciprocal centimeters. Because the lamella is one of the simpler forms for a sample the relations given above will be basic to the discussion of the measurements of optical constants.

C. Optical properties of simple classical systems.

16. The conductivity $\hat{\sigma}_1$ and dielectric constant $\hat{\varepsilon}_1$, as used in part B, were presumed to be functions of frequency. In this section a model will be presented to show the classical frequency dependence of these parameters. This model is presented because it characterizes the general frequency behavior of the optical constants of materials and because it forms a background of information that is helpful in understanding the interrelationships between the real and imaginary parts of the index of refraction. The deficiencies of this classical theory are being removed by quantum mechanical approaches to the problem; the general qualitative features remain sound, however. The model to be discussed will be exemplified by InSb and the figures which will be used to illustrate the frequency behavior for the optical parameters will be labeled with the values that are appropriate for this material. The excellent agreement of the classical theory for this material has been shown by SANDERSON [*5*].

The frequency dependence of $\hat{\sigma}_1$ and $\hat{\varepsilon}_1$ will be derived from the sinusoidal displacements of the charges in the material produced by a forcing sinusoidal electric field. The velocities of the charges produce a current density and determine the conductivity; the displacements of the charges produce a polarization and determine the dielectric constant. However the displacements and velocities are interdependent in the resulting sinusoidal motion, and, therefore, the motions of any group of charges contributes both to the conductivity and to the dielectric constant. It is convenient to obtain classical formulae for the frequency dependence of the free charge carrier contributions by calculating the conductivity and to find the lattice and atomic "bound charge" contributions by calculating the polarizability. The motions of both types of charges will have in phase and out of phase components and thus contribute to the effective values of $\hat{\sigma}_1$ and $\hat{\varepsilon}_1$.

I. Classical frequency dependence of the conductivity of the free charges.

16a. Let us consider the conductivity of the material to be modeled by a "gas" of free charge carriers in a viscous host material. Let the carriers be of charge q and have an effective mass m. The center of mass of all of the large number of free charges in a small volume will be displaced a vector distance $\mathbf{s}_q$ from its equilibrium position under the action of the electric field $\boldsymbol{E}$ of the sinusoidal electromagnetic wave. The center of mass motion is unaffected by the random thermal motions of the charges even though individually the random thermal energy of each charge is large compared to the average energy imparted by the

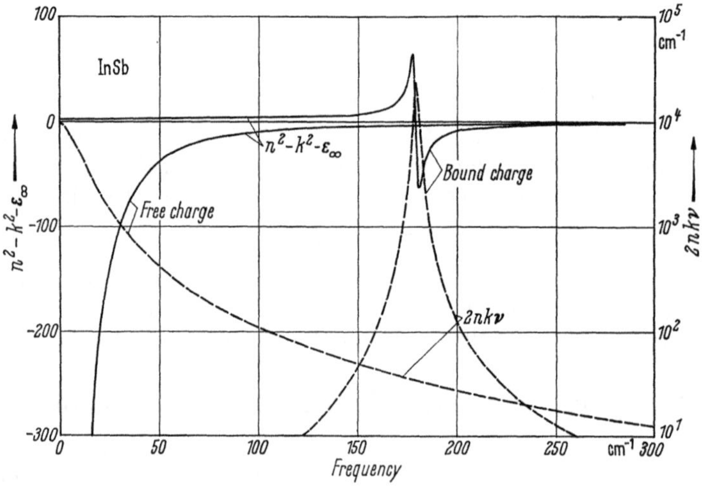

Fig. 5. The classical frequency dependence of the free charge and the bound charge (lattice) contributions to the dielectric constants ($\varepsilon' = n^2 - k^2$; $\varepsilon'' = 2nk$) for an InSb sample.

wave field. The dimensions of the volume under consideration are to be small compared to the wavelength of the radiation so that all of the charges will be under the action of the same electric field simultaneously. The volume is to be large enough, however, to contain a large number of charges so that the thermal motions may be considered to average to zero and so that the damping force on the charge collection can be considered to be represented by an average viscous force proportional to the speed of the center of mass motion.

The current density in the material is $N_q\, q\, \dot{\mathbf{s}}_q = \boldsymbol{j}$, where N_q is the number of free charges per unit volume and the dot notation is to indicate the derivative with respect to time. The center of mass motion is thus presumed to be given by

$$m\, \ddot{\mathbf{s}}_q = q\, \boldsymbol{E} - m\, \gamma_0\, \dot{\mathbf{s}}_q \tag{16.1}$$

where $m\gamma_0$ measures the effective viscous resistive force. With $\boldsymbol{E}$ having a time dependence $\boldsymbol{E}_0 \exp(-i\omega t) = \hat{\boldsymbol{E}}$ we find that the steady state sinusoidal solution for the displacement $\mathbf{s}_q$ is

$$\hat{\mathbf{s}}_q = -q\, \hat{\boldsymbol{E}}\, (i\,\omega\, m\, \gamma_0 + \omega^2\, m)^{-1} \tag{16.2}$$

so that

$$\hat{\boldsymbol{j}} = -i\,\omega\, \hat{\mathbf{s}}_q N_q\, q \tag{16.3}$$

and

$$\hat{\boldsymbol{j}} = \hat{\sigma}_1\, \hat{\boldsymbol{E}} = \hat{\boldsymbol{E}}\, i\,\omega\, N_q\, q^2 (\omega^2\, m + i\,\omega\, m\, \gamma_0)^{-1}. \tag{16.4}$$

As the frequency approaches zero, $\hat{\sigma}_1$ approaches σ_0, the direct current value, and

$$\sigma_1 = \sigma_0 \gamma_0 (\gamma_0 - i\omega)^{-1} \tag{16.5}$$

where $\sigma_0 = N_q q^2 (\gamma_0 m)^{-1}$.

The term in $\hat{n}^2$ containing $\hat{\sigma}_1$ is, see Eq. (3.10),

$$\left.\begin{aligned} i\,4\pi\hat{\sigma}_1\omega^{-1} &= -4\pi\sigma_0\gamma_0(\omega^2 - i\omega\gamma_0)^{-1} \\ &= -4\pi\sigma_0\gamma_0(\omega^2+\gamma_0^2)^{-1} + i\,4\pi\sigma_0\gamma_0(\omega^3+\omega\gamma_0^2)^{-1}. \end{aligned}\right\} \tag{16.6}$$

At sufficiently high frequencies, this contribution to $\hat{n}^2$ is negligible and the free charge carriers thus do not contribute to the high frequency values of the optical constants. At low frequencies the free charge carriers do contribute a frequency dependent portion to the optical constants. Fig. 5 shows the frequency dependence of the free charge contribution to the optical constants for InSb. There will also be lattice and atomic contributions to the frequency value of $\hat{n}^2$, so that the low frequency optical constants are not determined by the free charge carriers alone.

II. Classical frequency dependence of the dielectric constant of the bound charges.

17. The dielectric constant of a material arises from a number of different processes which couple the electromagnetic wave to the motions of the atoms and their constituents. For a classical picture of the frequency dependence of the dielectric constant only one classical mechanism will be described here. This description will center on a mechanism of importance in the infrared region for ionic crystals and will serve to typify the general frequency dependence of all the various mechanisms.

Consider a diatomic, ionic crystal with a positive ion lattice which can move against a negative ion lattice under the driving influence of an electromagnetic wave. Select a small portion of the lattice which contains a great many ion pairs but whose greatest dimension is small compared to the wavelength of the radiation. There are a number of positive ions each of mass m_+ and an equal number of negative ions each of mass m_- in this volume. The center of mass of the positive ions has a displacement $\mathbf{s}$ from the center of mass of the negative ions. With no electric wave the value of $\mathbf{s}$ is zero as the volume is chosen to have no resultant static dipole moment. Under the influence of the electromagnetic wave with an electric field $\boldsymbol{E}$, the equation of motion is

$$m_r\ddot{\mathbf{s}} = -m_r\omega_1^2\mathbf{s} - m_r\gamma_1\dot{\mathbf{s}} + q\boldsymbol{E} \tag{17.1}$$

where m_r is the reduced mass of the ion pair $m_r^{-1} = m_+^{-1} + m_-^{-1}$, $m_r\omega_1^2$ is a restoring force constant, and $m_r\gamma_1$ is an ad hoc damping constant and q is the magnitude of an effective charge for the ions which depends upon the polarizability of the ions. The natural undamped mechanical frequency of the system is ω_1 radians per second. For a sinusoidally varying field having the form $\boldsymbol{E}_0\exp(-i\omega t)$ at the ion pair site, a steady state solution of the displacement $\mathbf{s}$ in the resulting sinusoidal motion is

$$\hat{\mathbf{s}} = q\hat{\boldsymbol{E}}\,m_r^{-1}(\omega_1^2 - \omega^2 - i\omega\gamma_1)^{-1}. \tag{17.2}$$

The polarization, dipole moment per unit volume, of the material resulting from this ion motion is

$$\hat{\boldsymbol{P}} = N_1 q\hat{\mathbf{s}} + (\varepsilon_\infty - 1)(4\pi)^{-1}\hat{\boldsymbol{E}} \tag{17.3}$$

where the constant ε_∞ encompasses the effect of many other mechanisms such as the electric polarizability of the ions themselves and N_1 is the number of ion pairs per unit volume. Under the action of the electromagnetic wave the polarization becomes

$$\hat{\boldsymbol{P}} = [-N_1 q^2 m_r^{-1} (\omega^2 - \omega_1^2 + i\,\omega\,\gamma_1)^{-1} + (4\pi)^{-1} (\varepsilon_\infty - 1)]\,\hat{\boldsymbol{E}} \tag{17.4}$$

and the corresponding electric displacement vector is

$$\hat{\boldsymbol{D}} = \hat{\boldsymbol{E}} + 4\pi\,\hat{\boldsymbol{P}} = [\varepsilon_\infty - 4\pi\,N_1 q^2 m_r^{-1} (\omega^2 - \omega_1^2 + i\,\omega\,\gamma_1)^{-1}]\,\hat{\boldsymbol{E}} = \hat{\varepsilon}_1\,\boldsymbol{E}. \tag{17.5}$$

From this relation the value of $\hat{\varepsilon}_1$ can be found, and it can be put into the following from:

$$\hat{\varepsilon}_1 = \varepsilon_\infty + (\varepsilon_0 - \varepsilon_\infty)\,\omega_1^2 (\omega_1^2 - \omega^2 - i\,\omega\,\gamma_1)^{-1} \tag{17.6}$$

with ε_0 the value of the dielectric constant as the frequency approaches zero, and ε_∞ the value for very high frequencies. The dielectric constants at very high frequencies and the values at very low frequencies are related by

$$\varepsilon_0 - \varepsilon_\infty = 4\pi\,N_1 q^2 m_r^{-2} \omega_1^{-2}. \tag{17.7}$$

The contribution to the square of the complex index of refraction $\hat{n}^2$ from the attice and ionic polarizability, according to this particular model is

$$\hat{n}^2 = \varepsilon_\infty + \frac{(\varepsilon_0 - \varepsilon_\infty)(\omega_1^2 - \omega^2)\,\omega_1^2}{(\omega_1^2 - \omega^2)^2 + \omega^2\gamma_1^2} + i\,\frac{(\varepsilon_0 - \varepsilon_\infty)\,\omega_1^2\,\omega\,\gamma_1}{(\omega_1^2 - \omega^2)^2 + \omega^2\gamma_1^2}. \tag{17.8}$$

The frequency dependence of the optical constants from this lattice frequency absorption is strongest in the neighborhood of the natural undamped lattice frequency ω_1 but extends smoothly to high and low frequencies from this resonance frequency. Fig. 5 shows the frequency dependence of the bound charges contribution to the optical constants of InSb.

The dielectric constant of a material will depend upon all of the processes which allow an interaction of the electromagnetic wave with the material. The general dependence can be illustrated, however, by the combination of the frequency dependence of the conductivity and the frequency dependence of the polarizability which were modeled in the preceding sections. The value of $\hat{n}^2$ including both of these effects is

$$\hat{n}^2 = \hat{\varepsilon} = \varepsilon' + i\,\varepsilon'' = (n^2 - k^2) + i\,(2n\,k) = \hat{\varepsilon}_1 + i\,4\pi\,\hat{\sigma}_1\,\omega^{-1} \tag{17.9}$$

where we define the unsubscripted dielectric constants by this relation so that $\hat{\varepsilon}$ includes the effects of all processes which determine the amplitude and phase of the polarizability. The value of the dielectric constant, therefore, is equal to

$$\left.\begin{aligned} \hat{\varepsilon} = &\left\{\varepsilon_\infty + \frac{(\varepsilon_0 - \varepsilon_\infty)(\omega_1^2 - \omega^2)\,\omega_1^2}{(\omega_1^2 - \omega^2) + \omega^2\gamma_1^2} - \frac{4\pi\,\sigma_0\,\gamma_0}{\omega^2 + \gamma_0^2}\right\} + \\ &+ i\left\{\frac{(\varepsilon_0 - \varepsilon_\infty)\,\omega_1^2\,\omega\,\gamma_1}{(\omega_1^2 - \omega^2)^2 + \omega^2\,\gamma_1^2} + \frac{4\pi\,\sigma_0\,\gamma_0^2}{\omega\,(\omega^2 + \gamma_0^2)}\right\}. \end{aligned}\right\} \tag{17.10}$$

It should be noted that ε_0 is not the zero frequency dielectric constant because of the conductivity contribution. It is true, on the other hand, that ε_∞ is the high frequency dielectric constant with this model. In (17.10) the parameters ω, ω_1, γ_0, γ_1, and σ_0 all have the dimensions of radians per second. The equation would still be valid if those parameters were all put in the frequency units of cycles per centimeter, and for many purposes this would be very desirable.

III. Classical frequency dependence of the optical constants.

18. The optical constants n and k can be determined from (17.9) by following relations which are formally similar to (5.3) to (5.4):

$$n=\{\tfrac{1}{2}[(\varepsilon'^2+\varepsilon''^2)^{\frac{1}{2}}+\varepsilon']\}^{\frac{1}{2}}, \tag{18.1}$$

$$k=\{\tfrac{1}{2}[(\varepsilon'^2+\varepsilon''^2)^{\frac{1}{2}}-\varepsilon']\}^{\frac{1}{2}}=\varepsilon''(2n)^{-1}. \tag{18.2}$$

An example of the application of these classical relations to a real material is helpful as an illustration of the physical content of the relations and as an il-

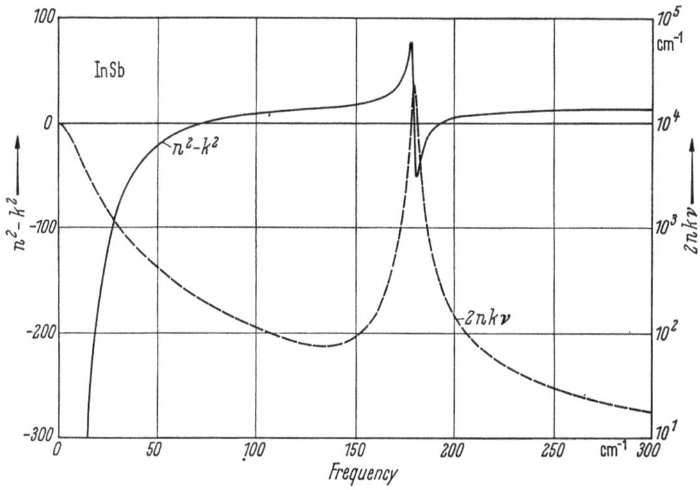

Fig. 6. The classical frequency dependence of the dielectric constants of the sample of InSb showing the combined result of the free charge and bound charge (lattice and atomic) contributions.

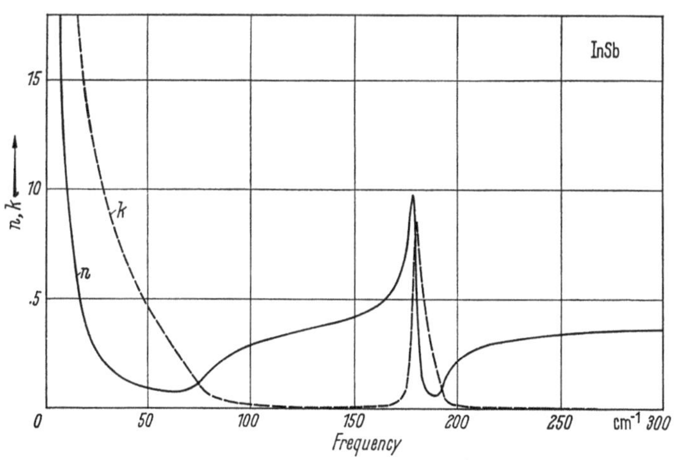

Fig. 7. The frequency dependence of the optical constants of the sample of InSb from the classical model.

lustration of the typical spectral behavior of the optical properties of solids. A careful measurement of the optical properties of InSb has been made by SANDERSON [5] and his data will be used to illustrate the frequency dependence of the optical constants. Figs. 5—8 show the frequency dependence of various parameters of interest for InSb at room temperature. The spectral frequency variable has been changed from ω to ν in cycles per centimeter for these figures. The values of the constants used in the calculations are:

$$\omega_1 = 179.1\ \text{cm}^{-1}, \qquad \varepsilon_0 = 17.72,$$
$$\gamma_1 = 2.86\ \text{cm}^{-1}, \qquad \varepsilon_\infty = 15.68,$$
$$\gamma_0 = 10.7\ \text{cm}^{-1}, \qquad 4\pi\sigma_0 = 9.6\times 10^3\ \text{cm}^{-1}.$$

These values give an excellent fit to the measured data. The dependence of the conductivity contribution on the free charge carrier concentration is shown by SANDERSON through the temperature dependence of the optical constants. The

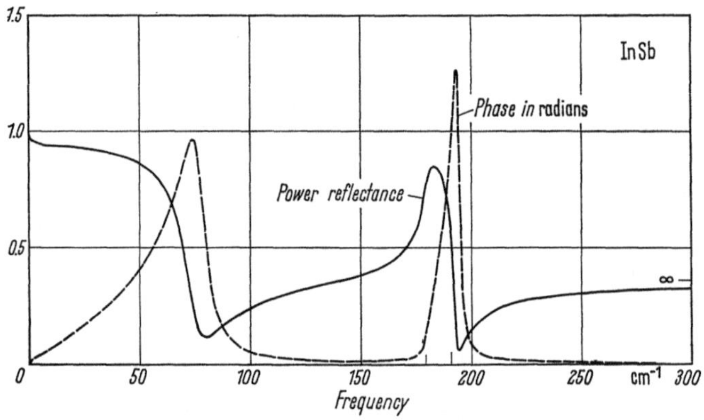

Fig. 8. The frequency dependence of the reflectance $\varrho = r^2$ and the phase of the reflectance $(\Phi_r - \pi)$ for the InSb sample. The reflectance for the sample at very high frequencies is shown by the ∞ mark on the right-hand side. The values of ω_1 and ω_L are marked on the frequency scale.

several frequency dependent terms of Eq. (17.10) are shown in Fig. 5, and one can see that, for this material, the free charge contributions and the bound charge contributions are separated in frequency. This separation, of course, depends upon the fact that the free charge concentration is low enough that the free charge effects do not overlay the lattice effects. The total $\varepsilon' = n^2 - k^2$ and $\varepsilon''\nu = 2n\,k\,\nu$ are shown in Fig. 6, and the values of the optical constants n and k are shown in Fig. 7. By following through this sequence, one can obtain a feeling for the important processes determining the optical behavior of the material throughout this spectral region. Particularly, the smooth interdependence of the real and the imaginary portions of the optical constants is seen to be a necessary consequence of the model which was used to predict the behavior. Some such interdependence must result from any mechanism which couples the electromagnetic wave to the material, however, so the general features of the interdependence of the optical constants should be expected for any material.

Fig. 8 shows the reflectance and the phase of the reflectance for the InSb sample calculated from the same data. The regions of high reflectance are related to the natural undamped frequencies in the model: frequency ω_1 for the lattice and zero frequency for the conductivity.

The value of ω_1 at 179.1 cm^{-1} is shown by a mark on the frequency axis of Fig. 8. The mark at 190.4 cm^{-1} is the longitudinal optical lattice frequency ω_L. At the frequency ω_L the lattice can have a longitudinal wave motion if the damping constant γ_1 is zero. In this motion the mechanical displacements of the charges produce the field $\boldsymbol{E}$ and a polarization $\boldsymbol{P}$ in opposition, so that $\boldsymbol{D}$ is zero.

From (17.5) or (17.6), with $\hat{\varepsilon}_1=0$ and $\gamma_1=0$, the frequency is found to be

$$\omega_L=(\varepsilon_0\,\varepsilon_\infty^{-1})^{\frac{1}{2}}\,\omega_1. \tag{18.3}$$

This is the LYDDANE, SACHS, TELLER [6] relation. Even if the damping constant is not zero, a longitudinal wave can be supported in a thin film by the oscillating charges on its surfaces.

The reflectance of a material with $\gamma_1=0$ is interesting as the limiting situation for real materials. The optical constants of such a material, as determined from (17.8) with $\gamma_1=0$, are such that no wave can be supported with a frequency between ω_1 and ω_L. Thus the material would reflect all of the incident radiation in this frequency range. For frequencies slightly higher than ω_L the reflectance would be very small. For materials with a small value of γ_1 the reflectance would be high for frequencies between ω_1 and ω_L and would have steep edges at these limiting frequencies.

The low frequency catastrophe for the optical constants for InSb illustrated in Fig. 7 is, of course, not the behavior for an insulating dielectric with zero conductivity. Omission of the free charge carrier terms from the effective dielectric constant (17.9) would let n approach $\varepsilon_0^{\frac{1}{2}}$ and k approach zero as the frequency approaches zero.

A full discussion of this classical model and the optical constant relations can be found in the book by BORN and HUANG [7].

a) Resonance circle diagram.

19. The behavior of the optical constants of InSb illustrates the fact that the optical constants throughout a range of spectral frequencies can be calculated from only a few parameters. The desired accuracy of the optical constants to be calculated will determine the number of necessary parameters and their precision. The concern in this article is with measurement methods applicable for low and moderately accurate optical constant determinations which are useful particularly in divining the nature of the material.

The simplicity of the classical model for the frequency dependence of the optical constants makes the determination of the parameters for such a model the first stage in a program of optical measurements on a material. Even though the quantum theories should show that the absorption process is somewhat different from the ad hoc damping assumed in the classical model, nevertheless the index of refraction n is not so seriously changed by the quantum theory. The index of refraction depends upon the absorption process through an integral relation, which will be explained in Sect. 20, that has the property of smoothing over the details of the absorption spectrum. The evaluation of the classical parameters for a material is therefore worth the effort. The classical parameters may be determined by just a few measurements and they lead to working values of the optical constants in a wide range of frequencies.

The ability of the classical model to predict the optical constants of a material may be judged by the fit of the experimental data on the material to a "resonance circle diagram". Such a circle diagram is shown for InSb in Fig. 9. The solid curve shows the calculated values for InSb for the frequency range shown in the Figs. 5 to 8. The cross marks on the curve show the frequencies at several positions along the curve. The dashed curve is a circle to show the approximation of the InSb data to the circle. For a classical model having only one natural frequency the optical data would fit on such a circle.

That such a circle would result from the lattice portion of the classical model may be seen by starting with Eq. (17.6) and putting it in the form

$$\omega(\hat{\varepsilon}-\varepsilon_\infty)\gamma_1(\omega_1^2(\varepsilon_0-\varepsilon_\infty))^{-1}=(i-(\omega_1^2-\omega^2)(\omega\gamma_1)^{-1})^{-1}. \tag{19.1}$$

Replacing $\hat{\varepsilon}$ with $(n+ik)^2$ the equation takes the form

$$(x+iy)d^{-1}=(i+s)^{-1} \tag{19.2}$$

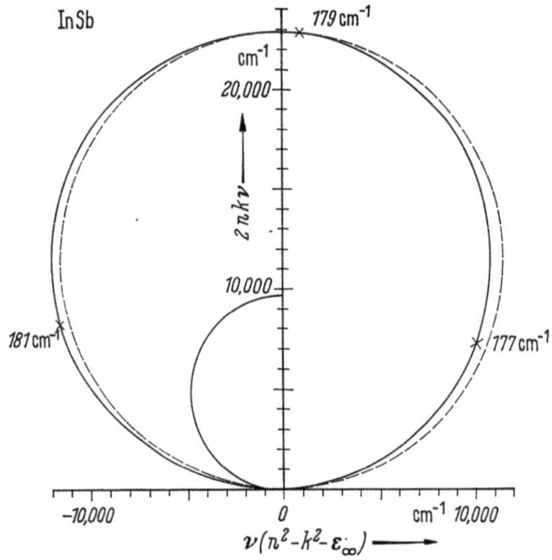

Fig. 9. A resonance circle diagram for the InSb sample calculated from the classical model. The dashed circle is for comparison. The cross marks on the solid line indicate the values at the frequencies labeled.

with s a function of frequency, x and y are functions of n and k, and d is a constant. In this form it is easily checked that $x/d=s(1+s^2)^{-1}$ and $y/d=-(1+s^2)^{-1}$, and then

$$x^2+(y+d/2)^2=(d/2)^2. \tag{19.3}$$

Thus the x and y functions of n and k are located on a circle of diameter d. Therefore a plot of

$$y=2\omega n k \tag{19.4}$$

and

$$x=\omega(n^2-k^2-\varepsilon_\infty) \tag{19.5}$$

will be a circle with a radius

$$(d/2)=\omega_1^2(\varepsilon_0-\varepsilon_\infty)(2\gamma_1)^{-1} \tag{19.6}$$

and a center at $(x, y)=(d/2,0)$. The angle Φ equal to $\arctan(y/x)$ on the diagram shown in the figure may be related to the frequency by

$$\tan\Phi=2nk(n^2-k^2-\varepsilon_\infty)^{-1}=(\omega_1^2-\omega^2)(\omega\gamma_1)^{-1}=-s^{-1}. \tag{19.7}$$

At the natural undamped frequency ω_1, the angle Φ is $\pi/2$. The change from ω to ν in reciprocal centimeters, as in Fig. 9, is convenient and simple.

IV. Dispersion relations.

20. The model presented in the previous section to show the frequency dependence of the complex refractive index also serves to illustrate the fact that there is an intimate relationship between the index of refraction and the extinction coefficient. Both the real and the imaginary parts of the complex refractive index were derived from the same physical model — the same differential equation. An interrelation between these two functions would exist for any model one may choose. Such a relationship is imposed by the linearity of the system and by causality — the fact that the response of the system to a driving force can not precede the application of the driving force. The relations between the real and imaginary portions of such response functions take the form of integral equations and are often called KRAMERS-KRONIG relations because of the work of H. A. KRAMERS and R. DE L. KRONIG who developed these relations for the dielectric constant and for the index of refraction. These relations are also called dispersion relations because they relate an absorptive process to a dispersive process. An account of these relationships and their physical significance may be found in the work of TOLL [8], and their application to optical constant measurements is discussed by STERN [9] and MOSS [10].

Several forms of these dispersion relations will be of use in the measurement of optical constants. The fundamental quantity derived from the classical model of a material was the dielectric constant as a function of frequency ω. The real and imaginary parts of this function are related by the dispersion relations:

$$\varepsilon'(\omega)-1=2\pi^{-1}\int_0^\infty \omega'\,\varepsilon''(\omega')\,(\omega'^2-\omega^2)^{-1}d\omega', \tag{20.1}$$

$$\varepsilon''(\omega)-4\pi\,\sigma_0\,\omega^{-1}=2\,\omega\,\pi^{-1}\int_0^\infty \varepsilon'(\omega')\,(\omega^2-\omega'^2)^{-1}d\omega'. \tag{20.2}$$

The integral must be evaluated as the CAUCHY principal value to properly include the singularity at $\omega'=\omega$. The CAUCHY principal value for all the necessary following integrals will be presumed without further notation to that effect.

The constant -1 on the left hand side of Eq. (20.1) is the high frequency limit of $\varepsilon'(\omega)$ representing the vacuum dielectric constant. The integral over $\varepsilon''(\omega')$ in (20.1) must be carried through all of the high frequency absorptions including the ultraviolet, X-ray, and nuclear resonances. This, of course, is not necessary in practice because the high frequency $\varepsilon''(\omega')$ contributions can be lumped together to produce a constant over the low frequency range for which the relation is to be used. It is true, to some approximation, that $\varepsilon''(\omega')$ has a value at low frequencies $\varepsilon''_{\text{low}}(\omega')$ due to low frequency absorptions and also has a value $\varepsilon''_{\text{high}}(\omega')$ at high frequencies and that there is a middle region of frequencies with no absorption contributions. The integral (20.1) can be separated, then, into two integrals: one containing only the low frequency contributions $\varepsilon''_{\text{low}}(\omega')$ and a second containing only the high frequency contributions $\varepsilon''_{\text{high}}(\omega')$. The integral over $\varepsilon''_{\text{high}}(\omega')$ with low frequency values of ω will give a constant independent of ω if the high frequency resonances are sufficiently far removed. Thus (20.1) can be put in the form

$$\varepsilon'(\omega)-\varepsilon_\infty=2\pi^{-1}\int_0^\infty \omega'\,\varepsilon''_{\text{low}}(\omega')\,(\omega'^2-\omega^2)^{-1}d\omega', \tag{20.3}$$

where the high frequency contributions have all been lumped into ε_∞. The integral in (20.3) still proceeds to infinity, but only the low frequency resonances are included in the integrand. Also the integration is simplified by the fact that $\omega'\,\varepsilon''_{\text{low}}(\omega')$ will approach zero at high frequencies so that the integral will converge. The high frequency region of the low frequency resonances, according to the classical model (17.10), will have $\omega'\,\varepsilon''_{\text{low}}(\omega')$ proportional to ω'^{-2}.

The integral (20.2) would not be affected by the addition of the constant ε_∞ to $\varepsilon'(\omega')$ since, for $\omega \neq 0$,

$$\int_0^\infty (\omega^2 - \omega'^2)^{-1} d\omega' = 0. \tag{20.4}$$

In the relations which follow it will be presumed that the $\varepsilon''(\omega')$ contains only the low frequency contributions and that the effects of the high frequency absorptions has been adequately included in the ε_∞ constant. Some judgement may be necessary in order to verify this for each situation in which these dispersion relations are used.

From the defining relations (17.9) for $\hat{\varepsilon}$ we see that (20.3) may be recast as relations between $n^2 - k^2$ and $2nk$. A useful relation derived from (20.3) is the value of the index of refraction at very low frequencies, $n^2(0)$ for non-conductors $(k(0) = 0)$. This is

$$n^2(0) - \varepsilon_\infty = 2\pi^{-1} \int_0^\infty 2n\, k\, \omega'^{-1} d\omega', \tag{20.5}$$

in which $2nk$ is a function of ω' resulting from the "low frequency" absorptions. A measurement of the low frequency dielectric constant, $n^2(0)$, gives information about the absorption $2nk$ and vice versa.

The Eqs. (20.2) and (20.3) may be integrated by parts to give

$$\varepsilon'(\omega) - \varepsilon_\infty = \pi^{-1} \int_0^\infty \frac{d\varepsilon''(\omega')}{d\omega'} \log\left(|\omega'^2 - \omega^2|^{-1}\right) d\omega' \tag{20.6}$$

and

$$\varepsilon''(\omega) - 4\pi\sigma_0\omega^{-1} = -\pi^{-1} \int_0^\infty \frac{d\varepsilon'(\omega')}{d\omega'} \log\left(|\omega' + \omega|\,|\omega' - \omega|^{-1}\right) d\omega'. \tag{20.7}$$

The logarithms in the integrands have strong peaks for frequencies in the neighborhood of ω and therefore behave like weighting functions to make the integral value very sensitive to the integrand values at frequencies near ω. The relations indicate that the value of $\varepsilon'(\omega)$ will be large in the spectral regions where the derivative of $\varepsilon''(\omega)$ is large and positive and $\varepsilon'(\omega)$ will be small when the derivative of $\varepsilon''(\omega)$ is large and negative, It is also seen that there is a corresponding behavior for $\varepsilon''(\omega)$ depending upon the derivative of $\varepsilon'(\omega)$. These ideas are easily verified for the particular example shown in the Figs. 5 and 6.

Not only are the real and imaginary parts of the effective dielectric constant related by dispersion relations but so are the optical constants interrelated. One relation, given by Moss [*10*], is

$$n(\omega) - 1 = 2\pi^{-1} \int_0^\infty \omega' k(\omega') (\omega'^2 - \omega^2)^{-1} d\omega'. \tag{20.8}$$

The integrand factor $\omega' k(\omega')$ in (20.8) is, except for constants, the absorption coefficient α of (15.22) and thus n and α are related by the dispersion relation (20.8).

For optical constant measurements, the previous dispersion relations have limited usefulness because of the difficulty of obtaining either the real or the imaginary part of the function for a sufficiently wide range of frequencies so that the integrals may be calculated. On the other hand, a corresponding relationship between the phase and the amplitude of the reflectance has been frequently used because reflectance data is more readily available over a wide spectral range.

The logarithm of the amplitude reflectance and the phase of the reflectance are the real and imaginary parts of the function $\log \hat{r}$. These parts are related by the dispersion relations:

$$\Phi_r(\omega) - \pi = 2\omega\, \pi^{-1} \int_0^\infty [\log r(\omega') - \log r(\omega)]\, (\omega^2 - \omega'^2)^{-1} d\omega', \qquad (20.9)$$

$$\log r(\omega) - \log r(\omega_0) = 2\pi^{-1} \int_0^\infty \omega' [\Phi_r(\omega') - \pi]\, [(\omega'^2 - \omega^2)^{-1} - (\omega'^2 - \omega_0^2)^{-1}]\, d\omega'. \qquad (20.10)$$

The relation (20.10) can be written in more useful forms if the phase function $\Phi_r(\omega)$ has a high frequency portion which is well separated from a low frequency portion and is zero in between. The high frequency portion then contributes a $\log r(\omega)$ value which is constant for all ω's from zero to ω_0. Thus (20.10) applies with the $\Phi_r(\omega')$ in the integrand being only the low frequency portion. If it is also true that ω_0 is well above the highest frequencies for which the low frequency phase function $\Phi_r(\omega') - \pi$ has non-zero values, then

$$\log r(\omega) - \log r(\omega_0) = 2\pi^{-1} \int_0^\infty \omega' [\Phi_r(\omega') - \pi]\, (\omega'^2 - \omega^2)^{-1} d\omega', \qquad (20.11)$$

where $r(\omega_0)$ is an extrapolated high frequency value of the reflectance.

Because the phase is not completely determined by the amplitude of the reflectance there are other possible spectral phase functions than (20.9). These other functions are not normally possible for the reflectivity of a solid according to STERN [9], since Φ_r is restricted to the range $\pi \leqq \Phi_r \leqq 2\pi$ for real materials. Discussion of these relations can be found in TOLL [8] and STERN [9].

The value of these dispersion relations in the measurement of optical constants comes from the fact that a measurement of the normal incidence reflectance of a "thick" sample over a wide frequency range allows the calculation of the phase by (20.9) and the optical constants from

$$n = (1 - r^2)\,(1 + r^2 + 2r\cos\Phi_r)^{-1}, \qquad (20.12)$$

$$k = (-2r\sin\Phi_r)\,(1 + r^2 + 2r\cos\Phi_r)^{-1} \qquad (20.13)$$

which are obtained from FRESNEL'S relations for normal incidence or from (10.21) and (10.22).

The constant $\log r(\omega)$ in the integrand of (20.9) does not contribute to the integral. The introduction of $\log r(\omega)$ into the integrand simplifies the computational problem at the singularity $\omega' = \omega$. The constant $-\pi$ in the left hand member of (20.9) is to keep Φ_r consistent with $\hat{r} = r \exp i\,\Phi_r$ as in the previous relations and especially with relations (20.12) to (20.13) which would be used to calculate the optical constants from the calculated phase of (20.9).

D. Determining optical constants from experimental data.

21. The problem of determining the optical constants from experimental data is also the problem of making the proper measurements so that the optical constants can be derived. Almost any measurement which can be made of the reflected or transmitted wave and of the incident wave will yield some information about the optical constants of the material at the frequency of the testing radiation wave. A high value of the reflectance obviously means that the complex refractive index is not of the magnitude of unity. Likewise a high value of the transmittance must mean that the extinction coefficient is small. Crude information of this kind is helpful in order to anticipate what kinds of further measurements are possible which will give a more accurate knowledge of the optical constants.

It will be assumed for this discussion that the experiments are made with the sample in air and that the precision of the measurements will not justify a correction for the index of refraction of air.

Measuring instruments, as such, will not be described in this article even though they limit the type and precision of the measurements. Instrumentation is constantly improving and previously difficult measurements may become simple with newer equipment. Remarks about instrumentation will therefore be very limited.

I. Measurement of the classical model parameters.

22. If the optical constants are known for a few frequencies and the material is known to have a simple spectrum — a single eigenfrequency, for example, like the classical model used in part C — then it will be possible to estimate the optical constants for other frequencies. Such an estimate may be all that is needed for some engineering work. The estimate may also be valuable in order to anticipate the range of values of the optical constants in the regions where more accurate measurements are to be made. One simple technique for such extrapolation is the use of the "resonance circle diagram". This diagram is illustrated in Fig. 9 for the InSb example of part C. The characteristics of this circle were described in Sect. 19 with the coordinates given by Eqs. (19.4) and (19.5). A similar diagram has been shown by FROEHLICH [*11*] for LiF and related diagrams have been shown by GEICK [*12*] for NaCl and [*13*] for CsBr.

The knowledge of the measured point (x, y) on the resonance circle diagram for a known frequency ω not too far from the resonance at ω_1 gives enough information to construct the circle through $(0, 0)$ and having a center on the y axis. The frequencies for all points are determined by Eq. (19.7). Thus a single measurement of n and k and a knowledge of ε_∞ is sufficient to determine the whole frequency dependence of the optical constants. Unfortunately a single measurement would not constitute sufficient evidence by itself to lend credence to the belief that a single resonance, classical model is adequate for the material investigated. Several measurements properly distributed on the circle, however, could lend faith to the acceptance of the optical constants as derived from the classical model.

Obviously the measured points will not fit the circle at frequencies which are influenced by other resonances. Notice, for example, the InSb data in Fig. 7 which shows the low frequency deviation because of the free carrier conductivity. It is important that the data used in producing such a "resonance circle diagram" should be obtained near the resonance peak. Such data is frequently difficult to obtain because of the strong absorption in this region. This is the reststrahlen region for ionic crystals and the strong reflectivity and strong absorption makes precision measurements difficult. The values of the classical parameters may be estimated from several kinds of measurements, however, and some comments about these measurements are of value.

The value of ε_∞ may be troublesome to obtain. Fortunately it does not have to be accurately known for use in otherwise approximate calculations — such as the resonance circle diagram. There may be no clear choice of a real finite frequency for which the value of ε_∞ will be the appropriate choice for use in the classical model of the material. Measurements of the optical properties of the material at frequencies higher than the resonant mode of interest should suggest an appropriate value for ε_∞.

The value of ε_0 may likewise have to be provided by poor, or indirect data. If the lowest resonant frequency is under investigation, then ε_0 may be obtained from radio frequency measurements. Here, again, the difficulty is in the attempt to isolate a single resonant optical mode in the dispersion model and to treat it as though all other modes are far enough away so that their effects may be combined into the ε_∞ and ε_0 constants.

Several relations also help in the evaluation of the ε_0 and ε_∞ constants. The difference $\varepsilon_0 - \varepsilon_\infty$, for example, is a measure of the strength of the lattice absorption according to the relation (17.7). Note the dispersion relation (20.5) in this regard also. Knowledge about the strength of the lattice absorption can give information about the low (or high) frequency dielectric constant. An example of this use of the dispersion relations is given by Moss [*10*] for PbTe. Another relation between ε_0 and ε_∞ is given by the LYDDANE, SACHS, TELLER relation (18.3) between the transverse optical frequency ω_1 and the longitudinal frequency ω_L. The values of ω_1 and ω_L can be estimated from the reststrahlen peak. The high frequency side of the reflectance peak falls to a minimum near the longitudinal optical lattice frequency. The low frequency edge of the peak is near the transverse optical lattice frequency. For a zero value of γ_1 the low and high frequency edges of the reflection maximum are ω_1 and ω_L respectively, following the argument given in Sect. 18. For the small values of γ_1 the reflectance peak is about this same width, not much influenced by the value of γ_1. An example is shown in Fig. 8 for InSb with ω_1 and ω_L marked on the frequency scale at 179.1 cm^{-1} and 190.4 cm^{-1} respectively. The width of the reststrahlen peak as measured between these two "edges" is approximately $\omega_L - \omega_1 = \omega_1 \left((\varepsilon_0\, \varepsilon_\infty^{-1})^{\frac{1}{2}} - 1\right)$. If the width of the reflectance peak can be estimated from the reflectance measurements, then this width sets some bounds upon the range of values of the dielectric constants ε_0 and ε_∞.

The value of the resonant frequency ω_1 neighbors the reststrahlen peak as seen in Fig. 8. The high reflectance, the high values of n and k, and the large phase of the reflectance all neighbor the resonant frequency but do not coincide with the resonant frequency. One of the simplest approximations for the resonant frequency from the experimental reflectance is the frequency of the reststrahlen peak. This value may be improved by the use of HAVELOCK's formula [*15*] which gives ω_1 in terms of the frequency ω_M of the maximum value of the reflectance as

$$\omega_1 = \omega_M \left(1 + (\varepsilon_0 - \varepsilon_\infty)(6\varepsilon_0 - 4)^{-1}\right)^{-\frac{1}{2}}. \tag{22.1}$$

This relation was derived from the classical dispersion model for the situation that $\gamma_1 \omega_1^{-1} \ll 1$, which holds if the reflectance peak is high. Other approximations have been discussed by JAFFE [*16*] and STERN [*9*].

A precise determination of the resonant frequency can be obtained from transmission measurements. For strong absorbers, however, the transmittance can be measured only with very thin samples. With ionic crystals, for example, it is necessary to produce evaporated films of the crystal material in order to obtain a narrow absorption maximum and accurate frequency measurement. The minimum transmittance occurs for the frequency for maximum value of $2n\,k\,\omega$ as will be seen later in Sect. 36. It is seen in the InSb example in Fig. 5 that $2n\,k\,\omega$ is a maximum at the lattice frequency ω_1.

A direct measurement of ω_L can be obtained by the method of BERREMAN [*14*]. BERREMAN observed the absorption associated with the longitudinal optical frequency in thin films of LiF both in transmission and in reflection. The radiation was p-polarized and the angles of incidence were non-zero but moderate. The more dramatic effect was in the reflectance of a thin film deposited on a silver substrate. The LiF film absorption by the transverse optical mode, with a lattice motion parallel to the surface, is prevented by the fact that the electric field component parallel to the silver metal surface must be zero. The longitudinal optical mode, on the other hand, has lattice motions perpendicular to the surface, is accompanied by surface polarization charges, and is excited by the incident radiation. The reflectance of the thin film on the silver substrate showed high values except at the longitudinal optical frequency, and the frequency ω_L was accurately measured.

The value of the damping constant γ_1 can be estimated from the peak value of the reflectance. For small values of γ_1, small damping, the reflectance peak is high; a large value of γ_1, on the other hand, produces a small value for the peak reflectance. A value for γ_1 can be obtained from the peak value of the reflectance ϱ_M by the method of MITRA [*17*]. MITRA utilizes the approximation that ϱ_M^{-1} is nearly proportional to $\gamma_1\,\omega_1^{-1}$. A calculation of the reflectance from the classical model optical constants (17.8) with reasonable values of γ_1 and for a range of frequencies in the region of the peak reflectance — or at the frequency given by HAVELOCK's formula (22.1) — will yield a maximum value for the reflectance for each γ_1 value. These calculated maximum reflectances establish the proportionality constant between ϱ_M^{-1} and $\gamma_1\,\omega_1^{-1}$. From this proportionality and the measured peak reflectance, γ_1 can be obtained.

A value of γ_1 can also be obtained from transmission measurements on a thin film of the material. The transmission of a thin film is discussed in a later section in which it will be shown, Eq. (36.4), that the absorption depends upon $n k$. The value of $n k$ depends upon γ_1 through the relation (17.8) to (17.9) and thus the measurement of the transmission of a thin film of the material determines γ_1. In particular, the minimum transmittance τ_m will be at the resonant frequency and will be related to γ_1 by $1-\tau_m=(\varepsilon_0-\varepsilon_\infty)\,\omega_1^2\, b\,(\gamma_1\,c)^{-1}$ in which b is the thickness of the film. Examples, with further elaboration of this technique, are given by JONES, MARTIN, MAWER, and PERRY [*18*] and GEICK [*13*].

The whole set of parameters for a single lattice frequency ε_0, ε_∞, ω_1, and γ_1, can be quickly adjusted to fit the measured reflectance peak spectrum with the use of electronic computers. The computer can be programmed to compute the optical constants from the chosen parameters with the equations like (14.8) to (14.9) and (15.3) to (15.4), and then the reflectance from (11.16), for example. The parameters are adjusted to match the measured reflectance to be accuracy warranted by the experimental data or by the simplicity of the theory. The parameters are chosen essentially with ω_1 as the parameter most influencing the low frequency edge of the reflectance peak, $\varepsilon_0\,\varepsilon_\infty^{-1}$ influencing the width, and γ_1 the reflectance peak height. Limitations of the theory and experimental data do not justify an elaborate fitting procedure and a reasonable fit may be quickly obtained. The application of such calculations to the determination of the optical properties of quartz is beautifully demonstrated by the work of SPITZER and KLEINMAN [*19*]. The same technique has been used by KLEINMAN and SPITZER [*20*] for GaP and by SPITZER, KLEINMAN, and WALSH [*21*] for SiC.

If the behavior of the optical constants were precisely determined by the classical model relations, then the determination of such classical parameters would be the end result of an optical measurements program. Real materials deviate from this simple state of affairs and other measurements are necessary to determine the precise spectral course of the optical constants. These classical parameters are only a first approximation — but a very useful approximation in that they give direct information about the microscopic nature of the material.

II. Specular reflection from a single surface.

23. One of the simplest configurations of a solid sample for use in optical constant measurements is that of a piece of the material having one plane surface, suitable for reflectance measurements. The available surface size will determine such instrumental conditions as the requisite solid angle of irradiation for adequate radiant power, the precision of the angle of incidence, the maximum angle of incidence, etc. The patience and skill of the experimenter are important for dealing with these limitations and for evaluating their importance.

A single measurement of the reflectance of the sample at a particular radiation frequency cannot determine, by itself, the values of the two optical constants n and k. If it is known that k is very small, however, then n is reasonably fixed by such a measurement. The precision of such a determination will depend upon the experimental conditions — e.g. the purity of the radiation. In the spectral regions close to an absorption frequency, both n and k are large and it is necessary to make at least two different measurements in order to calculate these optical constants. In a sense, it is the positions and strengths of the absorptions which give the material its optical properties, and it is these which one wishes to evaluate. The regions of strong absorption are, therefore, the most important regions for determining the classical model parameters.

It is desirable to chose two measurements that can be easily and accurately made and that are sufficiently independent to give a good determination of both optical constants. One could chose two reflectance measurements with different polarizations, or with different angles of incidence, or with a combination of these. Both instrumental and calculational difficulties will influence the choice of the pair of measurements.

The problem of calculating the optical constants from a pair of reflectance values at a particular radiant frequency is becoming simpler for most workers because of the availability of programmable high speed computers. Such computers must be fed with a program and with data which together will yield accurate results. The program will usually be one of choosing approximate optical constants and having the computer test these constants against the measured data, adjusting the constants for better fit, testing again, adjusting, etc., until the machine closes on a final result of sufficient accuracy. In order to chose the experimental conditions and program of calculation, it will be helpful to envision the routine by which the computer closes on the final result. It is desirable to know the sensitivity of the result to small changes in the reflectance data and to be sure that the closure is possible and rapid. For this purpose a simple graphical technique for the calculation can help in the visualization and appreciation of the process before the computer goes to work.

a) Graphical constructions for reflectance.

24. A very simple graphical relation for visualizing the relationship between the reflectance and the optical constants is shown in Fig. 10, and its construction will be described here. The reflectance of a single surface may be obtained from (10.7) and (10.13) as

$$\hat{r} = r \exp i\,\Phi_r = (1-\hat{\chi})\,(1+\hat{\chi})^{-1} \tag{24.1}$$

for either s- or p-polarization, accordingly as $\hat{\chi}_s$ or $\hat{\chi}_p$ is used in the formula. Let us consider both polarizations together by omitting the subscripts. This relation (24.1) is a linear fractional transformation (a bilinear or a Mobius transformation) which transforms circles in $\hat{r}$ space into circles in $\hat{\chi}$ space (and conversely) with lines as limiting cases of circles.

We wish to show that circles of constant reflectance amplitude in $\hat{r}$ space are circles in $\hat{\chi}$ space and that lines of constant reflectance phase in $\hat{r}$ space are circles in $\hat{\chi}$ space. From (24.1) it is seen that

$$r^2 = \hat{r}\,\hat{r}^* = (\hat{\chi}\,\hat{\chi}^* + 1 - (\hat{\chi} + \hat{\chi}^*))(\hat{\chi}\,\hat{\chi}^* + 1 + (\hat{\chi} + \hat{\chi}^*))^{-1}, \tag{24.2}$$

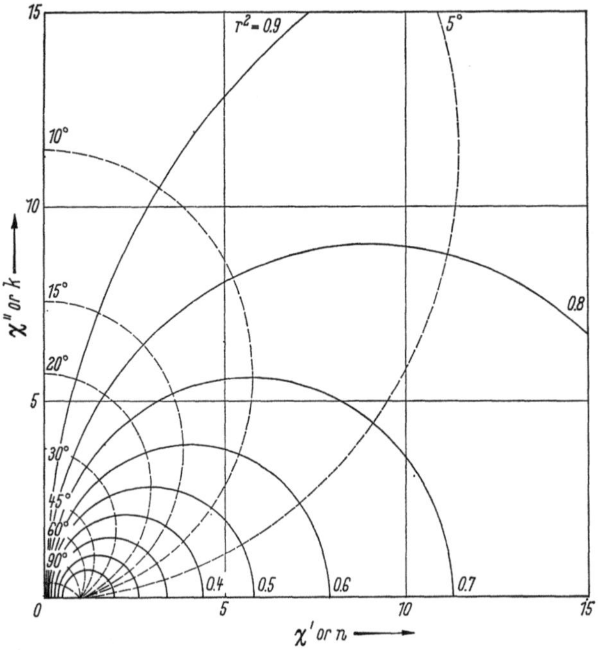

Fig. 10. The power reflectance $\varrho = r^2$ and the phase of the reflectance $(\Phi_r - \pi)$ related to the characteristic parameter $\hat{\chi}$. For a normal incidence reflection the abscissa and ordinates are equivalent to n and k.

in which the superscript * indicates the complex conjugate value. From this it follows that

$$\hat{\chi}\,\hat{\chi}^* + 1 - (\hat{\chi} + \hat{\chi}^*)(1 + r^2)(1 - r^2)^{-1} = 0; \tag{24.3}$$

and thus

$$[\hat{\chi} - (1 + r^2)(1 - r^2)^{-1}][\hat{\chi} - (1 + r^2)(1 - r^2)^{-1}]^* = [2r(1 - r^2)^{-1}]^2. \tag{24.4}$$

This last relation says that for any constant value of r the distance of all $\hat{\chi}$ points from the center at

$$C = (1 + r^2)(1 - r^2)^{-1} \tag{24.5}$$

is

$$(C^2 - 1)^{\frac{1}{2}} = 2r(1 - r^2)^{-1}. \tag{24.6}$$

That is, the locus of all points in $\hat{\chi}$ space corresponding to the same magnitude of the reflectance is a circle with center at $(1 + r^2)(1 - r^2)^{-1}$ and a radius of $2r(1 - r^2)^{-1}$. Several such circles are shown in Fig. 10. The sign convention for the phase shown in Fig. 10 corresponds to the convention established in Sect. 10 and follows from the conventions shown in Fig. 2 and 4.

Again from (24.1) it is seen that

$$\hat{r}\,\hat{r}^{*-1} = [1 - \hat{\chi}\,\hat{\chi}^* - (\hat{\chi} - \hat{\chi}^*)]\,[1 + \hat{\chi}\,\hat{\chi}^* + (\hat{\chi} - \hat{\chi}^*)]^{-1} \tag{24.7}$$

and therefore

$$\hat{\chi}\,\hat{\chi}^* - 1 - (\hat{r} + \hat{r}^*)\,(\hat{r} - \hat{r}^*)^{-1}\,(\hat{\chi} - \hat{\chi}^*) = 0. \tag{24.8}$$

However

$$(\hat{r} + \hat{r}^*)\,(\hat{r} - \hat{r}^*)^{-1} = -i \cot \Phi_r, \tag{24.9}$$

and thus,

$$(\hat{\chi} - i \cot \Phi_r)\,(\hat{\chi} - i \cot \Phi_r)^* = (\sin \Phi_r)^{-2},. \tag{24.10}$$

It is evident, therefore, that for a constant value of the phase of the reflectance Φ_r the locus of $\hat{\chi}$ values are those points at a distance $(\sin \Phi_r)^{-1}$ from the point $\hat{\chi} = i \cot \Phi_r$. Several such constant phase circles are shown in Fig. 10.

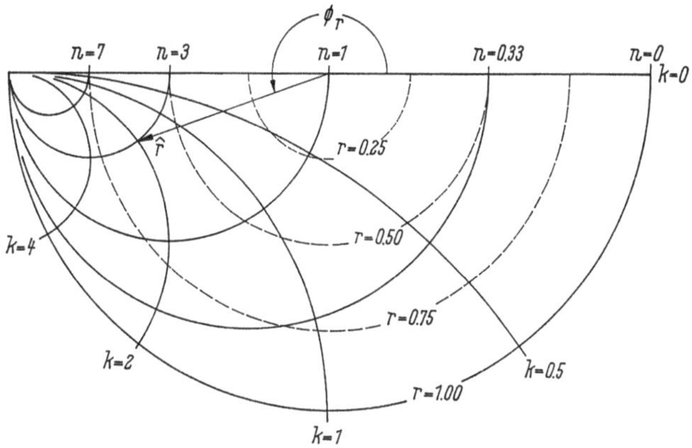

Fig. 11. Smith chart with labeling appropriate to optical reflectance. This chart shows the relationship between the amplitude reflectance $\hat{r} = r \exp i\,\Phi_r$ at normal incidence and the optical constant $\hat{n} = n + i k$. For non-normal incidence χ' and χ'' may be substituted for n and k on this chart.

These constant phase and constant reflectance circle relations are valid for each polarization, p and s, and any reflecting interface and incident angle. The values of $\hat{\chi}$ however are not simple relations of the complex indices of refraction of the interfacial materials. The most useful form of these relations is for normal incidence, $\Theta_0 = 0$, so that the p- and s-polarization distinction disappears, and for the incident wave in air, $n_0 = 1$. In this special situation, $\hat{\chi} = n + i\,k$, and the circle relations indicated in Fig. 10 are particularly simple. The formulae (20.12) and (20.13) may be used to obtain n and k from the normal incidence reflectance and phase.

It is interesting to note that the relationship between $\hat{r}$ and $\hat{\chi}$ is such that a similar set of circle relations must hold for the points in $\hat{r}$ space for constant magnitude of $\hat{\chi}$, and another set of circles for each constant phase of $\hat{\chi}$. These circle diagrams for this linear fractional transformation are useful in visualizing the interrelationships between small variations in $\hat{r}$ and $\hat{\chi}$ in some working region. The fact that the transformation is conformal makes small variations about a point in $\hat{\chi}$ space simply related to small variations in $\hat{r}$ in the neighborhood of the corresponding point. This conformal relationship is helpful in devising computational methods which require the calculation of incremental changes in the dependent variables for the choice of a better fit, next approximation. One can see from Fig. 10, for example, the manner in which small values of k can change the reflectance from the value approximated with k equal to zero.

The relationship between $\hat{r}$ and $\hat{\chi}$ may be presented with r and Φ_r as polar coordinates. The locus of constant χ' values and of constant χ'' values are an orthogonal set of circles in this polar coordinate system. Such a graphic presentation of the relationship is known as a SMITH [22] chart. SMITH charts are easily obtained because they are used in electrical transmission line problems. Such a chart is shown in Fig. 11 and has been adequately described in many electrical engineering texts as well as by SIMON [24]. This chart crowds the region of high values of χ' and χ'' and therefore serves as a good alternate for the rectangular presentation of Fig. 10, which crowds the low χ' and χ'' region.

25. The $\hat{\chi}$ to $\hat{r}$ transformation shown in Fig. 10 can not be used to obtain the two optical constants from a single measurement of reflectance except by adding some other additional information. Two measurements at two different angles of incidence could be used, but the calculation through the use of Fig. 10 would be difficult. A graphical routine for such a calculation, however, has been described by LINDQUIST and EWALD [23]. Their procedure depends upon the construction of the locus of possible values of $\hat{\chi}^2$ for each of the two measured reflectance values and finding the intersection of these two curves as the unique value of $\hat{\chi}^2$ for the material. The routine may be understood by considering the fact that Eq. (24.4) may be written in the form

$$\hat{\chi}=C+(C^2-1)^{\frac{1}{2}}\exp i\,\Phi \tag{25.1}$$

so that the circle in $\hat{\chi}$ space for a reflectance r is described by its center at C, Eq. (24.5), with a radius $(C^2-1)^{\frac{1}{2}}$, Eq. (24.6). The value of the construction angle Φ is, of course, a function of the unmeasured phase of the reflectance Φ_r. In $\hat{\chi}^2$ space, this circle has been transformed into an epicycloid with the form

$$\hat{\chi}^2=C^2+2C(C^2-1)^{\frac{1}{2}}\exp i\,\Phi+(C^2-1)\exp i\,2\Phi. \tag{25.2}$$

For s-polarized radiation in air, this can be transformed to $\hat{n}^2$ space through

$$\hat{\chi}_s^2=(\hat{n}^2-\sin^2\Theta_0)(\cos\Theta_0)^{-1}, \tag{25.3}$$

from (10.6) and (9.22). Thus we find,

$$\left.\begin{aligned}\hat{n}^2=&[C^2\cos\Theta_0+\sin^2\Theta_0]+\\&+[2C(C^2-1)^{\frac{1}{2}}\cos\Theta_0]\exp i\,\Phi+[(C^2-1)\cos\Theta_0]\exp i\,2\Phi,\end{aligned}\right\} \tag{25.4}$$

which is of the form

$$\hat{n}^2=A+B\exp i\,\Phi+D\exp i\,2\Phi. \tag{25.5}$$

The values of A, B, and D may be obtained from the measurement of the power reflectance $r^2=\varrho$ at an angle of incidence Θ_0, with s-polarized radiation from the relations:

$$C=(1+r^2)(1-r^2)^{-1}, \tag{25.6}$$

$$A=C^2\cos\Theta_0+\sin^2\Theta_0, \tag{25.7}$$

$$B=2C(C^2-1)^{\frac{1}{2}}\cos\Theta_0, \tag{25.8}$$

$$D=(C^2-1)\cos\Theta_0. \tag{25.9}$$

Two such measurements of r at different angles of incidence allows the construction of two different epicycloids in $\hat{n}^2$ space. The intersection of these two epicycloids is the value of $\hat{n}^2=\varepsilon'+i\,\varepsilon''$ for the material. The optical constants can then be calculated from the relations (18.1) and (18.2).

The graphical construction of such epicycloids is illustrated in Fig. 12. The construction proceeds in the following manner: first, lay off the vector $\boldsymbol{A}$ of length A along the ε' axis; second, construct the circle of radius B about the head of vector $\boldsymbol{A}$, locating the pivot point P; third, lay off the vector $\boldsymbol{B}$ at an arbitrary angle 2Φ from the ε' axis; fourth, lay off the vector $\boldsymbol{D}$ of length D from the head of vector $\boldsymbol{B}$ away from the circle along the line directed from the pivot point P to the head of $\boldsymbol{B}$, thus obtaining one point on the epicycloid at the head of vector $\boldsymbol{D}$; fifth, chose other angles 2Φ and construct other points on the epicycloid in the neighborhood of the $\hat{n}^2$ value as estimated from previous information or experience. Such a graphic construction can be as accurate as one wishes and surely as accurate as the experimental values of reflectance would justify.

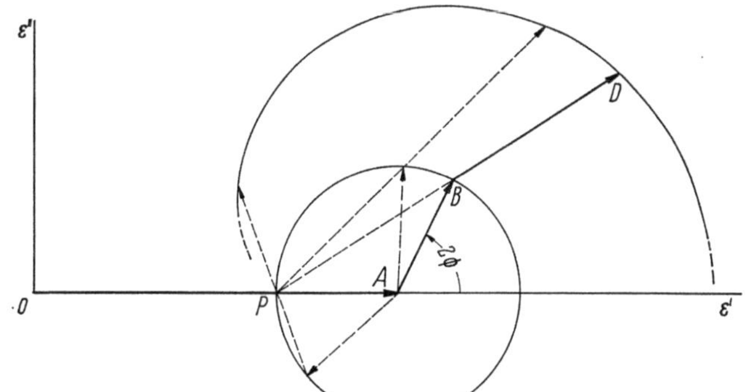

Fig. 12. Graphical construction of epicycloids of constant reflectance by the method of LINDQUIST and EWALD [23] as described in the text.

The construction for p-polarization can not proceed in the same way, unfortunately, because $\hat{\chi}_p^2$ is a different function of $\hat{n}$ and Θ_0. LINDQUIST and EWALD have shown, however, that a similar construction for the ratio $r_p^2\, r_s^{-2}$ can be used. This ratio is more sensitive to the properties of the material than r_s^2 and may also be easier to measure. From the Fresnel relation (10.26) for the situation $n_0=1$, we find

$$-\hat{r}_p \hat{r}_s^{-1} = (1-\hat{\chi}_r)\,(1+\hat{\chi}_r)^{-1}. \tag{25.10}$$

with $\hat{\chi}_r$ defined as

$$\hat{\chi}_r = (\hat{n}^2 - \sin^2\Theta_0)^{\frac{1}{2}}\,(\sin\Theta_0 \tan\Theta_0)^{-1}. \tag{25.11}$$

The relation corresponding to (25.5) has the coefficients for the polarization ratio measurement as follows:

$$C = (1 + r_p^2\, r_s^{-2})\,(1 - r_p^2\, r_s^{-2})^{-1}, \tag{25.12}$$

$$A = C^2 \sin^2\Theta_0 \tan^2\Theta_0 + \sin^2\Theta_0, \tag{25.13}$$

$$B = 2C(C^2-1)^{\frac{1}{2}} \sin^2\Theta_0 \tan^2\Theta_0, \tag{25.14}$$

$$D = (C^2-1)\sin^2\Theta_0 \tan^2\Theta_0. \tag{25.15}$$

Thus it is possible to construct epicycloids on the $\hat{n}^2$ plane for constant values of the measured ratio $r_p^2\, r_s^{-2}$ and these can be used to locate the value of $\hat{n}^2$ for the material.

Another manner of mapping the reflectivites in the dielectric constant plane has been described and used by GEICK [12]. The rectangular coordinates chosen by GEICK were $\log|\varepsilon|$ and the phase angle $\arctan(\varepsilon''/\varepsilon')$. In this coordinate system the isoflectance and the isophase curves as well as the isotransmittance of thin films were plotted. Many kinds of measurements — reflectance at several angles, with both polarizations, and transmittance of thin films — can be used to locate curves on this diagram. Since each pair of measurements determines a possible set of the optical constants, there can be inconsistencies in the optical constant values obtained. This diagram can be used to visualize the range of uncertainty in the optical constants determined from the various types of experimental data.

b) Precomputed charts.

26. It is possible to prepare charts relating the calculated results of measurements of the reflectance for a large set of values of optical constants. If the values chosen in the computations are the same as that of a measured sample then it would be possible to identify the optical constants from the measured reflectances. With a scheme of interpolation, it will not be necessary to have the exact value of the optical constants of the material represented by the particular values on the charts. If the grid of chart values is fine enough, then the interpolation can obviousy be a very good one. This method of determining the optical constants requires, however, that at least two charts be used for at least two different measurements of the material.

It is evident that a single measurement will yield a set of possible values of the two optical constants. A second measurement, related to the optical constants in a different manner, will need to be made and a second chart will need to be searched for a second set of possible values of the two optical constants. From these two sets of optical constants there should be one pair of optical constants in common — the optical constants of the material.

SIMON [*24*] has published charts for optical constant determinations from reflectance measurements and has described their use. These popular charts have been reproduced in a book by BRUGEL [*25*], in a review article by FAHRENFORT [*26*], and by LECOMTE [*27*] and are therefore easily obtained. The reflectance of the material must be measured at two angles of incidence and the polarization must be controlled. Because measuring instruments, spectrometers particularly, are not equally sensitive to all polarizations, it is necessary to account for this polarization in the calculation or to select the proper polarization with polarizers. SIMON has described a correction technique which may be used if the instrument has a known but incomplete polarization. This polarization problem is a hinderance to the effective use of the precomputed chart technique with power reflectance measurements. SIMON chose angles of incidence of 20 and 70° and both s- and p-polarizations for his charts. The choice of angles of incidence is discussed by BEATTIE [*28*] and HEILMANN [*29*]. A different form of such precomputed reflectance charts has been published by SASAKI and ISHIGARO [*30*].

The measurement of the ratio of the p- to s-polarized reflectances together with the determination of the relative phases, that is $r_p r_s^{-1}$ and $\Phi_{rp} - \Phi_{rs}$, will give the information needed to calculate both of the optical constants. The experimental technique for measuring these quantities has been described by BEATTIE [*28*] and used for the determination of the optical constants of metals. BEATTIE and CONN [*31*] have published a precomputed chart for use with these data. The highest accuracy in the use of this chart is for angles of incidence neighboring the principal angle and it might be expected that the principal angle measurement would as simple to measure. The calculation of the optical constants from principal angle measurements, however, is much simpler.

Other precomputed charts have been published by AVERY [32] for use with $r_p^2 r_s^{-2}$ measurements at several angles of incidence and by PRISHIVALKO [33] for use with Stokes parameter measurements of the reflected radiation.

PRISHIVALKO [34][1] has compared the Beattie technique, polarization ratio and phase angle, with the Simon technique, two angle measurement. He reports that with a power reflectance error of 0.005 his computations show the Simon technique to be better with optical constants in the range $0.4 < n < 2.4$ with $0.2 < n\,k < 3.0$. For $n > 3$ and $n\,k > 3$, he found the method of BEATTIE to be superior. The errors were large for both methods with $n\,k < 0.2$.

c) Special angles of reflectance determination.

27. The determination of the optical constants of a solid by measuring the critical angle for total internal reflection or by measuring BREWSTER's angle (13.1) have not been generally useful. The simple relations for these angles as a function of the optical constants hold only for non-absorbing dielectrics. Even for this restricted class of materials, the same information can be obtained by a careful normal incidence reflectance measurement. The measurement of these special angles has been more useful for the determination of optical constants of liquids than for solids. The special angle determinations do remove the difficulty of measuring quantitatively both the incident and the reflected radiation which would be required for a reflectance measurements. The special angles can be determined by locating a maximum or a minimum reflected power signal as the angle is scanned, and thus are somewhat simpler to determine than the reflectance itself. The angles, however, can not be located accurately if a small sample must be measured. The small sample must be measured with a large solid angle of incident radiation if there is to be sufficient radiant power for adequate sensitivity. Thus the small sample will have a large uncertainty in the value of the special angle of reflectance.

The critical angle (12.3) is not easily measured on solid materials unless they are very transparent and can be formed into the proper shape for such measurements. A material of these favorable characteristics can be formed into a prism and the minimum angle of deviation can be measured. Such a prism measurement of the index of refraction should have a high precision.

For BREWSTER's angle measurement, it is necessary to have polarizers. If such polarizers are available then the measurement of the principal angle of incidence and the principal azimuth can be made. The instrumental complications may be more than offset by the extra information which is obtained. These characteristic angles lead to the determination of both of the optical constants from (14.3) or (14.5) and (14.6). A description of such instrumentation and the errors in such characteristic angle measurements has been given by CONN and EATON [35] and [36]. The graphical method of LINDQUIST and EWALD, Sect. 20, may also be used to compute the optical constants from the measurements of $r_p^2 r_s^{-2}$.

For large values of $|\hat{n}|$ the principal angle of incidence is large. This is evident from (14.4) which shows that $|\hat{n}| = \sin\Theta_P \tan\Theta_P$ for $|\hat{n}| \gg 1$. Large angles of incidence are difficult in an experimental measurement because of the large-sized sample required.

---

[1] *Note added in proof.* An analysis of the errors in determining the optical constants from measurements of reflectance at various angles of incidence has also been given by HUNTER, W. R., in J. Opt. Soc. Am. **55**, 1197 (1965).

d) Dispersion relation calculations.

28. One of the most fruitful methods of determining the optical constants of solid materials in the infrared is the use of the dispersion relation (20.9) to obtain the phase of the reflectance from measured values of the amplitude of the reflectance over a wide range of frequencies. The power reflectance ϱ may be used in place of the amplitude reflectance in (20.9) by the replacement $\log r = \frac{1}{2}\log\varrho$. The knowledge of both r and Φ_r allows the computation of n and k from (20.12) and (20.13). The utility of this dispersion relation has been demonstrated by the results of many workers: ROBINSON's [*37*] measurements on polythene; GOTTLIEB's [*38*] measurements on LiF; GEICK's [*12*] measurement on NaCl; and PHILIP and TAFT's [*39*] measurement on Germanium are examples.

The difficulty of computation of the integral in the dispersion relation (20.9) is becoming a minor one because of the growing availability of high speed computers. Even without such a computer the integration can be obtained with reasonable speed and accuracy by a semigraphical method that has been described by BALKANSKI and BESSON [*40*]. Their method replaces the integral in (20.9) with a sum of $\log r(\omega')$ values and utilizes a graphical method to chose the sequence of ω' values so that the weighting function $(\omega^2-\omega'^2)^{-1}$ is included. The ω' values are close together where $(\omega^2-\omega'^2)^{-1}$ is large and conversely. In order to simplify the procedure, the $\log r(\omega)$ portion of the integrand is omitted.

Two problems of major concern in the use of the dispersion relations to determine the phase of the reflectance are the following: first, the influence of the reflectance at frequencies beyond the range of the measurements; and second, the influence of errors in the measured reflectance values. The reflectance in the spectral regions outside of the range of measurement may be approximated by various analytical expressions. Because of the weighting factor $(\omega^2-\omega'^2)^{-1}$ in the integrand of (20.9), the greatest influence on the calculated phase at any particular frequency will come from the reflectance values for closely neighboring frequencies. The integrated contribution from the infinite extent of distant frequencies can not be omitted, however, in the evaluation of the integral.

If the measured reflectance data covers a large range of frequencies and can be sensibly extrapolated to all frequencies, then the phase calculation can proceed with some degree of success. The replacement of the low frequency end, $0\leqq\omega\leqq\omega_a$, with a constant reflectance r_0 and the high frequency end, $\omega_b\leqq\omega\leqq\infty$ with a constant reflectance r_∞ is a reasonable approximation for many situations. This is particularly appropriate for measurements over isolated reststrahlen regions, for example.

In the region $\omega_a\leqq\omega\leqq\omega_b$ the phase $\Phi_r(\omega)$ can be calculated as the sum of the four portions

$$\Phi_r(\omega)=\Phi_r^{(0)}(\omega)+\Phi_r^{(1)}(\omega)+\Phi_r^{(\infty)}(\omega)+\pi, \tag{28.1}$$

with

$$\Phi_r^{(0)}(\omega)=\pi^{-1}(\log r_0-\log r(\omega))\left(\log(\omega_a+\omega)-\log|\omega_a-\omega|\right), \tag{28.2}$$

$$\Phi_r^{(1)}(\omega)=2\omega\,\pi^{-1}\int\limits_{\omega_a}^{\omega_b}(\log r(\omega)'-\log r(\omega))\,(\omega^2-\omega'^2)^{-1}\,d\omega', \tag{28.3}$$

$$\Phi_r^{(\infty)}(\omega)=\pi^{-1}(\log r_\infty-\log r(\omega))\left(\log|\omega_b-\omega|-\log(\omega_b+\omega)\right). \tag{28.4}$$

From the above formulae, it is seen that the limiting value of $\Phi_r^{(0)}(\omega)$ as ω approaches ω_a is $\Phi_r^{(0)}(\omega_a)=0$. Likewise, $\Phi_r^{(\infty)}(\omega_b)=0$. The machine computation of the integral in (28.3) can be carried out but the calculation must be programmed especially to provide for the undefined integrand value at $\omega=\omega'$. This can be done, for example, by assigning the average value for the integrand at the next neighboring points $\omega' = \omega - \Delta\omega$ and $\omega' = \omega + \Delta\omega$ for the point $\omega' = \omega$. This procedure is equivalent to evaluating the integrand by L'HOSPITAL's rule at this point.

If the high frequency reflectance can not be smoothly connected to a constant reflectance region, then other approximations will be necessary. A power series expansion for the phase correction produced by the unmeasured reflectance regions has been used by VELICKY [*41*]. If the phase were known by some other means, such as a transmission measurement, for example, then the use of the series expansion will help evaluate the correction to the phase calculated from the dispersion relations. Results obtained by the use of a single term power series expansion are given by SANDERSON [*5*]. Instructive detailed results with the use of a constant reflectance r_∞ are shown for NaCl by GEICK [*13*][1].

The second source of inaccuracy in the dispersion relation determined phase is the experimental error in the measured reflectance. This error is particularly important in the regions of low reflectance where the absolute value of $\log r(\omega)$ is large. The fractional error in the experimentally measured reflectance in such regions is large, and the computed phase is therefore subject to large errors. The low reflectance region on the high frequency side of a lattice band is such a region as can be seen in Fig. 8. Here, near the longitudinal lattice frequency, the reflectance is small; n is near unity; r is approximately $-i\,k/2$; and ϱ is approximately $k^2/4$. In this region it may be possible to measure k in transmission and then to compute a value for the reflectance. In this region, however, k is varying rapidly with the frequency and is not so small that thick lamellae can be used in transmission measurements. In general the errors of measurement of the reflectance lead to poor values of k when k is small. SPITZER and KLEINMAN [*19*] have described this difficulty with the small k regions in their work on quartz.

The important advantage in the use of the dispersion relations on reflectance data is the fact that normal incidence reflectance measurements with no polarization control can be used. The samples may be small, therefore, and the measuring instruments may be simple.

e) Direct measurement of the phase of the reflectance.

29. The advantage of the simplicity of the normal angle reflectance measurement and the dispersion relation phase angle calculation is degraded by the necessity of making measurements over a wide range of frequencies. If, at a particular frequency, a direct measure of the phase were available as well as the magnitude of the reflectance, then the optical constants would be determined for that frequency. Such phase measurements are possible in the far infrared region of the spectrum, and, in principle, can be obtained at higher frequencies.

One phase measuring technique, described by BELL [*42*], consists of the comparison of the reflectance from the sample with that from a standard mirror. The mirror is placed in one arm of the Michelson interferometer and the displacement of the fringes which is produced by exchanging the mirror with the sample is measured. This displacement determines the phase of the reflectance. The measurement need not be made with monochromatic radiation if a reasonable portion of the interferogram — the interferometer output signal as a function of the dis-

[1] *Note added in proof.* A method for optimizing the choice of ω_a and ω_b in (28.3) and fitting artificial wings to the reflectance peak is given by ANDERMANN, G. A., A. CARON, and D. A. DOWS in J. Opt. Soc. Am. **55**, 1210 (1965).

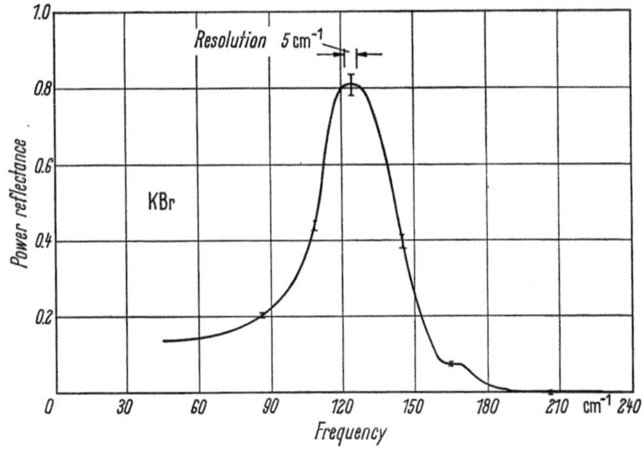

Fig. 13. Reflectance of KBr derived from interferograms as described in the text. The vertical lines on the curve indicate the limits of the range of values obtained from several different interferograms.

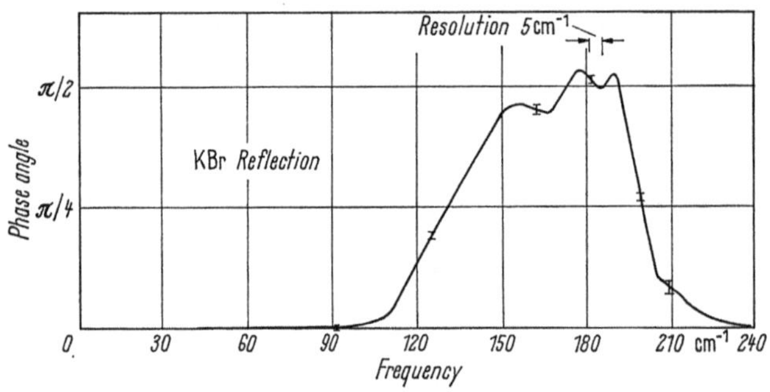

Fig. 14. The phase ($\Phi_r - \pi$) of the KBr reflectance obtained from interferograms by the method described in the text. The vertical lines on the curve indicate the limits of the range of values obtained from several different interferograms.

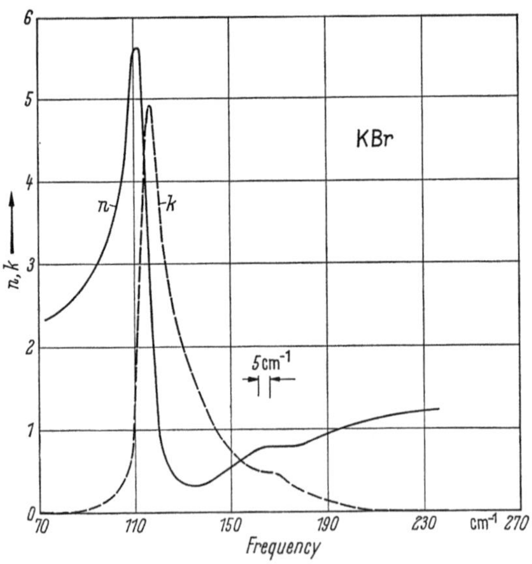

Fig. 15. The optical constants of KBr derived from the data shown in Figs. 13 and 14. The low resolution disturbs the peak values of n and k.

placement of the reference mirror in one arm of the interferometer — is measured. The interferogram obtained with the standard mirror will be symmetric about the central "white light" position, and a Fourier transformation of this interferogram will give the spectral power density of the radiation passing through the interferometer. The Fourier transformation of the sample interferogram, which is not symmetric, will give the spectral crosspower density with amplitude and phase information. By comparison of the two spectra, it is possible to obtain all of the spectral information about the reflectance of the sample both in amplitude and in phase.

The accuracy of the phase measurement is determined, of course, by the experimental techniques which have not been fully developed. Figs. 13 and 14 show the amplitude and the phase of the reflectance of the sample derived from such interferograms. The optical constants derived from this data through (20.12) and (20.13) are shown in Fig. 15. Because of the low resolution, the values of n and k in Fig. 15 can not be accurate at their peaks. The results given in these figures are from recent work by BELL and RUSSELL [*43*]. The simplicity of the laboratory instrument and the direct and intimate relation of the results to the optical constants of the sample make this technique very attractive. The precision of the determination, however, will be sensitive to the flatness of the reflecting surface since the surface position must be well defined for the phase to have meaning.

There are advantages in the interferogram technique in reducing the influence of detector noise in the spectral results. These advantages, which were pointed out by FELLGET [*44*], are obtained for the phase measurement also. The fact that the phase measurements are reasonable even in the low reflectance regions where the signal is small, lends encouragement to further development of this interferometer technique.

III. Measurements on a lamella.

30. A useful sample configuration is the plane parallel surfaced lamella. This configuration is the common one for transmission measurements on materials with small or moderate extinction coefficients. Even for samples which are highly absorbant, such a sample configuration is convenient if the lamella can be made thin enough for measurable transmission.

The emphasis in these sections about lamellae will be on transmission measurements because such measurements are the simplest to make. The angular orientation of the sample in the radiation beam is not as critical as it would be for reflection measurements, and information may be obtained even when the surfaces are too poor for reflectance measurements to be meaningful. A normal incidence measurement is the natural one to make, and it will be presumed that such is the situation in all of the discussion which follows unless otherwise noted.

It can generally be assumed that materials which are so transparent that the transmittance can be easily measured with a thick lamella sample will have a small extinction coefficient, and $k \ll n$. Thus the approximations, from (15.15) to (15.18), with $\omega\, c^{-1} = 2\pi\,\nu$,

$$\hat{a} = a \exp i\,\Phi_a = \exp(-2\pi\,\nu\,k\,b) \exp i(2\pi\,\nu\,n\,b), \tag{30.1}$$

$$\hat{r} = r \exp i\,\pi = -r = (1-n)(1+n)^{-1}, \tag{30.2}$$

$$n = (1+r)(1-r)^{-1}, \tag{30.3}$$

and

$$k = (2\pi\,\nu\,b)^{-1} \log(a^{-1}) \tag{30.4}$$

will usually be appropriate.

The relations (15.7) and (15.8) for the power transmittance and the power reflectance of lamellae are the basic relations which will be discussed and used in the following sections.

a) Resolved channeled spectra.

31. For lamellae which are sufficiently thin, which have a sufficiently small extinction coefficient, and which are measured with a spectrometer of sufficiently high resolution — narrow spectral band width —, the transmittance spectrum will show a channeled spectrum. Several such spectra are shown in Fig. 16. The major factor in producing this channeled spectrum appearance is the variation of the phase angle Φ_a, see (15.8), with spectral frequency. The variation of Φ_a depends upon $\hat{n}$, the thickness of the lamella b, and the frequency ν according to (15.9) or, with approximations, (30.1).

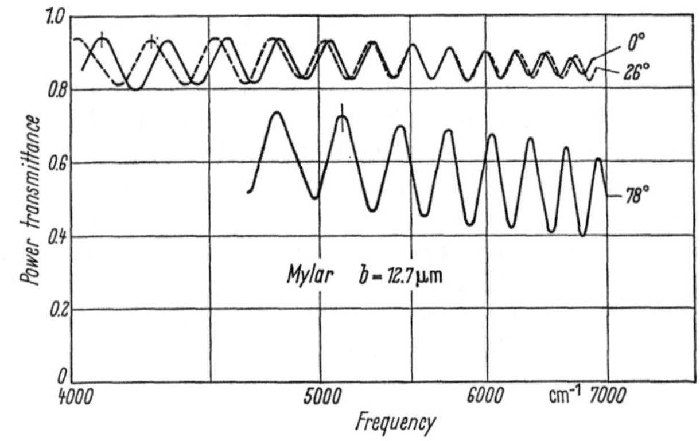

Fig. 16. The channeled spectra of Mylar film of 12.7 microns thickness measured with several angles of incidence, Θ_0. The vertical marks on the three spectra all correspond to peaks of the same order number.

32. Thick, transparent lamella. A thick, transparent lamella must have a small value of the extinction cofficient k and thus the phase angle for the reflectance from a single surface must be π radians according to FRESNEL's relations (10.21) and (10.22) or the relations (15.16). From the relation (15.8) for the power transmittance τ it is seen that the maximum transmittance will be for values of Φ_a equal to even multiples of $\pi/2$ and that the minimum value of the transmittance will be for Φ_a equal to odd multiples of $\pi/2$. Thus the maximum value of the power transmittance τ_M is

$$\tau_M = a^2(1-r^2)^2(1-a^2r^2)^{-2} \tag{32.1}$$

and the minimum value τ_m is

$$\tau_m = a^2(1-r^2)^2(1+a^2r^2)^{-2}. \tag{32.2}$$

For many situations the values of the optical constants are such slowly varying functions of the frequency that the values of a^2 and r^2 in (32.1) and (32.2) are the same for neighboring τ_M and τ_m frequencies. Thus these values of a^2 and r^2 can be computed from (32.1) and (32.2) as though they were constants. The relations for the computations are:

$$y = x[1+(1-x^{-2})^{\frac{1}{2}}] \tag{32.3}$$

with

$$y = r^{-2} \quad \text{for} \quad x = 2\tau_M\tau_m(\tau_M-\tau_m)^{-1}+1 \tag{32.4}$$

and

$$y = a^{-2}r^{-2} \quad \text{for} \quad x = 2\tau_m(\tau_M-\tau_m)^{-1}+1. \tag{32.5}$$

It is apparent that, for x large, y is approximately equal to $2x$, and this approximation is frequently as good as the experimental data justifies. Because $y^{-1}+y=2x$ it is necessary to check the computation to see whether y or y^{-1} has been computed from the experimental data. Poor resolution of the channeled spectrum will reduce the value of $\tau_M-\tau_m$ and thus give computed values of a^2 and r^2 larger than the correct values. From the values of a^2 and r^2, together with the lamella thickness and frequency, the values of n and k can be computed from (30.3) and (30.1).

33. The index of refraction may be obtained more accurately from the channeled spectrum fringe spacing. The phases for the maxima τ_M must be integral multiplies of π radians, that is

$$\Phi_a=2\pi\,\nu_M\,n\,b=M\pi \tag{33.1}$$

for M an integer, and ν_M the frequency for the M-th order maximum, The obvious method for measuring n is to measure the separation $\nu_{M+N}-\nu_M$ between the maxima of order $M+N$ and M and then to use

$$n=N\big(2b\,(\nu_{M+N}-\nu_M)\big)^{-1}. \tag{33.2}$$

For $N=1$ or -1, it is expected that (33.2) would give a value of n appropriate to the frequency neighborhood of ν_M. Since the channeled spectrum peaks are not sharp, their frequencies can not be precisely measured, and values of N greater than one may be needed to obtain an accurate value of n. Large values of N, however, yield values of n averaged over a large frequency range.

As pointed out by Moss [45], however, the use of (33.2) may lead to serious error if n is not a constant but, instead, has a value $n=n_\infty-A\nu^{-1}$, with A a constant, over the range of measurement. This can be a good approximation for some materials, with a non-trivial value of A. The relation corresponding to (33.1) becomes

$$2\nu_M\,n_\infty\,b=M+2A\,b \tag{33.3}$$

for this situation. The measurement of the frequency differences for the channeled spectrum peaks gives

$$n_\infty=N\big(2b\,(\nu_{M+N}-\nu_M)\big)^{-1}. \tag{33.4}$$

Thus the constant A and the true index of refraction n can not be determined from the fringe spacing in this case.

The utilization of the fringe formula (33.1) is good if the order M can be determined by some method. In the far infrared and with low values of M, it may be possible to deduce or "count" the order by a reasonable extrapolation of the channeled spectrum to zero frequency. The use of very thin sample may help to obtain low M values. In these cases (33.1) may be used directly to obtain values of n at the frequencies ν_M.

34. An attempt to determine M by using another sample of different thickness may be a help. For example, if a particular order M_1 for a lamella of thickness b_1 is at the same frequency as an order M_2 for a lamella of thickness b_2, then $(b_1-b_2)\,b_1^{-1}=(M_1-M_2)\,M_1^{-1}$. By having b_1-b_2 small enough so that M_1-M_2 is small, it is possible to identify the integer M_1-M_2 and thus determine M_1 and then the index of refraction from (33.1).

A similar result may be obtained by rotating the sample lamella in the beam to a non-normal orientation. The rotation of the lamella changes the "effective thickness" so that

$$\hat{a}=\exp i\,2\pi\,\nu\,b\,(\hat{n}^2-\sin^2\Theta_0)^{\frac{1}{2}}=\exp i\,2\pi\,\nu\,b\,(n\cos\Theta+i\,k/\cos\Theta) \tag{34.1}$$

according to (15.1), (9.21), and (15.14). The rotation from normal incidence produces the same change in the channeled spectrum fringe positions as though the lamella were made thinner, and it produces the same attenuation changes as though the lamella were made thicker. These changes are proportional to the cosine of the angle Θ of the radiation inside of the lamella. Fig. 16 shows a such channeled spectra from a lamella at normal incidence and in a rotated position for comparison. The rotation moves the maximum of any order number to a higher frequency and for a particular value of Θ_0 makes the maximum of order $M-1$ have the same frequency as the order M at a smaller angle.

One simple way to use this fringe shift with orientation of the lamella is to consider the situation with the approximation that k is negligible. Let Θ_M be the angle of incidence which has the maximum of order M at the frequency ν and let Θ_{M-1} be the angle which has the maximum of order $M-1$ at the same frequency, then

$$4\nu^2 b^2 (n^2 - \sin^2 \Theta_M) = M^2 \tag{34.2}$$

and a corresponding relation for angle Θ_{M-1}. From these two relations it can be found that

$$4\nu^2 b^2 (\sin^2 \Theta_{M-1} - \sin^2 \Theta_M) = 2M-1 . \tag{34.3}$$

Thus a knowledge of the angles, the thickness, and the frequency makes a unique determination of the order of the particular fringe maximum and, therefore, of all the neighboring fringe maxima. The precision of the alignment of the nearly sinusoidal fringes must be made with sufficient accuracy to be sure that a good measure of the order can result. The alignment must be more accurate than one-M-th of a cycle of the channeled spectrum. When this lamella rotation method can be used, it has the advantage of giving a value for the order by measurement at one frequency and thus does not suffer from the fringe spacing defect pointed out by Moss and given in Eq. (33.4). With high order numbers, the alignment may not be sufficiently accurate to use this method and with low orders the method may not be needed.

The effective change in the optical thickness of a lamella upon rotation has been used by CHAMBERLIN and GEBBIE [*46*] with a monochromatic laser source in the far infrared. The lamella measured by them was placed in one arm of a Michelson interferometer. The total path change in that arm produced by the insertion of the lamella is $2b\left((n^2 - \sin^2 \Theta_0)^{\frac{1}{2}} - \cos \Theta_0\right)$. The values of Θ_0 which produced integral wavelength changes in the path were observed, and from these angles and the thickness of the lamella the index of refraction was determined.

35. The channeled spectrum may be seen in reflection, of course, and similar relationships can be formulated to determine the optical constants from the reflectance.

HOHLS [*47*] has used reflectance channeled spectra to measure the indices of refraction of alkali halides. The use of a prism of these materials and a measurement of the minimum angle of deviation had given precise values of the indices, but HOHLS' method remains of interest for the absorbing spectral regions not measurable with a prism. The surfaces of the lamella were coated by HOHLS to increase their reflectivity and thereby enhance the multiple internal reflection phenomenon which produces the channeled spectrum. One surface was made completely reflecting by the deposited metal and the other only partially reflecting. The reflectance was measured from the partially reflecting side of the lamella.

A related method, used by HOHLS, was that of the measurement of TALBOT'S bands. TALBOT'S bands are produced by the insertion of the lamella in a portion of the collimated beam through the dispersing element of a spectrometer. The lamella produces an optical path change of $(n-1)\,b$ in part of the beam and thus produces periodic intensity variations in the spectrum of a continuous source. The lamella must be placed in the proper portion of the collimated beam, see DITCHBURN [48], to produce the bands.

The Talbot band method is not so demanding upon the quality of the lamella because it measures $(n-1)$ rather than n. The variations in the thickness of the lamella do not degrade the spectral pattern as seriously as they would in a channeled spectrum. The Talbot band measurements by HOHLS extended into the long wave absorption region even further than the channeled spectrum measurements.

36. Thin lamellae. Transmission measurements of absorbing materials must be made with thin lamella. The channeled spectrum order number M (33.1) will be small, therefore, and the spectral interval $\nu_{M+1}-\nu_M$ between orders will be large and perhaps not recognizable as a channeled spectrum. A transmission measurement on a lamella which is thin compared to the wavelength of radiation mainly serves to measure the quantity $2n\,k\,\nu$. In order to show this, consider a normal incidence measurement on a lamella of such thickness that the parameter defined by $\beta=\omega\,c^{-1}\,b=2\pi\,\nu\,b$ is small enough that $\hat{a}=\exp i\,\hat{n}\,\beta$ can be represented by $\hat{a}=1+i\,\hat{n}\,\beta-\hat{n}^2\,\beta^2/2$. This condition is equivalent to $2\pi\,\nu\,|\hat{n}|\,b\ll 1$ and terms containing β to higher powers than the square will be omitted in this discussion of thin films.

For normal incidence $\hat{r}=(1-\hat{n})\,(1+\hat{n})^{-1}$ and $\hat{r}^{-1}-\hat{r}=4\hat{n}(1-\hat{n}^2)$. From (15.4) we find

$$2\,\hat{T}^{-1}=-(1+\hat{n}^2)\,i\,\beta+2(1-\hat{n}^2\,\beta^2/2)+\cdots \tag{36.1}$$

so that the power transmittance is

$$\hat{T}\,\hat{T}^*=\tau=\{1+2n\,k\,\beta+[4n^2\,k^2+(n^2-k^2-1)^2]\,\beta^2/4\}^{-1}. \tag{36.2}$$

This relation may be used with measured values of τ for lamellae of several thickness to find the dielectric constant components $\varepsilon'=n^2-k^2$ and $\varepsilon''=2n\,k$. By placing the relation (36.2) in the form

$$2(\tau^{-1}-1)\,\beta^{-1}=2\varepsilon''+[\varepsilon''^2+(\varepsilon'-1)^2]\,\beta \tag{36.3}$$

it is seen that the left hand member can be graphed as a function of β from experimental data on the transmittance and that the dielectric constant components are then determined from the slope and intercept of the resulting straight line. Such a procedure has been used by GEICK [13] with evaporated films of CsBr.

For β small the power transmittance may be written in the form

$$\tau=1-2n\,k\,\beta+[3\,n^2\,k^2-(n^2-k^2-1)^2/4]\,\beta^2+\cdots \tag{36.4}$$

and one sees that the thin film absorptance $1-\tau$ measures $2n\,k\,\nu$.

Corresponding relations may be written for the reflectance of thin films. From (15.3) and (15.4) it is found that

$$\hat{R}\,\hat{T}^{-1}=(1-\hat{n}^2)\,i\,\beta/2 \tag{36.5}$$

in this approximation, and thus

$$\varrho\,\tau^{-1}=(\beta^2/4)\,[(n^2-k^2-1)^2+4n^2\,k^2]. \tag{36.6}$$

Together with (36.4) it is found that this relation yields the power reflectance of the thin film as

$$\varrho=(\beta^2/4)\,[(n^2-k^2-1)^2+4n^2\,k^2]+\cdots \tag{36.7}$$

It is seen that the thin film reflectance is not so useful as a direct measure of optical constants because it is so small — proportional to β^2.

The special measurement by BERREMAN [*14*] on thin films, described in Sect. 22, to determine the longitudinal optical frequency must be made with non-normal incidence and Eqs. (36.1) to (36.7) do not apply. The theory and techniques for the measurement of optical constants on thin evaporated films deposited on substrates is complicated by the fact that the optical constants of the deposited material may not be the same as the constants for the bulk material. The complications introduced by the addition of the substrate to the optical system also makes the analysis of the results more difficult. The work of ABELES [*49*] on "Methods for Determining Optical Parameters of Thin Films" and the article by ROUARD and BOUSQUET [*50*] on "Optical Constants of Thin Films" should be consulted for an insight into these problems.

b) Unresolved channeled spectra.

37. A frequently occuring experimental situation is the measurement of the transmittance of a lamella which does not exhibit a channeled spectrum appearance because the surfaces are not parallel and/or because the measuring instrument does not have sufficient spectral resolution. The requirement of a high signal to noise ratio for the measurement of accurate quantitative transmittance values is accomplished by obtaining high signal at the expense of spectral resolution. If any such degrading of the spectrum has washed-out the channel spectrum then the measured transmittance values are equivalent to those obtained by averaging the values which would be obtained with the channeled spectrum resolved over a complete cycle of the phase angle Φ_a.

The relations for the transmittance and reflectance of lamella (15.7) and (15.8) can be integrated over a cycle of Φ_a by the standard technique: the trigonometric function of Φ_a is replaced by a function of a complex variable and the resulting integrand is integrated around the origin on a circle of unit radius. The values of the integrals are determined by the residues of the singularities within the unit circle. The resultant average values of transmittance and reflectance are:

$$\bar{\tau}=(1-a^4 r^4)^{-1}[a^2(1+r^4)-2a^2 r^2 \cos 2\Phi_r], \tag{37.1}$$

$$\bar{\varrho}=(1-a^4 r^4)^{-1}[r^2(1+a^4)-2r^4 a^4 \cos 2\Phi_r]. \tag{37.2}$$

For lamella which are not thin and yet have non-zero transmittance, k must be small and Φ_r must be π radians for near normal incidence, (15.16). Thus for this very common situation,

$$\bar{\tau}=(1-a^4 r^4)^{-1} a^2 (1-r^2)^2 \tag{37.3}$$

and

$$\bar{\varrho}=(1-a^4 r^4)^{-1} r^2 (1+a^4-2\,a^4 r^2)=r^2(1+a^2\,\bar{\tau}) \tag{37.4}$$

are the appropriate expressions for the transmittance and reflectance.

The optical constants can be obtained from measurements on this thick lamella from (32.3) and (32.4) through the calculation of the value

$$(1-\bar{\varrho})^2\,\bar{\tau}^{-1}-\bar{\tau}=a^{-2}-a^2=2x \tag{37.5}$$

and thus the calculation of

$$a^2=x[(1+x^{-2})^{\frac{1}{2}}-1]=\exp(-4\pi\,\nu\,k\,b) \tag{37.6}$$

using (30.4) to obtain k.

The measurement of $\bar{\varrho}$, $\bar{\tau}$, ν and b can be used to determine x and thus to determine a^2 and k. Having determined a^2 it is possible to determine r^2 from (37.4) and n from (30.3). The measurements of $\bar{\varrho}$ and $\bar{\tau}$ for a moderately transparent lamella near normal incidence can thus be used to determine the optical constants.

For the situations in which the transmittance is small enough that $a^4 r^4 \ll 1$, it is useful to approximate the transmittance and reflectance by

$$\bar{\tau} = a^2(1-r^2)^2 \approx a^2(1-2r^2) \tag{37.7}$$

and

$$\bar{\varrho} = r^2\left(1 + a^4(1-2r^2)\right) \approx r^2(1+a^2\bar{\tau}) \approx r^2(1+a^4). \tag{37.8}$$

Such approximations may be used in many ways. For example, from a prior knowledge of r^2 a prepared curve of the relation

$$\log\bar{\tau} = 2\log(1-r^2) - 4\pi\nu k b \tag{37.9}$$

can be used to find $k b$ from the measured $\bar{\tau}$. Even if the value of r^2 is not known, the approximation may be known to be valid and transmission measurements on two thicknesses b_1 and b_2 giving transmittances $\bar{\tau}_1$ and $\bar{\tau}_2$ can be used to obtain k from the relation

$$\log(\bar{\tau}_2/\bar{\tau}_1) = 4\pi\nu(b_1 - b_2). \tag{37.10}$$

This is probably the most useful of the simple approximations which together with (30.3) gives the approximate optical constant of weakly absorbing materials.

c) Direct measurement of the phase of the transmittance.

38. From the discussion in Sect. 29 one sees that the phase of the radiation transmitted through a lamella can be determined from an interferogram measured with the lamella as a transmission sample in one arm of a Michelson interferometer. Such a use of the Fourier transform of an interferogram for the determination of the optical constants has been described by Chamberlin, Gibbs, and Gebbie [*51*] and by Bell [*42*]. If an interferogram without the sample is also made for reference, then the Fourier transform of the two interferograms gives the information for calculating both the phase and the amplitude of the transmittance. Fig. 17 shows a reference and a Mylar sample interferogram and Figs. 18 and 19 show the transmittance and the phase of the transmittance for a Mylar sample as measured by Bell and Russell [*43*].

If the index of refraction of the sample were a constant, then the phase for the transmitted partial wave which is not internally reflected in the sample would be $2\pi\nu n b$. The sample interferogram would be symmetric and would have its central "white light" peak shifted an amount $\Delta = (n-1)b$ from the central peak position of the reference interferogram. Because of the dispersion of the sample, the interferogram is not symmetric. The choice of a reasonable value for the zero shift Δ, however, allows the computed phase of the sample interferogram to have a smaller variation with frequency. Fig. 18 shows the computed phase for the Mylar sample with such a shift. The curvature remaining in the variation in phase with frequency results from the dispersion of the sample.

The phase Φ computed from the interferogram with a shift of amount Δ in the zero position, the phase Φ_T for the sample, and the amount of phase $2\pi\nu b$ for the radiation path replaced by the sample, are related by

$$\Phi + 2\pi\nu\Delta = \Phi_T - 2\pi\nu b. \tag{38.1}$$

A factor of two will need to be introduced into this equation, and the following ones, if the radiation passes through the sample twice in the interferometer arm.

The value of Φ_T may be approximated for transparent lamellae from (15.4) with the assumptions that $\hat{r}=-r$, $r\ll 1$, and $\hat{a}=a\exp i\,2\pi\nu n b$, so that

$$\hat{T}=\hat{a}\left(1-\hat{r}^2(1-\hat{a}^2)\right)=T\exp i\,\Phi_T \tag{38.2}$$

and

$$\Phi_T=2\pi\nu n b+\Psi. \tag{38.3}$$

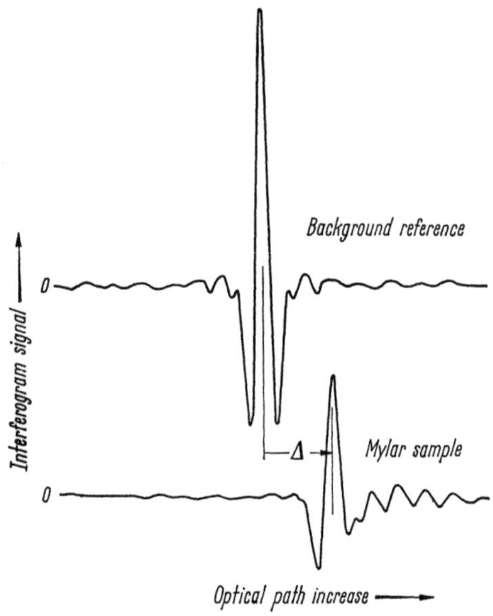

Fig. 17. The central portions of interferograms from a vacuum Michelson interferometer, BELL and RUSSELL [43]. The Mylar sample was inserted for a single passage of the radiation in one arm of the interferometer. The interferograms have been separated vertically for easy comparison.

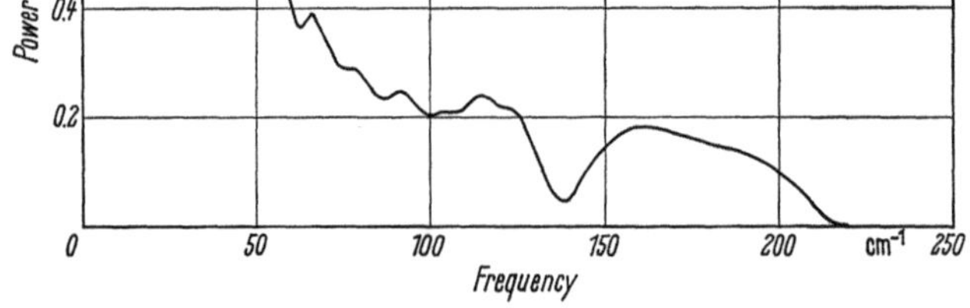

Fig. 18. The power transmittance of a Mylar film of 24.9 microns thickness as obtained from the interferogram pair shown in Fig. 17.

The phase angle Ψ is the channeled spectra addition and corresponds to the phase of the factor $\left(1-\hat{r}^2(1-\hat{a}^2)\right)=(1-\hat{r}^2+\hat{r}^2\hat{a}^2)$ in (38.2). The channeled spectra phase term is approximately[1]

$$\Psi=\left[\arc\sin\left(a^2r^2(1-r^2)^{-1}\right)\right]\sin 4\pi\nu n b. \tag{38.4}$$

[1] A simpler and more reasonable approximation is given in reference [43].

The channeled phase term is normally a small contribution a few cycles away from the zero frequency and can be estimated from approximate values of n and a if the channeled term is recognizable in the phase spectrum. The computed phase from the interferogram is, then,

$$\Phi = 2\pi\,\nu\,(n-1)\,b - 2n\,\nu\,\Delta + \Psi \tag{38.5}$$

and from this relation it is possible to obtain the index of refraction for the frequencies of radiation contributing to the interferogram. The dashed lines on Fig. 19 show the values of Φ in (38.5) for several constant values of n and for no channeled spectrum phase Ψ. The channeled spectrum term is seen contributing to the phase in the low frequency portions in Fig. 19, and, also, the channeled

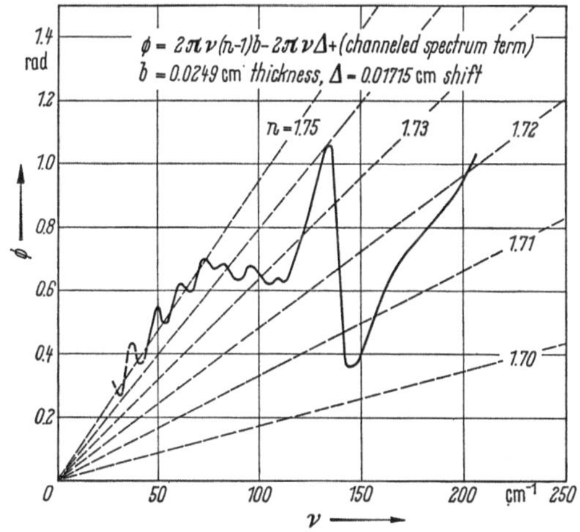

Fig. 19. The relative phase of the Fourier components of the interferograms shown in Fig. 17. The index of refraction of Mylar, neglecting the channeled spectrum contribution, is indicated by the dashed line coordinates on the diagram. Note, particularly, the channeled phase spectra at low frequencies and the anomalous dispersion features.

spectra can be seen in the corresponding portions in the power transmittance τ of the same sample shown in Fig. 18. The anomalous dispersion produced by the absorption of the sample is especially to be noted in this Mylar example. Even in the regions of low transmittance in the Mylar sample, the phase measurement is good and the index of refraction is accurately measured.

The advantages of this technique will surely be further exploited in the future because the apparatus is simple and the measurement precision is high. A high speed, programmable computer is an essential laboratory adjunct, however.

IV. Special techniques for optical constant measurements.

a) Measurement of emissivity.

39. It is possible to determine the absorptance of a body by measuring its emissivity because these quantities are intimately related according to KIRCHHOFF's law. Thus the optical constants of a material may be measured through the study of the emission process as well as through the study of the absorption process.

If a material in thermal equilibrium is absorbing all of the incident radiation, then it is also radiating an amount of power expressed by PLANCK's radiation law for blackbodies. A body which absorbs only a part of the incident radiation radiates only a fraction of the blackbody amount. This fraction is the emissivity of the body. The emissivity of a body is a function of the frequency of the radiation, being large for those frequencies which are well absorbed and small for those frequencies which are poorly absorbed.

For a lamellar sample, the average emissivity $\bar{\varepsilon}$ is related to the reflectance and transmittance by

$$\bar{\varepsilon}=1-\bar{\varrho}-\bar{\tau} \tag{39.1}$$

where the values averaged over the channeled spectra have been used. From the relations (37.3) and (37.4) it is found that

$$\bar{\varepsilon}=(1-a^2)(1-r^2)(1-a^2 r^2)^{-1}. \tag{39.2}$$

These relations have been presented in an interesting manner by McMahon [52] in graphical chart form so that a knowledge of any two of the quantities a^2, r^2, $\bar{\varrho}$, $\bar{\tau}$ or $\bar{\varepsilon}$ allow the determination of all of the others.

The emissivity is particularly simple for a very transparent sample having $1-a^2$ small compared to unity. Such a sample will have an emissivity equal to $1-a^2$ and will have a^2 well approximated by $\exp(-\alpha b)$. Thus the emissivity of a "transparent" lamella will be

$$\bar{\varepsilon}=\alpha b=4\pi \nu k b, \tag{39.3}$$

according to (15.19). A measurement of the emissivity of such a sample in order to determine its absorption coefficient or its extinction coefficient does not need correction for the reflectance of its surface. This is an important advantage of the emissivity measurement.

The relation (39.2) for a thick lamella having zero transmittance shows that the emissivity is $1-r^2$. The measurement of the emissivity of such a sample gives the same information as a measurement of the power reflectance r^2. In terms of this bulk emissivity,

$$\bar{\varepsilon}_B=1-r^2, \tag{39.4}$$

the relation (39.2) may be rewritten as

$$a^{-2}=1+\bar{\varepsilon}\,\bar{\varepsilon}_B(\bar{\varepsilon}_B-\bar{\varepsilon})^{-1}. \tag{39.5}$$

The measurement of the emissivity $\bar{\varepsilon}_B$ of a thick sample, or a knowledge of r^2, together with a measurement of the emissivity $\bar{\varepsilon}$ of a thin lamella, will allow the calculation of the optical constants from the values of r^2 and a^2 obtained from (39.4) and (39.5) through the use of (30.3) and (30.4).

The usefulness of these emissivity measurements to cover a full range of optical constants from very small values of αb through the reststrahlen peaks is shown in the work of Stierwalt and Potter [53]. The measuring apparatus, which was described by Stierwalt, Bernstein, and Kirk [54], takes advantage of a special magnetic tape memory, but the method does not depend upon this special apparatus for this effectiveness.

It should be realized that the optical constants of a material will be a function of the temperature so that all of the variations in emittance with temperature may not be the result of the blackbody radiation change but may be the result of a change in the optical constants of the material and a corresponding emissivity change.

40. For opaque samples the measurement of $\bar{\varepsilon}$ does not give information different from the measurement of $\bar{\varepsilon}_B$. It is possible to take advantage of the polarization of the emitted radiation, however, to increase the amount of information about the optical constants. By measuring the s- and p-polarized components of the radiation emitted from a surface at non-normal angles, the values of $1-r_s^2$ and $1-r_p^2$ are determined. The ratio of these polarized components can be measured by proper apparatus without the necessity of making a quantitative measurement of the radiant power. This ratio will depend upon the temperature of the sample only to the extent that the optical constants will depend upon the temperature. Such apparatus has been described by MARTIN, DUCHANE, and BLAU [55]. Their spectrometer looked alternately at the source through two polarizers adjusted to pass only the s-polarized radiation through one and the p-polarized radiation through the other. This alternating source signal to the spectrometer was passed through a rotatable polarizer, serving as an analyzer, that was adjusted to attenuate the two polarized signals to the same value and thereby produce no resulting alternating signal. The position of the analyzer measured the ratio of the emitted polarized components.

The ratio of these emissivities can be found from the relations (10.26) together with (10.27) and (10.28). The ratio is not a simple function of the optical constants of the material and the angle of observation. MARTIN, DUCHANE, and BLAU calculated the optical constants by preparing charts of the values of the polarization ratio for a range of values of n and k for the values of the angle used in the measurement. From such a tabulation all of the values of n and k which give a particular measured polarization ratio at the measured angle can be found. By making measurements at two angles, the unique pair of optical constants is determined. The measurement at a third angle then overdetermines the pair and gives a measure of the accuracy of the results. The experiments of MARTIN, DUCHANE, and BLAU on very hot metals indicated that this ratio could be measured to one part in a thousand. Their signals should have been large, however, because the sample sources were very hot.

b) Reflectance of an overcoated sample.

41. The use of normal incidence in measuring the reflectance of a sample is experimentally advantageous because no equipment need be provided to account for the polarization of the radiation and also because normal incidence allows the use of the smallest sample size. The fact is, however, that one such measurement at one frequency can not determine both n and k without other information. In order to keep the advantage of the polarization independence and yet obtain a second independent measurement, VINCENT-GEISSE and LECOMTE [56] have used an additional normal incidence reflection measurement on the sample with a dielectric coating. The non-absorbing dielectric coating is evaporated onto the surface of the sample. The thickness of the coating is determined by measuring a control deposit put on a separate substrate during the same evaporation. A graphic method for the calculation of the optical constants from these reflectances was described by VINCENT-GEISSE and LECOMTE.

An analytical expression for the reflectance of the coating sample can be obtained by the partial wave analysis as indicated in Sect. 15. For the coated sample the reflectance may be written in the form

$$\hat{R} = (\hat{r}_1 + \hat{a}^2 \hat{r}_2)(1 + \hat{a}^2 \hat{r}_1 \hat{r}_2)^{-1} \tag{41.1}$$

with the phase of the reflectance $\hat{R}$ referred to the air-coating surface; $\hat{r}_1$ is the amplitude reflectance of a single air-coating-air reflection; $\hat{r}_2$ is the amplitude reflectance of a single coating-material-coating reflection; and $\hat{a}$ is the attenuating factor (15.1) for the transmission through the coating. For a non-absorbing, dielectric coating of index of refraction n_1 and thickness b and a sample index of refraction $\hat{n}_2$, the parameters in the equation are given by:

$$\hat{r}_1 = (1-n_1)(1+n_1)^{-1} = -r_1, \tag{41.2}$$

$$\hat{a} = \exp i\, 2\pi \nu\, n_1\, b = \exp i\, \Phi_a \tag{41.3}$$

and

$$\hat{r}_2 = (n_1 - \hat{n}_2)(n_1 + \hat{n}_2)^{-1}. \tag{41.4}$$

The relation (41.1) can then be put in the form

$$|r_1 \hat{R}| = |r_1 \hat{a}^{-2} - \hat{r}_2|\, |r_1^{-1} \hat{a}^{-2} - \hat{r}_2|^{-1}. \tag{41.5}$$

The measurement of the power reflectance R^2 with the coating and the measurement of the power reflectance R_0^2 without the coating, that is with $a=1$, do not include the phase information. In each case, however, the possible values of the phases of $\hat{R}$ and $\hat{r}_2$ are such that $\hat{r}_2$ lies on a circle as shown in Fig. 20. This circle may be constructed in the following manner: first, construct the line OA and its extension at an angle Φ_a from the real axis; second, locate the points on OA at the distances r_1 and r_1^{-1} from the origin; third, locate the point B on OA such that the ratio of the distances from r_1 to B and from r_1^{-1} to B is $r_1 R$; fourth,

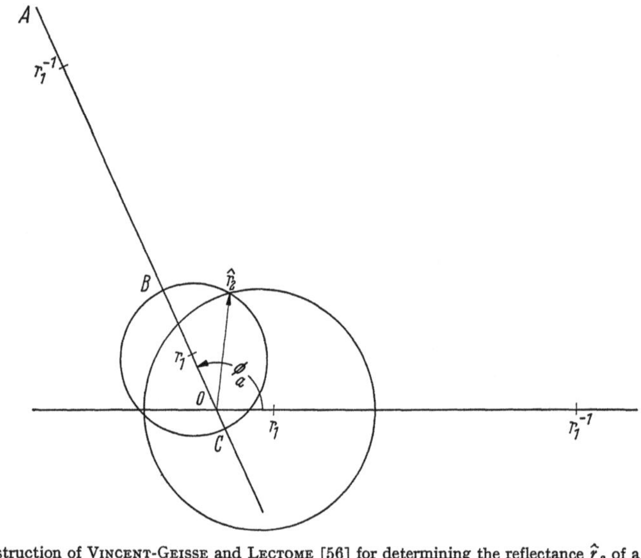

Fig. 20. The construction of Vincent-Geisse and Lectome [56] for determining the reflectance $\hat{r}_2$ of a dielectric sample interface from normal incidence measurements of the reflectance of a dielectric coated sample.

locate C on OA such that the distances r_1 to C and r_1^{-1} to C are also in the ratio $r_1 R$; fifth, construct the circle with its center on the line OA which passes through the points B and C, this circle is the locus of possible $\hat{r}_2$ values; sixth, repeat the process with the uncoated sample data, $\Phi_a = 0$ and R_0, to find another circle of possible $\hat{r}_2$ values. The intersection of the two $\hat{r}_2$ circles determines two possible values of $\hat{r}_2$. One of these values will be eliminated by a third measurement with a different thickness of dielectric coating. From the complex value of the $\hat{r}_2$, the values of the optical constants n_2 and k_2 can be calculated.

c) Attenuated total reflection.

The internal reflection from the interface between a highly refractive dielectric and a material will be total if the material is non-absorbing and the angle of incidence exceeds the critical angle. If the material is absorbing, then the reflectance can not be unity and there will be absorption in the material. The radiation wave in the absorbing material will be non-homogeneous and the power remains close to the interface — see Sect. 12. The spectrum of a material obtained in this way has the appearance of an absorption spectrum. The experimental advantages for obtaining characteristic spectra of non-transparent materials has made this technique very popular for material identification problems. The method is particularly useful with materials which have absorption coefficients too small to give good reflection spectra and yet too large to give good transmission spectra.

FAHRENFORT [57] pointed out the advantages to the attenuated total reflection technique for materials with intermediate values for their absorption coefficients. FAHRENFORT and VISSER [58] have used this attenuated total reflection phenomenon to obtain optical constants of liquids. Precomputed charts were prepared to relate the reflectance to the optical constants. A chart was prepared for each of a set of angles of incidence and for a particular high index reference dielectric. With measured values of the reflectance at two angles of incidence, the values of the optical constants can be obtained from the charts. Measurement at another angle serves to check the accuracy of the determination of the optical constant values.

The problem of mating the reference dielectric surface to a solid sample, the sensitivity of the reflectance to the incident angle, together with the necessity of polarization control makes attenuated total reflectance an unpopular method for determining the optical constants of solids.

Acknowledgements.

The author wishes to acknowledge the helpful discussions about optical constants measurements which he has had with Professor Dr. LUDWIG GENZEL and with Dr. RICHARD SANDERSON. He is also grateful for the aid of many students, particularly Dr. RAYMOND BROWN, Mr. EDGAR RUSSELL, and Mr. KENNETH JOHNSON.

References.

[1] BORN, M., and E. WOLF: Principles of Optics. New York: Pergamon Press 1959.
[2] LORENTZ, H. A.: Theory of Electrons. New York: Dover Publications, Inc. 1952.
[3] WHITTAKER, E. T.: A History of Theories of Aether and Electricity. The Classical Theories. New York: Thomas Nelson & Sons 1951.
[4] BARNES, R. B., and M. CZERNY: Phys. Rev. **38**, 338 (1931).
[5] SANDERSON, R. B.: J. Phys. Chem. Solids **26**, 803 (1965).
[6] LYDDANE, R. H., R. G. SACHS, and E. TELLER: Phys. Rev. **59**, 673 (1941).
[7] BORN, M., and K. HUANG: Dynamical Theory of Crystal Lattices. Oxford: Clarendon Press 1954.
[8] TOLL, J. S.: Phys. Rev. **104**, 1760 (1956).
[9] STERN, F.: Solid State Physics, vol. 15, p. 299 (editors: SEITZ and TURNBULL). New York: Academic Press 1963.
[10] MOSS, T. S.: Optical Properties of Semi-conductors. London: Butterworths & Co. 1961.
[11] FROEHLICH, D.: Z. Physik **169**, 114 (1962).
[12] GEICK, R.: Z. Physik **126**, 122 (1962).
[13] GEICK, R.: Z. Physik **163**, 499 (1961).
[14] BERRMAN, D. W.: Phys. Rev. **130**, 2193 (1963).
[15] HAVELOCK, T. H.: Proc. Roy. Soc. A (London) **105**, 488 (1924).

[16] JAFFE, G.: Handbuch der Experimental-Physik, Bd. 19, S. 64 und 190 (editors: W. WIEN and F. HARMS). Leipzig: Akademische Verlagsgesellschaft M. B. H. 1928.
[17] MITRA, S. S.: Crystallography and Crystal Perfection (editor RAMACHANDRAN). New York: p. 353 Academic Press 1963.
[18] JONES, G. O., D. H. MARTIN, P. A. MAWER, and C. H. PERRY: Proc. Roy. Soc. (London) A **261**, 10 (1961).
[19] SPITZER, W. G., and D. A. KLEINMAN: Phys. Rev. **121**, 1324 (1961).
[20] KLEINMAN, D. A., and W. G. SPITZER: Phys. Rev. **118**, 110 (1960).
[21] SPITZER, W. G., E. KLEINMAN, and D. WALSH: Phys. Rev. **113**, 127 (1959).
[22] SMITH, R. P.: Electronics **17**, 130, 318 (1944).
[23] LINDQUIST, R. E., and E. W. EWALD: J. Opt. Soc. Am. **53**, 247 (1963).
[24] SIMON, I.: J. Opt. Soc. Am. **41**, 336 (1951).
[25] BRUGEL, W.: Physik und Technik der Ultrastrahlung. Hannover: Curt R. Vincentz 1961.
[26] FAHRENFORT, J.: Infrared Spectroscopy and Molecular Structure (editor: DAVIS). London: Elsevier Publ. Co. 1963.
[27] LECOMTE, J.: Handbuch der Physik, Bd. 26, Licht und Materia II (editor: S. FLÜGGE). Berlin-Göttingen-Heidelberg: Springer 1958.
[28] BEATTIE, J. R.: Phil. Mag. **46**, 235 (1955).
[29] HEILMANN, G.: Z. Naturforsch. **16**A, 714 (1961).
[30] SASAKI, T., and K. ISHIGURO: Jap. J. Appl. Phys. **2**, 289 (1963).
[31] BEATTIE, J. R., and G. K. T. CONN: Phil. Mag. **46**, 222 (1955).
[32] AVERY, D. G.: Proc. Phys. Soc. (London) B **65**, 425 (1952).
[33] PRISHIVALKO, A. P.: Optics and Spect., English translation, **9**, 256 (1960).
[34] PRISHIVALKO, A. P.: Optics and Spect., English translation, **11**, 131 (1961).
[35] CONN, G. K. T., and G. K. EATON: J. Opt. Soc. Am. **44**, 477 (1954).
[36] CONN, G. K. T., and G. K. EATON: J. Opt. Soc. Am. **44**, 484 (1954).
[37] ROBINSON, T. S.: Proc. Phys. Soc. (London) B **25**, 910 (1952).
[38] GOTTLIEB, M.: J. Opt. Soc. Am. **50**, 343 (1960).
[39] PHILLIP, H. R., and A. E. TAFT: Phys. Rev. **113**, 1002 (1959).
[40] BALKANSKI, M., and J. M. BESSON: J. Opt. Soc. Am. **55**, 200 (1965).
[41] VELICKY, B.: Czechoslov. J. Phys. B **11**, 787 (1961).
[42] BELL, E. E.: Suppl. to Jap. J. Appl. Phys., Spring (1965).
[43] BELL, E. E., and E. RUSSELL: Submitted to Infrared Physics.
[44] FELLGET, P.: J. phys. radium **19**, 187 (1958).
[45] MOSS, T.S.: Proc. Phys. Soc. (London) B **70**, 778 (1957).
[46] CHAMBERLIN, J. E., and H. A. GEBBIE: Nature **206**, 602 (1965).
[47] HOHLS, H. W.: Ann. Physik **29**, 433 (1937).
[48] DITCHBURN, R. W.: Light. New York: Interscience Publ. 1953.
[49] ABELES, F.: Progress in Optics, vol. 2 (editor: E. WOLF). New York: John Wiley & Sons 1963.
[50] ROUARD, P., and P. BOUSQUET: Progress in Optics, vol. 4 (editor: E. WOLF). New York: John Wiley & Sons 1965.
[51] CHAMBERLIN, J. E., J. E. GIBBS, and H. A. GEBBIE: Nature **198**, 874 (1963).
[52] MCMAHON, M. O.: J. Soc. Am. **40**, 376 (1960).
[53] STIERWALT, D. L., and R. F. POTTER: J. Phys. Chem. Solids **23**, 99 (1962).
[54] STIERWALT, D. L., J. B. BERNSTEIN, and D. D. KIRK: Appl. Opt. **2**, 1169 (1963).
[55] MARTIN, W. S., E. M. DUCHANE, and H. H. BLAU jr.: Submitted to J. Opt. Soc. Am. (1965).
[56] VINCENT-GEISSE, J., et M. J. LECOMTE: J. phys. radium **20**, 841 (1958).
[57] FAHRENFORT, J.: Spectrochim. Acta **17**, 698 (1961).
[58] FAHRENFORT, J., and W. M. VISSER: Spectrochim. Acta **18**, 1108 (1962).

Phonons in Perfect Crystals.

By

W. COCHRAN and R. A. COWLEY.

With 56 Figures.

1. Introduction. In writing this article we have kept in mind the fact that two articles dealing with aspects of lattice dynamics have already appeared in Vol. VII of this series [*9*], [*10*]. The subject has however undergone very considerable development in the intervening decade. In particular, the experimental technique of neutron inelastic scattering has made possible the determination of the form of phonon dispersion curves for practically the first time, and unexpected features of interatomic forces have become apparent. Secondly, new mathematical techniques have made it possible to evaluate the influence of anharmonic effects on the physical properties of crystals in almost as much detail as these had been worked out formerly on the basis of the harmonic approximation. We have therefore devoted a considerable part of this article to these new developments. In order to make the account more nearly complete, we have included sections (notably A and C) which cover much the same ground as was covered in the articles by BLACKMAN [*9*] and LEIBFRIED [*10*]; where possible we have discussed the topics concerned from a different approach. In an Appendix we have summarised the principal features of the theory of thermodynamic GREEN'S functions, as applied to the evaluation of those physical properties of crystals which involve phonons.

A. Elementary lattice dynamics

2. The linear chain. Many of the features of the lattice dynamics of a crystal can be brought out by considering a simpler system, namely a linear chain of point masses [*7*], [*10*]. We take the masses m to be identical and to have a separation a, and represent the interaction between nearest neighbours by a spring of

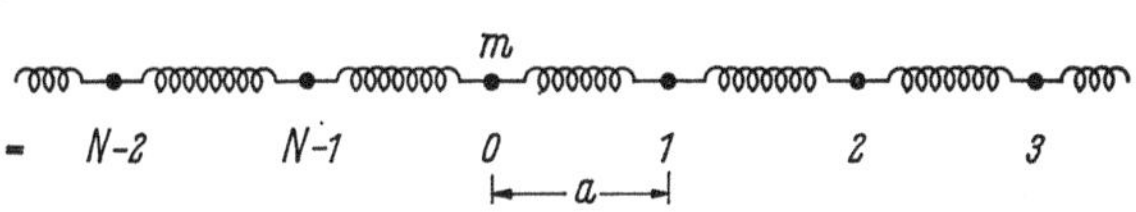

Fig. 1. A linear chain of masses m with forces between nearest neighbours.

force constant f, (Fig. 1). Various boundary conditions are clearly possible, but we shall consider only the cyclic boundary condition where one end of the chain is joined to the other so that the N-th mass is also the 0-th. The Hamiltonian of the system is then

$$H = T + \varphi_2 = \sum_{l=1}^{N} \left(\frac{p^2(l)}{2m} + \frac{1}{2} f \left(u(l+1) - u(l) \right)^2 \right) \tag{2.1}$$

where the displacement $u(l)$ is longitudinal, i.e. along the chain. The equilibrium position of the l-th mass is

$$X(l) = l a \tag{2.2}$$

and its equation of motion is

$$m\frac{\partial^2 u(l)}{\partial t^2} = f\{(u(l+1)-u(l)) + (u(l-1)-u(l))\}. \tag{2.3}$$

It is readily verified that the travelling wave

$$u(l) = \frac{|B(q)|}{\sqrt{N m}} \cos\left(q X(l) - \omega(q)t + \alpha(q)\right) \tag{2.4}$$

with amplitude and phase determined by the initial conditions, is a solution of Eq. (2.3), and substitution in Eq. (2.3) gives

$$\omega^2(q) = \frac{2f}{m}(1-\cos q a). \tag{2.5}$$

All values of the wavenumber q which satisfy the periodic boundary condition

$$u(l) = u(l+N) \tag{2.6}$$

are allowed, and these are

$$q = \frac{2\pi S}{N a} \tag{2.7}$$

where S is any integer, as can be verified from Eq. (2.4). It is also apparent that the wavenumbers

$$q,\ q \pm \frac{2\pi}{a},\quad q \pm \frac{4\pi}{a}, \ldots$$

do not produce different patterns of displacement, so we may take

$$-\tfrac{1}{2}(N-1) \leqq S \leqq \tfrac{1}{2}(N-1) \tag{2.8}$$

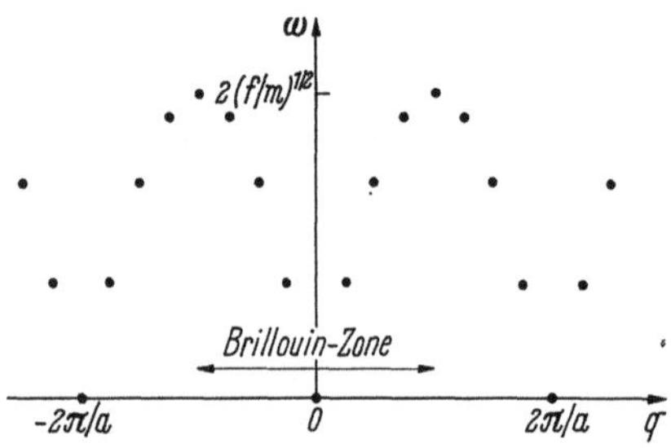

Fig. 2. Allowed wave numbers q and frequencies $\omega(q)$ for a linear chain of $N=8$ masses, with periodic boundary conditions. Those modes for which the values of q differ by a multiple of $2\pi/a$ are not physically distinct, there are therefore 8 independent normal modes.

with the convention that in Eq. (2.4), $\omega(q)$ is positive. The dispersion relation, Eq. (2.5), and the allowed values of q for the particular case $N=8$ are shown in Fig. 2. It is apparent that the unique values of q are contained in one unit cell of a one-dimensional reciprocal lattice whose spacing is $2\pi/a$. Note that in Eq. (2.4) there is no necessary relation between $|B(q)|$ and $|B(-q)|$, or between $\alpha(q)$ and $\alpha(-q)$. The general solution of Eq. (2.3) is therefore

$$u(l) = \frac{1}{\sqrt{N m}} \sum_q |B(q)| \cos\left(q X(l) - \omega(q)t + \alpha(q)\right) \tag{2.9}$$

where q can assume N distinct values lying between $-\pi/a$ and $+\pi/a$, and the amplitudes and phase angles are determined by $2N$ initial conditions.

The solution can be found somewhat more formally as follows. We make a transformation of coordinates by writing

$$u(l) = \frac{1}{\sqrt{N m}} \sum_q Q(q) \exp i q X(l). \tag{2.10}$$

Since $u(l)$ is real,

$$Q(-q) = Q^*(q). \tag{2.11}$$

The expression for the kinetic energy becomes

$$T = \frac{1}{2N} \sum_l \sum_{q q'} \dot{Q}(q)\, \dot{Q}(q') \exp i(q+q') X(l). \tag{2.12}$$

We now make use of the result that when N is large

$$\left.\begin{aligned}&\sum_l \exp iq\,X(l)=\sum_l \exp iq\,la=N\Delta(q),\\ &\text{where}\\ &\Delta(q)=1 \quad \text{for} \quad q=0,\frac{2\pi}{a}, \text{etc.},\\ &\Delta(q)=0 \quad \text{otherwise},\end{aligned}\right\} \tag{2.13}$$

so that

$$T=\tfrac{1}{2}\sum_q |\dot{Q}(q)|^2. \tag{2.14}$$

Similarly one finds that the potential energy is given by

$$\varphi_2=\tfrac{1}{2}\sum_q \omega^2(q)\,|Q(q)|^2 \tag{2.15}$$

so that in terms of the new coordinates,

$$H=\tfrac{1}{2}\sum_q \left(|\dot{Q}(q)|^2+\omega^2(q)\,|Q(q)|^2\right). \tag{2.16}$$

From the Lagrangian, $L=T-\varphi_2$, the momentum conjugate to $Q(q)$ is

$$P(q)=\frac{\partial L}{\partial \dot{Q}^*(q)}=\dot{Q}(q) \tag{2.17}$$

so that

$$H=\tfrac{1}{2}\sum_q \left(|P(q)|^2+\omega^2(q)\,|Q(q)|^2\right). \tag{2.18}$$

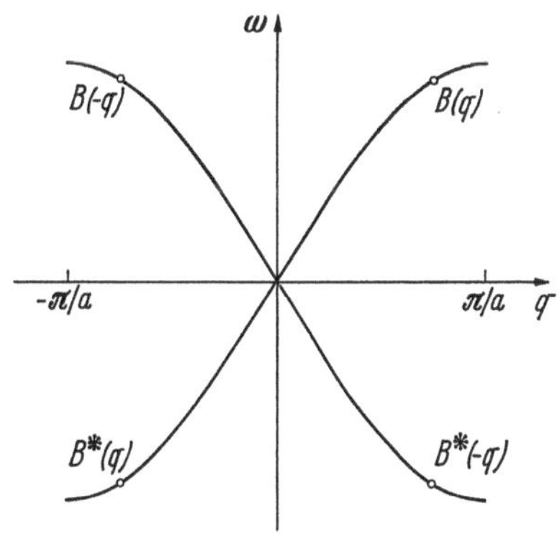

Fig. 3. $B(q)$ and $B^*(q)$ specify a wave travelling to the right, $B(-q)$ and $B^*(-q)$ a wave travelling to the left. $B(q)$ and $B(-q)$ are independent.

From the form of this expression we recognise the $Q(q)$ as normal coordinates of the linear chain, each of which describes an independent mode of vibration of the system. From HAMILTON'S equations it is found that the equation of motion of the coordinate $Q(q)$ is

$$\ddot{Q}(q)+\omega^2(q)\,Q(q)=0. \tag{2.19}$$

The relation between these complex normal coordinates and real travelling waves (Eq. 2.4) can be found by writing

$$B(q)=|B(q)|\exp i\alpha(q). \tag{2.20}$$

Comparison of Eqs. (2.4) and (2.10) then gives

$$\left.\begin{aligned}&Q(q)=\tfrac{1}{2}\left(B(q)\exp(-i\,\omega(q)\,t)+B^*(-q)\exp(i\,\omega(q)\,t)\right)\\ &\text{and therefore}\\ &P(q)=\tfrac{1}{2}i\,\omega(q)\left(B^*(-q)\exp(i\,\omega(q)\,t)-B(q)\exp(-i\,\omega(q)\,t)\right).\end{aligned}\right\} \tag{2.21}$$

It is not correct to associate $Q(q)$ with the wave travelling in the positive direction only; $Q(q)$ involves both $B(q)$ and $B(-q)$ which are independent, while $Q(q)$ and $Q(-q)$ are not. If the relevant part of the dispersion curve is redrawn so as to display both positive and negative values of $\omega(q)$, then the complex amplitudes $B(q)$ etc. are associated with the points shown in Fig. 3.

From the Hamiltonian (Eq. (2.16)], the total energy of the system is

$$E=\tfrac{1}{2}\sum_q \omega^2(q)\,|B(q)|^2 \tag{2.22}$$

i.e., it is the energy of N independent simple harmonic oscillators. For a system obeying the laws of classical physics and in thermal equilibrium with its surroundings

$$E(q) = \tfrac{1}{2}\,\omega^2(q)\,|B(q)|^2 = k_B T \tag{2.23}$$

and

$$E = N k_B T. \tag{2.24}$$

3. Quantisation of the harmonic oscillator. The Hamiltonian of a simple harmonic oscillator is

$$H = \frac{p^2}{2m} + \frac{1}{2}\, m\,\omega^2 u^2 \tag{3.1}$$

or

$$H = \tfrac{1}{2}(P^2 + \omega^2 Q^2) \tag{3.2}$$

on introducing

$$Q = \sqrt{m}\,u \tag{3.3}$$

and

$$P = \dot{Q}. \tag{3.4}$$

The corresponding Schrödinger equation is

$$\frac{1}{2}\left(-\hbar^2 \frac{\partial^2}{\partial Q^2} + \omega^2 Q^2\right)\Psi = E\Psi. \tag{3.5}$$

It is well known[1] that the wave functions which satisfy this equation are

$$\Psi_n(Q) = (2^n n!)^{-\frac{1}{2}} \left(\frac{\omega}{\hbar\pi}\right)^{\frac{1}{4}} \exp\left(\frac{-\omega Q^2}{2\hbar}\right) H_n\left(Q\left(\frac{\omega}{\hbar}\right)^{\frac{1}{2}}\right) \tag{3.6}$$

where $H_n(x)$ is the n-th Hermite polynomial. Alternatively we may identify the state of the system by $|n\rangle$.

The corresponding energy levels are

$$E_n = (n + \tfrac{1}{2})\,\hbar\,\omega. \tag{3.7}$$

It is convenient to introduce operators a and a^+ defined in terms of the quantum mechanical operators which correspond to Q and P by

$$Q = \left(\frac{\hbar}{2\omega}\right)^{\frac{1}{2}} (a^+ + a), \tag{3.8}$$

$$P = i\left(\frac{\hbar\omega}{2}\right)^{\frac{1}{2}} (a^+ - a). \tag{3.9}$$

The properties which justify the names creation and destruction operators for a^+ and a respectively are

$$a^+|n\rangle = (n+1)^{\frac{1}{2}}\,|n+1\rangle, \tag{3.10}$$

$$a|n\rangle = n^{\frac{1}{2}}\,|n-1\rangle. \tag{3.11}$$

These relations can be verified by operating on the wave function given by Eq. (3.6) with

$$a^+ = \left(\frac{\omega}{2\hbar}\right)^{\frac{1}{2}} Q - \left(\frac{\hbar}{2\omega}\right)^{\frac{1}{2}} \frac{\partial}{\partial Q} \tag{3.12}$$

and

$$a = \left(\frac{\omega}{2\hbar}\right)^{\frac{1}{2}} Q + \left(\frac{\hbar}{2\omega}\right)^{\frac{1}{2}} \frac{\partial}{\partial Q}. \tag{3.13}$$

---

[1] L. J. SCHIFF: Quantum Mechanics. New York: McGraw-Hill Book Co. 1949.

These latter equations are obtained by solving Eqs. (3.8) and (3.9) and remembering that, as an operator,

$$P = -i\hbar \frac{\partial}{\partial Q}. \tag{3.14}$$

Since P and Q satisfy the commutation relation

$$(QP - PQ)|n\rangle = i\hbar|n\rangle$$

written as

$$[Q, P] = i\hbar \tag{3.15}$$

it follows that

$$(a a^{+} - a^{+} a)|n\rangle = |n\rangle$$

i.e.,

$$[a, a^{+}] = 1. \tag{3.16}$$

In terms of a^{+} and a the Hamiltonian operator assumes the simple form

$$H = \tfrac{1}{2}\hbar\omega(a^{+}a + a a^{+}) = \hbar\omega(a^{+}a + \tfrac{1}{2}). \tag{3.17}$$

The eigenvalues of the operator $a^{+}a$ are the integers $n = 0, 1, 2 \ldots$.

4. Phonons. The results of the previous section can readily be extended to apply to the linear chain, since the normal coordinates of the latter behave like simple harmonic oscillators. We define operators $a(q)$ and $a^{+}(q)$ by means of the relations

$$Q(q) = \left(\frac{\hbar}{2\omega(q)}\right)^{\frac{1}{2}} \left(a^{+}(-q) + a(q)\right) \tag{4.1}$$

and

$$P(q) = i\left(\frac{\hbar\omega(q)}{2}\right)^{\frac{1}{2}} \left(a^{+}(-q) - a(q)\right). \tag{4.2}$$

The analogous equations in classical mechanics are Eqs. (2.21).

It can be shown [*4*] that these operators satisfy the commutation relations

$$\left.\begin{aligned} [Q^{*}(q), P(q')] &= i\hbar\Delta(q-q') \\ [Q(q), P^{*}(q')] &= i\hbar\Delta(q-q') \end{aligned}\right\} \tag{4.3}$$

from which it follows that the a's satisfy

$$[a(q), a^{+}(q')] = \Delta(q-q')$$

and

$$[a(q), a(q')] = [a^{+}(q), a^{+}(q')] = 0. \tag{4.4}$$

In terms of the a operators the Hamiltonian operator takes the form

$$H = \sum_{q} \hbar\omega(q)\left(a^{+}(q)\,a(q) + \tfrac{1}{2}\right). \tag{4.5}$$

The wave-function of the system is specified when we know the quantum number $n(q)$ of each mode q, being simply a product of the wave-functions of N independent simple harmonic oscillators. The energy of the system is

$$E = \sum_{q} \hbar\omega(q)\left(n(q) + \tfrac{1}{2}\right). \tag{4.6}$$

Denoting the state of the system by

$$|n(q_1), \quad n(q_2) \ldots n(q_N)\rangle,$$

the results analogous to Eqs. (3.10) and (3.11) are

$$a^+(q_1)\,|n(q_1),\, n(q_2)\,\ldots\, n(q_N)\rangle = \big(n(q_1)+1\big)^{\frac{1}{2}}\,|n(q_1)+1,\, n(q_2)\,\ldots\, n(q_N)\rangle, \tag{4.7}$$

$$a(q_1)\,|n(q_1),\, n(q_2)\,\ldots\, n(q_N)\rangle = \big(n(q_1)\big)^{\frac{1}{2}}\,|n(q_1)-1,\, n(q_2)\,\ldots\, n(q_N)\rangle. \tag{4.8}$$

It is worth noting that these formulae can be interpreted in two equivalent ways.

1. The system is considered as N independent simple harmonic oscillators which are distinguishable, each being identified by its wavenumber q. Each oscillator is in a stationary state of quantum number $n(q)$. The operator $a^+(q)$ excites the oscillator q and leaves all the other unaffected, similarly the operator $a(q)$ de-excites this oscillator only.

2. Alternatively we may describe the vibrations of the linear chain in terms of entities called phonons. The stationary states of a phonon are labelled by quantum numbers q, and the number of indistinguishable phonons in the state q is $n(q)$. The operator $a^+(q)$ creates an additional phonon in this state while $a(q)$ removes one. Phonons obey Bose-Einstein statistics and the total number of phonons is not conserved. For a system in equilibrium at temperature T,

$$\bar{n}(q) = \{\exp(\beta\hbar\,\omega(q)) - 1\}^{-1} \tag{4.9}$$

where

$$\beta = \frac{1}{k_B T}. \tag{4.10}$$

The dynamics of phonons has been considered in some detail by JENSEN[1].

5. Lattice vibrations in three dimensions. There is no formal difficulty in extending the results of foregoing sections so that they apply to a periodic array of point masses in three dimensions. At the same time we remove the restriction that there is only one mass per unit cell, and identify the different masses (atoms) by an index $\varkappa = 1, 2 \ldots r$. The position of the $\varkappa$-th atom in the l-th cell is

$$\boldsymbol{X}\begin{pmatrix} l \\ \varkappa \end{pmatrix} = \boldsymbol{X}(l) + \boldsymbol{X}(\varkappa) \tag{5.1}$$

and we shall use suffices α, β, γ to denote components in a Cartesian coordinate system. A basic assumption which we shall make meantime is that the change in potential energy when the atoms are given small displacements $\boldsymbol{u}\begin{pmatrix} l \\ \varkappa \end{pmatrix}$ is

$$\varphi_2 = \frac{1}{2} \sum_{\substack{l l' \\ \varkappa\varkappa' \\ \alpha\beta}} \varphi_{\alpha\beta}\begin{pmatrix} l l' \\ \varkappa\varkappa' \end{pmatrix} u_\alpha\begin{pmatrix} l \\ \varkappa \end{pmatrix} u_\beta\begin{pmatrix} l' \\ \varkappa' \end{pmatrix}. \tag{5.2}$$

The quantity $-\varphi_{\alpha\beta}\begin{pmatrix} l l' \\ \varkappa\varkappa' \end{pmatrix}$ is thus a force constant, it gives the force in the α-direction on the atom $\begin{pmatrix} l \\ \varkappa \end{pmatrix}$ when the atom $\begin{pmatrix} l' \\ \varkappa' \end{pmatrix}$ is given a (small) unit displacement in the β-direction. The restriction imposed in much of our discussion of the linear chain, that only nearest-neighbour masses interacted, has therefore also been removed. Since there can be no force on any atom as a result of a uniform translation of the crystal, it follows that

$$\varphi_{\alpha\beta}\begin{pmatrix} l l \\ \varkappa\varkappa \end{pmatrix} = -\sum_{l'\varkappa'}{}' \varphi_{\alpha\beta}\begin{pmatrix} l l' \\ \varkappa\varkappa' \end{pmatrix}. \tag{5.3}$$

---

[1] H. H. JENSEN, p. (1) of Ref. [8].

The Hamiltonian of the system is

$$H = \frac{1}{2} \sum_{\substack{ll'\\ \varkappa\varkappa'\\ \alpha\beta}} \left(m_\varkappa \dot{u}_\alpha^2 \binom{l}{\varkappa} + \varphi_{\alpha\beta} \binom{ll'}{\varkappa\varkappa'} u_\alpha \binom{l}{\varkappa} u_\beta \binom{l'}{\varkappa'} \right) \tag{5.4}$$

and the equation of motion of an individual atom,

$$m_\varkappa \frac{\partial^2 u_\alpha \binom{l}{\varkappa}}{\partial t^2} = - \sum_{l'\varkappa'\beta} \varphi_{\alpha\beta} \binom{ll'}{\varkappa\varkappa'} u_\beta \binom{l'}{\varkappa'} . \tag{5.5}$$

Guided by the results of Sect. 2, we look for a solution in terms of plane waves

$$u_\alpha \binom{l}{\varkappa} = \frac{u_\alpha(\varkappa|\boldsymbol{q})}{\sqrt{m_\varkappa}} \exp i \left(\boldsymbol{q} \cdot \boldsymbol{X} \binom{l}{\varkappa} - \omega(\boldsymbol{q}) t \right), \tag{5.6}$$

when only modes of a single wave vector are present. Substitution of Eq. (5.6) in the equations of motion (5.5) gives

$$\omega^2(\boldsymbol{q}) u_\alpha(\varkappa|\boldsymbol{q}) = \sum_{\varkappa'\beta} D_{\alpha\beta} \binom{\boldsymbol{q}}{\varkappa\,\varkappa'} u_\beta(\varkappa'|\boldsymbol{q}) \tag{5.7}$$

where

$$D_{\alpha\beta} \binom{\boldsymbol{q}}{\varkappa\,\varkappa'} = \frac{1}{\sqrt{m_\varkappa m_{\varkappa'}}} \sum_{l'} \varphi_{\alpha\beta} \binom{ll'}{\varkappa\varkappa'} \exp i\, \boldsymbol{q} \cdot \left(\boldsymbol{X} \binom{l'}{\varkappa'} - \boldsymbol{X} \binom{l}{\varkappa} \right). \tag{5.8}$$

We note that this quantity depends on $\boldsymbol{X}(l') - \boldsymbol{X}(l)$ [see Eq. (5.1)] and not on $\boldsymbol{X}(l')$ and $\boldsymbol{X}(l)$ separately, so that we could set $\boldsymbol{X}(l)$ equal to zero.

The number of equations of the type (5.7) is $3r$, the $u_\alpha(\varkappa|\boldsymbol{q})$ are arbitrary to the extent of a constant factor and we may replace them by quantities $e_\alpha(\varkappa|\boldsymbol{q})$ which, as we shall see shortly can be chosen to satisfy certain orthonormality and closure conditions. In matrix notation, Eq. (5.7) then become

$$\omega^2(\boldsymbol{q})\, \boldsymbol{e}(\boldsymbol{q}) = \boldsymbol{D}(\boldsymbol{q})\, \boldsymbol{e}(\boldsymbol{q}) \tag{5.9}$$

where $\boldsymbol{e}(\boldsymbol{q})$ is a column matrix:

$$\boldsymbol{e}(\boldsymbol{q}) = \begin{pmatrix} e_\alpha(1|\boldsymbol{q}) \\ e_\beta(1|\boldsymbol{q}) \\ e_\gamma(1|\boldsymbol{q}) \\ e_\alpha(2|\boldsymbol{q}) \\ \vdots \\ e_\gamma(r|\boldsymbol{q}) \end{pmatrix} \tag{5.10}$$

and $\boldsymbol{D}(\boldsymbol{q})$ is a square matrix containing $3r \times 3r$ elements:

$$\boldsymbol{D}(\boldsymbol{q}) = \begin{pmatrix} D_{\alpha\alpha}\binom{\boldsymbol{q}}{1\,1} D_{\alpha\beta}\binom{\boldsymbol{q}}{1\,1} D_{\alpha\gamma}\binom{\boldsymbol{q}}{1\,1} D_{\alpha\alpha}\binom{\boldsymbol{q}}{1\,2} \cdots D_{\alpha\gamma}\binom{\boldsymbol{q}}{1\,r} \\ D_{\beta\alpha}\binom{\boldsymbol{q}}{1\,1} D_{\beta\beta}\binom{\boldsymbol{q}}{1\,1} D_{\beta\gamma}\binom{\boldsymbol{q}}{1\,1} D_{\beta\alpha}\binom{\boldsymbol{q}}{1\,2} \cdots D_{\beta\gamma}\binom{\boldsymbol{q}}{1\,r} \\ \vdots \qquad\qquad\qquad\qquad\qquad\qquad \vdots \\ D_{\gamma\alpha}\binom{\boldsymbol{q}}{r\,1} D_{\gamma\beta}\binom{\boldsymbol{q}}{r\,1} D_{\gamma\gamma}\binom{\boldsymbol{q}}{r\,1} D_{\gamma\alpha}\binom{\boldsymbol{q}}{r\,2} \cdots D_{\gamma\gamma}\binom{\boldsymbol{q}}{r\,r} \end{pmatrix}. \tag{5.11}$$

The condition for the set of Eq. (5.9) to have a solution is that the determinant of the coefficients vanish,

$$|\boldsymbol{D}(\boldsymbol{q}) - \omega^2(\boldsymbol{q})\, \delta_{\alpha\beta}\, \delta_{\varkappa\varkappa'}| = 0. \tag{5.12}$$

We distinguish the $3r$ frequencies which satisfy this equation by subscripts j. Substituting a particular $\omega_j(\boldsymbol{q})$ into the Eq. (5.9) gives a set of $3r$ polarisation components $e_\alpha\left(\varkappa\middle|\begin{smallmatrix}\boldsymbol{q}\\ j\end{smallmatrix}\right)$ which specify the motion of the atoms in the mode whose wavevector is $\boldsymbol{q}$ and frequency $\omega_j(\boldsymbol{q})$. In fact the $\omega_j(\boldsymbol{q})$ and $e_\alpha\left(\varkappa\middle|\begin{smallmatrix}\boldsymbol{q}\\ j\end{smallmatrix}\right)$ are the eigenvalues and eigenvectors of the matrix $\boldsymbol{D}(\boldsymbol{q})$. As such, the eigenvectors can be chosen to satisfy the orthonormality and closure conditions

$$\sum_{\varkappa\alpha} e_\alpha^*\left(\varkappa\middle|\begin{matrix}\boldsymbol{q}\\ j\end{matrix}\right) e_\alpha\left(\varkappa\middle|\begin{matrix}\boldsymbol{q}\\ j'\end{matrix}\right)=\delta_{jj'}, \tag{5.13}$$

$$\sum_{j} e_\alpha\left(\varkappa\middle|\begin{matrix}\boldsymbol{q}\\ j\end{matrix}\right) e_\beta^*\left(\varkappa'\begin{matrix}\boldsymbol{q}\\ j\end{matrix}\right)=\delta_{\alpha\beta}\,\delta_{\varkappa\varkappa'}. \tag{5.14}$$

From the definition (4.9) we see that

$$D_{\alpha\beta}\begin{pmatrix}\boldsymbol{q}\\ \varkappa\varkappa'\end{pmatrix}=D_{\beta\alpha}^*\begin{pmatrix}\boldsymbol{q}\\ \varkappa'\varkappa\end{pmatrix}. \tag{5.15}$$

The matrix $\boldsymbol{D}(\boldsymbol{q})$ is therefore Hermitian. It follows that each $\omega_j^2(\boldsymbol{q})$ is entirely real, and if the system is stable against small displacements each $\omega_j^2(\boldsymbol{q})$ will also be non-negative. Since $\omega_j^2(\boldsymbol{q})$ is independent of the sign of $\boldsymbol{q}$ we can consistently choose either of

$$e_\alpha\left(\varkappa\middle|\begin{matrix}\boldsymbol{q}\\ j\end{matrix}\right)=\pm e_\alpha^*\left(\varkappa\middle|\begin{matrix}-\boldsymbol{q}\\ j\end{matrix}\right). \tag{5.16}$$

The choice of the positive sign is the more usual convention.

In terms of normal coordinates the displacement of an atom is

$$u_\alpha\begin{pmatrix}l\\ \varkappa\end{pmatrix}=\frac{1}{\sqrt{N m_\varkappa}}\sum_{\boldsymbol{q}j} e_\alpha\left(\varkappa\middle|\begin{matrix}\boldsymbol{q}\\ j\end{matrix}\right) Q\begin{pmatrix}\boldsymbol{q}\\ j\end{pmatrix}\exp i\boldsymbol{q}\cdot\boldsymbol{X}\begin{pmatrix}l\\ \varkappa\end{pmatrix}. \tag{5.17}$$

This outline follows fairly closely that of BORN and HUANG [*1*] and of MARADUDIN, MONTROLL and WEISS [*4*], we have however chosen to define the eigenvectors in such a way that $\exp i\boldsymbol{q}\cdot\boldsymbol{X}\begin{pmatrix}l\\ \varkappa\end{pmatrix}$ appears in Eq. (5.17), while these authors define the phases of the eigenvectors so that $\exp i\boldsymbol{q}\cdot\boldsymbol{X}(l)$ appears in the corresponding position. The advantage of the notation used here is that for a crystal such as an alkali halide where each atom is at a centre of symmetry, the matrix elements $D\begin{pmatrix}\boldsymbol{q}\\ \varkappa\varkappa'\end{pmatrix}$ and eigenvectors $\boldsymbol{e}\left(\varkappa\middle|\begin{smallmatrix}\boldsymbol{q}\\ j\end{smallmatrix}\right)$ are entirely real quantities.

6. Boundary conditions. Boundary conditions require further consideration since it is not possible to impose in three dimensions conditions that are periodic in the way defined in Sect. 2. Analogous conditions can however be imposed as follows.

Let the unit cell be defined by vectors $\boldsymbol{a}_1, \boldsymbol{a}_2, \boldsymbol{a}_3$, and let $\boldsymbol{b}_1, \boldsymbol{b}_2, \boldsymbol{b}_3$ define a reciprocal lattice such that

$$\boldsymbol{a}_i\cdot\boldsymbol{b}_j=2\pi\,\delta_{ij} \tag{6.1}$$

we then single out from the infinite crystal a volume defined by $N_1\boldsymbol{a}_1, N_2\boldsymbol{a}_2, N_3\boldsymbol{a}_3$ and impose the "periodic boundary condition" that $\boldsymbol{u}\begin{pmatrix}l\\ \varkappa\end{pmatrix}$ for atoms separated by a translation $N_i\boldsymbol{a}_i$, or a sum of such translations, must be the same. Possible values of $\boldsymbol{q}$ are then given by

$$\boldsymbol{q}=\sum_{i=1}^{3}\frac{S_i\boldsymbol{b}_i}{N_i} \tag{6.2}$$

where the S_i are integers. Each unit cell of the reciprocal lattice therefore contains $N_1 N_2 N_3 = N$ distinct values of $\boldsymbol{q}$, which are distributed uniformly and continuously in reciprocal space when N is large. The frequencies satisfy

$$\omega_j(\boldsymbol{q}) = \omega_j(\boldsymbol{q} + \boldsymbol{\tau}) \tag{6.3}$$

where

$$\boldsymbol{\tau} = \sum_{i=1}^{3} S_i \, \boldsymbol{b}_i \tag{6.4}$$

is any vector of the reciprocal lattice. It follows that all distinct modes are specified by wavevectors contained in the first Brillouin zone, the symmetrically-shaped unit cell of the reciprocal lattice which can be chosen to have its centre at the origin [7].

Provided that N is large, the imposition of particular boundary conditions cannot affect the values of $\omega_j(\boldsymbol{q})$ or of $e_\alpha\left(\varkappa \middle| \begin{smallmatrix} \boldsymbol{q} \\ j \end{smallmatrix}\right)$. This follows physically from the fact that these determine bulk properties of the solid which cannot be influenced by surface conditions. The point is discussed in more detail by Born and Huang [1].

7. Acoustic and optic modes. The condition that no forces are set up in the crystal by a uniform translation has already been expressed as Eq. (5.3). From this it follows that

$$\sum_{\varkappa'} D_{\alpha\beta}\begin{pmatrix} 0 \\ \varkappa\,\varkappa' \end{pmatrix} = 0 \tag{7.1}$$

and it is then readily shown from Eq. (5.9) that for $q = 0$ there are three frequencies $\omega_j(0)$ equal to zero. Those branches of the dispersion relation for which $\omega_j(\boldsymbol{q})$ tends linearly to zero as q tends to zero are called the acoustic branches. The remaining $3r - 3$ branches are called optic branches and for these $\omega_j(0)$ is finite.

For simple structures such as the alkali halides it is found that when q is parallel to an axis of symmetry, the determinantal equation for the $\omega_j^2(\boldsymbol{q})$ can be reduced to a product of three separate determinants. For example when $\boldsymbol{q}$ is parallel to [100] the determinant of order 6 factors into three determinants each of order 2, since each element $D_{\alpha\beta}\begin{pmatrix} \boldsymbol{q} \\ \varkappa\,\varkappa' \end{pmatrix}$ is zero for $\alpha \neq \beta$. We then have

$$|D| = |D_\alpha| \times |D_\beta| \times |D_\gamma| \tag{7.2}$$

where, for example

$$|D_\alpha| = \begin{vmatrix} D_{\alpha\alpha}\begin{pmatrix} \boldsymbol{q} \\ 1\,1 \end{pmatrix} & D_{\alpha\alpha}\begin{pmatrix} \boldsymbol{q} \\ 1\,2 \end{pmatrix} \\ D_{\alpha\alpha}\begin{pmatrix} \boldsymbol{q} \\ 2\,1 \end{pmatrix} & D_{\alpha\alpha}\begin{pmatrix} \boldsymbol{q} \\ 2\,2 \end{pmatrix} \end{vmatrix}. \tag{7.3}$$

The eigenvectors of this determinant are $\left(e_\alpha\left(1 \middle| \begin{smallmatrix} \boldsymbol{q} \\ 1 \end{smallmatrix}\right), e_\alpha\left(2 \middle| \begin{smallmatrix} \boldsymbol{q} \\ 1 \end{smallmatrix}\right)\right)$ and $\left(e_\alpha\left(1 \middle| \begin{smallmatrix} \boldsymbol{q} \\ 2 \end{smallmatrix}\right), e_\alpha\left(2 \middle| \begin{smallmatrix} \boldsymbol{q} \\ 2 \end{smallmatrix}\right)\right)$ and the frequencies are $\omega_1(\boldsymbol{q})$ and $\omega_2(\boldsymbol{q})$, corresponding respectively to the longitudinal acoustic and longitudinal optic modes. The other two determinants similarly give the polarisation properties and frequencies of the transverse modes. The equation

$$|\boldsymbol{D}_\alpha - \delta_{\varkappa\varkappa'}\, \omega^2| = 0 \tag{7.4}$$

is identical with that which holds for a linear chain with two masses in the repeat distance. In the crystal, when $\boldsymbol{q}$ is parallel to [100], successive planes of atoms play the role of the individual masses of the linear chain.

8. Symmetry conditions on the force constants. From the form of the Hamiltonian, Eq. (5.4), it is evident that

$$\varphi_{\alpha\beta}\begin{pmatrix} l\,l' \\ \varkappa\,\varkappa' \end{pmatrix} = \varphi_{\beta\alpha}\begin{pmatrix} l'\,l \\ \varkappa'\,\varkappa \end{pmatrix}. \tag{8.1}$$

In matrix notation, the relation between forces and displacements is

$$\boldsymbol{F}\begin{pmatrix} l \\ \varkappa \end{pmatrix} = -\boldsymbol{\Phi}\begin{pmatrix} l\,l' \\ \varkappa\,\varkappa' \end{pmatrix} \boldsymbol{u}\begin{pmatrix} l' \\ \varkappa' \end{pmatrix} \tag{8.2}$$

where $\boldsymbol{F}$ and $\boldsymbol{u}$ are column matrices and $\boldsymbol{\Phi}$ is a 3×3 matrix whose elements are the $\varphi_{\alpha\beta}\begin{pmatrix} l\,l' \\ \varkappa\,\varkappa' \end{pmatrix}$. Consider the relation between $\boldsymbol{\Phi}\begin{pmatrix} l\,l' \\ \varkappa\,\varkappa' \end{pmatrix}$ and $\boldsymbol{\Phi}\begin{pmatrix} l\,l'' \\ \varkappa\,\varkappa' \end{pmatrix}$ when the atom at $\boldsymbol{X}\begin{pmatrix} l'' \\ \varkappa' \end{pmatrix}$ is related to that at $\boldsymbol{X}\begin{pmatrix} l' \\ \varkappa' \end{pmatrix}$ by a symmetry operation without translation, $\boldsymbol{S}$, about $\boldsymbol{X}\begin{pmatrix} l \\ \varkappa \end{pmatrix}$ as origin. $\boldsymbol{S}$ then brings the atom at the origin into self-coincidence, and can be written as a 3×3 matrix. For example for inversion in a centre of symmetry,

$$\boldsymbol{S} = \begin{pmatrix} -1 & 0 & 0 \\ 0 & -1 & 0 \\ 0 & 0 & -1 \end{pmatrix}. \tag{8.3}$$

Since $\boldsymbol{\Phi}$ is a second-rank tensor it transforms as

$$\boldsymbol{\Phi}\begin{pmatrix} l\,l'' \\ \varkappa\,\varkappa' \end{pmatrix} = \boldsymbol{S}\,\boldsymbol{\Phi}\begin{pmatrix} l\,l' \\ \varkappa\,\varkappa' \end{pmatrix}\widetilde{\boldsymbol{S}} \tag{8.4}$$

where $\widetilde{\boldsymbol{S}}$ is the transpose of $\boldsymbol{S}$. Certain symmetry operations $\boldsymbol{S}_0$ will bring the atom at $\boldsymbol{X}\begin{pmatrix} l' \\ \varkappa' \end{pmatrix}$ into self-coincidence as well as that at $\boldsymbol{X}\begin{pmatrix} l \\ \varkappa \end{pmatrix}$. For these $l'' = l'$ and

$$\boldsymbol{S}_0\,\boldsymbol{\Phi}\begin{pmatrix} l\,l' \\ \varkappa\,\varkappa' \end{pmatrix}\widetilde{\boldsymbol{S}}_0 = \boldsymbol{\Phi}\begin{pmatrix} l\,l' \\ \varkappa\,\varkappa' \end{pmatrix}. \tag{8.5}$$

Such relations limit the number of independent force constants.

For example, consider the force constants between an atom at the origin and one in the body-centre position of a cubic unit cell. The symmetry operation

$$\begin{pmatrix} 0 & 1 & 0 \\ 1 & 0 & 0 \\ 0 & 0 & 1 \end{pmatrix}$$

(a mirror plane) leaves the atoms at (0, 0, 0) and $(\frac{1}{2}, \frac{1}{2}, \frac{1}{2})$ unaffected, as does

$$\begin{pmatrix} 0 & 1 & 0 \\ 0 & 0 & 1 \\ 1 & 0 & 0 \end{pmatrix}$$

(a three-fold axis). It is found on applying Eq. (8.5) that there can be only two independent force constants between these atoms. Appropriate application of Eq. (8.4) gives the force constants between the atom at $(\frac{1}{2}, \frac{1}{2}, \frac{1}{2})$ and its remaining seven nearest neighbours in terms of the same two parameters.

9. Phonons in a three-dimensional lattice. The results given in Sect. 4 apply in three dimensions with the addition of subscripts j to distinguish the different modes which have the same value of $\boldsymbol{q}$. For example Eq. (4.1) becomes

$$Q\begin{pmatrix} \boldsymbol{q} \\ j \end{pmatrix} = \left(\frac{\hbar}{2\omega_j(\boldsymbol{q})}\right)^{\frac{1}{2}} \left(a^+\begin{pmatrix} -\boldsymbol{q} \\ j \end{pmatrix} + a\begin{pmatrix} \boldsymbol{q} \\ j \end{pmatrix}\right) \tag{9.1}$$

and the first of the commutation relations, (4.4), becomes

$$\left[a\begin{pmatrix}\boldsymbol{q}\\ j\end{pmatrix}, a^{+}\begin{pmatrix}\boldsymbol{q}'\\ j'\end{pmatrix}\right] = \Delta(\boldsymbol{q}-\boldsymbol{q}')\,\delta_{jj'}. \tag{9.2}$$

Expressions for the Hamiltonian operator, the energy of a mode $\begin{pmatrix}\boldsymbol{q}\\ j\end{pmatrix}$, the effect of the operators $a\begin{pmatrix}\boldsymbol{q}\\ j\end{pmatrix}$ and $a^{+}\begin{pmatrix}\boldsymbol{q}\\ j\end{pmatrix}$, etc., are precisely analogous to those which apply in one dimension.

We may note that, given the form of the interaction between the atoms, the problem of determining the frequencies and polarisation properties of the normal modes of vibration does not involve the concepts of quantum physics. This is no longer true if the Hamiltonian is not of the form given as Eq. (5.4), that is if the harmonic approximation discussed in a later section does not apply.

10. The frequency distribution of the normal modes. We define $g(\omega)\,d\omega$ as the number of modes of vibration having a frequency between ω and $\omega+d\omega$. Evidently for a crystal of N unit cells with r atoms in the unit cell

$$\int_0^{\omega_M} g(\omega)\,d\omega = 3Nr \tag{10.1}$$

where ω_M is the maximum frequency. It is sometimes more convenient to consider the distribution of squared frequencies, $G(\omega^2)$. Clearly

$$g(\omega) = 2\omega\, G(\omega^2). \tag{10.2}$$

For the simple linear chain, since

$$g(\omega)\,d\omega = g(q)\,dq \tag{10.3}$$

and

$$g(q) = N\Big/\frac{2\pi}{a} \tag{10.4}$$

one finds on rewriting Eq. (2.5) as

$$\omega = \omega_M \sin\tfrac{1}{2}qa \tag{10.5}$$

that

$$g(\omega) = \frac{2N}{\pi}(\omega_M^2-\omega^2)^{-\frac{1}{2}}. \tag{10.6}$$

This distribution is illustrated in Fig. 4.

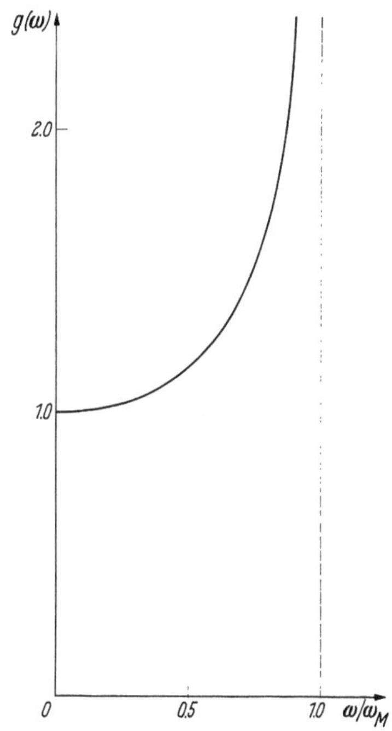

Fig. 4. The frequency distribution of the normal modes of vibration of a linear chain.

In three dimensions, we have

$$g(\omega)\,d\omega = \frac{Nv}{(2\pi)^3}\int d^3q \tag{10.7}$$

where the integral is extended over that region of the Brillouin zone for which the frequency $\omega(\boldsymbol{q})$ lies between ω and $\omega+d\omega$, the factor $\frac{Nv}{(2\pi)^3}$ being the density of points representing the wavevectors $\boldsymbol{q}$. Let dS be an element of area on the surface for which the frequency is equal to ω. The element of volume in reciprocal space between constant frequency surfaces ω and $\omega+d\omega$ is

$$d^3q = \frac{dS\,d\omega}{|\mathrm{grad}_{\boldsymbol{q}}\,\omega|}. \tag{10.8}$$

Thus, allowing for the different branches of the dispersion curve,

$$g(\omega)=\frac{Nv}{(2\pi)^3}\sum_{j=1}^{3r}\int\limits_{\omega=\omega_j(\boldsymbol{q})}\frac{dS}{|\mathrm{grad}_{\boldsymbol{q}}\,\omega_j(\boldsymbol{q})|}\,. \tag{10.9}$$

Further consideration of frequency distributions is left to later sections, we have given Eq. (10.9) because it renders plausible the conclusion that $g(\omega)$ should show characteristic features at frequencies for which $\mathrm{grad}_{\boldsymbol{q}}\,\omega_j(\boldsymbol{q})$ is zero. These critical points were discussed in detail by VAN HOVE[1] who showed that the existence of certain critical points could be deduced from the symmetry of the crystal, although others depend on the detailed shape of the dispersion curves and thus on the force constants. It is for instance usually the case that $\mathrm{grad}_{\boldsymbol{q}}\,\omega_j(\boldsymbol{q})=0$ for each $\omega_j(\boldsymbol{q})$ corresponding to $\boldsymbol{q}$ at, and perpendicular to, a surface of the Brillouin zone.

11. The harmonic and adiabatic approximations. The results given in the foregoing sections are exact for a system whose Hamiltonian is given by Eq. (5.4). This expression applies to a real crystal only as an approximation. The potential energy of a crystal can be written as a power series in the displacements of the atoms from their equilibrium positions,

$$\left.\begin{aligned}\varphi&=\varphi_0+\varphi_1+\varphi_2+\varphi_3\cdots\\&=\varphi_0+\sum_{l\varkappa\alpha}\varphi_\alpha\binom{l}{\varkappa}u_\alpha\binom{l}{\varkappa}+\frac{1}{2}\sum_{\substack{l\,l'\\\varkappa\varkappa'\\\alpha\beta}}\varphi_{\alpha\beta}\binom{l\;l'}{\varkappa\,\varkappa'}u_\alpha\binom{l}{\varkappa}u_\beta\binom{l'}{\varkappa'}+\\&\quad+\frac{1}{3!}\sum_{\substack{l\,l'\,l''\\\varkappa\varkappa'\varkappa''\\\alpha\beta\gamma}}\varphi_{\alpha\beta\gamma}\binom{l\;l'\;l''}{\varkappa\,\varkappa'\,\varkappa''}u_\alpha\binom{l}{\varkappa}u_\beta\binom{l'}{\varkappa'}u_\gamma\binom{l''}{\varkappa''}+\cdots\end{aligned}\right\} \tag{11.1}$$

where, for example,

$$\varphi_{\alpha\beta}\binom{l\;l'}{\varkappa\,\varkappa'}=\left(\frac{\partial^2\varphi}{\partial u_\alpha\binom{l}{\varkappa}\,\partial u_\beta\binom{l'}{\varkappa'}}\right)_0. \tag{11.2}$$

The equations of previous sections apply only in the harmonic approximation where the series is terminated at φ_2. Additional terms in φ_3 and φ_4 can be treated as a perturbation and are essential for a discussion of such properties as coefficient of expansion, thermal conductivity etc. In practice the harmonic approximation provides a fairly satisfactory basis for a discussion of the thermodynamic properties of crystals, particularly at low temperatures, [9], and of the interaction of X-rays, slow neutrons etc. with the crystal. The extension of the theory to include anharmonic effects is the subject matter of later sections.

Some justification is also necessary for treating the forces in a crystal as if they acted between the nuclei. The 'core' electrons no doubt move rigidly with the nuclei in the course of the lattice vibrations, but the wave-functions of the outer or valence electrons do not remain unaffected. It can usually be assumed however that the electrons make only virtual transitions to excited states. The system can be regarded as comprising ion cores and valence electrons, and its Hamiltonian written as

$$H=T_e+T_i+\varphi_{ie}(\boldsymbol{x},\boldsymbol{X})+\varphi_{ii}(\boldsymbol{X})+\varphi_{ee}(\boldsymbol{x}) \tag{11.3}$$

[1] L. VAN HOVE: Phys. Rev. 89, 1189 (1953).

the subscripts i and e denoting ions and electrons, while $\boldsymbol{X}$ and $\boldsymbol{x}$ indicate that ion and electron coordinates are involved respectively. The adiabatic approximation [1] is used to decouple the dynamical aspects of the ion and electron motion. Briefly, this consists in principle in finding the energy of the electrons $E_e(\boldsymbol{X})$ for a *fixed* configuration of the ions, using the Hamiltonian

$$H_e = T_e + \varphi_{ie}(\boldsymbol{x}, \boldsymbol{X}) + \varphi_{ee}(\boldsymbol{x}) . \tag{11.4}$$

The term $E_e(\boldsymbol{X})$ is then used to give the Hamiltonian for the motion of the ions as

$$H_i = T_i + \varphi_{ii}(\boldsymbol{X}) + E_e(\boldsymbol{X}) . \tag{11.5}$$

This approximation will be valid when excited states of the electrons are separated from the ground state by energies large compared with phonon energies, as in insulating crystals. Nevertheless the approximation should also be a valid one for the lattice dynamics of metals, since the great majority of the conduction electron transitions which are energetically allowable are forbidden by the Pauli exclusion principle. This point has been discussed by CHESTER[1].

The 'shell model' described in Sect. 22 was developed to take explicit account of the forces that result from the deformation of the wave-functions of the valence electrons. It also provides an example of the adiabatic approximation in that equations of motion can be set up involving both 'core' and 'shell' (i.e., valence electron) coordinates, but the latter are eliminated by the assumption that the electrons instantaneously take up an equilibrium configuration for any configuration of the cores. The valence electrons "belonging" to one core then have a configuration determined primarily by the position of this core, but also depending to some extent on the positions of other cores.

B. Experimental methods.

12. Elastic constants. Many of the properties of crystals are influenced by the lattice vibrations, and measurements of these properties yield information about the frequencies of the normal modes of vibration. Many of these properties however, for example the specific heat and thermal conductivity, depend on the normal modes only through some average over the whole spectrum, and it is difficult if not impossible to obtain $\omega_j(\boldsymbol{q})$ from such measurements. The most detailed and direct information is obviously provided by those properties which depend on the frequencies of normal modes with particular and well-defined wave vectors. In this section we shall concentrate on these latter measurements.

The long wavelength acoustic modes of vibration of a crystal describe macroscopic homogeneous deformations, and are identical with sound waves. The slopes of these dispersion curves for $\boldsymbol{q} \to 0$ are thus determined by the elastic constants of the crystal. Consider a homogeneous deformation of the crystal, $\boldsymbol{u}$. The equation of motion of a macroscopic portion of the crystal is [2]

$$\varrho \frac{\partial^2 u_\alpha}{\partial t^2} = \sum_\beta \frac{\partial T_{\alpha\beta}}{\partial X_\beta} \tag{12.1}$$

where ϱ is the density and $T_{\alpha\beta}$ the stress tensor which is given by

$$T_{\alpha\beta} = \frac{1}{2} \sum_{\gamma\delta} C_{\alpha\beta\gamma\delta} \left(\frac{\partial u_\gamma}{\partial X_\delta} + \frac{\partial u_\delta}{\partial X_\gamma} \right) . \tag{12.2}$$

[1] G. V. CHESTER: Advances in Phys. **10**, 357 (1961).
[2] C. KITTEL: Introduction to Solid State Physics. New York: John Wiley & Sons 1953.

The elastic constants, $C_{\alpha\beta\gamma\delta}$, are written in tensor notation. In the particular case of a plane wave $\boldsymbol{u}\exp i(\boldsymbol{q}\cdot\boldsymbol{X}-\omega t)$, Eq. (12.1) becomes

$$\varrho\,\omega^2 u_\alpha = \frac{1}{2}\sum_{\beta\gamma\delta} C_{\alpha\beta\gamma\delta}\,(q_\beta\, q_\gamma\, u_\delta + q_\beta\, q_\delta\, u_\gamma)\,. \tag{12.3}$$

This equation may be compared with Eq. (5.7) to obtain expressions for the elastic constants in terms of the force constants, as described by BORN and HUANG [1].

The experimental technique most frequently used to measure elastic constants is to excite a particular type of disturbance along a known direction in a single crystal specimen of the material, and to measure the velocity of this disturbance. For example in cubic crystals the velocity of a longitudinal disturbance along a [100] direction gives the elastic constant $C_{1111}(=C_{11})$, while the velocity of the corresponding transverse disturbance gives $C_{1212}(=C_{44})$. The measurements thus yield directly the slopes of the long wavelength acoustic dispersion curves in crystals. A detailed account of the experimental techniques and a review of many measurements has been given by HUNTINGDON[1].

13. One-phonon infra-red absorption and Raman scattering. The wavelength of both infra-red and visible light is about 1000 times larger than typical inter-atomic spacings. They can therefore be used to excite, under conditions in which momentum conservation is applicable, single phonons of very long wavelength only. Optical techniques thus give information about the long wavelength modes of vibration, in crystals for which the band gap is greater than the energy of the incident photons. They cannot therefore be used to study the lattice vibrations of metals or heavily doped semiconductors.

The detailed theory of the interaction between light and the lattice vibrations of crystals will be reviewed in other articles of this series [2], [3]. To describe the infra-red absorption spectrum of a harmonic crystal each normal mode may be treated independently and a simplified theory given. The incident radiation is described classically [1] by an oscillating electric field, $\boldsymbol{E}\exp(i\omega t)$. Its interaction with a normal mode $\binom{0}{j}$ is described by the dipole moment of the crystal associated with the normal mode, $\boldsymbol{M}\binom{0}{j}\,Q\binom{0}{j}$. The equation of motion for the amplitude of the normal mode $Q\binom{0}{j}$, [c.f. Eq. [2.19)] is then modified by the presence of the driving force $\boldsymbol{M}\binom{0}{j}\cdot\boldsymbol{E}\exp(i\omega t)$ to give

$$\frac{\partial^2 Q\binom{0}{j}}{\partial t^2} + \omega_j^2(0)\, Q\binom{0}{j} = \boldsymbol{M}\binom{0}{j}\cdot\boldsymbol{E}\exp(i\omega t)\,.$$

The dipole moment produced by the electric field is then

$$\frac{\boldsymbol{M}\binom{0}{j}\,\boldsymbol{M}\binom{0}{j}\cdot\boldsymbol{E}}{\omega_j^2(0)-\omega^2}\,\exp(i\omega t)$$

and the contribution of this mode to the dielectric constant $\varepsilon_{\alpha\beta}(\omega)$ is

$$\frac{4\pi}{N v}\,\frac{M_\alpha\binom{0}{j}\,M_\beta\binom{0}{j}}{\omega_j^2(0)-\omega^2} + \delta_{\alpha\beta}\,.$$

[1] H. B. HUNTINGDON: Solid State Physics 7, 214 (1958).

The dielectric constant shows a resonant behaviour near the frequency of the normal mode, $\omega_j(0)$, and strong absorption of the radiation occurs at this frequency.

This simplified theory shows that under conditions of normal incidence of the light only modes of long wavelength having a transverse electric dipole moment will be able to couple with the transverse electromagnetic radiation. The restrictions which this imposes as to the modes which can give absorption may be obtained with the use of group theory, as described by BIRMAN [3]. In alkali halides for example only the transverse optic mode for which $\boldsymbol{q} \to 0$ gives rise to absorption.

Optic modes which are associated with a longitudinal electric moment and for which $\boldsymbol{q} \to 0$ can also be studied by infra-red techniques, although their effect on the dielectric properties is different. In the absence of external fields, POISSON's equation gives $\nabla \cdot \boldsymbol{D} = 0$. However in a homogeneous medium $\boldsymbol{D} = \varepsilon \boldsymbol{E}$, so that $\varepsilon \nabla \cdot \boldsymbol{E} = 0$. This equation shows that longitudinal oscillations in the dipole moment will only occur for those frequencies for which the dielectric constant is zero; $\varepsilon(\omega) = 0$. Measurements of the dielectric constant as a function of frequency thus enable the frequencies of the longitudinal optic modes to be determined. An alternative way of observing longitudinal modes is to use oblique incidence in thin films[1]. However optical modes with no dipole moment cannot be investigated with infra-red techniques.

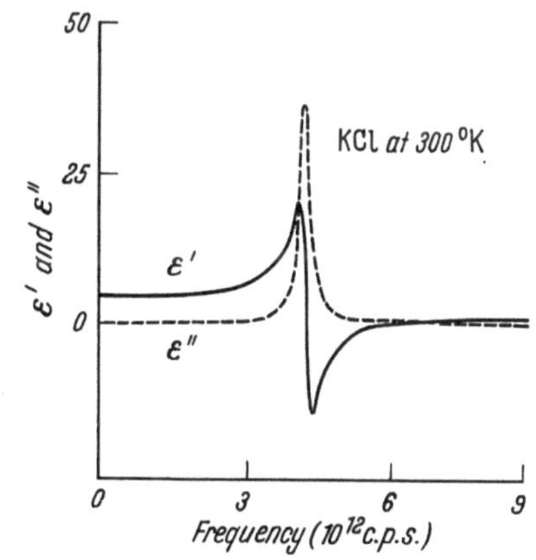

Fig. 5. The real and imaginary parts of the dielectric constant of KCl, as deduced from measurements of infra-red reflectivity and transmission.

The most detailed information is obtained by measuring both the reflectivity and transmission of specimens as a function of the infra-red frequency. The results can then be analysed[2] to give the real and imaginary parts of the dielectric constant. Some typical experimental results for potassium chloride[3] are shown in Fig. 5.

The Raman scattering of light arises from inelastic scattering by the normal modes of vibration with conservation of both energy and momentum. For a harmonic crystal the change in frequency of the light is then equal to the frequency of a long wavelength phonon. However the selection rules for Raman scattering differ from those for infra-red absorption, and different modes may be observed. The Raman scattering depends on the contribution of the mode to the polarisability of the crystal and detailed measurements of the different components of the polarisability tensor can yield information about the symmetry of the normal modes[4]. Examples of the complementary character of the infra-red absorption spectrum and the Raman spectrum are provided by alkali halides and by diamond. The optic modes of alkali halides can be studied by infra-red absorption but not by Raman scattering, while for diamond the optic mode is

[1] R. BERREMAN: Phys. Rev. **130**, 2193 (1963).

[2] D. H. MARTIN: Advances in Phys. **14**, 39 (1965).

[3] G. R. WILKINSON, and C. SMART: Private Comm., quoted by D. H. MARTIN in Advances in Phys. **14**, 39 (1965).

[4] R. LOUDON: Advances in Phys. **13**, 423 (1964).

Raman active but not infra-red active. Fig. 6 shows experimental measurements of the Raman scattering from quartz[1], for which some of the longitudinal and transverse optic modes are Raman active.

Light is also inelastically scattered by acoustic modes, when the process is known as Brillouin scattering. KRISHNAN[2] has made a study of this in diamond and has used it to measure the elastic constants. At present this technique is not so useful for materials with much smaller elastic constants than diamond, but interesting developments can be expected through the use of laser sources[3].

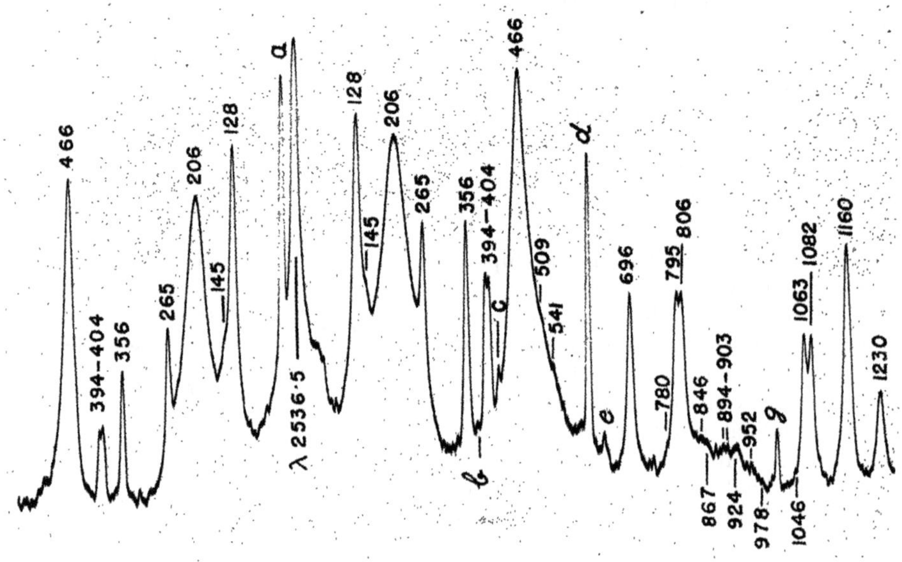

Fig. 6. The Raman spectra of quartz at room temperature.

14. X-ray scattering. X-rays are scattered by the lattice vibrations of crystals with conservation of energy and momentum. Since X-rays have much greater energy than lattice vibrations, the magnitudes of the incident wave vector, $\boldsymbol{k}_0$, and of the scattered wave vector, $\boldsymbol{k}$, are almost equal. The momentum transfer is given by $\hbar Q$ where

$$\boldsymbol{Q} = \boldsymbol{k} - \boldsymbol{k}_0 .$$

The scattering of X-rays by an atom is determined by the electron distribution, $\varrho(\boldsymbol{X})$. The amplitude of this scattering is determined by the form factor;

$$f(\boldsymbol{Q}) = \int\limits_{\text{atom}} \varrho(\boldsymbol{X}) \exp(i\,\boldsymbol{Q}\cdot\boldsymbol{X})\, d^3X .$$

The scattering from a crystal is given to a fair approximation by assuming that the electron distribution is the superposition of that of a series of atoms at $\boldsymbol{X}\binom{l}{\varkappa} + \boldsymbol{u}\binom{l}{\varkappa}$. If the electron distribution of each atom is the same in the crystal as for the free atom, the cross-section for scattering into an element of solid angle $d\mathscr{S}$ is

$$\frac{d\sigma}{d\mathscr{S}} = \sum_{l\varkappa}\sum_{l'\varkappa'} f_\varkappa(\boldsymbol{Q})\, f_{\varkappa'}(-\boldsymbol{Q}) \exp i\boldsymbol{Q}\cdot\left(\boldsymbol{X}\binom{l}{\varkappa} - \boldsymbol{X}\binom{l'}{\varkappa'}\right)\left\langle \exp i\boldsymbol{Q}\cdot\left(\boldsymbol{u}\binom{l}{\varkappa} - \boldsymbol{u}\binom{l'}{\varkappa'}\right)\right\rangle \tag{14.1}$$

[1] D. KRISHNAMURTI: Proc. Indian Acad. Sci. **47**, 276 (1958).

[2] R. S. KRISHNAN: Nature **159**, 740 (1947).

[3] G. B. BENEDEK, J. B. LASTOVKA, T. GREYTAK, and K. FRITSCH: J. Opt. Soc. Amer. **54**, 1286 (1964).

when the intensity is measured in units of the intensity scattered by a (classical) electron at the origin. The bracket $\langle \cdots \rangle$ represents the thermal average. The thermal expectation value can be evaluated with the aid of the relation [4]

$$\langle \exp i\boldsymbol{Q}\cdot\boldsymbol{u}\rangle = \exp\left(-\tfrac{1}{2}\langle(\boldsymbol{Q}\cdot\boldsymbol{u})^2\rangle\right). \tag{14.2}$$

The displacements can be expressed in terms of the normal modes [Eq. (5.17)] so that

$$\boldsymbol{u}\binom{l}{\varkappa} = \frac{1}{\sqrt{N m_\varkappa}} \sum_{\boldsymbol{q}j} \boldsymbol{e}\left(\varkappa\middle|\begin{matrix}\boldsymbol{q}\\ j\end{matrix}\right) Q\binom{\boldsymbol{q}}{j} \exp i\boldsymbol{q}\cdot\boldsymbol{X}\binom{l}{\varkappa}.$$

The expectation value in Eq. (14.1) then becomes

$$\begin{aligned}\left\langle \exp i\boldsymbol{Q}\cdot\left(\boldsymbol{u}\binom{l}{\varkappa}-\boldsymbol{u}\binom{l'}{\varkappa'}\right)\right\rangle = \exp\Bigg[-\frac{1}{2}N\Bigg\langle\Bigg|\sum_{\boldsymbol{q}j} Q\binom{\boldsymbol{q}}{j}&\frac{\left(\boldsymbol{Q}\cdot\boldsymbol{e}\left(\varkappa\middle|\begin{matrix}\boldsymbol{q}\\ j\end{matrix}\right)\right)}{\sqrt{m_\varkappa}}\exp\left(i\boldsymbol{q}\cdot\boldsymbol{X}\binom{l}{\varkappa}\right)-\\ &-\frac{\boldsymbol{Q}\cdot\boldsymbol{e}\left(\varkappa'\middle|\begin{matrix}\boldsymbol{q}\\ j\end{matrix}\right)}{\sqrt{m_{\varkappa'}}}\exp i\boldsymbol{q}\cdot\boldsymbol{X}\binom{l'}{\varkappa'}\Bigg|^2\Bigg\rangle\Bigg].\end{aligned}$$

Multiplying this expression out gives

$$\left.\begin{aligned}&\exp\left(-Y_\varkappa(\boldsymbol{Q})\right)\exp\left(-Y_{\varkappa'}(\boldsymbol{Q})\right)\exp\Bigg\{\frac{1}{2N}\sum_{\boldsymbol{q}j}\sum_{\boldsymbol{q}'j'}\left\langle Q\binom{\boldsymbol{q}}{j}Q\binom{\boldsymbol{q}'}{j'}\right\rangle\Big[\boldsymbol{Q}\cdot\boldsymbol{e}\left(\varkappa\middle|\begin{matrix}\boldsymbol{q}\\ j\end{matrix}\right)\times\\ &\quad\times\boldsymbol{Q}\cdot\boldsymbol{e}^*\left(\varkappa'\middle|\begin{matrix}\boldsymbol{q}'\\ j'\end{matrix}\right)\exp\left(i\boldsymbol{q}\cdot\left(\boldsymbol{X}\binom{l}{\varkappa}-\boldsymbol{X}\binom{l'}{\varkappa'}\right)\right)+\boldsymbol{Q}\cdot\boldsymbol{e}\left(\varkappa'\middle|\begin{matrix}\boldsymbol{q}'\\ j'\end{matrix}\right)\boldsymbol{Q}\cdot\boldsymbol{e}^*\left(\varkappa\middle|\begin{matrix}\boldsymbol{q}\\ j\end{matrix}\right)\times\\ &\quad\times\exp\left(i\boldsymbol{q}\cdot\left(\boldsymbol{X}\binom{l'}{\varkappa'}-\boldsymbol{X}\binom{l}{\varkappa}\right)\right)\Big]\Big/(m_\varkappa m_{\varkappa'})^{\frac{1}{2}}\Bigg\}\end{aligned}\right\}\tag{14.3}$$

where $\exp(-Y_\varkappa(\boldsymbol{Q}))$ is the Debye Waller factor for the $\varkappa$-th atom and

$$Y_\varkappa(\boldsymbol{Q}) = \frac{1}{2Nm_\varkappa}\sum_{\boldsymbol{q}j}\left|\boldsymbol{Q}\cdot\boldsymbol{e}\left(\varkappa\middle|\begin{matrix}\boldsymbol{q}\\ j\end{matrix}\right)\right|^2\frac{\hbar\,\omega_j(\boldsymbol{q})}{\omega_j^2(\boldsymbol{q})}\left(\bar{n}\binom{\boldsymbol{q}}{j}+\frac{1}{2}\right). \tag{14.4}$$

The scattering is then developed by expanding the exponential in { } in Eq. (14.3). The first term gives the elastic scattering, the second the one-phonon inelastic scattering and the higher terms the multi-phonon scattering. The cross section for one-phonon scattering is, in the harmonic approximation,

$$\left.\begin{aligned}\frac{d\sigma}{d\mathscr{S}} = \sum_{\boldsymbol{q}j}&\frac{N\hbar\omega_j(\boldsymbol{q})\left(\bar{n}\binom{\boldsymbol{q}}{j}+\frac{1}{2}\right)}{\omega_j^2(\boldsymbol{q})}\Delta(\boldsymbol{Q}-\boldsymbol{q})\Bigg|\sum_\varkappa f_\varkappa(\boldsymbol{Q})\exp\left(-Y_\varkappa(\boldsymbol{Q})\right)\times\\ &\times\frac{\boldsymbol{Q}\cdot\boldsymbol{e}\left(\varkappa\middle|\begin{matrix}\boldsymbol{q}\\ j\end{matrix}\right)}{\sqrt{m_\varkappa}}\exp\left(i(\boldsymbol{Q}-\boldsymbol{q})\cdot\boldsymbol{X}(\varkappa)\right)\Bigg|^2.\end{aligned}\right\}\tag{14.5}$$

The quantity $\Delta(\boldsymbol{Q}-\boldsymbol{q})$ is defined by Eq. (2.13),

$$\sum_l \exp\left(i\,\boldsymbol{Q}\cdot\boldsymbol{X}(l)\right) = N\,\Delta(\boldsymbol{Q})$$

so that $\boldsymbol{Q}-\boldsymbol{q}$ in Eq. (14.5) is equal to a reciprocal lattice vector, $\boldsymbol{\tau}$.

The intensity of the thermal diffuse scattering of X-rays is therefore related to the frequencies of the normal modes by (14.5). For simple crystals it is possible to derive the individual frequencies by measuring the intensities for several different $\boldsymbol{Q}$ which correspond to the same phonon wave vector, $\boldsymbol{q}$. For example if there is only one atom in the unit cell and the wave vector transfer, $\boldsymbol{Q}$, makes an angle α_j with the phonon polarisation vector $\boldsymbol{e}\binom{\boldsymbol{q}}{j}$, the intensity as a function of $\boldsymbol{Q}$ is

$$\frac{d\sigma}{d\mathscr{S}} = N|f(\boldsymbol{Q})|^2 Q^2 \exp\left(-2Y(\boldsymbol{Q})\right) \sum_j \frac{\cos^2\alpha_j \left(\bar{n}\binom{\boldsymbol{q}}{j} + \frac{1}{2}\right) \hbar\omega_j(\boldsymbol{q})}{m\omega_j^2(\boldsymbol{q})}. \tag{14.6}$$

When $\boldsymbol{q}$ is parallel to certain symmetry axes the directions of the polarisation vectors $\boldsymbol{e}\binom{\boldsymbol{q}}{j}$ are determined by symmetry to be parallel to $\boldsymbol{q}$ (longitudinal mode) or perpendicular to $\boldsymbol{q}$ (transverse modes). It is then possible to choose a value for

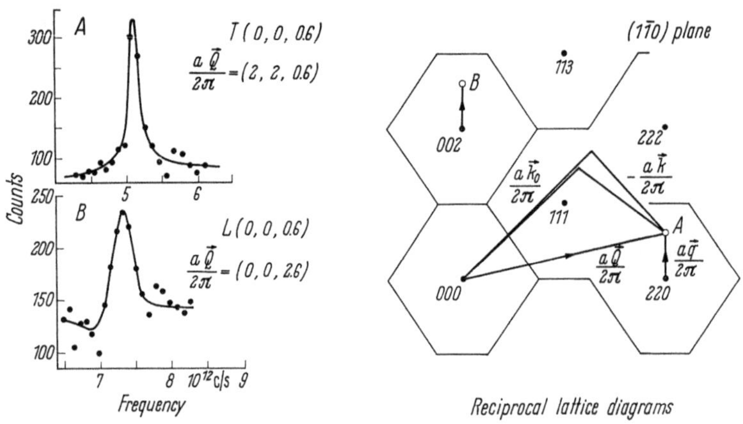

Fig. 7. The left-hand part of the figure shows two typical neutron groups in an experiment on nickel. The right-hand part shows at A the arrangement which gives scattering predominantly by a transverse mode for which $\boldsymbol{q}$ is parallel to [001], since the polarisation vector of this mode is nearly parallel to $\boldsymbol{Q}$. The two triangles at A show how the energy change of the scattered neutrons, proportional to $k^2 - k_0^2$, can be varied while keeping $\boldsymbol{Q}$ constant. The arrangement at B is such that only the longitudinal mode with $\boldsymbol{q}$ parallel to [001] gives scattered intensity.

$\boldsymbol{Q}$ such that for example $\alpha_1 = 0$ and $\alpha_2 = \alpha_3 = \pi/2$, so that the intensity depends only on $\omega_1(\boldsymbol{q})$. The corresponding situations for neutron scattering are illustrated in Fig. 7.

WALKER[1] has determined $\omega_j(\boldsymbol{q})$ for aluminum by this method, and CORBEAU[2] has successfully separated the scattering from acoustic and optic modes in silicon. BUYERS and SMITH[3] have determined $\omega_j(\boldsymbol{q})$ for certain acoustic and optic modes in NaF and have compared the results with those obtained by inelastic neutron scattering. However it is difficult to obtain accurate results by this method. The measured intensities must be corrected for the multiphonon contributions to the scattering, as well as for Compton scattering. Furthermore the method is not readily applicable to crystals with several atoms in the unit cell because the intensity cannot always be separated unambiguously into the contributions from the different modes.

15. Inelastic neutron scattering. The difficulty in studying lattice vibrations using electromagnetic radiation is that either the frequency is comparable with that of the lattice vibrations (infra-red), so that the phonon energy can be meas-

[1] C. B. WALKER: Phys. Rev. **103**, 547 (1956).
[2] J. CORBEAU: J. phys. radium **25**, 925 (1964).
[3] W. J. L. BUYERS, and T. SMITH: To be published.

ured directly, but only when the wave vector is very small, or the wave vector is comparable with that of lattice vibrations but the energy of the photon is too large for the energy change of the scattered photon to be detected (X-rays). The advantage of using thermal neutrons is that both the energy and the wave vector of the radiation are similar to those of lattice vibrations. They thus provide the most versatile and direct way of studying lattice vibrations at present.

Neutrons are scattered by their interaction with the nucleus of an atom. For slow neutrons this interaction is described by a Fermi pseudo-potential[1];

$$V(\boldsymbol{X}) = b\,\delta(\boldsymbol{X}),$$

where b, the scattering length, depends on the particular nucleus and the orientation of its spin relative to that of the neutron. The differential scattering cross section is given adequately by the first Born approximation;

$$\left.\begin{aligned}\frac{d^2\sigma}{d\mathscr{S}\,dE} = \frac{|\boldsymbol{k}|}{|\boldsymbol{k}_0|}\sum_{l\varkappa}\sum_{l'\varkappa'} b\binom{l}{\varkappa}\, b\binom{l'}{\varkappa'} \exp\left(i\boldsymbol{Q}\cdot\left(\boldsymbol{X}\binom{l}{\varkappa} - \boldsymbol{X}\binom{l'}{\varkappa'}\right)\right)\times\\ \times\left\langle\exp\left(i\boldsymbol{Q}\cdot\left(\boldsymbol{u}\binom{l}{\varkappa} - \boldsymbol{u}\binom{l'}{\varkappa'}\right)\right)\right\rangle\delta\left(E-\left(E_I - E_F\right)\right)\end{aligned}\right\} \tag{15.1}$$

where E is the difference between the final and initial energies of the neutron, so that

$$E = \frac{\hbar^2}{2m_n}\left(k^2 - k_0^2\right)$$

and E_I and E_F the initial and final energies of the crystal. The evaluation of the thermal expectation part of this expression can be made in a similar way to that for X-ray scattering (14.1). For example the cross-section for one-phonon scattering requires the evaluation of

$$\sum_{\boldsymbol{q}j}\sum_{\boldsymbol{q}'j'}\left\langle Q\binom{\boldsymbol{q}}{j} Q\binom{\boldsymbol{q}'}{j'}\right\rangle \delta\left(E - (E_I - E_F)\right).$$

For a harmonic crystal this expectation value is non-zero in a process involving one phonon only if $\binom{\boldsymbol{q}'}{j'} = \binom{-\boldsymbol{q}}{j}$, and $E = E_I - E_F = \pm\hbar\,\omega_j(\boldsymbol{q})$. If the plus sign is taken the expectation value involves $\bar{n}\binom{\boldsymbol{q}}{j}$, and if the minus sign is taken it involves $\left(\bar{n}\binom{\boldsymbol{q}}{j} + 1\right)$.

Since the scattering length may vary from one atom to another of the same index ($\varkappa$), because of different isotopes and spin states, it is convenient to divide the scattering into two parts. The coherent scattering is that part which depends upon the correlations between the atoms, while the incoherent part is the remainder. The coherent scattering length is defined by

$$\bar{b}_\varkappa = \frac{1}{N}\sum_l b\binom{l}{\varkappa}$$

and the incoherent scattering length by

$$b_\varkappa^{\mathrm{inc}} = \left|\frac{1}{N}\sum_l b\binom{l}{\varkappa}^2 - (\bar{b}_\varkappa)^2\right|^{\frac{1}{2}}.$$

[1] W. M. Lomer, and G. Low: In: Thermal Neutron Scattering, ed. P. A. Egelstaff. New York and London: Academic Press 1965.

The one-phonon coherent scattering cross-section for a mode $\binom{\boldsymbol{q}}{j}$ is then found to be given by

$$\left.\begin{aligned}\frac{d^2\sigma}{d\mathscr{S}\,dE}\binom{\boldsymbol{q}}{j} &= \frac{|\boldsymbol{k}|}{|\boldsymbol{k}_0|} N \left|\sum_{\varkappa} \bar{b}_{\varkappa} m_{\varkappa}^{-\frac{1}{2}} \boldsymbol{Q}\cdot\boldsymbol{e}\left(\varkappa\middle|\begin{matrix}\boldsymbol{q}\\ j\end{matrix}\right) \exp\left(i(\boldsymbol{Q}-\boldsymbol{q})\cdot\boldsymbol{X}(\varkappa)\right)\exp\left(-Y_{\varkappa}(\boldsymbol{Q})\right)\right|^2 \times \\ &\times \frac{\hbar\omega_j(\boldsymbol{q})}{2\omega_j^2(\boldsymbol{q})} \left\{\begin{matrix}\bar{n}\binom{\boldsymbol{q}}{j} \\ \bar{n}\binom{\boldsymbol{q}}{j}+1\end{matrix}\right\} \delta\left(-E \pm \hbar\,\omega_j(\boldsymbol{q})\right)\Delta(\boldsymbol{Q}-\boldsymbol{q}) \\ &\qquad\qquad \text{(compare Eq. (14.5))}.\end{aligned}\right\} \quad (15.2)$$

The factor $\bar{n}$ applies for scattering with neutron energy gain $(E=+\hbar\omega_j(\boldsymbol{q}))$, the factor $\bar{n}+1$ when $E=-\hbar\omega_j(\boldsymbol{q})$.

The incoherent one-phonon scattering depends only on the correlation of the atom $\binom{l}{\varkappa}$ with itself and the cross section for the mode $\binom{\boldsymbol{q}}{j}$ is

$$\left.\begin{aligned}\frac{d^2\sigma^{\text{inc}}}{d\mathscr{S}\,dE}\binom{\boldsymbol{q}}{j} &= \frac{|\boldsymbol{k}|}{|\boldsymbol{k}_0|} \sum_{\varkappa} \left| b_{\varkappa}^{\text{inc}} m_{\varkappa}^{-\frac{1}{2}} \boldsymbol{Q}\cdot\boldsymbol{e}\left(\varkappa\middle|\begin{matrix}\boldsymbol{q}\\ j\end{matrix}\right)\exp\left(-Y_{\varkappa}(\boldsymbol{Q})\right)\right|^2 \times \\ &\times \frac{\hbar\omega_j(\boldsymbol{q})}{2\omega_j^2(\boldsymbol{q})} \left\{\begin{matrix}\bar{n}\binom{\boldsymbol{q}}{j} \\ \bar{n}\binom{\boldsymbol{q}}{j}+1\end{matrix}\right\} \delta\left(-E \pm \hbar\,\omega_j(\boldsymbol{q})\right).\end{aligned}\right\} \quad (15.3)$$

This expression differs from the coherent scattering cross section (15.2) in that there is no momentum conservation condition, and in the form of the dependence on the eigenvectors. More detailed descriptions of the theory of neutron scattering by crystals are given by LOMER and LOW[1] and by SJOLANDER[2].

The coherent one-phonon scattering of a mode $\binom{\boldsymbol{q}}{j}$ is determined partly by the conditions for conservation of energy and of momentum:

$$-E = \frac{\hbar^2}{2m_n}(k_0^2 - k^2) = \pm\hbar\,\omega_j(\boldsymbol{q})$$

$$\boldsymbol{k} - \boldsymbol{k}_0 = \boldsymbol{Q} = \boldsymbol{\tau} + \boldsymbol{q}\,.$$

Experiments are made by scattering initially monoenergetic neutrons from a single crystal specimen, and determining the change in momentum and energy of the scattered neutrons. Several different techniques have been developed to do this. One of these uses the triple-axis crystal spectrometer[3]. Neutrons emerging from the reactor are Bragg reflected by a single crystal (the monochromator) to provide a monoenergetic beam of neutrons which is scattered by the specimen. (The beam is monoenergetic because only those neutrons of the appropriate wavelength (energy) will be Bragg reflected by the monochromator). The wavelength (energy) of the neutrons scattered through an angle determined by the collimating system is determined by Bragg reflection from another single crystal (the analyser) into a counter. The experimental technique most frequently employed is to vary the incident energy, angle of scattering, and specimen orientation, in such a way as to measure an energy distribution while keeping the

---

[1] See footnote 1, p. 77.

[2] A. SJÖLANDER, p. 76 of Ref. [8].

[3] B. N. BROCKHOUSE: In: Inelastic Scattering of Neutrons from Solids and Liquids. (Report of Vienna Conference.) Internat. Atomic Energy Agency, Vienna 1961. — P. K. IYENGAR: In: Thermal Neutron Scattering, ed. P. A. EGELSTAFF. New York and London: Academic Press 1965.

momentum transfer $\boldsymbol{Q}$, fixed at a preselected value[1]. The frequencies of the normal modes are then deduced from the peaks in the energy distributions. The preselected momentum transfer determines the phonon wave vector, $\boldsymbol{q}$, and the branch j is determined with the help of the cross section (15.2). Fig. 7 shows some typical results and also the way in which longitudinal and transverse modes may be distinguished. Other experimental techniques are described by BRUGGER[2] and by HARRIS et al.[3], and a comprehensive review of experimental results obtained on solids (up to 1965) is given by DOLLING and WOODS[4].

Coherent inelastic scattering of neutrons is at present the most flexible and powerful technique for the study of lattice vibrations. However there are considerable restrictions on its use. Although there are no restrictions arising from the metallic, semiconducting, or ionic character of the specimens, they must be available in the form of large single crystals and have small incoherent and absorption cross sections for thermal neutrons. Furthermore the resolution of the experiments is not nearly as good as can be obtained in optical experiments, and so the detailed study of certain normal modes is better done using optical techniques. For similar reasons the neutron scattering technique does not give as accurate values for the elastic constants of crystals as are obtained by ultrasonic measurements.

16. Phonon-assisted tunnelling of electrons. Some of the normal modes of vibration of semiconductors can be studied by examining the fine structure in the current-potential relation for narrow p-n junctions, at low temperatures. The material on one side of the junction is doped to be p-type and on the other to be n-type, so that well defined Fermi surfaces exist in the valence and conduction bands on opposite sides of the junction. These Fermi surfaces are in well defined regions of reciprocal space, whose wave vector is determined by the particular band structure of the semiconductor. A simple theory of the tunneling process is then that the electrons tunnel across the junction from either p-type to n-type material or vice versa, assisted by a phonon whose wave vector is equal to the wave vector difference between the top of the valence band and the minimum of the conduction band. The probability of tunneling at low temperatures involves

$$\sum_{\boldsymbol{q}j} \int f(E)\,[1 - f(E + eV - \hbar\,\omega_j(\boldsymbol{q}))]\,G\,dE,$$

where E is the electron energy and f the Fermi distribution function. The bias voltage across the junction is V and $G\,dE$ is the contribution to the current from electrons of energy E. The sum over the normal modes is restricted to those with the appropriate wave vectors. This equation shows there will be a discontinuity in dI/dV when $eV = \hbar\,\omega_j(\boldsymbol{q})$.

The experiments consist in measuring the current as a function of applied voltage, and the peaks in d^2I/dV^2 give[5] the frequencies of the normal modes with wave vector equal to the difference in wave vector between the top of the valence band and bottom of the conduction band. The technique has been

---

[1] See footnote 3, p. 78.

[2] R. M. BRUGGER: In: Thermal Neutron Scattering, ed. P. A. Egelstaff. New York and London: Academic Press 1965.

[3] D. H. C. HARRIS, S. J. COCKING, P. A. EGELSTAFF, and F. J. WEBB: In: Inelastic Scattering of Neutrons from Liquids and Solids, Vol. I. (Report of Chalk River Conference.) Internat. Atomic Energy Agency, Vienna 1963.

[4] G. DOLLING, and A. D. B. WOODS: In: Thermal Neutron Scattering, ed. P. A. EGELSTAFF. New York and London: Academic Press 1965.

[5] A. G. CHYNOWETH, R. A. LOGAN, and D. E. THOMAS: Phys. Rev. **125**, 877 (1962).

applied to determine the [111] zone boundary frequencies in germanium[1], (Fig. 8) and their pressure dependence[2] and also the longitudinal optic mode for which $\boldsymbol{q}\to 0$ of several polar semiconductors[3].

The experiments give great precision in energy, about 0,1% [4]. However the determination of the momentum transfer is more uncertain. The band structure must already be known and even then a detailed theory[5] shows that while momentum is conserved perpendicular to the tunnelling direction, it may not be conserved parallel to that direction. There is then some uncertainty about the value of $\boldsymbol{q}$ for the phonons involved. The range of materials for which this technique can be applied is very restricted, as it is limited to doped semiconductors at low temperature, which furthermore must be free from impurity levels.

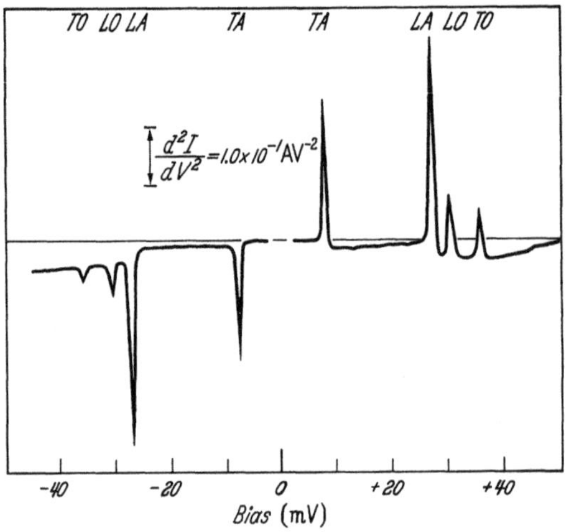

Fig. 8. The graph shows d^2I/dV^2 as a function of the voltage V applied across a $p-n$ junction of germanium at 4.2 °K. Peaks in d^2I/dV^2 give the frequencies of normal modes for which $\boldsymbol{q}$ is at the boundary of the Brillouin Zone in the [111] direction.

17. The frequency distribution.

α) *The one-phonon density of states.* The techniques described in this section yield information about the frequencies of the normal modes but not their wave vectors. The most direct information of this nature is provided by the frequency distribution

$$g(\omega)=\sum_{\boldsymbol{q}j}\delta\left(\omega_j(\boldsymbol{q})-\omega\right).$$

This function can be determined directly by the incoherent inelastic neutron scattering from a cubic monatomic crystal, in the form of a powder. Under these conditions the incoherent neutron scattering cross section for one-phonon processes with neutron energy gain is obtained from (15.3) as

$$\frac{d^2\sigma}{d\mathscr{S}d\omega}=\frac{|\boldsymbol{k}|}{|\boldsymbol{k}_0|}\,\frac{(b^{\mathrm{inc}})^2\,|\boldsymbol{Q}|^2\exp\left(-2Y(\boldsymbol{Q})\right)}{6m\,\omega^2}\left(\frac{\hbar\,\omega}{\exp\left(\frac{\hbar\,\omega}{k_B T}\right)-1}\right)g(\omega),$$

where $E=\hbar\omega$ is the neutron energy gain.

[1] R. N. HALL: In: Proc. Internat. Conf. on Semiconductors, Prague, 1960.
[2] R. T. PAYNE: Phys. Rev. Letters **13**, 53 (1964).
[3] R. N. HALL, J. H. RACETTE, and H. EHRENREICH: Phys. Rev. Letters **4**, 456 (1960).
[4] R. T. PAYNE: Phys. Rev. **139**, A 570 (1965).
[5] E. O. KANE: J. Phys. Chem. Solids **12**, 181 (1959).

The measured intensity as a function of energy must be corrected for the multi-phonon contributions, multiple scattering and the frequency-dependent factors in the cross section to obtain $g(\omega)$. The technique has been used to give the frequency distribution of vanadium[1] and of nickel[2] (see Sect. 28). Vanadium is one of the few elements for which the coherent scattering length of the naturally-occurring material is zero, and the sample of nickel was constructed to have an isotopic composition such that $\bar{b}=0$.

If the crystal is non-cubic or polyatomic the energy distribution of the scattered neutrons is not directly related to $g(\omega)$. However peaks in the distribution will usually be associated with peaks in $g(\omega)$, and sharp kinks with the critical points. The critical points of a frequency distribution are discussed in more detail in Sect. 27. The application of this technique is limited because few materials scatter neutrons entirely incoherently.

The tunnelling of electrons from a superconductor to a normal metal gives information about the one phonon density of states, through the energy dependence of the superconductivity gap. SCHRIEFFER[3] shows that if I_S is the current flowing across the barrier when the metal is superconducting, and I_N the current when it is normal then;

$$\frac{d I_{s/dV}}{d I_{N/dV}} = R_e \left[\frac{V}{\{V^2 - \Delta(E)\}^{\frac{1}{2}}} \right],$$

where V is the applied voltage and $\Delta(E)$ the energy-dependent superconductivity gap parameter. This parameter is related to $g(\omega)$ through the electron-phonon interaction. MCMILLAN and ROWELL[4] have solved the resulting integral equations to deduce a modified form of the density of states for lead. The function they obtain is

$$\sum_{\boldsymbol{q} j} \delta(\omega_j(\boldsymbol{q}) - \omega)\, \alpha \begin{pmatrix} \boldsymbol{q} \\ j \end{pmatrix},$$

where $\alpha \begin{pmatrix} \boldsymbol{q} \\ j \end{pmatrix}$ is the electron-phonon coupling constant averaged over the appropriate electrons. Although their measurements for lead do not give $g(\omega)$ they do exhibit several peaks and kinks which can be correlated with structure in the frequency distribution[5,6].

The one-phonon infra-red absorption and Raman scattering from homogeneous crystals give information only about the long wavelength modes. If defects are introduced into the crystals, then the momentum conservation condition is relaxed and one-phonon measurements involving many of the phonons are possible. The detailed shape of the spectra is very dependent upon the particular materials and the type of defect introduced. The intensity of the spectrum is proportional to

$$\sum_{\boldsymbol{q} j} \delta(\omega_j(\boldsymbol{q}) - \omega) \left| M \begin{pmatrix} \boldsymbol{q} \\ j \end{pmatrix} \right|^2$$

---

[1] K. C. TURBERFIELD, and P. A. EGELSTAFF: Phys. Rev. **127**, 1017 (1962). Refs. to earlier work are given there.

[2] N. A. CHERNOPLEKOV, M. G. ZEMLYANOV, A. G. CHICHERN, and B. G. LYCHSHENKO: In: Inelastic Scattering of Neutrons in Solids and Liquids, Vol. II. (Report of Chalk River Conference.) Internat. Atomic Energy Agency, Vienna 1963.

[3] J. R. SCHRIEFFER: Theory of Superconductivity. Benjamin 1964.

[4] W. MCMILLAN, and J. M. ROWELL: Phys. Rev. Letters **14**, 108 (1965).

[5] D. J. SCALAPINO, and P. W. ANDERSON: Phys. Rev. **133**, A 921 (1964).

[6] B. N. BROCKHOUSE, T. ARASE, G. CAGLIOTI, K. R. RAO, and A. D. B. WOODS: Phys. Rev. **128**, 1099 (1962).

where the factor $M\binom{q}{j}$ depends on the properties of the defect and of the normal mode. The defect may also introduce localised modes associated with the defect and these contribute strongly to the observed spectra. The theory of the effect has been discussed in detail by DAWBER and ELLIOTT[1] for the infra-red spectra of silicon with various impurities, and by MARADUDIN et al.[2] for the Raman scattering of alkali halide crystals. LOUDON[3] has investigated the selection rules which govern the allowed phonons in the infra-red spectra. One example of the use of defects is in the experiments of JONES and WOODFINE[4] to determine the one-phonon density of states in argon, by adding 1/2% zenon (see Sect. 20).

Similar properties can be deduced from the fine structure associated with electronic transitions of impurities in crystals, under conditions of weak electron-phonon coupling. In the same way as for the infra-red spectra, the results do not give the frequency distribution directly, but the peaks and kinks in the distribution can be correlated with structure in the frequency distribution. The detailed theory of the process is given by SATTEN[5], and LOUDON[3] has worked out some of the appropriate selection rules. An example of some experimental results is given by Er^{3+} in $SrTiO_3$[6].

β) The two- phonon density of states. In all of the techniques described for determining particular phonon frequencies and wave-vectors, there are not only contributions from one-phonon processes but also from processes involving two phonons. The theory of these processes is essentially similar to that of the one-phonon processes, except that for a given momentum transfer $\hbar\boldsymbol{Q}$, the conservation equation is

$$\boldsymbol{Q}=\boldsymbol{\tau}+\boldsymbol{q}_1+\boldsymbol{q}_2$$

while the energy transfer from the specimen is

$$\pm\hbar\,\omega_{j_1}(\boldsymbol{q}_1)\pm\hbar\,\omega_{j_2}(\boldsymbol{q}_2).$$

Since two-phonon processes are nearly always much weaker than one-phonon processes they have only been studied experimentally by the use of optical techniques, infra-red absorption and Raman scattering. In these cases, $\boldsymbol{Q}\approx 0$, and $\boldsymbol{q}_1\approx-\boldsymbol{q}_2$.

The detailed theory of these two-phonon processes has been discussed by BORN and HUANG [1] and by LAX and BURSTEIN[7] for infra-red spectra, and by COWLEY[8] for Raman spectra, while the selection rules for certain materials have been derived by BIRMAN[9] and by JOHNSON and LOUDON[10].

The theory of the effect shows that the two-phonon infra-red absorption or Raman scattering at a frequency ω depends on

$$\sum_{\boldsymbol{q}jj'}\left|M\begin{pmatrix}\boldsymbol{q} & -\boldsymbol{q}\\ j & j'\end{pmatrix}\right|^2\delta\left(\omega\pm\omega_j(\boldsymbol{q})\pm\omega_{j'}(\boldsymbol{q})\right).$$

[1] P. G. DAWBER, and R. J. ELLIOTT: Proc. Roy. Soc. (London) A **273**, 222 (1963); — Proc. Phys. Soc. (London) **81**, 453 (1963).

[2] NGUYEN XUAN XINH, A. A. MARADUDIN, and R. A. CALDWELL-HORSFALL: Journal de Physique **26**, 717 (1966).

[3] R. LOUDON: Proc. Phys. Soc. (London) **84**, 379 (1964).

[4] G. O. JONES, and J. M. WOODFINE: Proc. Phys. Soc. (London) **86**, 101 (1965).

[5] R. A. SATTEN: J. Chem. Phys. **40**, 1200 (1964).

[6] M. J. WEBER, and R. F. SCHAYFELE: Phys. Rev. **138**, A 1544 (1965).

[7] M. LAX, and E. BURSTEIN: Phys. Rev. **97**, 39 (1955).

[8] R. A. COWLEY: Proc. Phys. Soc. (London) **84**, 281 (1964).

[9] J. L. BIRMAN: Phys. Rev. **127**, 1093 (1962); **131**, 1489 (1963).

[10] F. A. JOHNSON, and R. LOUDON: Proc. Roy. Soc. (London) A **281**, 274 (1964).

The peaks in the measured distributions can then be correlated with the peaks in the density of states, and the kinks with the critical points.

Both in the measurements of the frequency distribution described in Sect. 17α, and of the two-phonon density of states, the kinks and peaks can be correlated with independent measurements of $\omega_j(\boldsymbol{q})$. Since the former measurements are frequently of greater precision than the latter, they can be used to improve the accuracy with which the frequencies of phonons are known[1]. However if the $\omega_j(\boldsymbol{q})$ have not been determined, for example by inelastic neutron scattering, then even with the help of selection rules, the assignment of the frequencies from the density of states spectra to particular wave vectors, $\boldsymbol{q}$, is extremely uncertain. In many cases in which the assignments have later been checked by measurement of $\omega_j(\boldsymbol{q})$, they have proved to be incorrect.

C. Interpretation of phonon dispersion curves.

18. $\boldsymbol{X}$-space and $\boldsymbol{q}$-space interpretations. We have already remarked that, within the framework of the harmonic approximation, and provided that interatomic force constants can be assumed, quantum concepts are not required in deriving the frequencies and polarisation properties of the normal modes of vibration. As a matter of practical convenience, throughout most of Section C we replace Eq. (5.17) by

$$u_\alpha\binom{l}{\varkappa}=\sum_{\boldsymbol{q}j} U_\alpha\left(\varkappa\middle|\begin{matrix}\boldsymbol{q}\\ j\end{matrix}\right)\exp i\left(\boldsymbol{q}\cdot\boldsymbol{X}\binom{l}{\varkappa}-\omega_j(\boldsymbol{q})\,t\right) \tag{18.1}$$

and the equations of motion by

$$m_\varkappa\,\omega^2(\boldsymbol{q})\,U_\alpha(\varkappa|\boldsymbol{q})=\sum_{\varkappa'\beta} M_{\alpha\beta}\binom{\boldsymbol{q}}{\varkappa\varkappa'}\,U_\beta(\varkappa'|\boldsymbol{q}), \tag{18.2}$$

where

$$M_{\alpha\beta}\binom{\boldsymbol{q}}{\varkappa\varkappa'}=\sum_{l'}\varphi_{\alpha\beta}\binom{l\,l'}{\varkappa\varkappa'}\exp i\boldsymbol{q}\cdot\left(\boldsymbol{X}\binom{l'}{\varkappa'}-\boldsymbol{X}\binom{l}{\varkappa}\right). \tag{18.3}$$

In matrix notation

$$\omega^2\,\boldsymbol{m}\,\boldsymbol{U}=\boldsymbol{M}\boldsymbol{U} \tag{18.4}$$

where ω^2, $\boldsymbol{M}$ and $\boldsymbol{U}$ are of course functions of $\boldsymbol{q}$, and $\boldsymbol{m}$ is a diagonal matrix whose $3r$ diagonal elements are $m_1, m_1, m_1, m_2, \ldots, m_r, m_r$. The values of $\omega^2(\boldsymbol{q})$ are then the eigenvalues of the matrix whose typical element is $(m_\varkappa m_{\varkappa'})^{-\frac{1}{2}} M_{\alpha\beta}\binom{\boldsymbol{q}}{\varkappa\varkappa'}$, which is identical with $D_{\alpha\beta}\binom{\boldsymbol{q}}{\varkappa\varkappa'}$ of Sect. 5.

When the potential $\varphi(X)$ between atoms depends only on the distance X between them, it may be more convenient to express the quantity $M_{\alpha\beta}\binom{\boldsymbol{q}}{\varkappa\varkappa'}$ in terms of the Fourier transform of the potential rather than in terms of the second derivatives of the potential itself. Let $\boldsymbol{Q}$, as before, be a general vector in reciprocal space (there should be no confusion with the normal coordinate used in Sect. 2, 3, 4). A function $\varphi(X)$ and its Fourier transform $\varphi(Q)$ are related by

$$\varphi(X)=\int\varphi(Q)\exp\left(-i\,\boldsymbol{Q}\cdot\boldsymbol{X}\right)\frac{d^3Q}{(2\pi)^3}. \tag{18.5}$$

We shall also make use of the result

$$\sum_l \exp\left(-i\,\boldsymbol{Q}\cdot\boldsymbol{X}(l)\right)=\frac{(2\pi)^3}{v}\sum_{\boldsymbol{\tau}}\delta(\boldsymbol{Q}-\boldsymbol{\tau}) \tag{18.6}$$

[1] See footnote 10, p. 82.

which applies when the sum over l includes a large number N of terms. As previously, $\boldsymbol{\tau}$ is a vector of the reciprocal lattice [Eq. (6.4)], v is the volume of the unit cell [so that $(2\pi)^3/v$ is in fact the volume of the unit cell in reciprocal space or Brillouin zone] and $\delta(Q)$ is the Dirac δ-function, which satisfies the conditions

$$\delta(\boldsymbol{Q})=0 \quad \text{for} \quad \boldsymbol{Q}\neq 0, \quad \int \delta(\boldsymbol{Q})\, d^3 Q = 1 \tag{18.7}$$

and

$$\int f(\boldsymbol{Q})\, \delta(\boldsymbol{Q}-\boldsymbol{\tau})\, d^3 Q = f(\boldsymbol{\tau}). \tag{18.8}$$

Clearly in the situation envisaged,

$$\varphi_{\alpha\beta}\begin{pmatrix} l\,l' \\ \varkappa\,\varkappa' \end{pmatrix} = -\left(\frac{\partial^2 \varphi \begin{pmatrix} X \\ \varkappa\,\varkappa' \end{pmatrix}}{\partial X_\alpha\, \partial X_\beta} \right)_{\boldsymbol{X}=\boldsymbol{X}\left(\substack{l'\\ \varkappa'}\right) - \boldsymbol{X}\left(\substack{l\\ \varkappa}\right)}. \tag{18.9}$$

Now define

$$N_{\alpha\beta}\begin{pmatrix} \boldsymbol{q} \\ \varkappa\,\varkappa' \end{pmatrix} = -\sum_{l'} \left(\frac{\partial^2 \varphi \begin{pmatrix} X \\ \varkappa\,\varkappa' \end{pmatrix}}{\partial X_\alpha\, \partial X_\beta} \exp i\boldsymbol{q}\cdot\boldsymbol{X} \right)_{\boldsymbol{X}=\boldsymbol{X}\left(\substack{l'\\ \varkappa'}\right) - \boldsymbol{X}\left(\substack{l\\ \varkappa}\right)}. \tag{18.10}$$

Comparison with the definition (18.3) shows that, at least for

$$\varkappa' \neq \varkappa, \quad M_{\alpha\beta}\begin{pmatrix} \boldsymbol{q} \\ \varkappa\,\varkappa' \end{pmatrix} = N_{\alpha\beta}\begin{pmatrix} \boldsymbol{q} \\ \varkappa\,\varkappa' \end{pmatrix}. \tag{18.11}$$

However in deriving $M_{\alpha\beta}\begin{pmatrix} \boldsymbol{q} \\ \varkappa\,\varkappa \end{pmatrix}$ one must remember that $\varphi_{\alpha\beta}\begin{pmatrix} l\,l \\ \varkappa\,\varkappa \end{pmatrix}$ is *not* the second derivative of a potential evaluated at $X=0$, but is given by Eq. (5.3). When this is taken into account, one finds

$$M_{\alpha\beta}\begin{pmatrix} \boldsymbol{q} \\ \varkappa\,\varkappa \end{pmatrix} = N_{\alpha\beta}\begin{pmatrix} \boldsymbol{q} \\ \varkappa\,\varkappa \end{pmatrix} - \sum_{\varkappa'} N_{\alpha\beta}\begin{pmatrix} 0 \\ \varkappa\,\varkappa' \end{pmatrix}. \tag{18.12}$$

Note that while the sum in Eq. (18.10), regarded as a definition of $N_{\alpha\beta}\begin{pmatrix} \boldsymbol{q} \\ \varkappa\,\varkappa \end{pmatrix}$, includes $l'=l$, this term contributes equally to $N_{\alpha\beta}\begin{pmatrix} \boldsymbol{q} \\ \varkappa\,\varkappa \end{pmatrix}$ and $N_{\alpha\beta}\begin{pmatrix} 0 \\ \varkappa\,\varkappa \end{pmatrix}$ and therefore cancels from the expression for $M_{\alpha\beta}\begin{pmatrix} \boldsymbol{q} \\ \varkappa\,\varkappa \end{pmatrix}$, Eq. (18.12).

Returning now to Eq. (18.10), we substitute in this from (18.5), obtaining

$$N_{\alpha\beta}\begin{pmatrix} \boldsymbol{q} \\ \varkappa\,\varkappa' \end{pmatrix} = \sum_{l'} \int Q_\alpha Q_\beta\, \varphi\begin{pmatrix} Q \\ \varkappa\,\varkappa' \end{pmatrix} \exp\left(-i(\boldsymbol{Q}-\boldsymbol{q})\right)\cdot\left(\boldsymbol{X}\begin{pmatrix} l' \\ \varkappa' \end{pmatrix} - \boldsymbol{X}\begin{pmatrix} l \\ \varkappa \end{pmatrix}\right) \frac{d^3 Q}{(2\pi)^3}. \tag{18.13}$$

On reversing the order of summation and integration, and using Eq. (18.6) and (18.8) this becomes

$$N_{\alpha\beta}\begin{pmatrix} \boldsymbol{q} \\ \varkappa\,\varkappa' \end{pmatrix} = \frac{1}{v} \sum_{\boldsymbol{\tau}} (\boldsymbol{\tau}+\boldsymbol{q})_\alpha (\boldsymbol{\tau}+\boldsymbol{q})_\beta\, \varphi\begin{pmatrix} |\boldsymbol{\tau}+\boldsymbol{q}| \\ \varkappa\,\varkappa' \end{pmatrix} \exp i\boldsymbol{\tau}\cdot\left(\boldsymbol{X}(\varkappa) - \boldsymbol{X}(\varkappa')\right). \tag{18.14}$$

Together with Eqs. (18.11) and (18.12), this gives the $\boldsymbol{Q}$-space representation of $M_{\alpha\beta}\begin{pmatrix} \boldsymbol{q} \\ \varkappa\,\varkappa' \end{pmatrix}$, which is particularly useful for the lattice dynamics of crystals in which there is Coulomb interaction.

19. Coulomb interaction. In this section we consider in more detail the important practical case of Coulomb interaction. We assume that the force constants of the crystal can be written as

$$\varphi_{\alpha\beta}\begin{pmatrix} l\,l' \\ \varkappa\,\varkappa' \end{pmatrix} = \varphi^{(R)}_{\alpha\beta}\begin{pmatrix} l\,l' \\ \varkappa\,\varkappa' \end{pmatrix} + \varphi^{(C)}_{\alpha\beta}\begin{pmatrix} l\,l' \\ \varkappa\,\varkappa' \end{pmatrix}, \tag{19.1}$$

where the superscript (R) indicates an interaction of short range, and the superscript (C) indicates Coulomb interaction between ions carrying charges $Z_\varkappa$ and $Z_{\varkappa'}$, so that in fact

$$\varphi_{\alpha\beta}^{(C)}\binom{l\,l'}{\varkappa\,\varkappa'} = -Z_\varkappa Z_{\varkappa'}\left(\frac{\partial^2|\boldsymbol{X}|^{-1}}{\partial X_\alpha\,\partial X_\beta}\right)_{\boldsymbol{X}=\boldsymbol{X}\binom{l'}{\varkappa'}-\boldsymbol{X}\binom{l}{\varkappa}} \tag{19.2}$$

except, as usual, for $\binom{l'}{\varkappa'}=\binom{l}{\varkappa}$.

Corresponding to Eq. (19.1) we write

$$M_{\alpha\beta}\binom{\boldsymbol{q}}{\varkappa\,\varkappa'} = R_{\alpha\beta}\binom{\boldsymbol{q}}{\varkappa\,\varkappa'} + Z_\varkappa Z_{\varkappa'}\, C_{\alpha\beta}\binom{\boldsymbol{q}}{\varkappa\,\varkappa'}, \tag{19.3}$$

where

$$R_{\alpha\beta}\binom{\boldsymbol{q}}{\varkappa\,\varkappa'} = \sum_{l'} \varphi_{\alpha\beta}^{(R)}\binom{l\,l'}{\varkappa\,\varkappa'} \exp i\boldsymbol{q}\cdot\left(\boldsymbol{X}\binom{l'}{\varkappa'} - \boldsymbol{X}\binom{l}{\varkappa}\right) \tag{19.4}$$

and

$$Z_\varkappa Z_{\varkappa'}\, C_{\alpha\beta}\binom{\boldsymbol{q}}{\varkappa\,\varkappa'} = \sum_{l'} \varphi_{\alpha\beta}^{(C)}\binom{l\,l'}{\varkappa\,\varkappa'} \exp i\boldsymbol{q}\cdot\left(\boldsymbol{X}\binom{l'}{\varkappa'} - \boldsymbol{X}\binom{l}{\varkappa}\right). \tag{19.5}$$

The latter definition gives $C_{\alpha\beta}\binom{\boldsymbol{q}}{\varkappa\,\varkappa'}$ in terms of a series which is only conditionally convergent, as we shall see shortly $C_{\alpha\beta}\binom{\boldsymbol{q}}{\varkappa\,\varkappa'}$ does not in fact tend to a unique value as $q\to 0$, but to a value which depends on the relative directions of $\boldsymbol{q}$ and the electrical polarisation associated with the mode of vibration concerned [*1*].

A transformation discovered by EWALD[1] enables $C_{\alpha\beta}\binom{\boldsymbol{q}}{\varkappa\,\varkappa'}$ to be evaluated, and is described in some detail by KELLERMANN [*12*]. The following derivation is more general in that it applies to a crystal of any symmetry. Let $\varphi^{(G)}\binom{X}{\varkappa\,\varkappa'}$ be the potential energy of a point charge $Z_\varkappa$ at a point $\boldsymbol{X}$ in a three-dimensional Gaussian charge distribution, $Z_{\varkappa'}\left(\frac{p}{\pi}\right)^{\frac{3}{2}} \exp(-pX^2)$, which contains a total charge $Z_{\varkappa'}$. The energy of two point charges $Z_\varkappa$ and $Z_{\varkappa'}$ separated by $\boldsymbol{X}$ would be

$$\varphi^{(C)}\binom{X}{\varkappa\,\varkappa'} = \frac{Z_\varkappa Z_{\varkappa'}}{X}. \tag{19.6}$$

Now write

$$\begin{aligned}\varphi^{(C)} &= \varphi^{(G)} + (\varphi^{(C)} - \varphi^{(G)})\\ &= \varphi^{(G)} + \varphi^{(H)}, \text{ say.}\end{aligned} \tag{19.7}$$

The last term can be evaluated from elementary electrostatics and is given by

$$\varphi^{(H)}\binom{X}{\varkappa\,\varkappa'} = \frac{2Z_\varkappa Z_{\varkappa'}}{X}\left(\frac{p}{\pi}\right)^{\frac{1}{2}} \int\limits_X^\infty \exp(-p\,x^2)\,dx. \tag{19.8}$$

This quantity diverges at $X=0$, but it decreases rapidly with X and therefore behaves in a lattice like a short-range potential, provided p is suitably chosen. $\varphi^{(G)}\binom{X}{\varkappa\,\varkappa'}$ is not divergent, but has the characteristics of the Coulomb potential at large distances and is therefore unsuitable for evaluating lattice sums. Its contribution to $Z_\varkappa Z_{\varkappa'} C_{\alpha\beta}$ can however readily be evaluated using the $\boldsymbol{Q}$-space formulation described in Sect. 18. Using POISSON's theorem together with the result

$$\left(\frac{p}{\pi}\right)^{\frac{3}{2}} \int \exp(-pX^2)\exp i\boldsymbol{Q}\cdot\boldsymbol{X}\,d^3X = \exp\left(-\frac{Q^2}{4p}\right) \tag{19.9}$$

---

[1] P. P. EWALD: Ann. Phys. (Lpz.) (4) **64**, 253 (1921).

one finds that the Fourier transform of $\varphi^{(G)}\binom{X}{\varkappa\varkappa'}$ is

$$\varphi^{(G)}\binom{Q}{\varkappa\varkappa'} = \frac{4\pi Z_\varkappa Z_{\varkappa'}}{Q^2}\exp\left(-\frac{Q^2}{4p}\right). \tag{19.10}$$

Using the results of Sect. 18 we then have, for $\varkappa\neq\varkappa'$,

$$Z_\varkappa Z_{\varkappa'} C_{\alpha\beta}\binom{\boldsymbol{q}}{\varkappa\varkappa'} = N_{\alpha\beta}^{(H)}\binom{\boldsymbol{q}}{\varkappa\varkappa'} + N_{\alpha\beta}^{(G)}\binom{\boldsymbol{q}}{\varkappa\varkappa'} \tag{19.11}$$

where

$$N_{\alpha\beta}^{(H)}\binom{\boldsymbol{q}}{\varkappa\varkappa'} = -\sum_{l'}\left(\frac{\partial^2\varphi^{(H)}\binom{X}{\varkappa\varkappa'}}{\partial X_\alpha\,\partial X_\beta}\exp i\,\boldsymbol{q}\cdot\boldsymbol{X}\right)_{\boldsymbol{X}=\boldsymbol{X}\binom{l'}{\varkappa'}-\boldsymbol{X}\binom{l}{\varkappa}} \tag{19.12}$$

is to be evaluated as it stands, but the corresponding $N_{\alpha\beta}^{(G)}\binom{\boldsymbol{q}}{\varkappa\varkappa'}$ is expressed as a sum in reciprocal space,

$$N_{\alpha\beta}^{(G)}\binom{\boldsymbol{q}}{\varkappa\varkappa'} = \frac{4\pi Z_\varkappa Z_{\varkappa'}}{v}\sum_{\boldsymbol{\tau}}\frac{(\boldsymbol{\tau}+\boldsymbol{q})_\alpha(\boldsymbol{\tau}+\boldsymbol{q})_\beta}{|\boldsymbol{\tau}+\boldsymbol{q}|^2}\exp\left(-\frac{|\boldsymbol{\tau}+\boldsymbol{q}|^2}{4p}\right)\exp i\boldsymbol{\tau}\cdot((\boldsymbol{X}(\varkappa)-\boldsymbol{X}(\varkappa')). \tag{19.13}$$

For $\varkappa=\varkappa'$ one must again remember that $\varphi_{\alpha\beta}^{(H)}\binom{l\,l}{\varkappa\varkappa}$ is *not* a second derivative of $\varphi^{(H)}\binom{X}{\varkappa\varkappa'}$ at $X=0$, but is found using Eq. (5.3). One then obtain

$$Z_\varkappa^2 C_{\alpha\beta}\binom{\boldsymbol{q}}{\varkappa\varkappa} = N_{\alpha\beta}^{(H)}\binom{\boldsymbol{q}}{\varkappa\varkappa} + N_{\alpha\beta}^{(G)}\binom{\boldsymbol{q}}{\varkappa\varkappa} - \sum_{\varkappa'}\left\{N_{\alpha\beta}^{(H)}\binom{0}{\varkappa\varkappa'} + N_{\alpha\beta}^{(G)}\binom{0}{\varkappa\varkappa'}\right\}. \tag{19.14}$$

The divergent term arising from $\binom{l'}{\varkappa'}=\binom{l}{\varkappa}$ in taking Eq. (19.12) to be the definition of $N_{\alpha\beta}^{(H)}\binom{\boldsymbol{q}}{\varkappa\varkappa}$ can be ignored since it is cancelled by an identical contribution to $N_{\alpha\beta}^{(H)}\binom{0}{\varkappa\varkappa}$ in the final result, Eq. (19.14).

Table 1. *Values of* $vC_{\alpha\beta}\binom{\boldsymbol{q}}{\varkappa\varkappa'}$ *for* $\boldsymbol{q}$ *parallel to* [100] *in the sodium chloride lattice.*

By symmetry, $C_{\alpha\beta}\binom{\boldsymbol{q}}{1\,1} = C_{\alpha\beta}\binom{\boldsymbol{q}}{2\,2}$. For $\boldsymbol{q}$ in this direction we also have $C_{\beta\beta}\binom{\boldsymbol{q}}{\varkappa\varkappa'} = C_{\gamma\gamma}\binom{\boldsymbol{q}}{\varkappa\varkappa'}$, and $C_{\alpha\beta}\binom{\boldsymbol{q}}{\varkappa\varkappa'}=0$ for $\alpha\neq\beta$.

$q_\alpha \div \frac{2\pi}{a}$	$vC_{\alpha\alpha}\binom{q_\alpha}{1\,1}$	$vC_{\beta\beta}\binom{q_\alpha}{1\,2}$	$vC_{\alpha\alpha}\binom{q_\alpha}{1\,2}$	$vC_{\beta\beta}\binom{q_\alpha}{1\,2}$
0.0	8.38	−4.19	8.38	−4.19
0.2	8.01	−4.00	8.99	−4.49
0.4	6.99	−3.51	10.62	−5.31
0.6	5.78	−2.89	11.68	−6.34
0.8	4.74	−2.37	14.39	−7.19
1.0	4.33	−2.16	15.04	−7.52

A value for p of the order $(1/X_0)^2$, where X_0 is the nearest neighbour distances gives rapid convergence of both series, (19.12) and (19.13). Explicit expressions for $N_{\alpha\beta}^{(H)}\binom{\boldsymbol{q}}{\varkappa\varkappa'}$ are given by KELLERMANN [*12*] for the sodium chloride structure. In Table 1 the dimensionless functions $vC_{\alpha\beta}\binom{\boldsymbol{q}}{\varkappa\varkappa'}$ are tabulated for $\boldsymbol{q}$ in the [100] direction of a crystal having the sodium chloride structure.

The equations of motion for an ionic crystal can be obtained in another way, introducing the effective field acting on an atom, so that

$$m_\varkappa \frac{\partial^2 u_\alpha\binom{l}{\varkappa}}{\partial t^2} = -\sum_{l'\varkappa'\beta} \varphi_{\alpha\beta}^{(\mathrm{R})}\binom{l\ l'}{\varkappa\,\varkappa'} u_\beta\binom{l'}{\varkappa'} + Z_\varkappa E_\alpha\binom{l}{\varkappa}. \tag{19.15}$$

Now by analogy with

$$u_\alpha\binom{l}{\varkappa} = U_\alpha(\varkappa|\boldsymbol{q}) \exp i\left(\boldsymbol{q}\cdot\boldsymbol{X}\binom{l}{\varkappa} - \omega(\boldsymbol{q})\,t\right), \tag{19.16}$$

we introduce

$$E_\alpha\binom{l}{\varkappa} = E_\alpha(\varkappa|\boldsymbol{q}) \exp i\left(\boldsymbol{q}\cdot\boldsymbol{X}\binom{l}{\varkappa} - \omega(\boldsymbol{q})\,t\right), \tag{19.17}$$

and Eq. (19.15) reduces to

$$m_\varkappa \omega^2(\boldsymbol{q})\, U_\alpha(\varkappa|\boldsymbol{q}) = \sum_{\varkappa'\beta} R_{\alpha\beta}\binom{\boldsymbol{q}}{\varkappa\,\varkappa'} U_\beta(\varkappa'|\boldsymbol{q}) - Z_\varkappa E_\alpha(\varkappa|\boldsymbol{q}), \tag{19.18}$$

on using the definition (19.4). Comparison of Eq. (19.18) with Eq. (18.2) and using the definition (19.3) shows that

$$E_\alpha(\varkappa|\boldsymbol{q}) = -\sum_{\varkappa'\beta} Z_{\varkappa'} C_{\alpha\beta}\binom{\boldsymbol{q}}{\varkappa\,\varkappa'} U_\beta(\varkappa'\boldsymbol{q}). \tag{19.19}$$

The displacement of charges $Z_\varkappa$ by a wave of amplitude $\boldsymbol{U}(\varkappa|\boldsymbol{q})$ produces a wave of electrical polarisation whose amplitude is

$$\boldsymbol{P}(\varkappa|\boldsymbol{q}) = \frac{1}{v} Z_\varkappa \boldsymbol{U}(\varkappa|\boldsymbol{q}). \tag{19.20}$$

It follows from Eq. (19.19) that

$$E_\alpha(\varkappa|\boldsymbol{q}) = -v \sum_{\varkappa'\beta} C_{\alpha\beta}\binom{\boldsymbol{q}}{\varkappa\,\varkappa'} P_\beta(\varkappa'|\boldsymbol{q}). \tag{19.21}$$

The elements of the matrix $\boldsymbol{C}(\boldsymbol{q})$ therefore relate the amplitude of the polarisation produced by the displacement of ions of one type to the amplitude of the field produced at another. Our definition of $C_{\alpha\beta}\binom{\boldsymbol{q}}{\varkappa\,\varkappa}$ is such that it is the field *at the displaced positions* of the ions of type $\varkappa$, which is given by Eq. (19.21).

Returning now to the entries in Table 1, which refer to the sodium chloride structure with $\boldsymbol{q} = (q_\alpha, 0, 0)$, one finds that for $q \to 0$

$$E_\alpha(1|0) = E_\alpha(2|0) = -\frac{8\pi}{3}\left(P_\alpha(1|0) + P_\alpha(2|0)\right), \tag{19.22}$$

while

$$E_\beta(1|0) = E_\beta(2|0) = +\frac{4\pi}{3}\left(P_\beta(1|0) + P_\beta(2|0)\right). \tag{19.23}$$

In other words, in a transverse optic wave of long wavelength the effective field is the Lorentz field $\frac{4\pi}{3}P$ but in a longitudinal wave it is the Lorentz field minus $4\pi P$. The field $-4\pi P$ associated with a longitudinal wave can be shown not to be a local field but a macroscopic field whose average value is the same over a unit cell [*1*]. In terms of the $\boldsymbol{Q}$-space representation, it arises from the term corresponding to $\tau = 0$ in Eq. (19.13). This term contributes $4\pi/v$ to $C_{\alpha\alpha}\binom{q_\alpha}{\varkappa\,\varkappa'}$ for small values of $\boldsymbol{q}$, but makes zero contribution to $C_{\beta\beta}\binom{q_\alpha}{\varkappa\,\varkappa'}$.

It is this field which is responsible for the different frequencies of the longitudinal optic and transverse optic modes for which $q \to 0$, in an ionic crystal (see Figs. 12, 13). In this context, $q \to 0$ implies a wave whose wavelength is still small compared with a dimension of the crystal, but large compared with a unit cell dimension. For smaller values of q than this, one must take explicit account of retardation of the Coulomb interaction [1], which the above discussion neglects, and of the shape of the crystal[1].

D. Calculation of phonon dispersion curves, and comparison with experiment.

20. Inert-gas solids. The physics of inert-gas solids has been reviewed by DOBBS and JONES[2] and by POLLACK[3].

Inert gases, which are one of the classic solid types, crystallise in the face centred cubic structure. Solution of the quantum mechanical problem for atoms with closed electron configurations, leading to an internuclear potential energy, scarcely seems possible at present and more or less empirical potentials have to be assumed. The commonly-used form for inert gases is a central two-body potential

$$\varphi(X) = \frac{-\varepsilon}{n-6}\left[n\left(\frac{X_0}{X}\right)^6 - 6\left(\frac{X_0}{X}\right)^n\right]. \quad (20.1)$$

The Lennard-Jones 6, 12 potential is the special case $n=12$, for which $\varphi(X)$ can be written

$$\varphi(X) = -4\varepsilon\left[\left(\frac{\sigma}{X}\right)^6 - \left(\frac{\sigma}{X}\right)^{12}\right], \quad (20.2)$$

where ε is the minimum value of the potential, $X_0 = 2^{\frac{1}{6}}\sigma$ and $\varphi(\sigma)=0$.

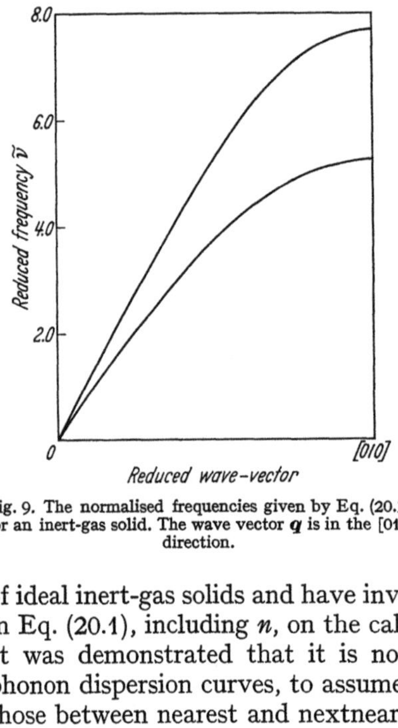

Fig. 9. The normalised frequencies given by Eq. (20.3), for an inert-gas solid. The wave vector $\boldsymbol{q}$ is in the [010] direction.

HORTON and LEECH[4] have made a thorough study of the statistical mechanics of ideal inert-gas solids and have investigated the effect of varying the parameters in Eq. (20.1), including n, on the calculated values of various physical properties. It was demonstrated that it is not a good approximation, in calculating the phonon dispersion curves, to assume that the only significant force constants are those between nearest and nextnearest neighbours.

GRINDLAY and HOWARD[5] have used Eq. (20.2) for the potential to calculate the phonon dispersion curves in such a way that, with appropriate choice of the parameters σ and ε, the results are applicable to any inert-gas solid.

Results were evaluated numerically in the form of a reduced frequency

$$\tilde{\nu}(\boldsymbol{q}) = \frac{1}{2}a_0\left(\frac{m}{8\varepsilon}\right)^{\frac{1}{2}}\omega(\boldsymbol{q}). \quad (20.3)$$

[1] T. H. K. BARRON: Phys. Rev. **123**, 1995 (1961). — J. HARDY: Phil. Mag. **7**, 315 (1962). — A. A. MARADUDIN, and G. H. WEISS: Phys. Rev. **123**, 1968 (1961).

[2] E. R. DOBBS, and G. O. JONES: Repts. Progr. in Phys. **20**, 516 (1957).

[3] G. L. POLLACK: Rev. Mod. Phys. **36**, 748 (1964).

[4] G. K. HORTON, and J. W. LEECH: Proc. Phys. Soc. (London) **82**, 816 (1963).

[5] J. GRINDLAY, and R. HOWARD, p. 129 of Ref. [11].

This is shown in Fig. 9 for $\boldsymbol{q}$ in the [100] direction. The two transverse modes are degenerate by symmetry.

There are unfortunately at the time of writing no direct measurements of $\omega_j(\boldsymbol{q})$ or of the elastic constants for any inert gas solid. GRINDLAY and HOWARD[1]

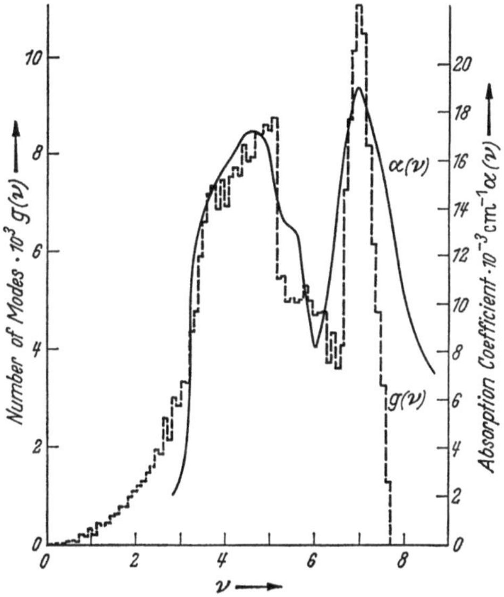

Fig. 10. The one-phonon infra-red absorption induced by defects in argon, compared with the calculated frequency distribution.

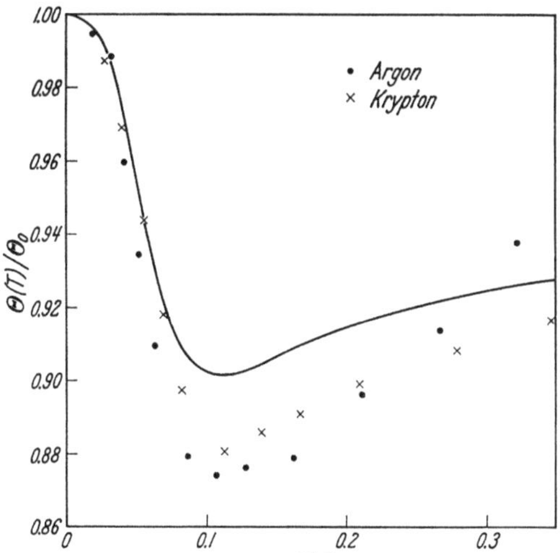

Fig. 11. Measured and calculated values of the Debye temperature $\Theta_D(T)$ for solid argon and krypton, scaled to $\Theta_D(0)=1$.

have evaluated $g(\nu)$ by the sampling method (Sect. 28) using approximately 10^5 values of $\boldsymbol{q}$ uniformly distributed in the Brillouin zone. The histogram obtained is shown by the dotted line in Fig. 10. The histogram was used to evaluate the Debye temperature as a function of temperature, $\Theta_D(T)$, for argon and krypton,

[1] See footnote 5, p. 88.

the parameters ε and σ being chosen to give agreement with experimental values of $\Theta_D(0)$ and a_0 at 0 °K. The agreement between measured and calculated values of $\Theta_D(T)$ is fairly good (Fig. 11), this is not however a sensitive test of the correctness of the dispersion curves or of $g(\omega)$. The defect-induced one-phonon infra-red absorption (Sect. 17) in argon[1] is in qualitative agreement with $g(\omega)$ calculated by GRINDLAY and HOWARD (Fig. 10). For both argon and krypton, calculated cohesive energies and zero-point energies agree with measured values to within 3% and 20% respectively.

First-neighbour force constants in argon are some fifty times less than in an alkali halide, so that the dipoles induced on the atoms by short range forces, which have a marked effect on the dispersion curves of alkali halides (see Sect. 22) are likely to be less important for inert-gas solids although they may not be negligible. Anharmonic effects are relatively greater than in the alkali halides.

21. Ionic crystals. The alkali halides provide another example of a class of solids for which a simple model potential approximates the actual situation. In BORN's model of an alkali halide the potential between two ions is taken to be

$$\varphi\begin{pmatrix} X \\ \varkappa\varkappa' \end{pmatrix} = \frac{Z_\varkappa Z_{\varkappa'}}{X} + b_{\varkappa\varkappa'} \exp(-X/\varrho), \tag{21.1}$$

where $Z_\varkappa = +e$ for cations, $-e$ for anions. The characteristic distance ϱ is such that the short range potential is inappreciable beyond next-nearest neighbours. The general theory of the cohesion of ionic solids, including methods of determining $b_{\varkappa\varkappa'}$ and ϱ, has been reviewed by TOSI[2]. We write the potential as

$$\varphi = \varphi^{(C)} + \varphi^{(R)}, \tag{21.2}$$

where superscripts C and R denote COULOMB and short-range interactions respectively, as before.

A calculation of $\omega_j(\boldsymbol{q})$ by KELLERMANN [*11*] for NaCl provided one of the first examples of a fairly realistic calculation of dispersion curves for lattice vibrations. The short-range interaction was taken to extend only to nearest neighbours. Taking the α-direction to be that joining neighbouring Na^+ and Cl^- ions identified by $\binom{l}{1}$ and $\binom{l}{2}$, we may write

$$-\varphi^{(R)}_{\alpha\alpha}\begin{pmatrix} l\,l' \\ 1\,2 \end{pmatrix} = \left(\frac{\partial^2 \varphi^{(R)}(X)}{\partial X_\alpha^2}\right)_{X=X_0} = \frac{e^2}{2v} A \tag{21.3}$$

and

$$-\varphi^{(R)}_{\beta\beta}\begin{pmatrix} l\,l' \\ 1\,2 \end{pmatrix} = \frac{1}{X_0}\left(\frac{\partial \varphi^{(R)}(X)}{\partial X_\alpha}\right)_{X=X_0} = \frac{e^2}{2v} B. \tag{21.4}$$

X_0 being the distance between nearest neighbours, and $v = 2X_0^3$ the volume of the primitive unit cell. The factor $\frac{e^2}{2v}$ is introduced simply to make the parameters A and B of a convenient order of magnitude. The energy of the static crystal per unit cell is

$$U_0 = -\alpha_M \frac{e^2}{X_0} + 6\varphi^{(R)}(X_0) \tag{21.5}$$

and the equilibrium condition

$$\left(\frac{\partial U}{\partial X}\right)_0 = 0 \tag{21.6}$$

leads to

$$B = -\tfrac{2}{3}\alpha_M = -1.165, \tag{21.7}$$

[1] D. H. MARTIN: Advances in Phys. **14**, 39 (1965).
[2] M. TOSI: Solid State Physics **16**, 1 (1964).

where α_M is the Madelung constant [1]. The compressibility β is given by

$$\frac{1}{\beta}=\frac{1}{18X_0}\left(\frac{\partial^2 U}{\partial X^2}\right)_0=\frac{e^2}{12X_0^4}(A+2B). \tag{21.8}$$

In KELLERMANN's calculation the parameter A was chosen to give agreement with the measured compressibility (at room temperature). The results of the calculation are therefore independent of the precise form of the short range interaction, provided this does not extend beyond nearest neighbours and is

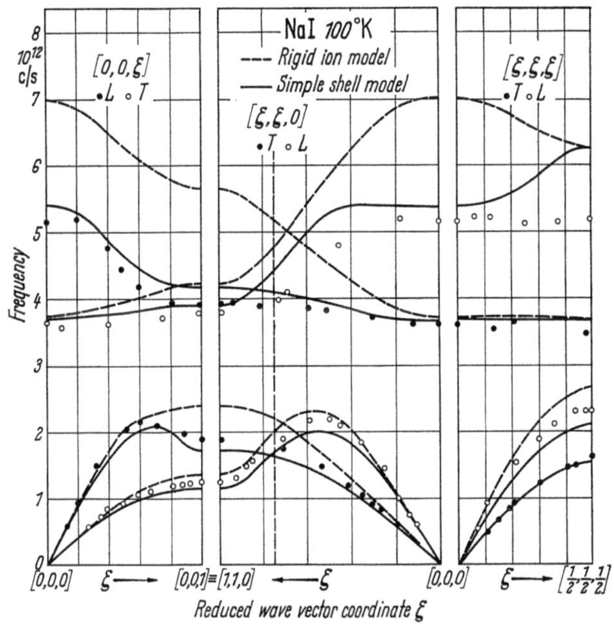

Fig. 12. Points show the measured values of $\omega_j(\boldsymbol{q})$ for NaI with $\boldsymbol{q}$ in each of three symmetry directions. The dotted line gives the results of a calculation based on the rigid ion model, corresponding results for the simplest version of the shell model are shown by the full lines.

central. We have already seen that in general $M_{\alpha\beta}\binom{\boldsymbol{q}}{\varkappa\varkappa'}$ can be divided into short-range and Coulomb contributions, and have discussed how the latter can be evaluated. Use of Eq. (8.4) with the appropriate symmetry elements of the sodium chloride structure gives all short-range force constants in terms of the parameters A and B, and it is found for example that

$$R_{\alpha\alpha}\binom{\boldsymbol{q}}{12}=-\frac{e^2}{v}\{A\cos q_\alpha X_0+B(\cos q_\beta X_0+\cos q_\gamma X_0)\}. \tag{21.9}$$

The equation corresponding to (18.4) is now

$$\omega^2\boldsymbol{m}\boldsymbol{U}=(\boldsymbol{R}+\boldsymbol{Z}\boldsymbol{C}\boldsymbol{Z})\boldsymbol{U}, \tag{21.10}$$

where $\boldsymbol{Z}$ is a 6×6 diagonal matrix whose diagonal elements are Z_1, Z_1, Z_1, Z_2, Z_2, Z_2. In KELLERMANN's work $Z_1=-Z_2=+e$.

KELLERMANN's calculation for NaCl has been repeated by IONA[1] for KCl and by KARO for the Li and Na halides[2] and for those K, Rb and Cs halides which

[1] M. IONA: Phys. Rev. 60, 822 (1941).
[2] A. M. KARO: J. Chem. Phys. 31, 1489 (1959).

have the NaCl structure[1]. Fairly detailed measurements of the dispersion curves are available for NaI and KBr. A comparison of the experimental results with those obtained using KELLERMANN's "rigid ion" model is made in Figs. 12 and 13[2]. The extent of the agreement is good when it is remembered that the theory contains only one parameter, A of Eq. 21.3, which was fitted to the elastic constant C_{11} in these instances, and not to the results obtained by neutron inelastic scattering. There are however considerable discrepancies, particularly for the longitudinal optic modes.

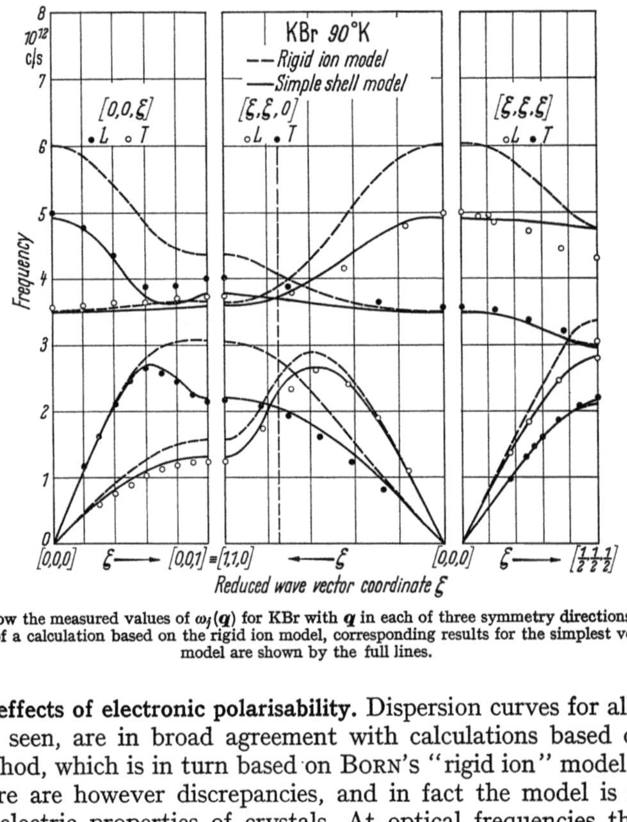

Fig. 13. Points show the measured values of $\omega_j(\boldsymbol{q})$ for KBr with $\boldsymbol{q}$ in each of three symmetry directions. The dotted line gives the results of a calculation based on the rigid ion model, corresponding results for the simplest version of the shell model are shown by the full lines.

22. The effects of electronic polarisability. Dispersion curves for alkali halides, as we have seen, are in broad agreement with calculations based on KELLERMANN's method, which is in turn based on BORN's "rigid ion" model of an alkali halide. There are however discrepancies, and in fact the model is inconsistent with the dielectric properties of crystals. At optical frequencies the electrons, but not the ions, respond to an electric field, the dielectric constant $\varepsilon(\infty)$ being equal to the square of the optical refractive index (extrapolated to long wavelength). For the rigid ion model $\varepsilon(\infty)$ is necessarily unity, the observed value for an alkali halide is typically between 2.0 and 3.0. If in the theory of the lattice dynamics of these crystals one attempts to remedy this defect by giving each ion a dipole moment

$$\boldsymbol{p}\binom{l}{\varkappa}=\alpha_\varkappa \boldsymbol{E}\binom{l}{\varkappa}, \qquad (22.1)$$

where $\alpha_\varkappa$ is the polarisability of the ion concerned and as before $\boldsymbol{E}\binom{l}{\varkappa}$ is the effective field, the agreement between theory and experiment becomes considerably

[1] A. M. KARO: J. Chem. Phys. **33**, 7 (1960).

[2] A. D. B. WOODS, B. N. BROCKHOUSE, R. A. COWLEY, and W. COCHRAN: Phys. Rev. **131**, 1025 (1963).

worse[1]. Furthermore the assumption that the electronic dipole moment of an ion depends only on the local field leads to certain inconsistencies. It was shown by SZIGETI[2] that when condition (22.1) is satisfied, the static dielectric constant $\varepsilon(0)$ should be given by

$$\varepsilon(0)=\varepsilon(\infty)+\frac{4\pi e^2(\varepsilon(\infty)+2)^2}{9v\mu\omega_t^2}, \tag{22.2}$$

where $\mu=\frac{m_1 m_2}{m_1+m_2}$ is the reduced mass of the ions and ω_t is the frequency of the transverse optic mode for $q\to 0$. This relation is satisfied only if the electronic charge e is reduced to an "effective charge" e' of about $0.75\,e$ for sodium chloride, although the cohesive energies and elastic constants of alkali halides are certainly more nearly consistent with the assumption that the ions are fully charged. SZIGETI[2] pointed out that when the two Bravais arrays of positive and negative ions are relatively displaced, as in an electric field, the configuration of the outer electrons of an ion will depend not only on the field, but also on the positions of at least nearest-neighbours, since the short-range or overlap potential must act primarily on the outer electrons. In other words there are two mechanisms of polarisation, and for the negative ion (which is usually the more polarisable) they act in opposite directions. This is illustrated schematically in Fig. 14 in which the outer electrons of the negative ions are represented by spherically-symmetric shells. This 'shell-model' has been used by DICK and OVERHAUSER[3] and by HANLON and LAWSON[4] to account quantitatively for the dielectric properties of the alkali halides, and the model has been used to calculate dispersion curves in NaI.[5] The shell model however really does no more than provide a convenient nomenclature and a link with the standard theory of lattice dynamics (given in Sect. 5). The equations of motion are the same as are obtained by taking the potential energy of a crystal in the harmonic approximation to be a function of the nuclear displacements and the electronic dipole moments, thus

$$\left.\begin{aligned}\varphi_2=\frac{1}{2}\sum_{l\varkappa\alpha}\Big\{\sum_{l'\varkappa'\beta}\Big[&\varphi_{\alpha\beta}^{(R)}\binom{l\,l'}{\varkappa\,\varkappa'}u_\alpha\binom{l}{\varkappa}u_\beta\binom{l'}{\varkappa'}+\\&+Y_{\varkappa'}^{-1}\varphi_{\alpha\beta}^{(T)}\binom{l\,l'}{\varkappa\,\varkappa'}u_\alpha\binom{l}{\varkappa}p_\beta\binom{l'}{\varkappa'}+\\&+Y_{\varkappa}^{-1}\varphi_{\beta\alpha}^{(T)}\binom{l'\,l}{\varkappa'\,\varkappa}p_\alpha\binom{l}{\varkappa}u_\beta\binom{l'}{\varkappa'}+\\&+Y_{\varkappa}^{-1}Y_{\varkappa'}^{-1}\varphi_{\alpha\beta}^{(S)}\binom{l\,l'}{\varkappa\,\varkappa'}p_\alpha\binom{l}{\varkappa}p_\beta\binom{l'}{\varkappa'}\Big]+\\&+\alpha_{\varkappa}^{-1}p_\alpha^2\binom{l}{\varkappa}-\Big[p_\alpha\binom{l}{\varkappa}+Z_\varkappa u_\alpha\binom{l}{\varkappa}\Big]E_\alpha\binom{l}{\varkappa}\Big\}.\end{aligned}\right\} \tag{22.3}$$

In this expression those terms not involving the dipole moments give the energy of the rigid ion model. The other terms are introduced to take account of the energy of the dipoles, which arises from their short-range interactions with one another and with displaced ions, and their energy of interaction with the field. An expression of this form was first suggested by TOLPYGO[6]. The quantities $Y_\varkappa$

---

[1] R. H. LYDANNE, and K. F. HERZFELD: Phys. Rev. **54**, 846 (1938).
[2] B. SZIGETI: Trans. Faraday Soc. **45**, 155 (1959); — Proc. Roy. Soc. (London) A **204**, 51 (1950).
[3] B. J. DICK, and A. W. OVERHAUSER: Phys. Rev. **112**, 90 (1958).
[4] J. E. HANLON, and A. W. LAWSON: Phys. Rev. **113**, 472 (1959).
[5] A. D. B. WOODS, W. COCHRAN, and B. N. BROCKHOUSE: Phys. Rev. **119**, 980 (1960).
[6] K. B. TOLPYGO: J. Exptl. Theoret Phys. (U.S.S.R.) **20**, 497 (1950).

are redundant, for example $(Y_\varkappa Y_{\varkappa'})^{-1}\varphi^{(S)}_{\alpha\beta}\binom{l\,l'}{\varkappa\,\varkappa'}$ could be written as a single parameter. The advantage of this notation is that it enables the concept of a force constant to be retained, $-\varphi^{(S)}_{\alpha\beta}\binom{l\,l'}{\varkappa\,\varkappa'}$ is a force constant between two shells carrying charges $Y_\varkappa$ and $Y_{\varkappa'}$ respectively. In terms of the shell model, the dipole moment is given by

$$\boldsymbol{p}\binom{l}{\varkappa}=Y_\varkappa\boldsymbol{w}\binom{l}{\varkappa}, \tag{22.4}$$

where $\boldsymbol{w}\binom{l}{\varkappa}$ is the relative displacement of a shell and a core of the same ion.

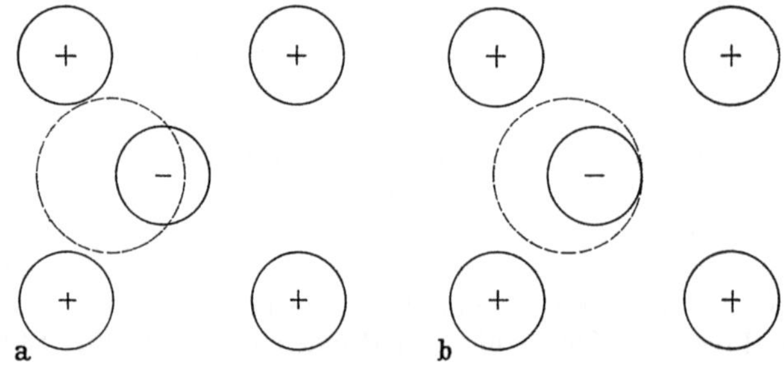

Fig. 14. a Shows schematically the effect on the outer electrons belonging to the negative ions of the electric field set up by a relative displacement of positive ions (to the right) and negative ions (to the left). The outer electrons are represented by a "shell", shown dotted. b Illustrates the fact that this polarisation of the negative ions will be reduced by short-range repulsive forces between the shell and the positive ions.

The equations of motion for the system are[1]

$$m_\varkappa\frac{\partial^2}{\partial t^2}u_\alpha\binom{l}{\varkappa}=-\frac{\partial\varphi_2}{\partial u_\alpha\binom{l}{\varkappa}} \tag{22.5}$$

and

$$\frac{\partial\varphi_2}{\partial p_\alpha\binom{l}{\varkappa}}=0. \tag{22.6}$$

The second of these equations expresses the adiabatic approximation (see Sect. 11).

Introducing

$$w_\alpha\binom{l}{\varkappa}=W_\alpha(\varkappa|\boldsymbol{q})\exp i\left(\boldsymbol{q}\cdot\boldsymbol{X}\binom{l}{\varkappa}-\omega(\boldsymbol{q})t\right) \tag{22.7}$$

which is analogous to Eq. (18.1),

$$T_{\alpha\beta}\binom{\boldsymbol{q}}{\varkappa\,\varkappa'}=\sum_{l'}\varphi^{(T)}_{\alpha\beta}\binom{l\,l'}{\varkappa\,\varkappa'}\exp i\boldsymbol{q}\cdot\left(\boldsymbol{X}\binom{l'}{\varkappa'}-\boldsymbol{X}\binom{l}{\varkappa}\right), \tag{22.8}$$

and

$$S_{\alpha\beta}\binom{\boldsymbol{q}}{\varkappa\,\varkappa'}=\sum_{l'}\varphi^{(S)}_{\alpha\beta}\binom{l\,l'}{\varkappa\,\varkappa'}\exp i\boldsymbol{q}\cdot\left(\boldsymbol{X}\binom{l'}{\varkappa'}-\boldsymbol{X}\binom{l}{\varkappa}\right) \tag{22.9}$$

which are analogous to Eq. (19.4), it is found that the equations of motion reduce to

$$\left.\begin{aligned}\omega^2(\boldsymbol{q})\,m_\varkappa U_\alpha(\varkappa|\boldsymbol{q})=\sum_{\varkappa'\beta}\Big[&\Big(R_{\alpha\beta}\binom{\boldsymbol{q}}{\varkappa\,\varkappa'}+Z_\varkappa Z_{\varkappa'}C_{\alpha\beta}\binom{\boldsymbol{q}}{\varkappa\,\varkappa'}\Big)U_\beta(\varkappa'|\boldsymbol{q})+\\&+\Big(T_{\alpha\beta}\binom{\boldsymbol{q}}{\varkappa\,\varkappa'}+Z_\varkappa Y_{\varkappa'}C_{\alpha\beta}\binom{\boldsymbol{q}}{\varkappa\,\varkappa'}\Big)W_\beta(\varkappa'|\boldsymbol{q})\Big],\end{aligned}\right\} \tag{22.10}$$

$$\left.\begin{aligned}0=\sum_{\varkappa'\beta}\Big[&\Big(T^*_{\beta\alpha}\binom{\boldsymbol{q}}{\varkappa\,\varkappa'}+Y_\varkappa Z_{\varkappa'}C_{\alpha\beta}\binom{\boldsymbol{q}}{\varkappa\,\varkappa'}\Big)U_\beta(\varkappa'|\boldsymbol{q})+\\&+\Big(\mathscr{S}_{\alpha\beta}\binom{\boldsymbol{q}}{\varkappa\,\varkappa'}+Y_\varkappa Y_{\varkappa'}C_{\alpha\beta}\binom{\boldsymbol{q}}{\varkappa\,\varkappa'}\Big)W_\beta(\varkappa'|\boldsymbol{q})\Big],\end{aligned}\right\} \tag{22.11}$$

[1] V. S. MASHKEVICH i K. B. TOLPYGO: J. Exptl. Theoret. Phys. (U.S.S.R.) **32**, 520 (1957).

where

$$\mathcal{S}_{\alpha\beta}\binom{\boldsymbol{q}}{\varkappa\varkappa'} = S_{\alpha\beta}\binom{\boldsymbol{q}}{\varkappa\varkappa'} + \delta_{\alpha\beta}\,\delta_{\varkappa\varkappa'}\,\alpha_\varkappa^{-1}\,Y_\varkappa^2. \tag{22.12}$$

The amplitude of the electric field is given by

$$E_\alpha(\varkappa|\boldsymbol{q}) = -\sum_{\varkappa'\beta} C_{\alpha\beta}\binom{\boldsymbol{q}}{\varkappa\varkappa'}\left(Z_\varkappa\,U_\beta(\varkappa'|\boldsymbol{q}) + Y_\varkappa\,W_\beta(\varkappa'|\boldsymbol{q})\right). \tag{22.13}$$

These equations can be written in matrix notation as

$$\omega^2 \boldsymbol{m}\boldsymbol{U} = (\boldsymbol{R}+\boldsymbol{Z}\boldsymbol{C}\boldsymbol{Z})\,\boldsymbol{U} + (\boldsymbol{T}+\boldsymbol{Z}\boldsymbol{C}\boldsymbol{Y})\,\boldsymbol{W}, \tag{22.14}$$

$$0 = (\tilde{\boldsymbol{T}}^* + \boldsymbol{Y}\boldsymbol{C}\boldsymbol{Z})\,\boldsymbol{U} + (\mathcal{S} + \boldsymbol{Y}\boldsymbol{C}\boldsymbol{Y})\,\boldsymbol{W} \tag{22.15}$$

and

$$\boldsymbol{E} = -\boldsymbol{C}(\boldsymbol{Z}\boldsymbol{U} + \boldsymbol{Y}\boldsymbol{W}). \tag{22.16}$$

On eliminating $\boldsymbol{W}$ between Eqs. (22.14) and (22.15) one finally obtains

$$\omega^2 \boldsymbol{m}\boldsymbol{U} = \{(\boldsymbol{R}+\boldsymbol{Z}\boldsymbol{C}\boldsymbol{Z}) - (\boldsymbol{T}+\boldsymbol{Z}\boldsymbol{C}\boldsymbol{Y})\,(\mathcal{S}+\boldsymbol{Y}\boldsymbol{C}\boldsymbol{Y})^{-1}(\tilde{\boldsymbol{T}}^*+\boldsymbol{Y}\boldsymbol{C}\boldsymbol{Z})\}\,\boldsymbol{U}. \tag{22.17}$$

Writing this result in the standard form

$$\omega^2 \boldsymbol{m}\boldsymbol{U} = \boldsymbol{M}\boldsymbol{U} \tag{22.18}$$

the values of $\omega_j^2(\boldsymbol{q})$ are then the eigenvalues of the matrix whose typical element is $(m_\varkappa\, m_{\varkappa'})^{-\frac{1}{2}} M_{\alpha\beta}\binom{\boldsymbol{q}}{\varkappa\varkappa'}$.

Dispersion curves have been calculated for NaI and KBr,[1] with the following simplifying assumptions

(i) Only the negative ion is polarisable

(ii) The short-range interaction is between the positive ion and the shell of the negative ion, and not the core.

Other assumptions are as in KELLERMANN'S work, i.e. short-range forces extend only to nearest neighbours, and the forces are central. The theory then involves three parameters, which were chosen to agree with the macroscopic quantities C_{11}, $\varepsilon(0)$ and $\varepsilon(\infty)$. The extent of the agreement is shown by the full lines in Figs. 12 and 13. The agreement is better than for the rigid ion model, the general shape of the experimental curves is quite well reproduced, but discrepancies remain, particularly for the longitudinal optic mode with $\boldsymbol{q}$ at the Brillouin zone boundary in the [111] direction.

Calculations were subsequently made[2] in which both ions were polarisable, the interactions between nearest neighbours was not necessarily central, neighbouring anions interacted through a short-range potential, and the ionic charges were not necessarily $\pm e$. The short-range potential was however assumed to act entirely through the shells. Nine parameters are then involved. Good agreement with the measured values of $\omega_j(\boldsymbol{q})$, the elastic constants C_{11}, C_{12} and C_{44} and the dielectric constants $\varepsilon(0)$ and $\varepsilon(\infty)$ could be obtained by suitable choice of the parameters, but the values for some of them (such as the polarisability of the positive ion in NaI) did not appear entirely reasonable. It is possible that some factor, such as quadrupole moments on the negative ions, is more important than some of those taken into account in improving on the simplest version of the model.

---

[1] See footnote 2, p. 92.

[2] R. A. COWLEY, W. COCHRAN, B. N. BROCKHOUSE, and A. D. B. WOODS: Phys. Rev. 131, 1030 (1963).

A very similar theory has been worked out by HARDY[1] based on the concept of "deformation dipoles" introduced by BORN and HUANG [1]. For alkali halides the equations of motion are not essentially different from those of the shell model, although somewhat different assumptions were made about the magnitudes of the parameters involved. The theory worked out by LUNDQUIST et al.[2] is more influenced by LÖWDIN's work [13]. Detailed calculations for sodium iodide have been made by TOLPYGO and collaborators[3]. Their procedure is in practice not essentially different from that outlined in this section. The relationships of the different calculations based on the "dipole approximation" has been discussed by COWLEY et al.[4].

23. Covalent and ionic-covalent crystals. There is no known approximate formula for the potential energy of a covalent crystal which would give the interatomic force constants, and there has not to date been any attempt to derive the phonon frequencies from a first-principles calculation, for example by repeating energy-band calculations with atom cores displaced in the pattern of a normal mode of vibration. The procedure adopted in the first calculation of dispersion curves for diamond and germanium by SMITH[5] and HSIEH[6] respectively was to assume that the forces are of short range, and to choose force constants in agreement with the values of such other physical constants as were available, for example the elastic constants and the Raman frequency. Calculations were made by HSIEH[6] of the phonon frequencies and the frequency distribution for germanium with two parameters chosen to fit the elastic constants. The results are shown in Fig. 15 for q in the [100] direction, and are compared with experimental results subsequently obtained using the technique of neutron inelastic scattering[7]. The calculated frequencies are generally much too high, and BROCKHOUSE and IYENGAR[7] showed that the agreement is not significantly improved by introducing force constants between second-nearest neighbours. Subsequently HERMAN[8] showed that to account for the experimental results it was necessary to postulate force constants extending at least to fifth neighbours. Fifteen independent force constants are then involved. MASHKEVICH and TOLPYGO[9] had already drawn attention to the importance of dipole-dipole interaction, the

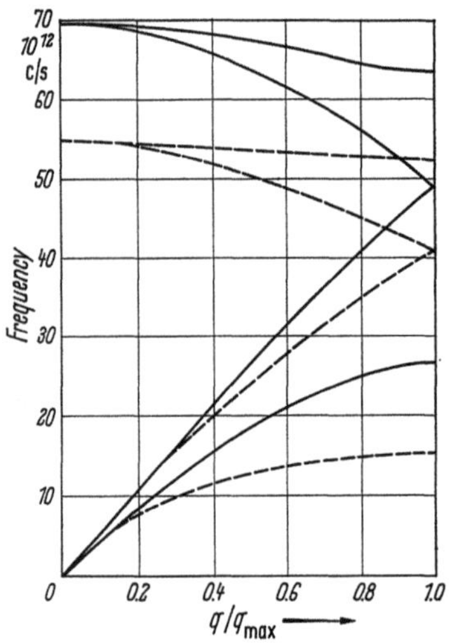

Fig. 15. Comparison of certain measured values of $\omega_j(\mathbf{q})$ for germanium (dotted lines) with those calculated on the assumption that there are forces only between nearest-neighbouratoms, neglecting polarisability (full lines).

[1] J. HARDY: Phil. Mag. **4**, 1278 (1959); **7**, 315 (1962).
[2] S. O. LUNDQUIST, V. LUNDSTRÖM, E. TENERZ, and J. WALLER: Arkiv Fysik **15**, 193 (1959).
[3] A. A. DEMIDENKO, Z. A. DEMIDENKO i K. B. TOLPYGO: Uspekhi Fiz. Zh. **3**, 728 (1958). — Z. A. DEMIDENKO i K. B. TOLPYGO: Fiz. Tverdoga Tela **3**, 3455 (1961).
[4] See footnote 2, p. 95.
[5] H. M. J. SMITH: Phil. Trans. Roy. Soc. London A **241**, 105 (1948).
[6] Y. C. HSIEH: J. Chem. Phys. **22**, 306 (1954).
[7] B. N. BROCKHOUSE, and P. K. IYENGAR: Phys. Rev. **111**, 747 (1958).
[8] F. HERMAN: J. Phys. Chem. Solids **8**, 405 (1959).
[9] See footnote 1, p. 94.

dipoles of course being absent from the static crystal but induced by the lattice vibrations. Their conclusion was that the energy should be expressed by Eq. (22.3), the ionic charges of course being set equal to zero (their analysis also did not include terms involving $p_\alpha\binom{l}{\varkappa}p_\beta\binom{l'}{\varkappa'}$). This suggests that the shell model can be applied to covalent crystals. It clearly lacks intuitive appeal when applied to a material with strongly directed covalent bonds, however it must be remembered that the assumptions on which Eq. (22.17) are derived are merely those of the dipole approximation, expressed by Eq. (22.3). Putting $\boldsymbol{Z}=0$, Eq. (22.17) becomes

$$\omega^2 m \boldsymbol{U} = \{\boldsymbol{R} - \boldsymbol{T}(\mathcal{S} + \boldsymbol{Y}\boldsymbol{C}\boldsymbol{Y})^{-1}\widetilde{\boldsymbol{T}}^*\}\boldsymbol{U}. \tag{23.1}$$

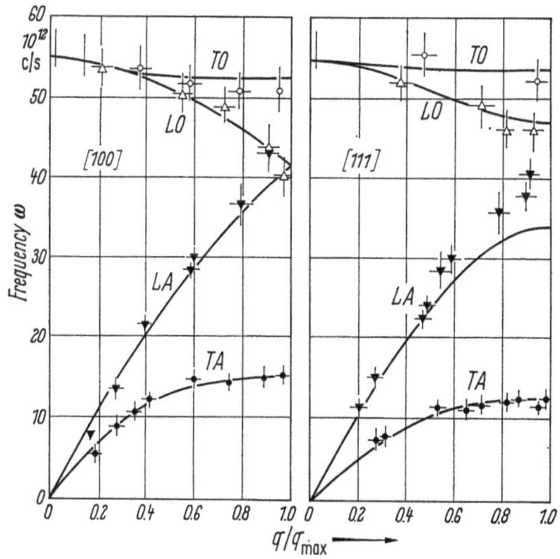

Fig. 16. Comparison of certain measured values of $\omega_j(\boldsymbol{q})$ for germanium with those calculated from the simplest version of the shell model.

If in germanium non-Coulomb forces between the dipoles, and the forces between the atoms, extend only to nearest neighbours atoms, each of the matrices $\boldsymbol{R}$, $\boldsymbol{T}$ and $\boldsymbol{S}$ involves two parameters (for $\boldsymbol{R}$ for example they are the independent force constants $\varphi_{\alpha\alpha}\binom{l\,l}{1\,2}$ and $\varphi_{\alpha\beta}\binom{l\,l}{1\,2}$. The remaining parameters are the charge Y on a shell and the isotropic force constant which links the shell and core of the same atom. Of these eight parameters, one is redundant since two parameters have been used to specify a dipole when only one is required (Eq. 22.4). The number can be reduced to five by assuming that the matrices $\boldsymbol{T}$ and $\boldsymbol{S}$ are proportional to $\boldsymbol{R}$, the remaining parameters were then chosen to be in agreement with the elastic constants, the high-frequency dielectric constant, and as far as possible with the measured values of $\omega_j(\boldsymbol{q})$ [1]. The general shape of the experimental curves was fairly well reproduced (Fig. 16).

Accurate measurements of $\omega_j(\boldsymbol{q})$ for $\boldsymbol{q}$ in the directions [100], [110] and [111] are also available for silicon. It has been shown by DOLLING [2] that to account satisfactorily for the experimental results it is necessary to remove some of the restrictions placed on the first-neighbour force constants in the calculation for

[1] W. COCHRAN: Proc. Roy. Soc. (London) A **253**, 260 (1959).
[2] G. DOLLING: In: Inelastic Scattering of Neutrons in Solids and Liquids. Internat. Atomic Energy Agency 1963.

germanium discussed above, and to include second-neighbour interactions. The extent of the agreement obtained with a model having eleven parameters is shown in Fig. 17. The agreement with the experimental value of $\varepsilon(\infty)$ was also satisfactory. Similar conclusions about the range of the force constants in the dipole approximation have been reached by DEMIDENKO, KUCHER and TOLPYGO[1].

Neutron measurements are available for the III—V compound GaAs. Independent evidence for the extent of the ionic character of this material is lacking, although the bonding is generally believed to be largely covalent. The neutron measurements and measurements of infra-red reflectivity[2] show that the longitudinal optic and transverse optic modes are not degenerate for $q\to 0$, as they

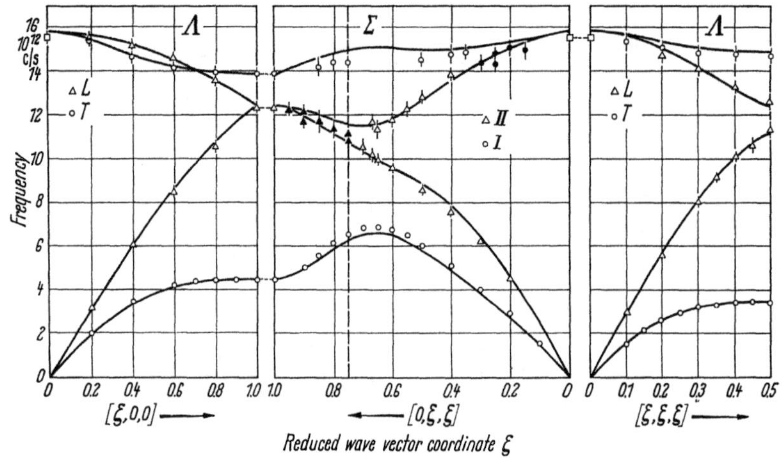

Fig. 17. Comparison of certain measured values of $\omega_j(q)$ for silicon with those calculated from a more complicated version of the shell model.

necessarily are in a diamond-type crystal. This proves the existence of Coulomb interaction, which could however arise solely from the difference in polarisabilities of the two atoms—in other words the SZIGETI effective charge[3] must be non-zero, although the ionic charge may be zero. Calculations made by WAUGH and DOLLING[4] are consistent with an ionic charge of not more than $0.04e$; shell models including second-nearest neighbour interaction did not fit the experimental data completely and it was suggested that any further elaboration of the theory should involve quadrupole-quadrupole interactions.

KAPLAN and SULLIVAN[5] have summarised the experimental evidence which is relevant for the lattice dynamics of other crystals with the zinc blende structure, including ZnS, AlSb and InSb. The data is rather meagre, consisting at most of the three elastic constants, the one-phonon infra-red absorption frequency $(\omega_t(q\to 0)]$, the dielectric constants $\varepsilon(0)$ and $\varepsilon(\infty)$ and the frequencies of certain features of the infra-red absorption spectrum involving two-phonon processes. In favourable circumstances the latter can be used to determine $\omega_j(\boldsymbol{q})$ for certain values of $\boldsymbol{q}$, generally at the Brillouin zone boundary in symmetry directions [2].

[1] Z. A. DEMIDENKO, T. J. KUCHER i K. B. TOLPYGO: Fiz. Tverdoga Tela 3, 2482 (1961); 4, 104 (1962).

[2] F. A. JOHNSON, and W. COCHRAN: Proc. Internat. Conference Physics Semiconductors (Phys. Soc. London) p. 498 (1962).

[3] See footnote 2, p. 93.

[4] J. WAUGH, and G. DOLLING: Phys. Rev. 132, 2410 (1963).

[5] H. KAPLAN, and J. J. SULLIVAN: Phys. Rev. 130, 120 (1963).

A range of parameters were found to be in agreement with the available data in each instance, and in particular it was not possible to come to a definite conclusion about the ionic charge of ZnS.

24. Discussion of dispersion curves for insulating crystals. In what follows we attempt to summarise the situation concerning the calculation and interpretation of phonon dispersion curves for non-metallic crystals. No very fundamental calculations have been made at the time of writing (1965). BORN's theory involving the concept of force constants has been extended to allow for the effects of the polarisability of the valence electrons, in the dipole approximation. Attempts to provide a quantum-mechanical basis for this approximation have been made[1], but the basis of Eq. (22.3) remains at least partly intuitive. The formalism of the shell model provides a link with the earlier theory. A simple version of this model involving only nearest-neighbour short-range interactions and therefore a small number of parameters which can be chosen to agree with macroscopic measurements, explains the dispersion curves of alkali halides fairly well, and those of more polarisable materials less well. The experimental results can be completely accounted for on the basis of the dipole approximation, without involving interactions (other than Coulomb interaction) which extend further than second-nearest neighbours. The number of parameters is then usually at least ten and these have to be chosen on an ad hoc basis. For crystals of more complicated structure, for example uranium dioxide[2] (which has the CaF_2-type of structure) it has been found that the set of parameters which accounts for the measured values of $\omega_j(\boldsymbol{q})$ is not unique. A *reliable* a priori calculation of the phonon frequencies and of the frequency distribution is not possible unless the crystal is one of a group of materials such as the alkali halides for which measurements are already available for one or more members of the group.

It might be thought to be a weakness of the theory that the Coulomb interaction is taken to involve point charges and dipoles of infinitesimal extent. The theoretical work of NOZIERES and PINES[3] and of ADLER[4] for instance shows that in a highly polarisable material like germanium the valence electrons are acted on by a field which more nearly approximates the macroscopic field than the Lorentz field. The Q-space formulation described in Sects. 18 and 19 enables one to calculate the $C_{\alpha\beta}\begin{pmatrix}\boldsymbol{q}\\ \varkappa\varkappa'\end{pmatrix}$ appropriate to extended charge distributions without difficulty, but when force constants derived from a postulated short-range interaction have to be treated as disposable parameters, there seems little point in attempting to allow for the finite extent of the "shells". The reason is that the Coulomb interactions between spherically symmetric charge distributions which overlap is the same as that between point charges, supplemented by a short-range interaction whose range is simply that of the overlap. Another way of expressing this is that terms such as $\boldsymbol{R}$ and $\boldsymbol{ZCZ}$ always occur together in the dynamical matrix. $\boldsymbol{ZCZ}$ can be changed without altering the calculated frequencies provided $\boldsymbol{R}$ is altered to keep the sum $\boldsymbol{R}+\boldsymbol{ZCZ}$ unchanged. It is significant that in materials such as Ge and PbTe[5] [which have $\varepsilon(\infty)$ equal to 16.0 and 31.8 respectively] and in which the valence electrons must be much less localised than in KBr for

---

[1] See for example R. A. COWLEY: Proc. Roy. Soc. (London) A **268**, 109 (1962).
[2] G. DOLLING, R. A. COWLEY, and A. D. B. WOODS: Can. J. Phys. **43**, 1397 (1965).
[3] P. NOZIÈRES, and D. PINES: Phys. Rev. **109**, 762 (1958).
[4] S. L. ADLER: Phys. Rev. **126**, 413 (1962).
[5] W. COCHRAN, R. A. COWLEY, G. DOLLING, and M. M. ELCOMBE: Proc. Roy. Soc. (London) A **293**, 433 (1966).

example [for which $\varepsilon(\infty)=2.3$], second neighbour short-range force constants are found to be relatively greater.

Frequencies measured experimentally have usually not been corrected for anharmonic effects before being interpreted on the basis of the harmonic approximation. The corrected frequencies in an alkali halide would be a few percent higher (Sect. 33). Where a parameter of the theory is fitted to the value of a macroscopic constant which was also measured at a finite temperature [for example A of Eq. (21.3) can be derived from the value of the elastic constant C_{11}] a partial correction for anharmonicity is made, provided that all $\omega_j(\boldsymbol{q})$ have the same temperature dependence. This is seldom likely to be the case. In certain materials, of which $SrTiO_3$ is an example[1], certain frequencies are anomalously temperature dependent and explicit allowance for anharmonic effects must be made (Sect. 36).

In degenerate semi-conductors the collective movement of the conduction electrons produces an elementary excitation, the plasmon, whose frequency may be comparable with that of the longitudinal optic mode. Coupling of the two modes through the macroscopic electric field associated with the longitudinal optic mode then alters the frequencies of both[2]. It is necessary to allow for this effect in interpreting the dispersion curves of PbTe, which has the NaCl-type structure. Coupling between phonons and spin waves in magnetic materials also occurs but there is at present no direct experimental evidence for this effect except for acoustic phonons of comparatively low frequency (10^9 c.p.s.)[3].

25. Force-constant models for metals. Coherent inelastic scattering of neutrons is almost the only method of determining $\omega_j(\boldsymbol{q})$ for a metallic crystal, although as we saw in Sects. 16 and 17 information about $g(\omega)$ can in some instances be obtained from electron tunnelling experiments and in other ways. At the time of writing, neutron experiments have been made for some sixteen metallic elements and a few alloys[4]. When the results are interpreted in terms of force constants, forces with a range of several unit cells must usually be postulated. We have already seen (Sect. 23) that this is also characteristic of such materials as germanium. These results were somewhat unexpected, it had generally been assumed in earlier calculations that the Coulomb interaction between the ions in a metal would be very effectively screened by the conduction electrons.

Results obtained for sodium, which has the body-centred-cubic structure at the temperature of the experiment, 90 °K, are shown in Fig. 18. The dispersion curves are comparatively smooth, indicating the absence of forces of very long range. The general shape of the curves can be well reproduced using general force constants extending to second-nearest neighbours, but to fit the results within experimental error force constants extending to fifth neighbours are required[5].

The dispersion curves for lead[6] are shown in Fig. 19. Fairly abrupt changes of slope are apparent in these, for example in the longitudinal mode for $\boldsymbol{q}$ just less than $\left(\frac{1}{2}, \frac{1}{2}, \frac{1}{2}\right)\frac{2\pi}{a}$. It was pointed out by KOHN[7] that in a free-electron metal an

[1] R. A. COWLEY: Phil. Mag. **11**, 673 (1965).

[2] R. A. COWLEY, and G. DOLLING: Phys. Rev. Letters **14**, 549 (1965).

[3] E. H. JACOBSEN, p. 505 of Ref. [8].

[4] G. DOLLING, and A. D. B. WOODS: In: Thermal Neutron Scattering, ed. P. A. EGELSTAFF. New York and London: Academic Press 1965.

[5] A. D. B. WOODS, B. N. BROCKHOUSE, R. H. MARCH, A. T. STEWART, and R. BOWERS: Phys. Rev. **128**, 1112 (1962).

[6] B. N. BROCKHOUSE, T. ARASE, G. CAGLIOTI, K. R. RAO, and A. D. B. WOODS: Phys. Rev. **128**, 1099 (1962).

[7] W. KOHN: Phys. Rev. Letters **2**, 393 (1959).

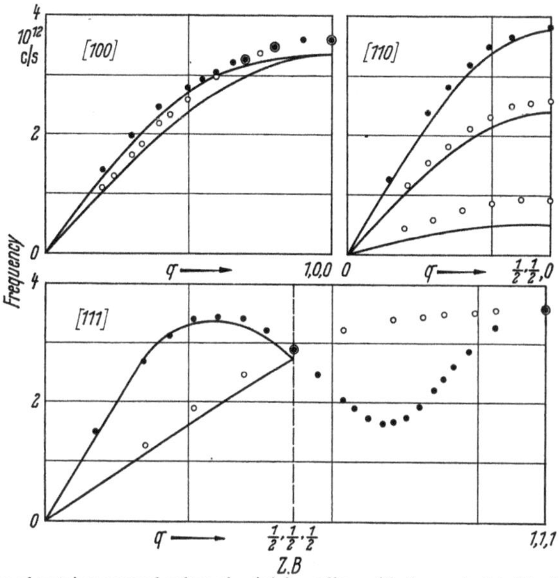

Fig. 18. Comparison of certain measured values of $\omega_j(\boldsymbol{q})$ for sodium with those calculated by TOYA (see Sect. 26).

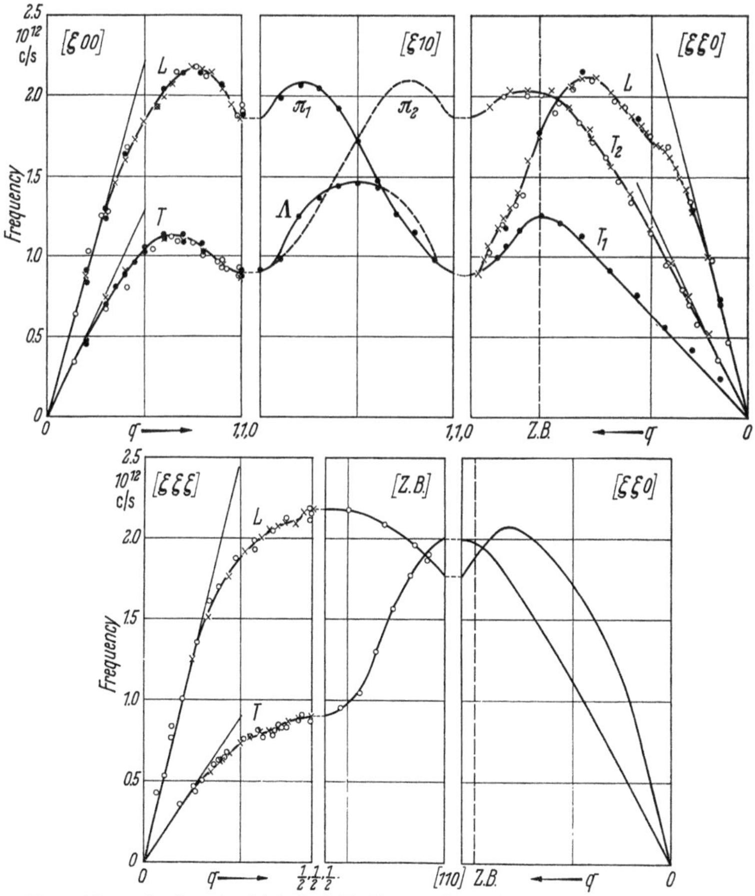

Fig. 19. Measured values of $\omega_j(\boldsymbol{q})$ for lead. The lines are drawn through the experimental points.

abrupt change in the screening power of the electrons, leading to a logarithmic infinity in the gradient of $\omega_j(\boldsymbol{q})$, occurs when

$$|\boldsymbol{\tau}+\boldsymbol{q}|=2k_F, \tag{25.1}$$

k_F being the radius of the Fermi sphere. The "anomalies" which occur in the dispersion curves of lead are in agreement with accepted dimensions of the Fermi surface. No choice of force constants was found to give an adequate fit to the measured frequencies. An analysis in terms of interplanar force constants

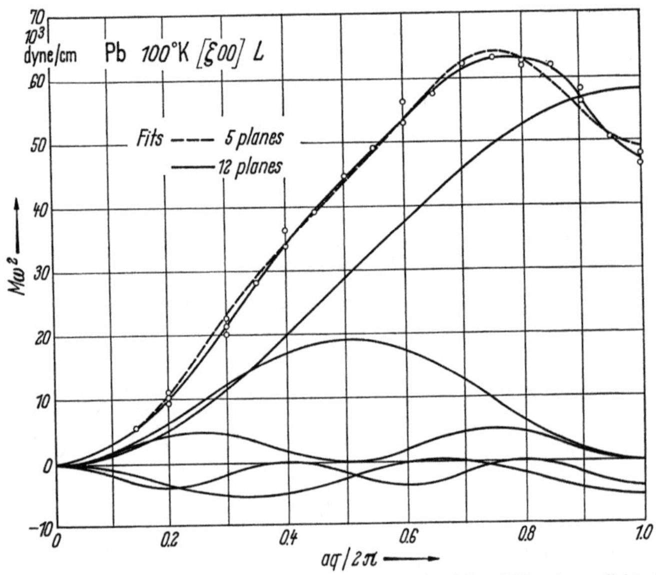

Fig. 20. Fourier analysis of $\omega_j(\boldsymbol{q})$ for the longitudinal mode in lead for which $\boldsymbol{q}$ is parallel to [100].

was made as follows. The determinant $|\boldsymbol{M}(\boldsymbol{q})|$ can be factorised when $\boldsymbol{q}$ is in certain symmetry directions. For example when $\boldsymbol{q}$ is parallel to [100] the equation

$$|\boldsymbol{M}-\boldsymbol{m}\,\omega^2|=0 \tag{25.2}$$

reduces to

$$(M_{\alpha\alpha}-m\,\omega^2)\,(M_{\beta\beta}-m\,\omega^2)\,(M_{\gamma\gamma}-m\,\omega^2)=0. \tag{25.3}$$

These separate equations are the same as for a linear chain, successive planes of the crystal playing the part of successive masses of the chain. We therefore have, for the longitudinal modes for instance

$$m\,\omega^2(q)=\sum_n \Phi_n(1-\cos n\pi q/q_m)\,. \tag{25.4}$$

Φ_n being the force constant (per atom) between n-th neighbour planes, and q_m the value of q at the boundary of the Brillouin zone. Fig. 20 shows the results of a Fourier analysis which gives Φ_n for the longitudinal mode with $\boldsymbol{q}$ parallel to [100]. Successive values of Φ_n are shown in Fig. 21, measurable forces extend over a distance of more than 20 Å.

Measurements made on nickel provide an example of an instance where more than one set of force constants accounts adequately for the data[1]. It may readily

[1] R. J. Birgenau, J. Cordes, G. Dolling, and A. D. B. Woods: Phys. Rev. **136**, A 1359 (1964).

be shown that if values of the eigenvalues $\omega_j^2(\boldsymbol{q})$ and eigenvectors $e_\alpha\binom{\boldsymbol{q}}{j}$ are available for all values of $\boldsymbol{q}$ (in the first Brillouin zone), the force constants $\varphi_{\alpha\beta}(l l')$ can be derived uniquely. In practice the polarisation properties of the modes are usually unknown (except for $\boldsymbol{q}$ in certain special directions, when they

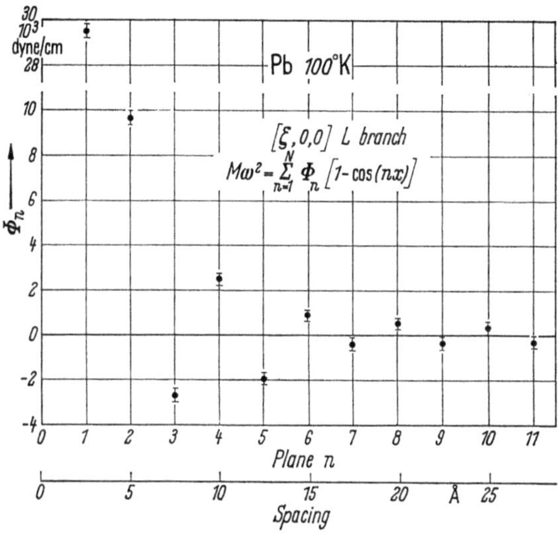

Fig. 21. The amplitudes Φ_n of successive terms in the Fourier analysis of $\omega^2(\boldsymbol{q})$ for the longitudinal mode in lead for which $\boldsymbol{q}$ is parallel to [100].

Table 2. *Two sets of force constants for nickel which give a satisfactory fit to measured frequencies* $\omega\binom{\boldsymbol{q}}{j}$.

Position of atom l'	Force constants	General forces (dyne cm⁻¹)	Axially-symmetric forces (dyne cm⁻¹)
$(\frac{1}{2}, \frac{1}{2}, 0)\,a$	$\alpha_1\ \gamma_1\ 0$ $\gamma_1\ \alpha_1\ 0$ $0\ 0\ \beta_1$	$\alpha_1 = 17178$ $\beta_1 = -26$ $\gamma_1 = 19316$	$\alpha_1 = 17720$ $\beta_1 = -1015$ $\gamma_1 = \alpha_1 - \beta_1$
$(1, 0, 0)\,a$	$\alpha_2\ 0\ 0$ $0\ \beta_2\ 0$ $0\ 0\ \beta_2$	$\alpha_2 = 880$ $\beta_2 = -519$	$\alpha_2 = 1148$ $\beta_2 = -998$
$(1, \frac{1}{2}, \frac{1}{2})\,a$	$\alpha_3\ \gamma_3\ \gamma_3$ $\gamma_3\ \beta_3\ \delta_3$ $\gamma_3\ \delta_3\ \beta_3$	$\alpha_3 = 626$ $\beta_3 = 320$ $\gamma_3 = 453$ $\delta_3 = -173$	$\alpha_3 = 940$ $\beta_3 = 182$ $\gamma_3 = \frac{2}{3}(\alpha_3 - \beta_3)$ $\delta_3 = \frac{1}{3}(\alpha_3 - \beta_3)$
$(1, 1, 0)\,a$	$\alpha_4\ \gamma_4\ 0$ $\gamma_4\ \alpha_4\ 0$ $0\ 0\ \beta_4$	$\alpha_4 = 275$ $\beta_4 = -160$ $\gamma_4 = 424$	$\alpha_4 = 459$ $\beta_4 = -153$ $\gamma_4 = \alpha_4 - \beta_4$
$(\frac{3}{2}, \frac{1}{2}, 0)\,a$	$\alpha_5\ \delta_5\ 0$ $\delta_5\ \beta_5\ 0$ $0\ 0\ \gamma_5$	Not determined	$\alpha_5 = -363$ $\beta_5 = 100$ $\gamma_5 = \frac{1}{8}(9\beta_5 - \alpha_5)$ $\delta_5 = \frac{3}{8}(\alpha_5 - \beta_5)$

are determined by symmetry) and measurements of $\omega_j(\boldsymbol{q})$ are generally, as was the case for nickel, made over a limited range of $\boldsymbol{q}$. It is therefore not surprising to find more than one set of force constants compatible with the data available. Table 2 shows two sets of force constants for nickel which give a satisfactory fit to values of $\omega_j(\boldsymbol{q})$ shown in Fig. 22. In Col. 3 general force constants extend to fourth neighbours, in Col. 4 the force constants extend to fifth neighbours but are axially symmetric[1]. By this is meant that the force constants are derivable from a potential $\varphi(X)$. There are then necessarily only two independent force

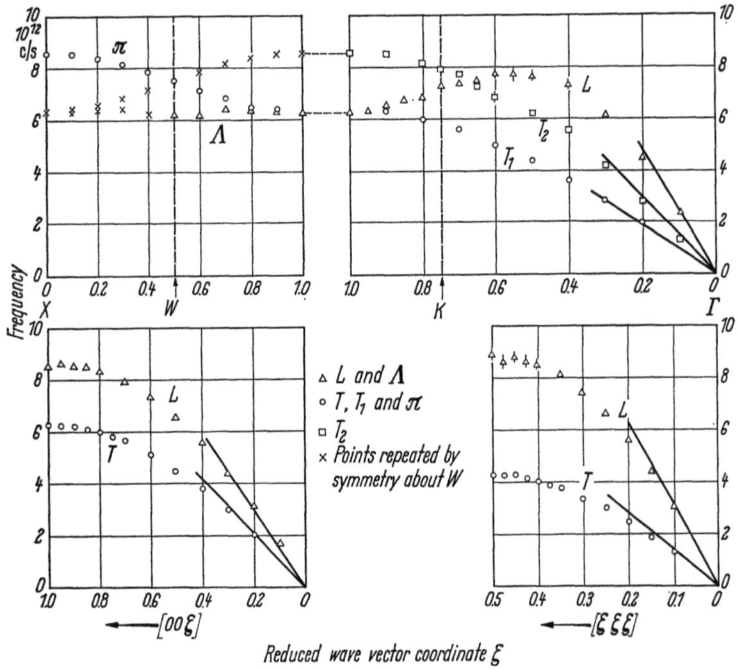

Fig. 22. Measured values of $\omega_j(\boldsymbol{q})$ for nickel. The lines drawn near the origins give the initial slopes of the dispersion curves derived from the elastic constants.

constants between any pair of atoms:

$$\left.\begin{aligned}-\varphi(l\,l')_{\text{radial}} &= \left(\frac{\partial^2\varphi(X)}{\partial X^2}\right)_{\boldsymbol{X}=\boldsymbol{X}(l')-\boldsymbol{X}(l)},\\ -\varphi(l\,l')_{\text{tangential}} &= \left(\frac{1}{X}\frac{\partial\varphi(X)}{\partial X}\right)_{\boldsymbol{X}=\boldsymbol{X}(l')-\boldsymbol{X}(l)}.\end{aligned}\right\} \tag{25.5}$$

The crystal is however not taken to be in equilibrium under the action of this potential only, the cohesive energy for example is not merely $\frac{1}{2}\sum_{l\neq l'}\varphi(|\boldsymbol{X}(l')-\boldsymbol{X}(l)|)$ but involves a term which is constant as long as the volume remains constant. The forces are therefore not central in the sense in which the term is usually used and the calculated elastic constants do not satisfy $C_{12}=C_{44}$.

The dispersion curves for niobium[2] show some unusual features (Fig. 23). For example for $\boldsymbol{q}$ parallel to [100], $\omega_j(\boldsymbol{q})$ for the longitudinal mode shows an extra maximum and minimum, and the slope of the curve for the transverse mode is anomalous near $q=(0.2, 0, 0)\,\frac{2\pi}{a}$, while the two branches cross at $(0.7, 0, 0)\,\frac{2\pi}{a}$. The full lines in Fig. 23 are the result of a calculation based on general force constants extending to eighth neighbours, and involving twenty three parameters.

[1] G. W. LEHMAN, T. WOLFRAM, and R. E. DEWAMES: Phys. Rev. **128**, 1593 (1962).
[2] Y. NAKAGAWA, and A. D. B. WOODS: Phys. Rev. Letters **11**, 271 (1964).

Small but significant discrepancies between measured and calculated values can be seen. For this metal, as for lead, an analysis in terms of force constants is clearly not very meaningful.

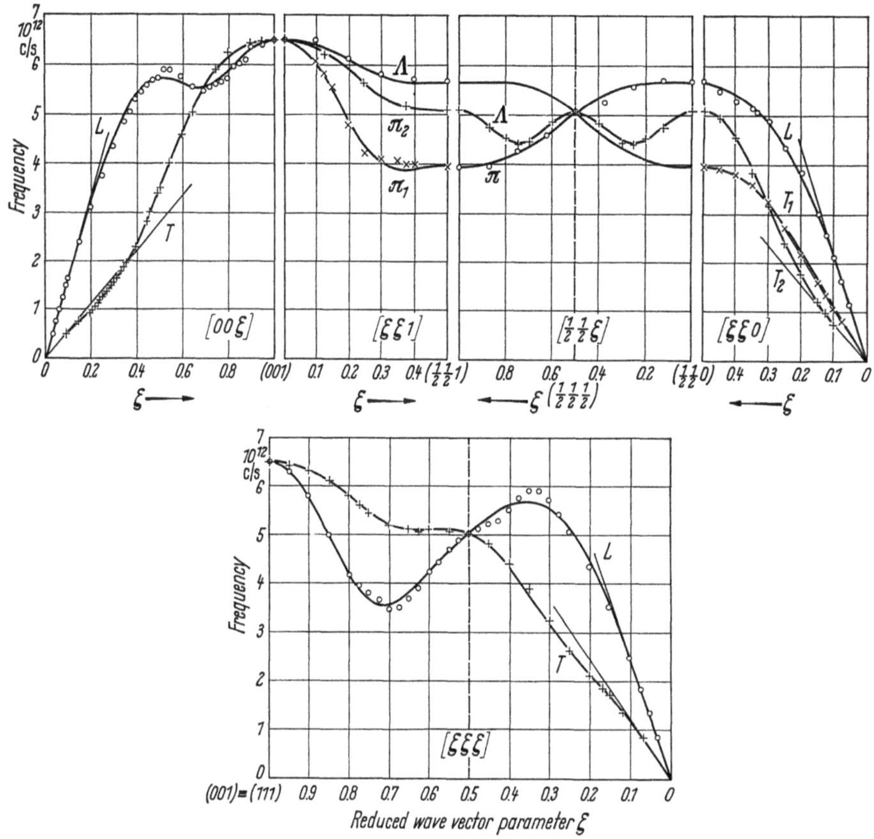

Fig. 23. Measured values of $\omega_j(\boldsymbol{q})$ for niobium. The full lines are the result of a calculation based on general force constants extending to eighth neighbours.

26. Ion-electron-ion interaction in metals. An alkali metal such as sodium provides perhaps the best situation for an interpretation of $\omega_j(\boldsymbol{q})$ from first principles. In sodium the ion cores are sufficiently far apart that the only important direct interaction between them is the Coulomb interaction[1]. A comparatively simple treatment of the screening of this interaction by the conduction electrons should be possible since sodium is known to be a good approximation to a "free electron metal" with a Fermi surface which is closely spherical. The following elementary account is perhaps only relevant to the alkali metals, but serves to introduce some of the concepts which are used in interpreting $\omega_j(\boldsymbol{q})$ without invoking force constants.

We consider a free electron gas, with wave functions

$$\Psi(\boldsymbol{k}, \boldsymbol{x}) = \exp(i\,\boldsymbol{k}\cdot\boldsymbol{x}) \tag{26.1}$$

normalised to unit volume. A small positive test charge having a charge distribution $Z(x)$ such that

$$\int Z(x)\, d^3x = Z \tag{26.2}$$

[1] S. H. Vosko: Phys. Letters **13**, 97 (1964).

is introduced. If $Z(Q)$ is the Fourier transform of $Z(x)$, the transform of the unscreened potential arising from $Z(x)$ is $4\pi Z(Q)/Q^2$. It was shown by BARDEEN[1] that in the Hartree or self-consistent-field approximation $Z(x)$ will attract a screening charge such that the Fourier transform of the screened potential is

$$V(Q)=4\pi Z(Q)/Q^2 \varepsilon(Q) \tag{26.3}$$

where $\varepsilon(Q)$ is usually referred to as the dielectric function. It is a special case of the function $\varepsilon(Q, \omega)$ discussed, for example, by PINES[2]. For the screened potential to be given by this expression, the distribution $Z(x)$ must be surrounded by a distribution of negative charge given by

$$Z_e(x)=-\int Z(Q)\left(1-\frac{1}{\varepsilon(Q)}\right)\exp(-i\boldsymbol{Q}\cdot\boldsymbol{x})\,\frac{d^3Q}{(2\pi)^3}. \tag{26.4}$$

The function $\varepsilon(Q)$ can be evaluated from the following considerations. The perturbed wave functions are

$$\psi(\boldsymbol{k},\boldsymbol{x})=\exp i\,\boldsymbol{k}\cdot\boldsymbol{x}+\sum_{\boldsymbol{Q}}\frac{e\,V(\boldsymbol{Q})}{E(\boldsymbol{k}+\boldsymbol{Q})-E(\boldsymbol{k})}\exp i(\boldsymbol{k}+\boldsymbol{Q})\cdot\boldsymbol{x}. \tag{26.5}$$

However we must have

$$Z_e(x)=-e\sum_{\boldsymbol{k}}\left(|\psi(\boldsymbol{k},\boldsymbol{x})|^2-1\right) \tag{26.6}$$

and this, together with the result

$$\sum_{k<k_F}\left(E(\boldsymbol{k}+\boldsymbol{Q})-E(\boldsymbol{k})\right)^{-1}=\frac{3n}{4E_F}\left(\frac{1}{2}+\frac{4k_F^2-Q^2}{8k_F Q}\log\left|\frac{2k_F+Q}{2k_F-Q}\right|\right), \tag{26.7}$$

where n is the electron density and E_F the Fermi energy, leads to Eq. (26.3) with

$$\varepsilon(Q)=1+\frac{6\pi n e^2}{Q^2 E_F}\left(\frac{1}{2}+\frac{4k_F^2-Q^2}{8k_F Q}\log\left|\frac{2k_F+Q}{2k_F-Q}\right|\right). \tag{26.8}$$

The function $\varepsilon(Q)$ varies as Q^{-2} for small Q, so that potentials which vary slowly in space are completely screened. A rapidly-varying potential however is unscreened since $\varepsilon(Q)\to 1$ for $Q\to\infty$. At $Q=2k_F$ the dielectric function has a logarithmic infinity in its slope. It can be shown to follow from this that at large values of x the screening charge $Z_e(x)$, and therefore the screened potential around $Z(x)$, varies approximately as $x^{-3}\cos(2k_F x)$, and is therefore of comparatively long range.

Before these results can be regarded as relevant to a metal we must attempt to justify the replacement of the core by an extended charge distribution and the representation of the conduction electrons by simple plane waves. This finds some justification in the concept of a pseudo-potential for electron-ion interaction, extensively discussed by HARRISON[3] and others. The pseudo-potential is strictly speaking not a localised potential but may be approximated by one. Outside the ion core it is equal to the Coulomb potential. A confined charge distribution $Z(x)$ has this property, and we may identify our $4\pi Z(Q)/Q^2$ with the Fourier transform of the (localised) pseudo-potential. For internal consistency we must set the electron density n which appears in $\varepsilon(Q)$ equal to $\frac{1}{v}\left(\frac{Z}{e}\right)$, and assume that $Z(\boldsymbol{\tau})\approx 0$ since any other result is inconsistent with the original assumption of free electrons.

---

[1] J. BARDEEN: Phys. Rev. **52**, 688 (1937).
[2] D. PINES: Elementary excitations in solids. New York: Benjamin 1963.
[3] W. HARRISON: Pseudopotentials in the Theory of Metals. New York: Benjamin 1965.

The Fourier transform of the effective interatomic potential is then

$$\varphi(Q)=\frac{4\pi Z^2}{Q^2}-\frac{4\pi Z^2(Q)}{Q^2}\left(1-\frac{1}{\varepsilon(Q)}\right) \tag{26.9}$$

the first term arising from the Coulomb interaction of point charges Z and the second from the interaction of the effective charge distribution $Z(x)$ of one ion with the screening charge distribution $Z_e(x)$ of another. Values of $\omega_j(\boldsymbol{q})$ can be evaluated from Eq. (26.9) using the Ewald method to deal with the Coulomb contribution, and the $\boldsymbol{Q}$-space method (described in Sect. 18) to find the contribution of the second term. From the form of Eq. (18.14) we see that when $|\boldsymbol{\tau}+\boldsymbol{q}|=2k_F$ the "anomalous" behaviour of $\varepsilon(Q)$ and therefore of $\varphi(Q)$ at $Q=2k_F$ will produce a corresponding anomaly in $\omega_j(\boldsymbol{q})$. If however the form factor $Z(Q)$ is small at $Q=2k_F$ the discontinuity in slope will be inappreciable, as appears to be the case in practice for sodium and aluminium, but not lead.

The problem of course is the determination of $Z(Q)$. In principle the pseudo-potential is determined by the wave functions and energy levels of the core electrons. It has been shown by several authors[1] that the pseudo-potential operator can be approximated in its effect by a localised potential, for which an expression involving two parameters was suggested[2]. These parameters can be chosen to give a good fit to the measured $\omega_j(\boldsymbol{q})$ for aluminium[3].

In calculations made by SHAM[4] for sodium the pseudo-potential was calculated from first principles, taking the true ion potential to be the Prokofjew potential which reproduces correctly the energy levels of an isolated ion. Good agreement with the measured $\omega_j(\boldsymbol{q})$ was obtained.

The calculations made by TOYA[5] for sodium in fact preceded the experimental measurements. The concept of a pseudo-potential was not used, although as usual the interaction was separated into the direct ion-ion interaction and the ion-electron-ion interaction, including the effect of electron-electron interaction. The change of energy of the conduction electrons on introduction of a lattice wave of wave-vector $\boldsymbol{q}$ was evaluated from perturbation theory. This involved the matrix element for electron-phonon interaction, which was taken to be as evaluated by BARDEEN[6] using the cellular method of WIGNER and SEITZ. Since the conduction electrons were represented everywhere by plane waves, for sodium this amounts to setting

$$Z(Q)=eG(Qr_s) \tag{26.10}$$

where

$$G(y)=3y^{-3}(\sin y-y\cos y) \tag{26.11}$$

and r_s is the radius of the Wigner-Seitz sphere. This calculation, which made no use of adjustable parameters, gave results in very fair agreement with experiment (Fig. 18).

E. The frequency distribution of the normal modes.

27. Critical points. From the discussion in Sect. 10, it follows that for a crystal containing r atoms per unit cell, and of volume Nv, the frequency distribution

[1] See for instance L. SHAM: Proc. Phys. Soc. (London) **78**, 895 (1961). — B. J. AUSTIN, V. HEINE, and L. SHAM: Phys. Rev. **127**, 276 (1962).
[2] W. HARRISON: Phys. Rev. **131**, 2433 (1963).
[3] W. HARRISON: Phys. Rev. **136**, A 1107 (1964).
[4] L. SHAM: Proc. Roy. Soc. (London) A **283**, 33 (1965).
[5] T. TOYA: J. Research Inst. Catalysis, Sapporo **6**, 183 (1958).
[6] See footnote 1, p. 106.

is given by

$$g(\omega)\,d\omega = \frac{N v}{(2\pi)^3}\int d^3q \tag{27.1}$$

where the integral is extended over the regions of the Brillouin zone for which any $\omega_j(\boldsymbol{q})$ lies between ω and $\omega+d\omega$. (These regions overlap, and in the course of constructing $g(\omega)$ any element of volume will be included $3r$ times.) An alternative form of this expression was given in Sect. 10,

$$g(\omega) = \frac{N v}{(2\pi)^3}\sum_{j=1}^{3r}\int\limits_{\omega=\omega_j(q)}\frac{dS}{|\mathrm{grad}_{\boldsymbol{q}}\,\omega_j(\boldsymbol{q})|}\,. \tag{27.2}$$

The frequency distribution of the linear chain [Eq. (10.5)] has an infinite discontinuity at the maximum frequency. MONTROLL[1] has proved that the frequency distribution of a two-dimensional lattice has logarithmic infinities. In three dimensions however there are no divergencies in the frequency distribution, although they may be introduced by some of the approximate methods which have been used to evaluate $g(\omega)$. Infinite discontinuities of slope do however occur, as has been shown by VAN HOVE[2] and by PHILLIPS[3]. The following discussion follows that of WANNIER[4]. We move the origin to the point in q-space for which $\mathrm{grad}_{\boldsymbol{q}}\,\omega_j(\boldsymbol{q})$ is zero [Eq. (27.2) shows that this will be a critical point], and choose a Cartesian coordinate system such that we may write, for example near a maximum of $\omega_j(\boldsymbol{q})$,

$$\omega = \omega_0 - (p_\alpha^2 + p_\beta^2 + p_\gamma^2)\,. \tag{27.3}$$

The element of volume $dp_\alpha\,dp_\beta\,dp_\gamma$ is then proportional to d^3q, and a transformation to another coordinate system which makes ω one of the variables of integration may be possible. The "local contribution" to $g(\omega)$ is then got by dropping the integration over ω.

The point $\omega=0$ of $g(\omega)$ is not a critical point but the same method applies. At the bottom of an acoustic branch $\omega_0=0$, and following the above procedure we have

$$\omega = (p_\alpha^2 + p_\beta^2 + p_\gamma^2)^{\frac{1}{2}}\,. \tag{27.4}$$

Surfaces of constant ω for the branch are ellipsoids in $\boldsymbol{q}$-space. The appropriate transformation is to polar coordinates

$$\left.\begin{aligned} p_\alpha &= \omega\sin\vartheta\cos\varphi\,,\\ p_\beta &= \omega\sin\vartheta\sin\varphi\,,\\ p_\gamma &= \omega\sin\vartheta \end{aligned}\right\} \tag{27.5}$$

so that

$$d^3q = \omega^2\cos\vartheta\,d\vartheta\,d\varphi\,d\omega \tag{27.6}$$

and therefore

$$g(\omega)\propto\omega^2\,. \tag{27.7}$$

As is well known, this feature is reproduced by the Debye distribution.

At any other minimum we have

$$\omega-\omega_0 = p_\alpha^2 + p_\beta^2 + p_\gamma^2 \tag{27.8}$$

---

[1] E. W. MONTROLL: Am. Math. Monthly **61**, 46 (1954).

[2] L. VAN HOVE: Phys. Rev. **89**, 1189 (1953).

[3] J. C. PHILLIPS: Phys. Rev. **104**, 1263 (1956).

[4] G. H. WANNIER: Elements of Solid State Theory. Cambridge: Cambridge Univ. Press 1959.

and a transformation to

$$\left.\begin{aligned} p_\alpha &= (\omega-\omega_0)^{\frac{1}{2}} \sin\vartheta \cos\varphi\,,\\ p_\beta &= (\omega-\omega_0)^{\frac{1}{2}} \sin\vartheta \sin\varphi\,,\\ p_\gamma &= (\omega-\omega_0)^{\frac{1}{2}} \sin\vartheta \end{aligned}\right\} \tag{27.9}$$

leads to

$$g(\omega) \propto (\omega-\omega_0)^{\frac{1}{2}}. \tag{27.10}$$

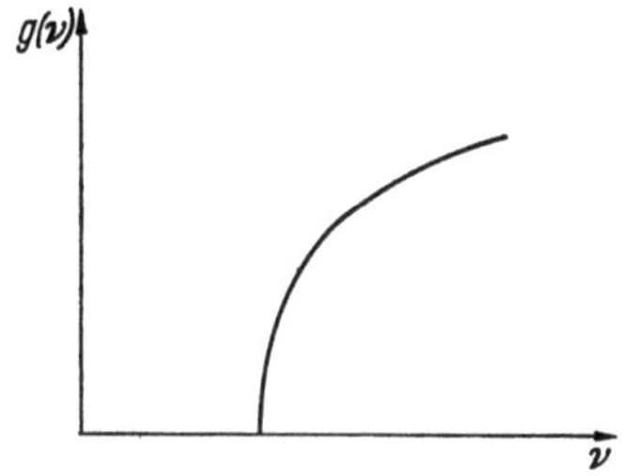

Fig. 24. A critical point in $g(\omega)$ resulting from a minimum in $\omega_j(\boldsymbol{q})$.

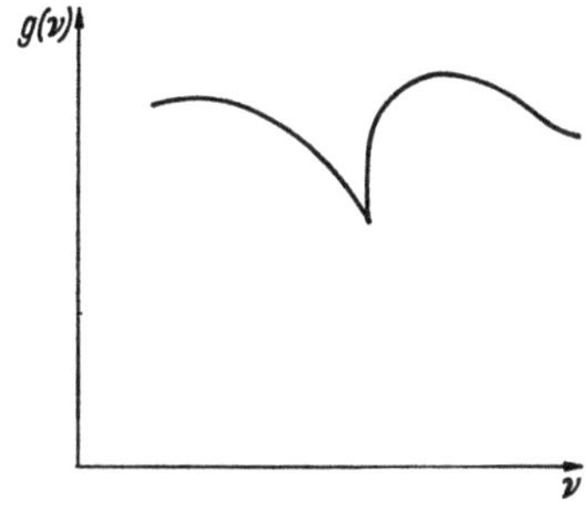

Fig. 25. The same critical point as in Fig. 24, but occurring on a background produced by other branches of the dispersion relation.

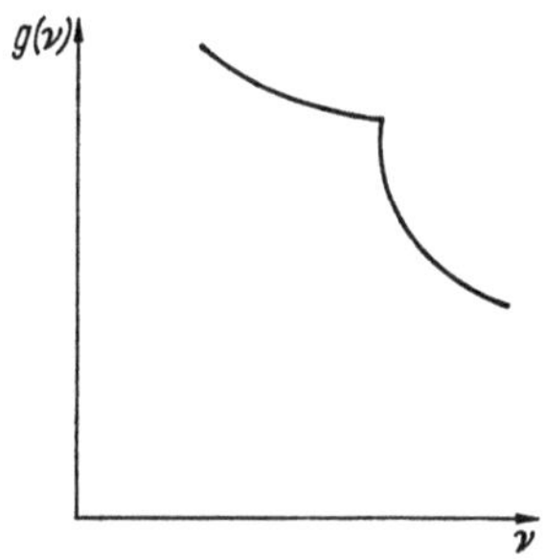

Fig. 26. A critical point in $g(\omega)$ resulting from a saddle point in $\omega_j(\boldsymbol{q})$, at which the frequency decreases in two directions and increases in a third.

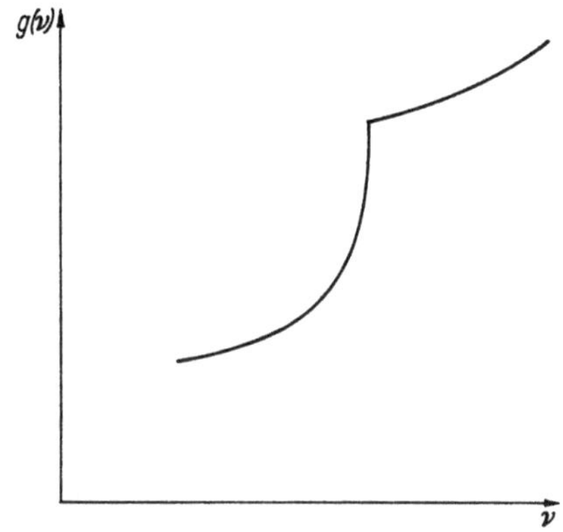

Fig. 27. A critical point in $g(\omega)$ resulting from a saddle point in $\omega_j(\boldsymbol{q})$, at which the frequency increases in two directions and decreases in a third.

This is shown in Fig. 24. The discontinuity can occur on a normal background, in which case the appearance is as in Fig. 25.

At a maximum

$$g(\omega) \propto (\omega_0-\omega)^{\frac{1}{2}}. \tag{27.11}$$

The coordinate transformations appropriate to the saddle points

$$\omega = \omega_0 - p_\alpha^2 - p_\beta^2 + p_\gamma^2 \tag{27.12}$$

and

$$\omega = \omega_0 + p_\alpha^2 + p_\beta^2 - p_\gamma^2 \tag{27.13}$$

are given by WANNIER[1]. For a saddle point given by (27.12) the slope of $g(\omega)$ changes from a finite value below ω_0 to an infinite negative value at the critical point (Fig. 26). For a saddle point given by (27.13) the slope is infinite at the critical point, and finite above it (Fig. 27).

[1] See footnote 4, p. 108.

VAN HOVE[1] proved from topological considerations that the frequency distribution must always exhibit these features. Simply because of the periodic character of $\omega_j(\boldsymbol{q})$ in the reciprocal lattice, critical points occur at frequencies corresponding to particular values of $\boldsymbol{q}$ on the surface of the Brillouin zone. Other critical points depend on the detailed shape of the dispersion curves for a particular material, and are not associated with the symmetry properties of $\omega_j(\boldsymbol{q})$.

28. The sampling method. None of the semianalytical methods for the calculation of $g(\omega)$ which has been proposed [*4*] is entirely adequate, as has been emphasised by BLACKMAN [*9*]. The method which has proved most satisfactory in practice is the sampling method, which has become much more powerful with the

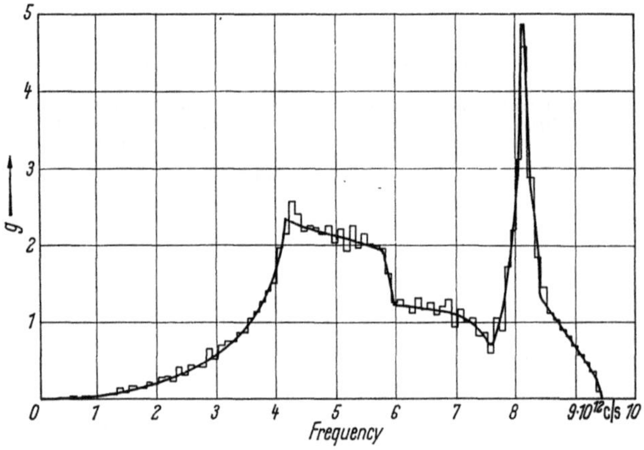

Fig. 28. The frequency distribution $g(\omega)$ for aluminium calculated by the sampling method and based on measurements of X-ray scattering.

development of fast computers. In this method, a force-constant or other model is found which accounts for the values of $\omega_j(\boldsymbol{q})$ determined by coherent inelastic scattering of neutrons or by some other reliable method. Such measurements are often confined to symmetry directions ($\boldsymbol{q}$ parallel to [100], [110] or [111]) and it is assumed that the model correctly reproduces $\omega_j(\boldsymbol{q})$ for general values of $\boldsymbol{q}$. This has been checked in some investigations[2]. Note that one need not then have confidence in the physical significance of the model, or ensure that the force constants are unique, since in this application the model is used only to provide a means of extrapolation and interpolation. (To the extent that reliable calculation of $\omega_j(\boldsymbol{q})$ is possible in the absence of experimental measurements on the material concerned, it is of course possible to make an à priori calculation of $g(\omega)$ by the Sampling method). Frequencies are obtained by finding the eigenvalues of the matrix $D(\boldsymbol{q})$ [Eq. (5.11)] for a large number of values of $\boldsymbol{q}$, uniformly distributed in the unique volume of the Brillouin zone. (By this is meant the volume which when operated on by the point group symmetry elements of the crystal, produces the complete Brillouin zone. For structures such as Na, Cu, Ge, NaCl, CsCl this unique volume is $\frac{1}{48}$ of the Brillouin zone.) A histogram is then made of the frequencies.

The sampling method was used by WALKER[3] to derive $g(\omega)$ for aluminium. The force constants were chosen to fit the dispersion curves obtained by measure-

[1] See footnote 2, p. 108.
[2] See footnote 2, p. 92.
[3] C. B. WALKER: Phys. Rev. **103**, 547, 558 (1956).

ment of the intensity of inelastic X-ray scattering (Sect. 14) and the matrix $D(\boldsymbol{q})$ derived for approximately 1.5×10^5 points in the Brillouin zone. The frequency distribution (Fig. 28) is noticeably a histogram, although adequate in resolution for the calculation of thermodynamic properties.

DIXON et al.[1] derived a frequency distribution for sodium using a force constant model with force constants extending to fifth neighbours, and giving good agreement with experimental measurements (see Sect. 25). Approximately

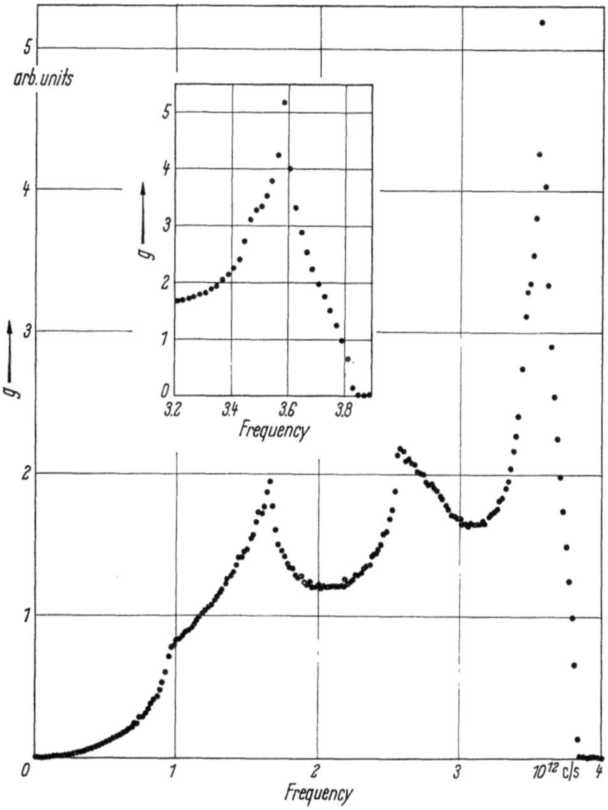

Fig. 29. The frequency distribution $g(\nu)$ for sodium derived by the sampling method. The inset shows part of the diagram on a different scale, to make one of the critical points more obvious.

8.5×10^6 values of $\boldsymbol{q}$ were involved, the histogram shown in Fig. 29 is almost a smooth curve on which critical points are quite evident. That at 0.93×10^{12} c.p.s. for example is associated with a transverse mode for which $\boldsymbol{q}$ is at the point $\left(\frac{1}{2}, \frac{1}{2}, 0\right)\frac{2\pi}{a}$ and for which $\omega_j(\boldsymbol{q})$ is a maximum in the [110] direction and a minimum in the two perpendicular directions (see also Fig. 18).

GILAT and DOLLING[2] have described a method by which $D(\boldsymbol{q})$ is diagonalised on a comparatively coarse mesh of points in $\boldsymbol{q}$-space, and frequencies on a finer mesh are obtained by linear extrapolation, using perturbation theory. Frequency distributions involving as many as 1.6×10^7 points in the Brillouin zone have been constructed in this way using only a few minutes of computer time. The

[1] A. E. DIXON, A. D. B. WOODS, and B. N. BROCKHOUSE: Proc. Phys. Soc. (London) **81**, 973 (1963).
[2] G. GILAT, and G. DOLLING: Phys. Rev. Letters **8**, 304 (1964).

frequency distribution of nickel shown in Fig. 30 was derived[1] by this variant of the sampling method, using the force constant model specified in Col. 3 of Table 2. Fig. 30 also shows the results of three independent experiments. Experi-

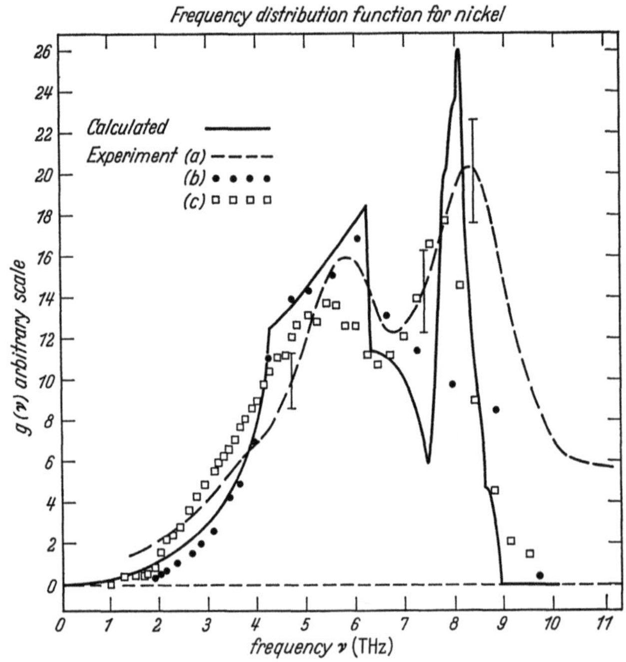

Fig. 30. The frequency distribution $g(\nu)$ for nickel derived by the sampling method, compared with the results of three independent experimental methods described in the text (Sect. 28).

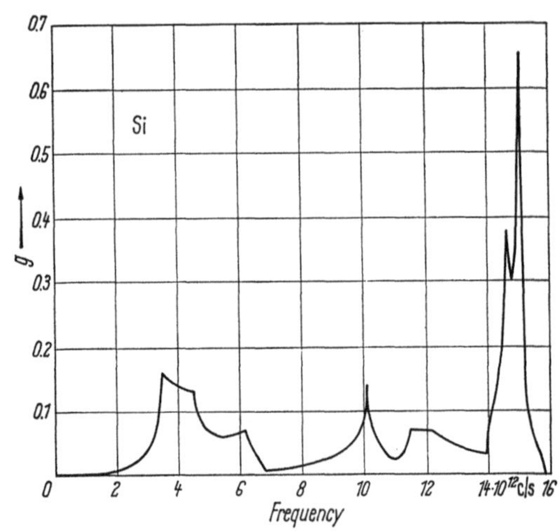

Fig. 31. The frequency distribution $g(\nu)$ for silicon derived by the sampling method.

ment (a)[1] was one in which an energy analysis was made of neutrons scattered incoherently from a suitable mixture of isotopes of nickel (Sect. 17). In experiments (b) and (c)[1] the observations were of the total inelastic scattering from a

[1] See footnote 1, p. 102.

single crystal of nickel of ordinary isotopic composition as a function of crystal orientation. None of the experimental methods gave sufficient resolution to show the interesting features of the distribution.

Fig. 31 shows the frequency distribution for silicon, based on a shell model with parameters chosen to fit experimental measurements of $\omega_j(q)$ (see Sect. 23).

An extensive calculation of the frequency distributions of alkali halides of the NaCl-type has been made by KARO and HARDY[1], using the dipole approximation.

The force constants or other parameters used in deriving frequency distributions are usually chosen to fit experimental values of $\omega_j(\boldsymbol{q})$ which have not been corrected for anharmonic effects. The distribution is therefore that of the quasi-harmonic frequencies (Sect. 32) at a particular temperature.

F. Anharmonic interactions.

29. The anharmonic potential and thermal expansion. In the earlier sections the harmonic approximation has been used to discuss the properties of the normal modes of vibration of crystals. Many properties of real crystals cannot be explained within this approximation; the thermal expansion, and the temperature dependence of the elastic and dielectric constants are examples of properties which depend on the interactions between the normal modes.

The potential of the crystal was expanded in Sect. 11 as a power series in the displacements of the atoms from their equilibrium positions. In the harmonic approximation only the quadratic terms in this expansion are retained, and the anharmonic interactions result from higher order terms such as:

$$H_A = \frac{1}{3!} \sum_{\alpha\beta\gamma} \sum_{\substack{l_1 l_2 l_3 \\ \varkappa_1 \varkappa_2 \varkappa_3}} \varphi_{\alpha\beta\gamma}\begin{pmatrix} l_1 & l_2 & l_3 \\ \varkappa_1 & \varkappa_2 & \varkappa_3 \end{pmatrix} u_\alpha\begin{pmatrix} l_1 \\ \varkappa_1 \end{pmatrix} u_\beta\begin{pmatrix} l_2 \\ \varkappa_2 \end{pmatrix} u_\gamma\begin{pmatrix} l_3 \\ \varkappa_3 \end{pmatrix}. \tag{29.1}$$

The displacements of the atoms are given in terms of the creation and destruction operators for the phonons as explained in Sect. 3 and Sect. 9. Since these operators always appear as a sum, it is more useful to introduce new operators;

$$A\begin{pmatrix} \boldsymbol{q} \\ j \end{pmatrix} = a\begin{pmatrix} \boldsymbol{q} \\ j \end{pmatrix} + a^+\begin{pmatrix} -\boldsymbol{q} \\ j \end{pmatrix}, \tag{29.2}$$

so that

$$\boldsymbol{u}\begin{pmatrix} l \\ \varkappa \end{pmatrix} = \sum_{\boldsymbol{q}j} \left(\frac{\hbar}{2Nm_\varkappa\,\omega_j(\boldsymbol{q})}\right)^{\frac{1}{2}} \boldsymbol{e}\left(\varkappa \middle| \begin{matrix} \boldsymbol{q} \\ j \end{matrix}\right) A\begin{pmatrix} \boldsymbol{q} \\ j \end{pmatrix} \exp i\boldsymbol{q}\cdot\boldsymbol{X}\begin{pmatrix} l \\ \varkappa \end{pmatrix}.$$

The expansion (29.1) can then be written as

$$\left.\begin{aligned} H_A = &\sum_{\substack{\boldsymbol{q}_1 \boldsymbol{q}_2 \boldsymbol{q}_3 \\ j_1 j_2 j_3}} V\begin{pmatrix} \boldsymbol{q}_1 & \boldsymbol{q}_2 & \boldsymbol{q}_3 \\ j_1 & j_2 & j_3 \end{pmatrix} A\begin{pmatrix} \boldsymbol{q}_1 \\ j_1 \end{pmatrix} A\begin{pmatrix} \boldsymbol{q}_2 \\ j_2 \end{pmatrix} A\begin{pmatrix} \boldsymbol{q}_3 \\ j_3 \end{pmatrix} + \\ &+ \sum_{\substack{\boldsymbol{q}_1 \boldsymbol{q}_2 \boldsymbol{q}_3 \boldsymbol{q}_4 \\ j_1 j_2 j_3 j_4}} V\begin{pmatrix} \boldsymbol{q}_1 & \boldsymbol{q}_2 & \boldsymbol{q}_3 & \boldsymbol{q}_4 \\ j_1 & j_2 & j_3 & j_4 \end{pmatrix} A\begin{pmatrix} \boldsymbol{q}_1 \\ j_1 \end{pmatrix} A\begin{pmatrix} \boldsymbol{q}_2 \\ j_2 \end{pmatrix} A\begin{pmatrix} \boldsymbol{q}_3 \\ j_3 \end{pmatrix} A\begin{pmatrix} \boldsymbol{q}_4 \\ j_4 \end{pmatrix}, \end{aligned}\right\} \tag{29.3}$$

where a further term has now been included.

---

[1] A. M. KARO, and J. HARDY: Phys. Rev. **129**, 2024 (1963).

A typical coefficient in this expression is given by [1] as

$$\left.\begin{aligned} V\begin{pmatrix} q_1\, q_2\, q_3 \\ j_1\, j_2\, j_3 \end{pmatrix} = \frac{1}{3!} \sum_{\alpha\varkappa_1} \sum_{\beta l_2 \varkappa_2} \sum_{\gamma l_3 \varkappa_3} \left(\hbar^3/8\,\omega_{j_1}(q_1)\,\omega_{j_2}(q_2)\,\omega_{j_3}(q_3)\,N\right)^{\frac{1}{2}} \times \\ \times (m_{\varkappa_1} m_{\varkappa_2} m_{\varkappa_3})^{-\frac{1}{2}} \varphi_{\alpha\beta\gamma}\begin{pmatrix} l_1\, l_2\, l_3 \\ \varkappa_1\, \varkappa_2\, \varkappa_3 \end{pmatrix} e_\alpha\left(\varkappa_1 \middle| \begin{matrix} q_1 \\ j_1 \end{matrix}\right) e_\beta\left(\varkappa_2 \middle| \begin{matrix} q_2 \\ j_2 \end{matrix}\right) e_\gamma\left(\varkappa_3 \middle| \begin{matrix} q_3 \\ j_3 \end{matrix}\right) \times \\ \times \exp\left(i\left(q_1 \cdot X\begin{pmatrix} l_1 \\ \varkappa_1 \end{pmatrix} + q_2 \cdot X\begin{pmatrix} l_2 \\ \varkappa_2 \end{pmatrix} + q_3 \cdot X\begin{pmatrix} l_3 \\ \varkappa_3 \end{pmatrix}\right)\right). \end{aligned}\right\} \quad (29.4)$$

(It is convenient to set $l_1 = 0$.)

The wave vectors of the phonons in the coefficients of (29.3) obey momentumconservation:

$$q_1 + q_2 + q_3 = \tau. \quad (29.5)$$

Besides this direct interaction between the normal modes, there are also effects resulting from the thermal expansion of the crystal. The thermal expansion alters the equilibrium distances between the atoms and hence the forces between them. The homogeneous deformation created by the thermal expansion is described by strain parameters $u^T_{\alpha\beta}$ [1]. If the equilibrium distance between two ions in the harmonic approximation is

$$\left(\bar{X}\begin{pmatrix} l\; l' \\ \varkappa\, \varkappa' \end{pmatrix}\right),$$

the distance in the deformed crystal is

$$X_\alpha\begin{pmatrix} l\; l' \\ \varkappa\, \varkappa' \end{pmatrix} = \sum_\beta (\delta_{\alpha\beta} + u^T_{\alpha\beta})\, \bar{X}_\beta\begin{pmatrix} l\; l' \\ \varkappa\, \varkappa' \end{pmatrix}. \quad (29.6)$$

This deformation adds additional terms to the expression for the potential energy of the crystal, which can be obtained by substituting the displacement given by (29.6) into equation (29.1). The additional potential energy can then be expanded in a double power series of the phonon coordinates, $A\begin{pmatrix} q \\ j \end{pmatrix}$, and the deformation parameters, $u^T_{\alpha\beta}$. For a crystal with every atom situated on a centre of symmetry this expansion gives

$$\left.\begin{aligned} H_S = \sum_{\alpha\beta} \sum_{q j j'} V_{\alpha\beta}\begin{pmatrix} q & -q \\ j & j' \end{pmatrix} A\begin{pmatrix} q \\ j \end{pmatrix} A\begin{pmatrix} -q \\ j' \end{pmatrix} u^T_{\alpha\beta} + \sum_{\alpha\beta\gamma\delta} V_{\alpha\beta\gamma\delta}\, u^T_{\alpha\beta}\, u^T_{\gamma\delta} + \\ + \sum_{\alpha\beta\gamma\delta} \sum_{q j j'} V_{\alpha\beta\gamma\delta}\begin{pmatrix} q & -q \\ j & j' \end{pmatrix} u^T_{\alpha\beta}\, u^T_{\gamma\delta}\, A\begin{pmatrix} q \\ j \end{pmatrix} A\begin{pmatrix} -q \\ j' \end{pmatrix}, \end{aligned}\right\} \quad (29.7)$$

where a typical term is

$$\left.\begin{aligned} V_{\gamma\delta}\begin{pmatrix} q & -q \\ j & j' \end{pmatrix} = \frac{1}{2} \sum_{\alpha\varkappa_1} \sum_{\beta\varkappa_2 l_2} \sum_{\varkappa_3 l_3} \left(\frac{\hbar^2}{4\omega_j(q)\,\omega_{j'}(q)\, m_{\varkappa_1} m_{\varkappa_2}}\right)^{\frac{1}{2}} \varphi_{\alpha\beta\gamma}\begin{pmatrix} l_1\, l_2\, l_3 \\ \varkappa_1\, \varkappa_2\, \varkappa_3 \end{pmatrix} X_\delta\begin{pmatrix} l_1\, l_3 \\ \varkappa_1\, \varkappa_3 \end{pmatrix} \times \\ \times e_\alpha\left(\varkappa_1 \middle| \begin{matrix} q \\ j \end{matrix}\right) e_\beta\left(\varkappa_2 \middle| \begin{matrix} -q \\ j' \end{matrix}\right) \exp\left(i q \cdot X\begin{pmatrix} l_1\, l_2 \\ \varkappa_1\, \varkappa_2 \end{pmatrix}\right). \end{aligned}\right\} \quad (29.8)$$

The thermal expansion coefficient may be deduced by differentiating Eq. (29.7) with respect to the strain[1]. The equilibrium condition is then, for the leading terms of (29.7), given by:

$$0 = \sum_{q j} V_{\alpha\beta}\begin{pmatrix} q & -q \\ j & j \end{pmatrix} \left(2\bar{n}\begin{pmatrix} q \\ j \end{pmatrix} + 1\right) + 2 \sum_{\gamma\delta} V_{\alpha\beta\gamma\delta}\, u^T_{\gamma\delta}.$$

[1] A. A. MARADUDIN: Phys. Stat. Solidi 2, 1493 (1962).

We have made use of the expectation value of $A\binom{\boldsymbol{q}}{j}A\binom{-\boldsymbol{q}}{j'}$ for a harmonic crystal. The $V_{\alpha\beta\gamma\sigma}$ can be identified with the elastic constants [1], and for cubic crystals the thermal strain is

$$u_{\alpha\alpha}^{T}=-\left(N v\left(C_{11}+2C_{12}\right)\right)^{-1}\sum_{\boldsymbol{q}j}V_{\alpha\alpha}\begin{pmatrix}\boldsymbol{q} & -\boldsymbol{q}\\ j & j\end{pmatrix}\left(\bar{n}\binom{\boldsymbol{q}}{j}+\frac{1}{2}\right). \tag{29.9}$$

The thermal expansion can then be obtained from this equation as $\dfrac{\partial u_{\alpha\alpha}^{T}}{\partial T}$.

30. The one-phonon GREEN's function. The techniques of many-body perturbation theory can be used with profit to discuss the effects of anharmonicity on the normal modes of vibration of crystals. There are several excellent accounts of these techniques as applied to both fermion and boson systems available[1] [5], and consequently they are not reproduced in detail here. A summary of the relevant theory is given in the Appendix. The unusual features which occur in applying these techniques to anharmonicity are firstly that the system consists of weakly interacting bosons (the phonons) with no conservation of particle number, and secondly that the phonon energies are nearly always comparable with $k_B T$, so that it is essential to use finite temperature theory from the beginning. The approach adopted here is to follow the development of MATSUBARA[2] and others as described by ABRIKOSOV et al. [5].

The theory describes a system in terms of the GREEN's functions or propagators for the excitations of the system, which are usually written in the Heisenberg representation [5]. Whereas in the Schrödinger representation the operators for an isolated system are time independent and the wave functions time dependent, in the Heisenberg representation the operators are time dependent and the wave functions independent of time. The transformations are given by:

$$\Psi_H=\exp\left(i\,Ht/\hbar\right)\Psi_S,$$

and

$$O_H=\exp\left(i\,H\,t/\hbar\right)O_S\exp\left(-i\,H\,t/\hbar\right).$$

The most important GREEN's function in the theory of an anharmonic crystal is the one-phonon thermodynamic GREEN's function defined by

$$G(\boldsymbol{q}j,\boldsymbol{q}'j',t)=\left\langle P\left\{A\left(\begin{matrix}\boldsymbol{q}\\ j\end{matrix}\middle|t\right)A^{+}\left(\begin{matrix}\boldsymbol{q}'\\ j'\end{matrix}\middle|0\right)\right\}\right\rangle \tag{30.1}$$

where $\langle\ldots\rangle$ represents the thermal average and the operators A are written in the Heisenberg representation. Since the crystal is assumed to be translationally invariant the expectation value of this GREEN's function is zero unless $\boldsymbol{q}=\boldsymbol{q}'$. The expression will therefore be contracted to $G(\boldsymbol{q}jj',t)$. The operator P is the time ordering operator;

$$\left.\begin{aligned}G(\boldsymbol{q}jj',t)&=\left\langle A\left(\begin{matrix}\boldsymbol{q}\\ j\end{matrix}\middle|t\right)A^{+}\left(\begin{matrix}\boldsymbol{q}\\ j'\end{matrix}\middle|0\right)\right\rangle \quad t>0\\ &=\left\langle A^{+}\left(\begin{matrix}\boldsymbol{q}\\ j'\end{matrix}\middle|0\right)A\left(\begin{matrix}\boldsymbol{q}\\ j\end{matrix}\middle|t\right)\right\rangle \quad t<0.\end{aligned}\right\} \tag{30.2}$$

The significance of these GREEN's functions is that if an additional displacement described by $A^{+}\binom{\boldsymbol{q}}{j'}$ is created at time 0, then a measure of the probability that

[1] D. J. THOULESS: The Quantum Mechanics of Many Body Systems. London and New York: Academic Press 1961. — D. PINES: The Many-Body Problem. New York: Benjamin 1962.

[2] T. MATSUBARA: Progr. Theoret. Phys. (Kyoto) 14, 351 (1955).

there will be an additional displacement $A\binom{\boldsymbol{q}}{j'}$ at a later time, t, is given by $G(\boldsymbol{q}jj', t)$. Alternatively if a displacement $A\binom{\boldsymbol{q}}{j}$ is removed at time t, then $G(\boldsymbol{q}jj', t)$ gives a measure of the probability that a displacement $A^+\binom{\boldsymbol{q}}{j'}$ is still absent at a later time, 0.

If the temperature dependence is not of interest, the theory may be developed from the GREEN's functions introduced above. However, when this is not the case we need to exploit the similarity in the way in which the time and the temperature enter into Eq. (30.2). This similarity is more explicit if (30.2) is written in full, as

$$\frac{1}{Z}\,\mathrm{Tr}\left[\exp(-\beta H)\,P\left\{\exp(i\,H\,t/\hbar)\,A\binom{\boldsymbol{q}}{j}\exp(-i\,H\,t/\hbar)\,A^+\binom{\boldsymbol{q}}{j'}\right\}\right].$$

The modification required to these functions to make use of this similarity is to write them in terms of the imaginary time $\tau = it$. The time ordering operator is then redefined to operate on the imaginary part of the time. The GREEN's functions are then

$$\left.\begin{aligned} G(\boldsymbol{q}jj', \tau) &= (1/Z)\,Tr\left[\exp(-\beta H + \tau H/\hbar)\,A\binom{\boldsymbol{q}}{j}\exp(-\tau H/\hbar)\,A^+\binom{\boldsymbol{q}}{j'}\right], \quad \tau > 0 \\ &\quad (1/Z)\,Tr\left[\exp(-\beta H)\,A^+\binom{\boldsymbol{q}}{j'}\exp(\tau H/\hbar)\,A\binom{\boldsymbol{q}}{j}\exp(-\tau H/\hbar)\right], \quad \tau < 0. \end{aligned}\right\} \tag{30.3}$$

Comparing these two expressions and using the cyclic property of the trace gives the periodicity condition

$$G(\boldsymbol{q}jj', \tau + \beta\hbar) = G(\boldsymbol{q}jj', \tau) \qquad 0 > \tau > -\beta h. \tag{30.4}$$

These GREEN's functions are therefore periodic in the complex time direction provided that $\beta\hbar > \tau > -\beta\hbar$. Since they are periodic they may be expanded in a Fourier series;

$$G(\boldsymbol{q}jj', \tau) = \sum_{n=-\infty}^{\infty} G(\boldsymbol{q}jj', i\omega_n)\exp(i\omega_n\,\tau), \tag{30.5}$$

where

$$\omega_n = 2\pi\, n/\beta\hbar, \tag{30.6}$$

and

$$G(\boldsymbol{q}jj', i\omega_n) = \frac{1}{2\beta\hbar}\int_{-\beta\hbar}^{\beta\hbar} G(\boldsymbol{q}jj', \tau)\exp(-i\,\omega_n\,\tau)\,d\tau. \tag{30.7}$$

Now let us apply these expressions to a harmonic crystal. The definition of the A operators gives (30.3a) as

$$G(\boldsymbol{q}jj', \tau) = \left\langle \exp(+\tau H/\hbar)\left(a\binom{\boldsymbol{q}}{j} + a^+\binom{-\boldsymbol{q}}{j}\right)\exp(-\tau H/\hbar)\left(a^+\binom{\boldsymbol{q}}{j'} + a\binom{-\boldsymbol{q}}{j}\right)\right\rangle.$$

For a harmonic crystal the expectation value is zero unless $j = j'$, when the expression reduces to

$$g(\boldsymbol{q}j, \tau) = \bar{n}\binom{\boldsymbol{q}}{j}\exp(|\tau|\,\omega_j(\boldsymbol{q})) + \left(\bar{n}\binom{\boldsymbol{q}}{j} + 1\right)\exp(-|\tau|\,\omega_j(\boldsymbol{q})). \tag{30.8}$$

The coefficients of the Fourier transform may then be deduced from (30.7), and the expression for the occupation number, to give

$$g(\boldsymbol{q}j, i\,\omega_n) = \frac{1}{\beta\hbar}\left(\frac{1}{\omega_j(\boldsymbol{q}) + i\omega_n} + \frac{1}{\omega_j(\boldsymbol{q}) - i\omega_n}\right) = \frac{2\omega_j(\boldsymbol{q})}{\beta\hbar\,(\omega_j^2(\boldsymbol{q}) + \omega_n^2)}. \tag{30.9}$$

This expression in part explains why the transformation to imaginary time and the use of the cyclic boundary condition is so powerful a technique. Apart from the constant factor $1/\beta\hbar$ the whole of the dependence upon temperature has now been eliminated, and the theory can be developed in a very similar way to that for the ground state at absolute zero.

As yet the Fourier coefficients in Eq. (30.5) have been defined only for values of $i\omega_n$ at the infinite set of points (30.6) on the imaginary frequency axis. Many of the physical properties of crystals can be calculated from these coefficients, as shown in the appendix, when they are analytically continued to the real frequency axis by letting

$$i\omega_n \to \Omega + i\varepsilon\,(\varepsilon > 0,\ \varepsilon \to 0). \tag{30.10}$$

In general this analytic continuation from a function defined at an infinite set of discrete points to one defined over the whole complex plane is not unique. Nevertheless in various special cases[1] [5], which nearly always occur in practice, the continuation can be performed. The physical properties of the crystal are then deduced from these GREEN's functions continued to the real frequency axis, as given by (30.10).

The anharmonic interactions are expected to alter the frequency and to produce a finite lifetime for the normal modes. If the shift in frequency is Δ and the inverse lifetime Γ then the one-phonon GREEN's function, analytically continued to the real frequency axis, might be expected to have the form

$$G(\boldsymbol{q}jj, \Omega + i\varepsilon) = \frac{2\omega_j(\boldsymbol{q})}{\beta\hbar\left(\omega_j^2(\boldsymbol{q}) - \Omega^2 + 2\omega_j(\boldsymbol{q})(\Delta - i\Gamma)\right)}.$$

In practice this equation is an oversimplification because the anharmonic interactions may couple the modes j and j'. The equation giving the GREEN's function is deduced in the appendix as[2,3]:

$$\left.\begin{aligned} &\sum_{j'}\left[(\omega_j^2(\boldsymbol{q}) - \Omega^2)\,\delta_{jj'} + 2\omega_j(\boldsymbol{q})\,\left(\Delta(\boldsymbol{q}jj'|\Omega) - i\Gamma(\boldsymbol{q}jj'|\Omega)\right)\right] \\ &G(\boldsymbol{q}j'j'', \Omega + i\varepsilon) = 2\omega_j(\boldsymbol{q})\,\delta_{jj''}. \end{aligned}\right\} \tag{30.11}$$

This equation is the basic equation for describing the effects of anharmonicity on the one-phonon properties of crystals. In the harmonic approximation both the shift in frequency and the inverse lifetime are zero, and we recover Eq. (30.9). The matrix $\Delta(\boldsymbol{q}jj'|\Omega)$ is the Hermitian part of the effect of the anharmonicity on the normal modes, and gives rise to a change in the frequency of the normal modes. The matrix $\Gamma(\boldsymbol{q}jj'|\Omega)$ is anti-Hermitian and gives rise to the lifetime effects.

The matrix on the left hand side of (30.11) plays a comparable role in the theory of anharmonic crystals to that of the dynamical matrix in the harmonic approximation. The anharmonic interactions couple the modes j and j', although they are orthogonal in the harmonic approximation. This coupling will occur whenever the modes have similar symmetry properties, or more exactly when they transform according to the same irreducible representation of the space group of the crystal. Another feature of (30.11) is that the shift and inverse lifetime depend not only on temperature, but also on the applied frequency, Ω, and may vary as this applied frequency varies. In order to discuss both of these features in more detail let us evaluate the lowest-order contributions to the shift and inverse lifetime.

---

[1] G. BAYM, and N. D. MERMIN: J. Math. and Phys. **2**, 232 (1961).
[2] A. A. MARADUDIN, and A. E. FEIN: Phys. Rev. **128**, 2589 (1962).
[3] R. A. COWLEY: Advances in Phys. **12**, 421 (1963).

The contributions to the shift and inverse lifetime are most easily evaluated by means of diagrams. In the diagrams the phonons are represented by lines and the anharmonic interactions by vertices. In one-phonon diagrams the line entering on the right is associated with the $A\begin{pmatrix}\boldsymbol{q}\\ j'\end{pmatrix}$ operator, and that leaving on the left with the $A^+\begin{pmatrix}\boldsymbol{q}\\ j\end{pmatrix}$ operator. The diagrams which contribute to $\Delta(\boldsymbol{q}jj'|\Omega)$ and $\Gamma(\boldsymbol{q}jj'|\Omega)$ are proper self-energy diagrams [5], which are those which cannot be broken into two by cutting any single phonon line other than an external line. The three lowest-order diagrams which contribute to the one-phonon Green's function are shown in Fig. 32, for crystals in which the atoms are at special positions in the unit cell. Their contributions may be evaluated by use of the rules given in the appendix[1,2]. The results for the Hermitian part are[1,2]:

$$\left.\begin{aligned}\Delta(\boldsymbol{q}jj'|\Omega) = {} & \frac{2}{\hbar}\sum_{\alpha\beta} V_{\alpha\beta}\begin{pmatrix}\boldsymbol{q} & -\boldsymbol{q}\\ j & j'\end{pmatrix} u_{\alpha\beta}^{T} + \frac{12}{\hbar}\sum_{\boldsymbol{q}_1 j_1} V\begin{pmatrix}\boldsymbol{q} & -\boldsymbol{q} & \boldsymbol{q}_1 & -\boldsymbol{q}_1\\ j & j' & j_1 & j_1\end{pmatrix}\left(2\bar{n}\begin{pmatrix}\boldsymbol{q}_1\\ j_1\end{pmatrix}+1\right) - \\ & - \frac{18}{\hbar^2}\sum_{\boldsymbol{q}_1\boldsymbol{q}_2 j_1 j_2} V\begin{pmatrix}\boldsymbol{q} & -\boldsymbol{q}_1 & -\boldsymbol{q}_2\\ j & j_1 & j_2\end{pmatrix} V\begin{pmatrix}-\boldsymbol{q} & \boldsymbol{q}_1 & \boldsymbol{q}_2\\ j' & j_1 & j_2\end{pmatrix} R(\Omega),\end{aligned}\right\}\quad(30.12)$$

where

$$\left.\begin{aligned}R(\Omega) = {} & (\bar{n}_1+\bar{n}_2+1)\,[1/(\omega_1+\omega_2+\Omega)_p + 1/(\omega_1+\omega_2-\Omega)_p] + (\bar{n}_2-\bar{n}_1)\times \\ & \times[1/(\omega_1-\omega_2+\Omega)_p + 1/(\omega_1-\omega_2-\Omega)_p].\end{aligned}\right\}\quad(30.13)$$

The abbreviations, $\bar{n}_1=\bar{n}\begin{pmatrix}\boldsymbol{q}_1\\ j_1\end{pmatrix}$ and $\omega_1=\omega_{j_1}(\boldsymbol{q}_1)$, have been used, subscript p denotes the principal part. The first term in Eq. (30.12) arises from diagram (a) of Fig. 32 and represents the effect of thermal expansion on the phonon frequency, while the second term is shown by diagram (b). Both of these contributions are independent of the applied frequency, Ω. The third contribution, diagram (c) of Fig. 32, depends on the applied frequency. The anti-Hermitian inverse lifetime is given by:

$$\Gamma(\boldsymbol{q}jj'|\Omega) = \frac{18\pi}{\hbar^2}\sum_{\boldsymbol{q}_1\boldsymbol{q}_2 j_1 j_2} V\begin{pmatrix}\boldsymbol{q} & -\boldsymbol{q}_1 & -\boldsymbol{q}_2\\ j & j_1 & j_2\end{pmatrix} V\begin{pmatrix}-\boldsymbol{q} & \boldsymbol{q}_1 & \boldsymbol{q}_2\\ j' & j_1 & j_2\end{pmatrix} S(\Omega), \quad (30.14)$$

where

$$\left.\begin{aligned}S(\Omega) = {} & (\bar{n}_1+\bar{n}_2+1)\,[\delta(\omega_1+\omega_2-\Omega) - \delta(\omega_1+\omega_2+\Omega)] + \\ & + (\bar{n}_2-\bar{n}_1)\,[\delta(\omega_1-\omega_2-\Omega) - \delta(\omega_1-\omega_2+\Omega)].\end{aligned}\right\}\quad(30.15)$$

Now consider the effect of these results in some special cases. Initially suppose the interactions do not intermix the different normal modes. An example of this type is provided by the long wavelength optical vibrations in alkali halides for which the eigenvectors are entirely determined by symmetry. Under these conditions the Green's function (30.11) is of the form:

$$G(\boldsymbol{q}jj,\Omega+i\varepsilon) = 2\omega_j(\boldsymbol{q})/\beta\hbar\left(\omega_j^2(\boldsymbol{q}) - \Omega^2 + 2\omega_j(\boldsymbol{q})\,[\Delta(\boldsymbol{q}jj|\Omega) - i\Gamma(\boldsymbol{q}jj|\Omega)]\right). \quad (30.16)$$

Provided that Δ and Γ are far smaller than $\omega_j(\boldsymbol{q})$ this equation shows that they can be interpreted as the shift in frequency and the inverse lifetime of the normal mode, respectively. However both the shift and inverse lifetime depend on the frequency, Ω. Fig. 33 shows calculations of the third (frequency-dependent) part of Δ (see Eq. (30.12)], and of Γ, (30.14), for the long wavelength transverse optic modes of KBr, as a function both of the applied frequency and of temperature[2,3]. The results show that the frequency dependence can be quite marked.

[1] See footnote 2, p. 117.
[2] See footnote 3, p. 117.
[3] E. R. Cowley, and R. A. Cowley: Proc. Roy. Soc. (London) A **287**, 259 (1965).

More generally the GREEN's function is given by the matrix Eq. (30.11) and depends on the intermixing of the normal modes. As the coefficient of the intermixing depends both on frequency and temperature the behaviour changes with both of these. This can be illustrated by the behaviour of the transverse optic modes of long wavelength in strontium titanate[1]. In this case there are 3 modes belonging to the same irreducible representation of the space group which can interact with one another. If we neglect the anti-Hermitian part of the matrix, the left hand side of (30.11) can be diagonalized to give a set of normal modes for each temperature and frequency. In Table 3 are given the components of the eigenvectors, in terms of those for $\Omega=0$, for various applied frequencies at 296 °K. Similar results are obtained as the temperature is varied. The results are probably more striking in the case of strontium titanate

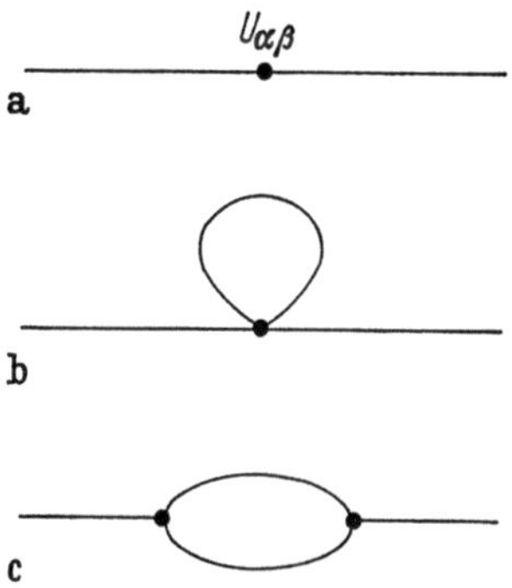

Fig. 32a—c. Diagrams which contribute to the self-energy of a phonon in lowest order, for crystals in which the atoms are at centres of symmetry in the unit cell. Diagram (a) shows the effect of thermal expansion, while (b) and (c) are the result of interaction with other phonons.

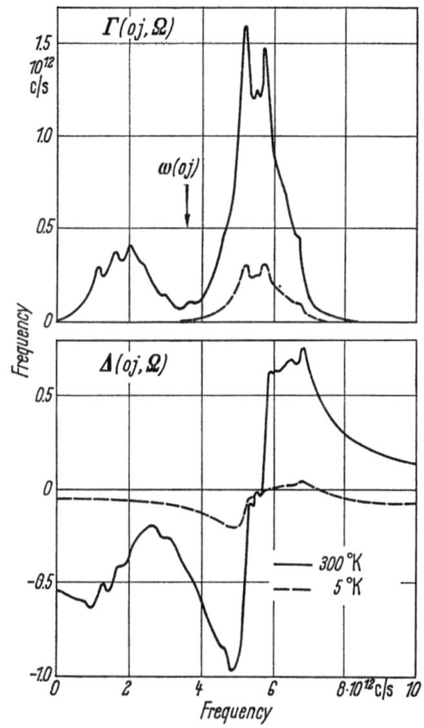

Fig. 33. The real part $\Delta(0j, \Omega)$, and imaginary part $\Gamma(0j, \Omega)$ of contributions from diagram (c) of Fig. 32 to the self energy of those transverse optic modes of potassium bromide for which $q\approx 0$.

than for most materials because the effect of the anharmonicity on the modes is particularly large (see Sect. 34), however these results show how significant the effect may be.

In this section we have shown that the GREEN's functions in an anharmonic crystal are governed by a matrix Eq. (30.11), which is very similar to the dynamical matrix of the harmonic approximation.

The additional complications are:

a) The matrix is no longer Hermitian, giving rise to lifetime effects.

b) The elements of the matrix depend on temperature.

c) The elements of the matrix depend on frequency.

31. Dielectric and elastic properties. One of the most direct ways of studying a material is to apply an external perturbation and to measure the response of the material. The static dielectric constant, for example, is determined by applying an electric field to a specimen and measuring the resultant polarisation.

[1] R. A. COWLEY: Phil. Mag. **11**, 673 (1964).

Table 3: *Frequencies are measured in units of* 10^{12} *c.p.s.*

Applied frequency Ω	Eigenvector I	Eigenvector II	Eigenvector III	Frequency $\omega_T\binom{0}{j}$
		Transverse optic mode I		
0	1	0	0	2.83
2.7	0.9999	−0.0117	0.0020	2.94
5.5	0.9864	−0.1631	0.0210	3.73
10.25	0.9971	0.0742	−0.0143	1.65
13.75	0.7951	0.5976	0.1035	6.92
16.5	0.9003	−0.4341	0.0322	4.40
20.5	0.8571	0.4948	0.1437	8.06
44.0	0.6898	0.7219	0.0540	6.25
		Transverse optic mode II		
0	0	1	0	5.18
2.7	0.0117	0.9999	0.0009	5.31
5.5	0.1628	0.9865	0.0153	5.45
10.25	−0.0743	0.9972	−0.0043	5.34
13.75	−0.5984	0.8007	−0.0259	4.89
16.3	0.4330	0.9006	0.0374	5.72
20.5	−0.4958	0.8672	−0.0182	4.96
44.0	−0.7220	0.6915	−0.0229	4.75
		Transverse optic mode III		
0	0	0	1	16.4
2.7	−0.0020	−0.0008	1.0000	16.1
5.5	−0.0232	−0.0117	0.9997	16.25
10.25	0.0139	0.0054	0.9999	16.0
13.75	−0.0984	−0.0414	0.9943	17.9
16.5	−0.0452	−0.0197	0.9988	16.5
20.5	0.1336	−0.0558	0.9895	17.2
44.0	−0.0539	0.0232	0.9983	16.6

If the external perturbation excites normal modes of vibration, while the response is also dependent upon these normal modes, the susceptibility of the crystal will be determined by the thermodynamic GREEN's functions, as shown in the appendix[1] [*5*].

In the case of an incident light wave, it is adequate to treat the light as a plane electric wave and to neglect the finite wavelength of the light [*1*]. If the dipole moment operator is $\boldsymbol{M}$, the dielectric susceptibility of the crystal at a frequency Ω is given by

$$\chi_{\alpha\beta}(\Omega)=\chi^E+\beta G(M_\beta M_\alpha, \Omega+i\varepsilon), \qquad (\varepsilon>0,\ \varepsilon\to 0) \tag{31.1}$$

where χ^E is the electronic contribution to the susceptibility. The GREEN's function is defined analogously to the one-phonon GREEN's function, namely as the analytically-contineud Fourier transform of $G(M_\beta M_\alpha, \tau)$ where

$$\left.\begin{aligned} G(M_\beta M_\alpha, \tau) &= \langle M_\beta(\tau)\, M_\alpha(0)\rangle, & \tau>0,\\ &= \langle M_\alpha(0)\, M_\beta(\tau)\rangle, & \tau<0. \end{aligned}\right\} \tag{31.2}$$

The dielectric properties can now be obtained by expanding the dipole moment operators in a phonon series:

$$\boldsymbol{M}=\sum_j \boldsymbol{M}\binom{0}{j} A\binom{0}{j}+\sum_{\boldsymbol{q}jj'} \boldsymbol{M}\begin{pmatrix}\boldsymbol{q} & -\boldsymbol{q}\\ j & j'\end{pmatrix} A\binom{\boldsymbol{q}}{j} A\binom{-\boldsymbol{q}}{j'}+\cdots. \tag{31.3}$$

[1] D. N. ZUBAREV: Uspekhi Fiz. Nauk **71**, 71 (1960).

The GREEN's functions and hence the dielectric susceptibility are then obtained by use of the diagram technique.

The diagrams of lowest order which contribute to the dielectric susceptibility are shown in Fig. 34. Diagram (a) is the one-phonon contribution to the susceptibility which is

$$\beta \sum_{jj'} M_\alpha\binom{0}{j} M_\beta\binom{0}{j'} G(0jj', \Omega + i\varepsilon). \tag{31.4}$$

In the case of many simple materials there is only one mode j which has the appropriate symmetry to contribute to (31.4). Under these conditions Eq. (30.16) can be used for the GREEN's function to give

$$\left.\begin{aligned} &\chi_{\alpha\beta}(\Omega) = \chi^E + \\ &+ \frac{2\omega_j(0) M_\alpha\binom{0}{j} M_\beta\binom{0}{j}}{\hbar\left[\omega_j^2(0) - \Omega^2 + 2\omega_j(0)\left(\Delta(0jj|\Omega) - i\Gamma(0jj|\;\;)\right)\right]}. \end{aligned}\right\} \tag{31.5}$$

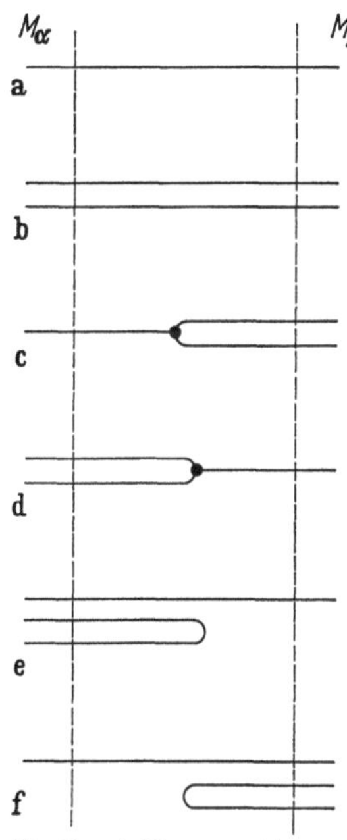

Fig. 34 a—f. Diagrams which contribute to the dielectric susceptibility of crystals in which the atoms are at centres of symmetry in the unit cell.

This equation shows that the centre of the one-phonon peak will be at the frequency which is the solution of the equation

$$\Omega^2 = \omega_j^2(0) + 2\omega_j(0)\,\Delta(0jj|\Omega), \tag{31.6}$$

and the half-width of the peak will be

$$\Gamma(0jj|\Omega).$$

The frequency $\omega_T\binom{0}{j}$ which is the solution of Eq. (31.6) will be referred to subsequently as the quasi-harmonic frequency of the mode $\binom{0}{j}$.

The frequency dependence of both the shift and inverse lifetime gives rise to structure in both the real and imaginary parts of the susceptibility. Fig. 35 shows these effects in some calculations for alkali halides[1, 2].

In many materials there is more than one infra-red active optical mode and Eq. (31.4) shows that there will be interaction between the different modes. Eq. (30.11) must then be solved exactly by inverting the self-energy matrix. However the form of the solution can be seen by a perturbation-theory expansion of (30.11) to give a contribution to the susceptibility of

$$\left.\begin{aligned} &\frac{1}{\hbar}\sum_{jj'} M_\alpha\binom{0}{j} M_\beta\binom{0}{j'} G(0jj|\Omega+i\varepsilon)\, G(0j'j'|\Omega+i\varepsilon)\left(\Delta(0jj'|\Omega+i\varepsilon) - \right. \\ &\qquad \left. - i\Gamma(0jj'|\Omega+i\varepsilon)\right), \end{aligned}\right\} \tag{31.7}$$

where the GREEN's functions are taken to be of the same form as in Eq. (31.5). The anomalous behaviour of this term is seen by considering the imaginary part of the susceptibility. The contribution from terms like (31.5) is peaked in a roughly Lorentzian shape about the frequencies given by (31.6). However there is an imaginary contribution from (31.7) multiplied by the real part of $G(0jj, \Omega+i\varepsilon)$ which is asymmetric about this frequency. This contribution will therefore tend to shift the centre of the absorption away from the frequency of the direct one-phonon part as given by (31.6). In a similar way the real part of (31.7) contains a

[1] See footnote 3, p. 117.
[2] See footnote 3, p. 118.

contribution which is peaked symmetrically around $\omega_j(0)$ instead of being asymmetric in this region.

The contribution from diagram (b) of Fig. 34 is the two-phonon contribution to the susceptibility. Its contribution to the susceptibility is

$$\chi_{\alpha\beta}(\Omega)=\frac{2}{\hbar}\sum_{\boldsymbol{q}jj'} M_\alpha\begin{pmatrix}\boldsymbol{q} & -\boldsymbol{q}\\ j' & j\end{pmatrix} M_\beta\begin{pmatrix}\boldsymbol{q} & -\boldsymbol{q}\\ j & j'\end{pmatrix}[R(\Omega)+i\pi S(\Omega)], \tag{31.8}$$

where $R(\Omega)$ and $S(\Omega)$ are defined by Eqs. (30.13) and (30.15) respectively. The contribution of this term to the dielectric susceptibility is shown in Fig. 35 for the

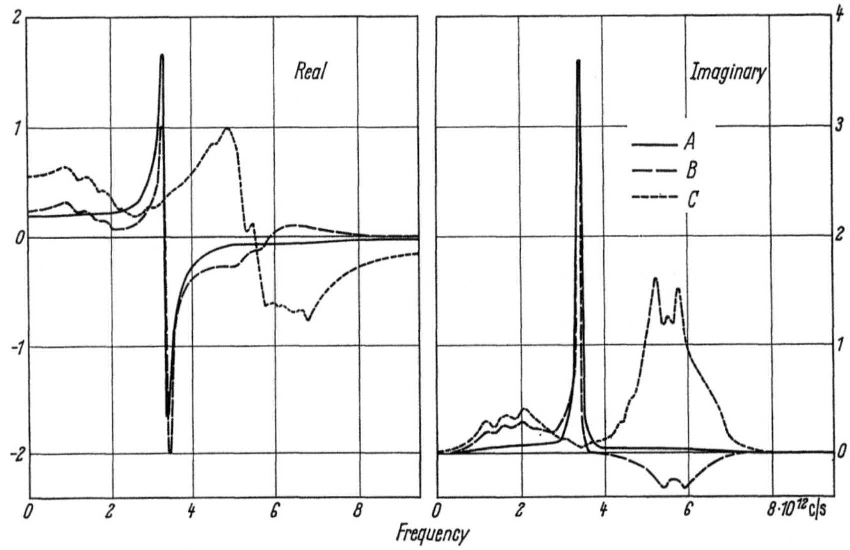

Fig. 35. Contributions to the dielectric susceptibility of potassium bromide at 300 °K. A arises from diagram (a) of Fig. 34, B from diagrams (c) and (d) (and is multiplied by 100), while C is from diagram (b) (and is multiplied by 263).

alkali halide, KBr. In the case of crystals with only one type of atom, germanium and silicon for example, this is the only term shown in Fig. 34 which contributes to the susceptibility.

Diagrams (c) and (d) of Fig. 34 show diagrams with one and two-phonon interactions with the electric field and also an anharmonic vertex. The contribution to the susceptibility of diagram (c) is, if there is only one infra-red active mode $\begin{pmatrix}0\\ j\end{pmatrix}$:

$$\frac{6}{\hbar^2}\sum_{\boldsymbol{q}j_1j_2}\frac{M_\alpha\begin{pmatrix}0\\ j\end{pmatrix} V\begin{pmatrix}0 & \boldsymbol{q} & -\boldsymbol{q}\\ j & j_1 & j_2\end{pmatrix} M_\beta\begin{pmatrix}\boldsymbol{q} & -\boldsymbol{q}\\ j_2 & j_1\end{pmatrix}[R(\Omega)+i\pi S(\Omega)]}{\omega_j^2(0)-\Omega^2+2\omega_j(0)\,(\Delta(0jj|\Omega)-i\Gamma(0jj|\Omega))}. \tag{31.9}$$

This contribution is peaked near the peak of the one-phonon contribution (31.5), however it is readily seen that the imaginary part of this contribution contains an asymmetric part to the intensity near the one-phonon peak. In a similar way to the term arising from the interaction between the modes this will give rise to a shift in the peak of the absorption spectrum. The contribution of these terms to the susceptibility of KBr is shown in Fig. 35.

The last two contributions, diagrams (e) and (f) of Fig. 34, modify the intensity of the one-phonon contribution. The contribution from diagram (e) is

$$3\beta\sum_{jj'\boldsymbol{q}j_1} M_\alpha\begin{pmatrix}0 & \boldsymbol{q} & -\boldsymbol{q}\\ j & j_1 & j_1\end{pmatrix} M_\beta\begin{pmatrix}0\\ j'\end{pmatrix} G(0jj',\Omega+i\varepsilon)\left(2\bar{n}\begin{pmatrix}\boldsymbol{q}\\ j_1\end{pmatrix}+1\right). \tag{31.10}$$

These contributions give the lowest-order contributions to the dielectric susceptibility of centrosymmetric crystals.

The effect of anharmonicity on the experimental measurement of phonon frequencies by infra-red absorption can now be discussed. The anharmonicity gives rise to a width of the one-phonon absorption peaks and to a shift in their frequency with temperature. The detailed shape of the peaks is determined by the frequency-dependent width and shift functions, and these do not necessarily give rise to simple Lorentzian peaks shapes. There is an interaction with the two-phonon absorption, through Eq. (31.9), which adds an asymmetric contribution to the absorption. This contribution, whose size is dependent upon the particular phonons involved, will tend to shift the measured frequencies away from the solutions of Eq. (31.6). In crystals where there is more than one infra-red active mode the coupling between the modes complicates the detailed structure of the spectrum, in a way which is dependent upon the dipole moments associated with each mode. The dielectric properties of crystals are discussed further by BILZ and WEHNER [2].

The elastic properties of crystals may be described by precisely the same formalism as was used for the dielectric properties. The theory is somewhat simpler in that the applied frequency is usually so much less than most of the phonon frequencies that it can be taken as zero. This approximation necessarily neglects ultrasonic attenuation, the theory of which is difficult because it requires a detailed evaluation of the spectral function for every normal mode of the crystal.

The distinction between adiabatic and isothermal elastic constants is important in practice. The elastic constants obtained experimentally by measuring the velocity of sound waves are the adiabatic elastic constants [6], while theoretical treatments give more directly the isothermal elastic constants. The difference between the two sets, for cubic crystals, is given by well-known thermodynamic relations

$$\left.\begin{aligned} C_{11}^{\mathrm{ad}} - C_{11}^{\mathrm{is}} = \frac{C_p - C_v}{K^{\mathrm{is}}\, C_v} = C_{12}^{\mathrm{ad}} - C_{12}^{\mathrm{is}} \\ C_{44}^{\mathrm{ad}} - C_{44}^{\mathrm{is}} = 0 \end{aligned}\right\}, \tag{31.11}$$

where C_p and C_v are the specific heats at constant pressure and at constant volume respectively, and K^{is} is the isothermal compressibility.

In the harmonic approximation the elastic constant is given by the term $V_{\alpha\beta\gamma\delta}$ of Eq. (29.7). Anharmonic interactions augment this by amounts which can be derived from the diagrams shown in Fig. 36 and which in total contribute the following to the isothermal elastic constant:

$$\left.\begin{aligned} &\sum_{\boldsymbol{q}j} V_{\alpha\beta\gamma\delta}\begin{pmatrix}\boldsymbol{q} & -\boldsymbol{q}\\ j & j\end{pmatrix}\left(2\bar{n}\begin{pmatrix}\boldsymbol{q}\\ j\end{pmatrix}+1\right) + \frac{8}{\hbar}\sum_{\boldsymbol{q}j\neq j'} V_{\alpha\beta}\begin{pmatrix}\boldsymbol{q} & -\boldsymbol{q}\\ j & j'\end{pmatrix} V_{\gamma\delta}\begin{pmatrix}-\boldsymbol{q} & \boldsymbol{q}\\ j & j'\end{pmatrix}\times \\ &\times\frac{\omega_j(\boldsymbol{q})\left(2\bar{n}\begin{pmatrix}\boldsymbol{q}\\ j'\end{pmatrix}+1\right)}{\omega_{j'}^2(\boldsymbol{q})-\omega_j^2(\boldsymbol{q})} - \sum_{\boldsymbol{q}j} V_{\alpha\beta}\begin{pmatrix}\boldsymbol{q} & -\boldsymbol{q}\\ j & j\end{pmatrix} V_{\gamma\delta}\begin{pmatrix}-\boldsymbol{q} & \boldsymbol{q}\\ j & j\end{pmatrix}\left[\frac{2}{\hbar\,\omega_j(\boldsymbol{q})}\left(2\bar{n}\begin{pmatrix}\boldsymbol{q}\\ j\end{pmatrix}+1\right)+\right. \\ &\left.+4\beta\left(\bar{n}\begin{pmatrix}\boldsymbol{q}\\ j\end{pmatrix}+\bar{n}^2\begin{pmatrix}\boldsymbol{q}\\ j\end{pmatrix}\right)\right]. \end{aligned}\right\} \tag{31.12}$$

The effect of thermal expansion on the elastic constants must also be considered. For simple crystals the correction can best be made by evaluating the expression given by the microscopic theory for the elastic constants in the harmonic approximation, and differentiating with respect to the lattice parameter[1].

[1] M. BLACKMAN: Proc. Phys. Soc. (London) **84**, 371 (1964).

For crystals of more complicated structure it is however necessary to use the theory which applies for finite strains [1], and to take considerable care to ensure that the elastic constants have the correct symmetry[1].

The numerical results of calculations of these corrections to the elastic constants of potassium bromide are shown in Fig. 37[1].

The techniques described in this section can equally readily be applied to obtain the piezo-electric constants, and non-linear properties of crystals. A further description of these techniques and of those employed by other authors is given in references [5][2,3].

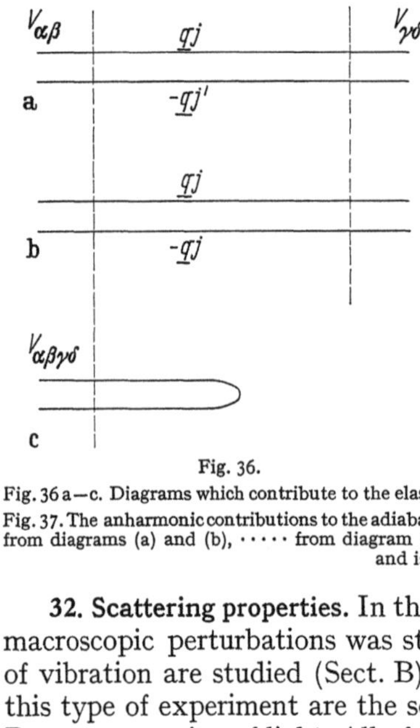

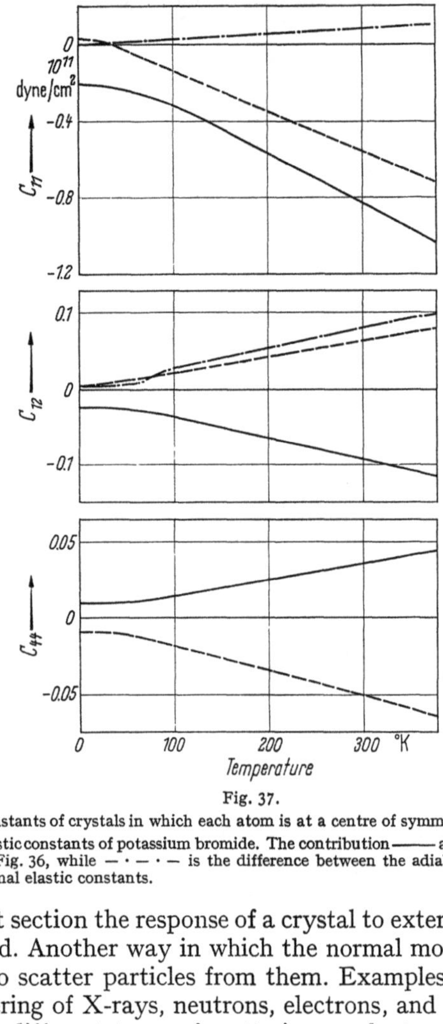

Fig. 36. Fig. 37.

Fig. 36a—c. Diagrams which contribute to the elastic constants of crystals in which each atom is at a centre of symmetry.

Fig. 37. The anharmonic contributions to the adiabatic elastic constants of potassium bromide. The contribution ——— arises from diagrams (a) and (b), ····· from diagram (c) of Fig. 36, while — · — · — is the difference between the adiabatic and isothermal elastic constants.

32. Scattering properties. In the last section the response of a crystal to external macroscopic perturbations was studied. Another way in which the normal modes of vibration are studied (Sect. B) is to scatter particles from them. Examples of this type of experiment are the scattering of X-rays, neutrons, electrons, and the Raman scattering of light. All of these different types of scattering can be treated by the same formalism: the scattering of plane waves from a state $\boldsymbol{k}_0$ to a state $\boldsymbol{k}$ in the first Born approximation, with a corresponding energy change of $\hbar\Omega$. The differential scattering cross-section is given in each instance by an equation which is similar to (15.1). Such differences as arise do so because whereas the atomic potential seen by the plane wave is independent of the atom's position for neutron scattering, in the other processes the potential $V_\varkappa(\boldsymbol{X})$ may depend on the atomic displacement and therefore on the normal modes. If the form factors of the atoms of type $\varkappa$ are all the same, $W_\varkappa(Q)$, then

$$W_\varkappa(Q) = \int_{\text{atom}} \exp(i\boldsymbol{Q}\cdot\boldsymbol{X})\, V_\varkappa(\boldsymbol{X})\, d^3\boldsymbol{X}, \tag{32.1}$$

[1] See footnote 3, p. 118.
[2] See footnote 3, p. 117.
[3] See footnote 1, p. 120.

and the differential cross section is then

$$\left.\begin{aligned}\frac{d^2\sigma}{d\mathscr{S}\,d\Omega}=\hbar\frac{|\boldsymbol{k}|}{|\boldsymbol{k}_0|}\sum_{\substack{l\varkappa\\ l'\varkappa'}}\Big\langle W_\varkappa(Q)\exp\Big(i\boldsymbol{Q}\cdot\Big(\boldsymbol{X}\binom{l}{\varkappa}+\boldsymbol{u}\binom{l}{\varkappa}\Big)\Big)\\ \times W_{\varkappa'}(Q)\exp\Big(-i\boldsymbol{Q}\cdot\Big(\boldsymbol{X}\binom{l'}{\varkappa'}+\boldsymbol{u}\binom{l'}{\varkappa'}\Big)\Big)\Big\rangle\,\delta(E_I-E_F-\hbar\Omega).\end{aligned}\right\}\quad(32.2)$$

Eq. (32.2) is very similar to that for the spectral function of two operators O_1 and O_2 [5] [Eq. (A.10) of the appendix], which can be obtained in terms of the imaginary part of the thermodynamic GREEN'S functions [Eq. (A.24)]. The differential scattering cross-section is then determined by the same GREEN'S functions as these introduced in Sect. 30.

We shall begin by considering coherent neutron scattering for which the form factors are constants, given by the scattering lengths, (Sect. 15). The GREEN'S functions for the scattering are obtained by expanding the exponential operators in terms of the phonon coordinates, $A\binom{\boldsymbol{q}}{j}$. In general the diagrams consist of three different regions, shown in Fig. 38. Region A is joined solely to the left hand side of the diagram, region B to the right hand side, and region C to both sides.

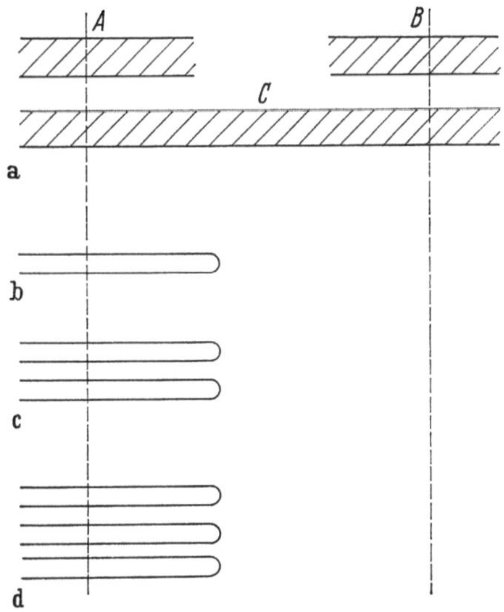

Fig. 38a–d. Diagram (a) shows the general form of a scattering diagram. Diagrams (b), (c) and (d) show the three lowest contributions to diagram A for a harmonic crystal.

Now consider the three simplest contributions to region A in the harmonic approximation, as shown in Fig. 38. These three diagrams are clearly the first members of a series whose general term is:

$$\left(-\frac{1}{2}\right)^m\frac{1}{m!}\left(\frac{\hbar}{2Nm_\varkappa}\sum_{\boldsymbol{q}j}\frac{\left|\boldsymbol{e}\left(\varkappa\left|\begin{matrix}\boldsymbol{q}\\ j\end{matrix}\right.\right)\cdot\boldsymbol{Q}\right|^2}{\omega_j(\boldsymbol{q})}\left(2\bar{n}\binom{\boldsymbol{q}}{j}+1\right)\right)^m.\quad(32.3)$$

Eq. (32.3) is the general term in the expansion of an exponential

$$\exp(-Y_\varkappa(Q)),$$

where

$$Y_\varkappa(Q)=\frac{1}{4Nm_\varkappa}\sum_{\boldsymbol{q}j}\frac{\left|\boldsymbol{e}\left(\varkappa\left|\begin{matrix}\boldsymbol{q}\\ j\end{matrix}\right.\right)\cdot\boldsymbol{Q}\right|^2}{\omega_j^2(\boldsymbol{q})}\left(2\bar{n}\binom{\boldsymbol{q}}{j}+\frac{1}{2}\right)\hbar\,\omega_j(\boldsymbol{q}).\quad(32.4)$$

This is the well-known Debye-Waller factor in the harmonic approximation, and regions A and B of the diagrams can now be identified as the Debye-Waller factors. The anharmonic interactions influence the Debye-Waller factors as shown by the diagrams of Fig. 39. Diagrams (a) and (b) are similar to the harmonic diagrams but the phonons contain self-energy parts. These diagrams alter the magnitude of the $Y_\varkappa(Q)$ factors. Diagrams (c) and (d) give rise to new effects however. Their contributions to the Debye-Waller factor are proportional to

Q^3 and Q^4 respectively. These contributions have different symmetry properties from the harmonic Debye-Waller factors, in particular they may give rise to an anisotropic factor in cubic crystals. Although calculations[1] suggest these terms are small, recent measurements[2] have shown anisotropic Debye-Waller factors in uranium dioxide and calcium fluoride.

The elastic scattering from crystals arises from diagrams for which region C is absent. The differential cross-section of Eq. (32.2) then reduces to

$$N^2\delta(\Omega)\Delta(\boldsymbol{Q})|F(\boldsymbol{Q})|^2,$$

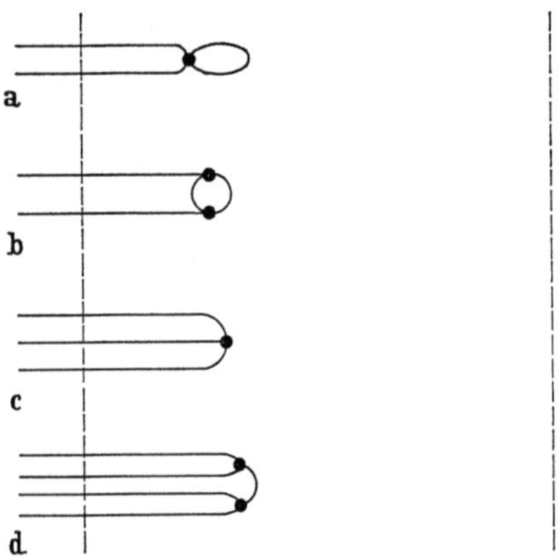

Fig. 39a—d. Diagrams for the anharmonic contributions to the Debye-Waller factor. Diagrams (a) and (b) are self-energy diagrams of the harmonic Debye-Waller factor but (c) and (d) are more complicated.

where $\Delta(\boldsymbol{Q})$ is zero unless $\boldsymbol{Q}$ is a reciprocal lattice vector $\boldsymbol{\tau}$, and the structure factor for elastic scattering is:

$$\left.\begin{aligned} F(\boldsymbol{Q}) = \sum_{\varkappa} W_{\varkappa}(Q)\exp(-Y_{\varkappa}(Q)) \times \\ \times \exp(i\boldsymbol{Q}\cdot\boldsymbol{X}(\varkappa)). \end{aligned}\right\} \quad (32.5)$$

This result shows that anharmonicity does not broaden the elastic scattering in either frequency or wave-vector. The only effect is to alter the intensity through the Debye-Waller factors.

The inelastic scattering arises from diagrams in which region C is present, and the basic one-phonon diagram is shown in Fig. 40a. The expansion of the exponential and Eq. (A.24) shows that the one-phonon cross-section is given by

$$\left.\begin{aligned} \sum_{\boldsymbol{q}jj'} \frac{|\boldsymbol{k}|}{|\boldsymbol{k}_0|} \Delta(\boldsymbol{Q}-\boldsymbol{q}) F(\boldsymbol{Q}|\boldsymbol{q}j) F(-\boldsymbol{Q}|-\boldsymbol{q}j') \lim_{\varepsilon\to 0} \frac{\beta\hbar}{2\pi i} (4\omega_j(\boldsymbol{q})\,\omega_{j'}(\boldsymbol{q}))^{-\frac{1}{2}} \times \\ \times (1-\exp(\beta\hbar\Omega))^{-1}[G(\boldsymbol{q}jj',\Omega+i\varepsilon) - G(\boldsymbol{q}jj',\Omega-i\varepsilon)], \end{aligned}\right\} \quad (32.6)$$

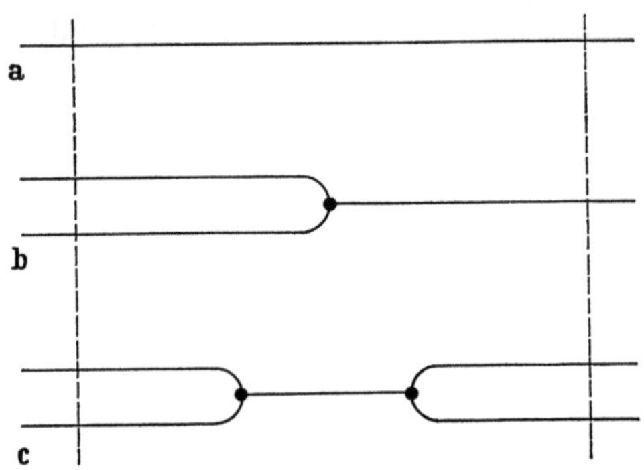

Fig. 40a—c. Diagrams for one-phonon scattering.

where

$$F(\boldsymbol{Q}|\boldsymbol{q}j) = \sum_{\varkappa} \left(\frac{\hbar}{2m_{\varkappa}}\right)^{\frac{1}{2}} \boldsymbol{Q}\cdot\boldsymbol{e}\left(\varkappa\middle|\begin{matrix}\boldsymbol{q}\\ j\end{matrix}\right) \exp(-Y_{\varkappa}(Q)) \exp(i(\boldsymbol{Q}-\boldsymbol{q})\cdot\boldsymbol{X}(\varkappa)). \quad (32.7)$$

---

[1] A. A. MARADUDIN, and P. A. FLINN: Phys. Rev. **129**, 2529 (1963).

[2] B. T. M. WILLIS: Proc. Roy. Soc. (London) A **274**, 122, 134 (1963); — Acta Cryst. **18**, 75 (1965). See also G. DOLLING, R. A. COWLEY, and A. D. B. WOODS: Can. J. Phys. **43**, 1397 (1965).

This quantity may be regarded as the structure factor for inelastic scattering. The GREEN's functions are given by Eq. (30.11), and in the simple case where there is no interaction between modes j and j', the part of the expression (32.6) inside the limit becomes;

$$\frac{1}{2\pi}\left[\exp(\beta\hbar\Omega)-1\right]^{-1}\frac{4\,\omega_j(\boldsymbol{q})\,\Gamma(\boldsymbol{q}jj|\Omega)}{(\omega_j^2(\boldsymbol{q})+2\omega_j(\boldsymbol{q})\,\Delta(\boldsymbol{q}jj|\Omega)-\Omega^2)^2+4\omega_j^2(\boldsymbol{q})\,\Gamma^2(\boldsymbol{q}jj|\Omega)}\,. \tag{32.8}$$

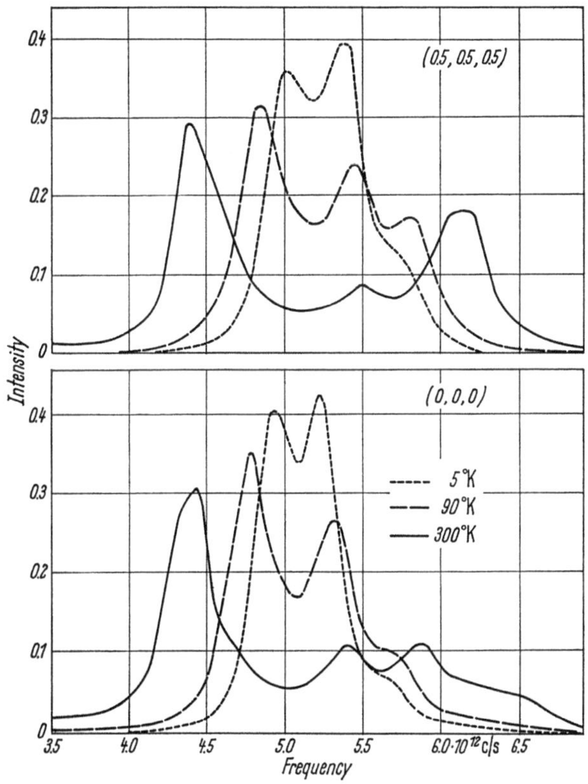

Fig. 41. The shapes of some one-phonon spectral functions for two longitudinal optic phonons in sodium iodide, having $\boldsymbol{q}\approx 0$ and $q=\left(\frac{1}{2},\frac{1}{2},\frac{1}{2}\right)\frac{2\pi}{a}$ respectively.

A comparison of Eqs. (31.4) and (31.5) for the one-phonon part of the dielectric susceptibility with these equations shows that the shape of the imaginary part of the dielectric susceptibility is very similar to the inelastic one-phonon scattering cross-section. The only difference between (31.5) and (32.8) is the population factor $[\exp(\hbar\beta\Omega)-1]^{-1}$ in (32.8).

The effect of anharmonicity on the one-phonon peaks (32.8) is therefore to give rise to a frequency-dependent width and lifetime for the normal modes. Figs. 41 and 42 show calculations of the shape of the spectral functions for some modes in sodium iodide and potassium bromide[1]. Fig. 41 shows some anomalously large effects for optical modes in sodium iodide. The origin of the structure in these spectral functions is illustrated in Fig. 43 which shows both the real and imaginary parts of the self-energy as a function of frequency.

[1] See footnore 3, p. 118.

When there is more than one mode belonging to a particular irreducible representation of the space group, the off-diagonal parts of the GREEN's functions

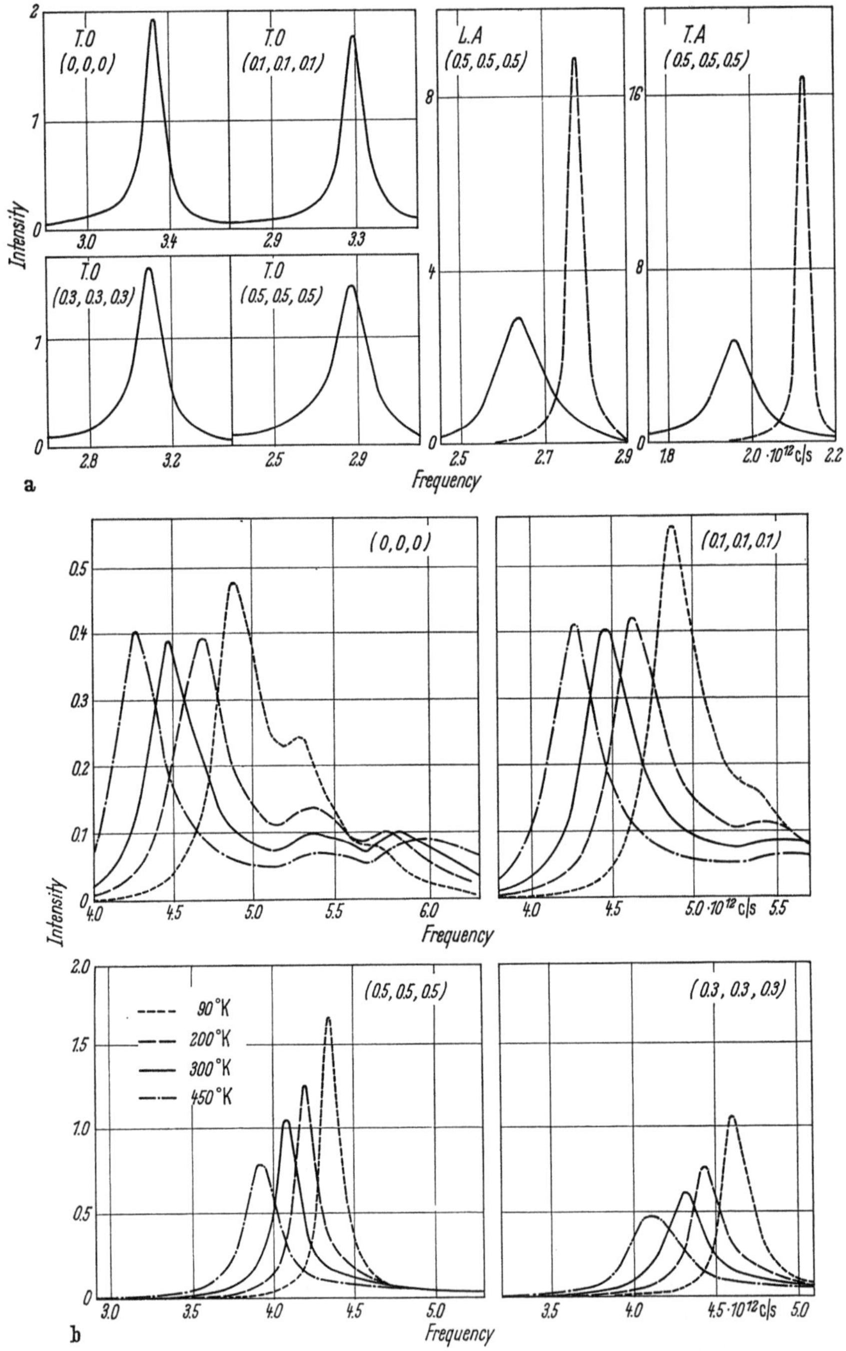

Fig. 42 a and b. The shapes of some one-phonon spectral functions in potassium bromide. A shows some acoustic and transverse optic phonons at —— 300 °K and – – – 90 °K. B shows some longitudinal optic phonons at four different temperatures. The values of q in units of $\frac{2\pi}{a}$ are given in brackets.

give rise to asymmetric contributions to the inelastic scattering cross-section at the frequencies of the phonons, as described for the dielectric susceptibility. However since the magnitude of this interference depends on the structure factor $F(\boldsymbol{Q}|\boldsymbol{q}j)$ for inelastic scattering, and on the $M_\alpha(0j)$ for the dielectric properties, the asymmetric term will contribute differently for the two properties. Furthermore since it is frequently possible to observe inelastic scattering under conditions such that different structure factors are involved, the size of the asymmetric

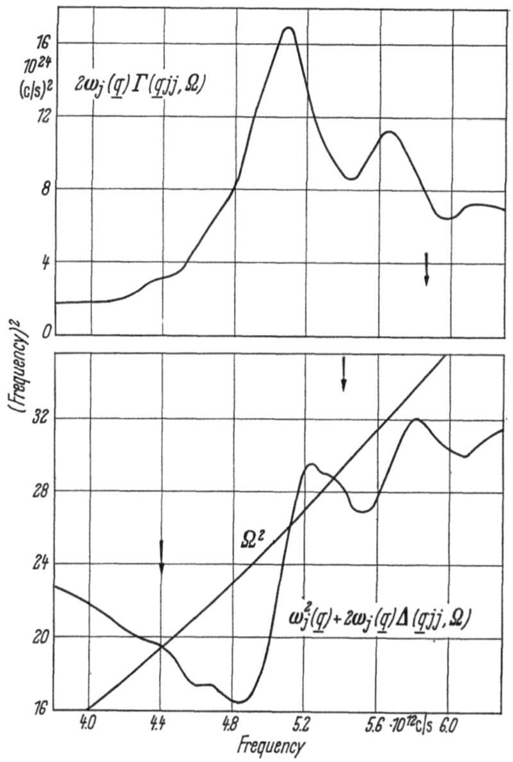

Fig. 43. The shapes of the real part of the self energy $\omega_j^2(\boldsymbol{q})+2\omega_j(\boldsymbol{q})\Delta(\boldsymbol{q}jj,\Omega)$ (Ω^2 is also shown), and of the imaginary part $2\omega_j(\boldsymbol{q})\,\Gamma(qjj,\Omega)$ for the longitudinal optic mode for which $q\simeq 0$. The crystal is sodium iodide at 300 °K.

term will vary according to the experimental conditions. The conclusions therefore are that the shape of the one-phonon peaks is the same for infra-red absorption and scattering processes if only one mode belongs to the irreducible representation. If this is not the case the shape of the curves may vary considerably from one experiment to another.

More complicated diagrams influence the one-phonon scattering cross-section however. Consider diagrams (b) and (c) of Fig. 40. These show the interaction of the one-phonon peak with the multiphonon background. Diagram (b) is similar to diagrams (c) and (d) of Fig. 34 for the dielectric susceptibility, and its contribution is similar to Eq. (31.8) for the susceptibility. This contribution also contains a contribution which is asymmetric about the one-phonon frequency and this contribution will in general differ for different experimental conditions[1,2].

---

[1] V. Ambegaokar, J. Conway, and G. Baym, p. 261 of Ref. [*11*].
[2] V. Ambegaokar, and A. A. Maradudin: Phys. Rev. **135**, A 107 (1964).

A further feature of anharmonicity is illustrated by diagram (c) of Fig. 40. This diagram contributes to the one-phonon type of scattering but its intensity is not governed by the structure factors $F(\boldsymbol{Q}|\boldsymbol{q}j)$. Since the identification of the modes is usually made with the help of these factors (Sect. 15), these interactions may make the interpretation of scattering results more difficult.

At present the resolution of neutron scattering measurements is not adequate to detect many of the phenomena described above. The experiments can normally yield only an approximate lifetime and a change in frequency with temperature. If we neglect both the interaction of the normal modes with the background and the off-diagonal parts of Eq. (30.11), the frequencies are given by the solutions of the equation

$$\left.\begin{aligned}\Omega^2 &= \omega_j^2(\boldsymbol{q}) + 2\omega_j(\boldsymbol{q})\,\Delta(\boldsymbol{q}jj|\Omega)\\ &= \omega_{\mathrm{T}}^2\binom{\boldsymbol{q}}{j}, \text{ say.}\end{aligned}\right\} \tag{32.9}$$

The inverse life-time is given by

$$\Gamma(\boldsymbol{q}jj|\Omega) \text{ evaluated at } \Omega = \omega_{\mathrm{T}}\binom{\boldsymbol{q}}{j}.$$

These frequencies are not only those measured experimentally but also provide the most useful set of quasi-harmonic frequencies with which to describe the normal modes[1]. They can also be used to some extent to describe the thermodynamic properties of crystals[2]. (These are not the quasi-harmonic frequencies of other authors [6].)

The effect of anharmonicity on the Raman scattering of crystals differs from the situation in neutron scattering because $\boldsymbol{Q}=0$ and the whole of the interaction with the normal modes arises through the form factor. The effect of the Debye-Waller factor is then more difficult to calculate, as it is no longer possible to write down the general term in the series (32.3). The evidence suggests however that two and three-phonon processes are far less important than for neutron scattering. The effect of anharmonicity on the one-phonon peaks of Raman scattering is exactly comparable to the effect on neutron scattering peaks. Different frequencies may however be observed because the relative magnitudes of the structure factors for the different modes and for the multi-phonon processes differ. A detailed discussion of the Raman spectra of diamond, silicon and germanium has been given by COWLEY[3].

The effect of anharmonicity on electron tunneling experiments (Sect. 16) is exactly comparable. Multi-phonon processes will occur as well as one-phonon processes, and anharmonicity will allow these to intermix.

X-ray scattering measurements of dispersion curves rely on measuring the intensity of the scattering (Sect. 14). Since anharmonicity intermixes the one and multi-phonon contributions to the intensity it is impossible to sort out these processes entirely unambiguously. It is also impossible to determine the detailed line shape of any modes. The only possibility is to determine the quasi-harmonic frequency when the lifetime is sufficiently long for the frequency to be well determined. These quasi-harmonic frequencies are the same as those obtained by neutron scattering measurements.

All single-phonon measurements are therefore influenced by anharmonicity in a similar way. The detailed shape of the one-phonon peak is identical no matter what type of measurement is performed, provided that interactions between

[1] See footnote 3, p. 117.
[2] T. H. K. BARRON, p. 247 of Ref. [11].
[3] R. A. COWLEY: J. phys. radium 26, 659 (1965).

modes of the same symmetry and with multi-phonon processes can be neglected. When these are included even the peak frequencies observed by the different techniques may differ, and the detailed shapes of the spectra may be quite different. Anharmonicity may also alter the intensity of the one-phonon peaks so that they no longer occur under the same experimental conditions, as when anharmonicity is neglected.

33. Anharmonic forces and comparison with experiment. The formalism for discussing the effect of anharmonicity on the physical properties of crystals, as described in the preceding sections, is dependent upon expansions in terms of phonon coordinates $A\binom{\boldsymbol{q}}{j}$. For example the potential energy (29.3), its strain dependence (29.7), and the dipole moment (31.3) are all expanded in such a power series. Without doubt the major difficulty in calculating anharmonic effects at present lies in the determination of the coefficients of these expansions. The most satisfactory approach would be to calculate them from first principles, which would require a knowledge of the electronic band structure of the distorted crystal. However the phonon dispersion curves of only a few simple metals have been obtained in this way even in the harmonic approximation (see Sect. D), so that as yet few attempts[1] have been made to calculate anharmonic effects from first principles.

The procedure usually adoped is similar to that for the harmonic approximation, described in Sect. D. Parameters are introduced to describe the anharmonic interactions, and these are fixed either by making some ad hoc assumptions or by comparison with experiment. For example consider the coefficients in the expansion (29.1). Translational invariance of the crystal leads to the condition [6],

$$\sum_{l_1 \varkappa_1} \varphi_{\alpha\beta\gamma}\begin{pmatrix} l_1 & l_2 & l_3 \\ \varkappa_1 & \varkappa_2 & \varkappa_3 \end{pmatrix} = 0,$$

Furthermore the coefficients are clearly symmetric in the indices $(\alpha\, l_1\, K_1)$, $(\beta\, l_2\, K_2)$ and $(\gamma\, l_3\, K_3)$ and in a similar way the space group symmetry of the crystal can be used to give further relations between these coefficients [3].

Even with the help of symmetry there are usually a large number of independent parameters needed to describe the coefficients, unless some explicit form of potential can be assumed. One form which is frequently adopted is an axially symmetric two-body interaction. The coefficients then always involve two parameters [6], for example

$$\varphi_{\alpha\beta\gamma}\begin{pmatrix} 0 & 0 & l \\ \varkappa_1 & \varkappa_1 & \varkappa_2 \end{pmatrix} = A\, X_\alpha X_\beta X_\gamma + B\,(X_\alpha \delta_{\beta\gamma} + X_\beta \delta_{\alpha\gamma} + X_\gamma \delta_{\alpha\beta}) \tag{33.1}$$

where

$$A = \left[\left(\frac{\partial^3 \varphi}{\partial r^3}\right)_{r=X} - \frac{3}{X}\left(\frac{\partial^2 \varphi}{\partial r^2}\right)_{r=X} + \frac{3}{X^2}\left(\frac{\partial \varphi}{\partial r}\right)_{r=X}\right] \Big/ 6X^3,$$

and

$$B = \left[\left(\frac{\partial^2 \varphi}{\partial r^2}\right)_{r=X} - \frac{1}{X}\left(\frac{\partial \varphi}{\partial r}\right)_{r=X}\right] \Big/ 6X^2,$$

and

$$X = \left|\boldsymbol{X}\begin{pmatrix} 0 & l \\ \varkappa_1 & \varkappa_2 \end{pmatrix}\right|.$$

Similar relations can be obtained for the higher-order coefficients.

The main advantage in the use of axially symmetric forces is that for each interatomic interaction only one parameter is introduced for each higher coef-

[1] T. Toya: J. Research Inst. Catalysis 9, 178 (1961).

ficient in the expansions. The number of parameters is then quite small if forces are restricted to nearest neighbour atoms. Furthermore if we assume a functional form for the potential energy, for example

$$\varphi(r) = b_{\varkappa\varkappa'} \exp(-r/\varrho) \tag{33.2}$$

only the two constants $b_{\varkappa\varkappa'}$ and ϱ are required.

Similar considerations can be used to limit the number of parameters in the dipole moment expansion (31.3), and in the expansion of the polarisability to obtain the Raman spectra. Alternatively it is possible to deduce these coefficients from those of the potential energy by using the shell model. The electronic dipole moment produced on the ions is $Y\boldsymbol{W}$ (Sect. 22). If we introduce anharmonic interactions into the equation of motion of the electronic dipoles then the coefficients of these expansions can be obtained. The further assumption that the anharmonic forces act entirely through the shells (Sect. 22) enables the coefficients of the dipole moment and polarisability expansions to be obtained in terms of the parameters of the potential energy expansion[1-3].

The calculation of anharmonic effects in crystal involves always the frequencies and eigenvectors of the normal modes of vibration. Since experimental measurements of these, and reasonably accurate lattice dynamical models, have only recently become available, many calculations have made use of approximations concerning the harmonic frequencies and eigenvectors, as for example in the work of LEIBFRIED and LUDWIG [6]. Unfortunately it is then impossible to decide whether a discrepancy between experiment and the results of such calculations arises from these approximations or from the form assumed for the anharmonic interactions.

A practical difficulty is that the evaluation of many of the properties requires summations over the whole Brillouin zone. The calculation of these summations without simplifying the form of the interaction coefficients is very lengthy and laborious even on modern electronic computers. Because of this several simplifications have been attempted in the form of the coefficients of Eqs. (29.4) and (29.8). The simplest is to assume the coefficients are proportional to the frequencies of the normal modes of vibration;

$$V\begin{pmatrix} \boldsymbol{q}_1 & \boldsymbol{q}_2 & \boldsymbol{q}_3 \\ j_1 & j_2 & j_3 \end{pmatrix} = K\left(\hbar^3\, \omega_{j_1}(\boldsymbol{q}_1)\, \omega_{j_2}(\boldsymbol{q}_2) \omega_{j_3}(\boldsymbol{q}_3)\right)^{\frac{1}{2}} N^{-\frac{1}{2}} \tag{33.3}$$

GRUNEISEN'S approximation is equivalent to a similar relation:

$$V_{\alpha\beta}\begin{pmatrix} \boldsymbol{q} & -\boldsymbol{q} \\ j & j \end{pmatrix} = C\hbar\omega_j(\boldsymbol{q}). \tag{33.4}$$

This coefficient gives the volume derivative of frequencies directly and GRUNEISEN'S parameter,

$$\gamma_j(\boldsymbol{q}) = -\frac{1}{2}\,\frac{\partial \ln \omega_j^2(\boldsymbol{q})}{\partial \ln V} \tag{33.5}$$

is a constant if Eq. (33.4) is valid. The results of a calculation of the Gruneisen constants for germanium[4] are shown in Fig. 44 and certainly do not support this conclusion.

Another possible simplification is in the form of the potential. If it is assumed that the highest derivative of the potential entering in any particular term is

[1] See foontote 3, p. 117.

[2] See footnote 3, p. 130.

[3] J. S. LANGER, A. A. MARADUDIN, and R. F. WALLIS, p. 411 of Ref. [11].

[4] G. DOLLING, and R. A. COWLEY: Proc. Phys. Soc. (London) **88**, 463 (1966).

much larger than the others [for example $\partial^3 \varphi/\partial r^3$ in A of Eq. (33.1)] then the calculations are simplified. MITSKEVICH[1] and COWLEY and COWLEY[2] have made use of this approximation. In practice the leading derivative is frequently 7 or 8 times larger than the others and this approximation is probably no worse than the neglect of anharmonicity in the electrostatic forces and in the polarisability of the atoms.

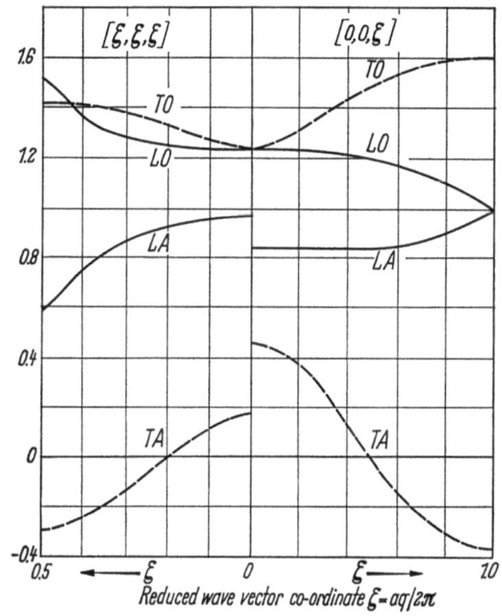

Fig. 44. The Gruneisen constants of modes in two symmetry directions for germanium.

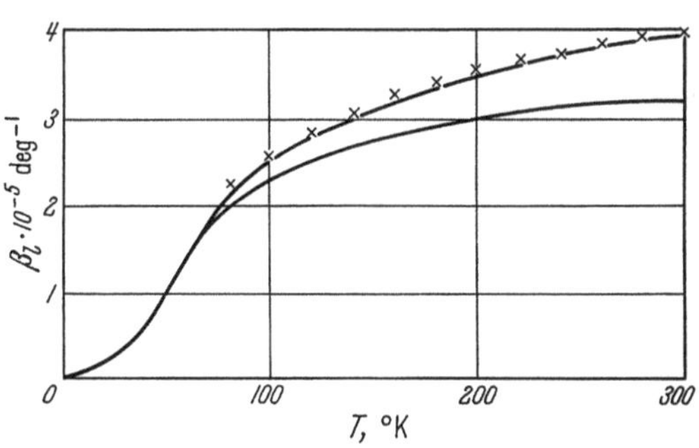

Fig. 45. The thermal expansion of sodium chloride, as calculated by MITSKEVITCH, and compared with experiment. The upper curve takes account of variation of compressibility with temperature.

Comprehensive calculations of anharmonic effects in crystals have mostly been made for alkali halides. LEIBFRIED and LUDWIG [6] used approximate ways of averaging over the normal modes and approximate forms for the coefficients. MITSKEVICH[1] has used the frequencies and eigenvectors of a fairly realistic model[3] and has calculated the elastic, thermal and dielectric properties of sodium chloride. He has made use of both of the approximations discussed above to simplify the coefficients. In Fig. 45 are shown his results for the thermal

[1] V. V. MITSKEVICH: Fiz. Tverdogo Tela 3, 3022, 3036 (1962).
[2] See footnote 3, p. 118.
[3] I. G. ZASLAVSKAYA, i K. B. TOLPYGO: Uspekhi Fiz. Zh. 1, 226 (1956).

expansion of NaCl. COWLEY and COWLEY[1] have used the models deduced from the measured dispersion curves of sodium iodide and potassium bromide to derive general frequencies and eigenvectors. The anharmonic interaction was assumed to involve an axially-symmetric two-body potential given by Eq. (33.2). Calculations were made of the elastic, thermal, dielectric and scattering properties of the crystals. The calculated elastic and dielectric properties are shown in Fig. 46 where they are compared with experimental results. MARADUDIN[2] and his

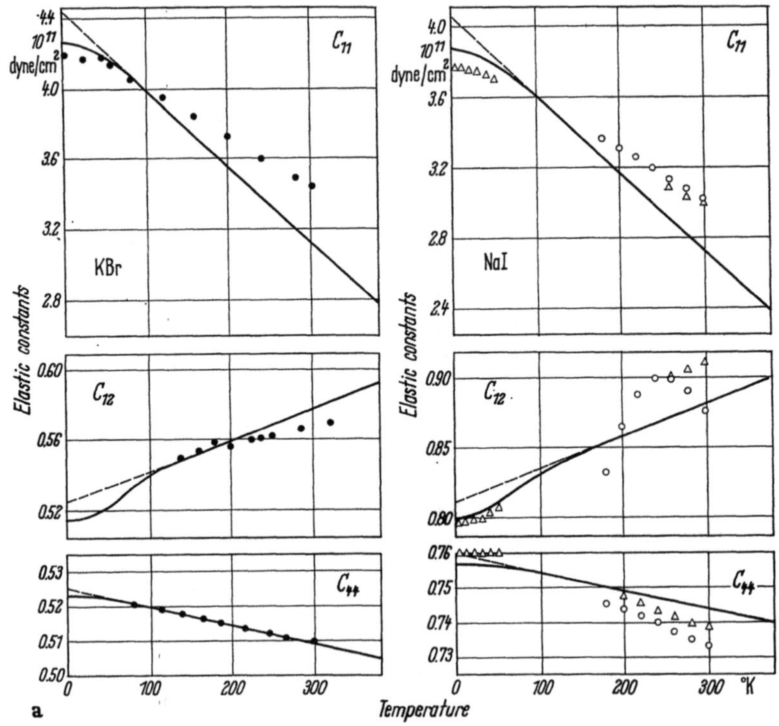

Fig. 46a and b. a shows the calculated and experimental elastic constants of sodium iodide and potassium bromide, while b shows the infra-red reflectivities.

collaborators have evaluated in detail many of the anharmonic properties of a simple model of a face-centred cubic crystal with interactions between nearest neighbours.

Most of the calculations give tolerable agreement with experiment. This is in part because there are few measurements which directly and accurately give information about the anharmonic interactions. The specific heat and Debye-Waller factor depend sensitively on the result given by the harmonic approximation, while the dielectric and scattering properties are more difficult to measure with adequate accuracy. A further difficulty is that frequently the different calculated contributions to the properties almost exactly cancel one another.

Most of the properties mentioned above depend on summations over the normal modes, and these are not very suitable for testing the validity of different anharmonic potentials. If disagreement is obtained at the end of the calculations,

[1] See footnote 3, p. 118.

[2] A. A. MARADUDIN, P. A. FLINN, and R. A. CALDWELL-HORSFALL: Ann. Phys. 15, 337, 360 (1961). See also footnotes 2, p. 117; 1, p. 126; 2, p. 129.

it is by no means easy to judge whether this arises from an inadequacy in the numerical work, or in the potential. Even if the error appears certainly to arise from the potential it is difficult to see how to improve the agreement.

The most direct information about anharmonic interactions is obtained from those properties which do not involve summations over all the normal modes. Unfortunately there are comparatively few of these. The pressure dependence of

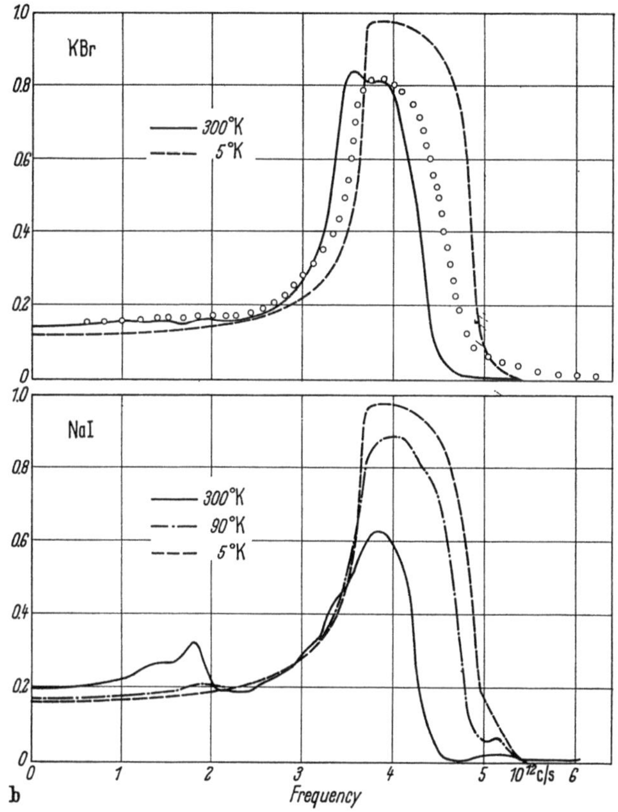

Fig. 46b.

normal mode frequencies, the non-linear dielectric constants and the pressure dependence of the dielectric constants are among the few, and these cannot be measured very accurately. It is difficult to see how much progress can be made without a better knowledge of the form of the anharmonic interaction. This will probably only be obtained by measuring the direct phonon-phonon interaction for many normal modes, in experiments which are similar in principle to those performed by TAYLOR and ROLLINS[1] for ultrasonic waves.

G. Lattice dynamics of ferroelectric crystals.

34. Introduction. A ferroelectric material is one in which the *E*—*D* (field-displacement) characteristic exhibits hysteresis. The name ferroelectric is derived by analogy with ferromagnetism, the analogy however extends only to phenomenological relationships and at the atomic level the mechanisms are quite

[1] L. H. TAYLOR, and F. R. ROLLINS: Phys. Rev. **136**, 591, 597 (1964).

dissimilar. Ferroelectrics of very varied physical and chemical properties are now known[1], the best known and the one having the simplest crystal structure is $BaTiO_3$. We shall largely confine our attention to this and other materials of similar crystal structure, such as $SrTiO_3$, since they are as yet the only materials for which the relation between electrical and dynamical properties is well understood.

The crystal structure of $BaTiO_3$ at temperatures above about 120 °C is shown in Fig. 47. This is the structure of the "paraelectric" phase, in which the static

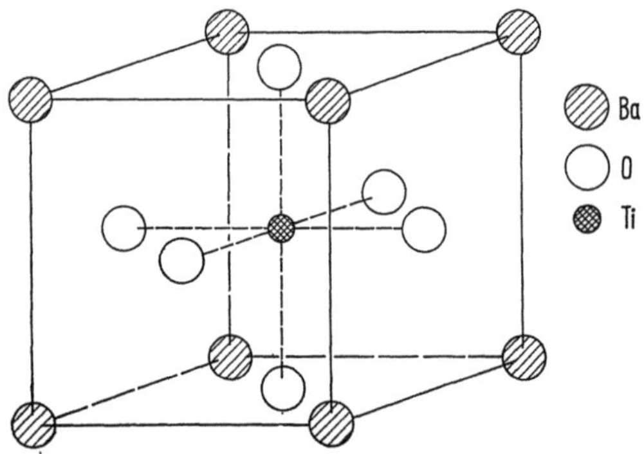

Fig. 47. The crystal structure of $BaTiO_3$ in the cubic phase.

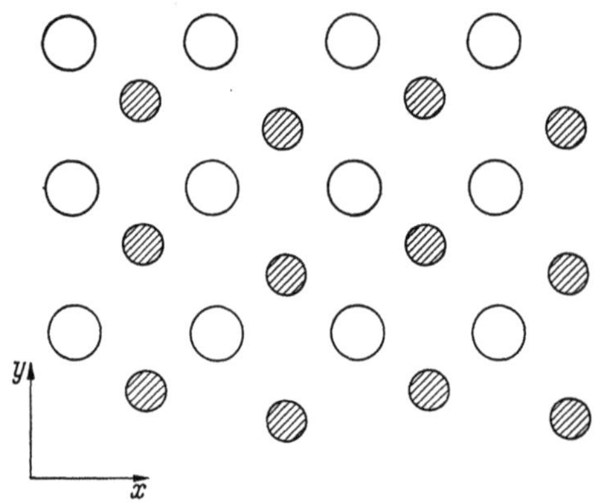

Fig. 48. The atomic displacements in a (hypothetical) antiferroelectric crystal structure.

dielectric constant follows a Curie-Weiss law

$$\frac{\varepsilon(0)-1}{4\pi}=\frac{\mathscr{C}}{T-T_c}\,. \tag{34.1}$$

As the temperature is decreased, a few degrees above the Curie temperature T_c the crystal deforms spontaneously to a tetragonal phase and becomes spontaneously polarised parallel to a cube axis — this is therefore a ferroelectric phase. The departure from cubic symmetry is small, $c/a=1.01$[1]. At a temperature of approximately 0 °C the spontaneous polarisation switches to a [110] direction (of the pseudo-cubic unit cell) and the unit cell becomes orthorhombic. Below −100 °C there is a phase with trigonal symmetry in which the polarisation is parallel to a cube diagonal. The transition temperatures and the mechanical and dielectric properties of the different phases are well correlated by DEVONSHIRE's phenomenological theory, which is briefly discussed in Sect. 38. X-ray and neutron

[1] F. JONA, and G. SHIRANE: Ferroelectric Crystals. Oxford: Pergamon 1962.

diffraction experiments[1] have shown that at the cubic-tetragonal transition, taking the barium atoms to be essentially unchanged in position, each titanium atom moves by approximately 0.07 Å, and the oxygen atoms by about the same amount in the opposite direction, leaving the octahedral arrangement of oxygen atoms almost undistorted. It is of course the nuclear positions which are determined by neutron diffraction, and X-ray diffraction is insensitive to the configuration of the polarisable valence electrons so that it is not possible experimentally to separate the spontaneous polarisation (amounting to 26 μ coul./cm² at 20 °C) unambiguously into ionic and electronic contributions, since the ionic charges are not known.

Strontium titanate has a Curie temperature of 32 °K, the exact nature of the transition or transitions occurring below 120 °K is not clear and we shall be concerned here only with the cubic paraelectric phase.

Certain other materials whose structures are distortions of the cubic structure shown in Fig. 47 have anti-ferroelectric phases in which for example oppositely-directed atomic displacements occur in alternate unit cells of the cubic structure. This is shown schematically in Fig. 48, the structures of antiferroelectric crystals such as $CaTiO_3$ are in practice more complicated. Transitions in which the atomic displacements are small compared with the unit cell dimensions are referred to as displacive; those we shall consider are in this category.

35. Lattice dynamics and dielectric properties of simple ionic crystals. We begin by discussing briefly a cubic diatomic ionic crystal in which the ions lie on 3-fold axes, for example the NaCl or the CsCl structure. So far as is known, no ferroelectric crystal has this structure in its paraelectric phase, but there is in principle no reason why this should not be possible, and the diatomic crystal is convenient for illustrative purposes[2].

The equations of the dipole approximation (or shell model) given in Sect. 22 can be simplified because of the high symmetry of the crystal, and the following results are found to apply. The frequencies ω_t and ω_l of the transverse and longitudinal optic modes for which $q\to 0$ are given by

$$\mu\,\omega_t^2 = R_0' - \frac{4\pi(\varepsilon(\infty)+2)\,(Z')^2}{9v}, \tag{35.1}$$

$$\mu\,\omega_l^2 = R_0' + \frac{8\pi(\varepsilon(\infty)+2)\,(Z')^2}{9v\,\varepsilon(\infty)}. \tag{35.2}$$

μ is the reduced mass $m_1 m_2/(m_1+m_2)$. Primed quantities can be defined in terms of the elements of the matrices $\boldsymbol{R}$, $\boldsymbol{T}$, $\boldsymbol{S}$ and $\boldsymbol{Z}$ but it is simpler to define them physically. When the positive and negative ions are displaced a relative distance $\boldsymbol{u}$ in the configuration of the transverse optic mode, the short-range restoring force is $R_0'\boldsymbol{u}$ per ion, and the polarisation is

$$\boldsymbol{P} = \frac{(\varepsilon(\infty)+2)}{3v} Z'\boldsymbol{u}.$$

In the corresponding longitudinal optic mode, the short range force is the same, and the polarisation is

$$\boldsymbol{P} = \frac{(\varepsilon(\infty)+2)}{3\varepsilon(\infty)\,v} Z'\boldsymbol{u}.$$

The corresponding effective fields are $\frac{4\pi}{3}\boldsymbol{P}$ in the transverse optic mode and $\frac{-8\pi}{3}\boldsymbol{P}$ in the longitudinal optic mode (see Sect. 19). Z' is therefore the "effective

[1] B. C. Frazer, H. Danner, and R. Pepinsky: Phys. Rev. **100**, 745 (1955).

[2] *Note added in proof.* See however G. S. Pawley, W. Cochran, R. A. Cowley, and G. Dolling: Phys. Rev. Letters, **17**, 753 (1966).

charge" defined by SZIGETI[1]. By considering the ratio of polarisation to effective field it is found that the polarizability for static fields is

$$\alpha(0) = \frac{(Z')^2}{R_0'} + \alpha(\infty). \tag{35.3}$$

The high-frequency polarisability $\alpha(\infty)$ can also be expressed in terms of the parameters of the shell model. The polarisabilities are related to the corresponding dielectric constants by the Clausius-Mossotti relation

$$\frac{\varepsilon - 1}{\varepsilon + 2} = \frac{4\pi\alpha}{3v}. \tag{35.4}$$

The quantities introduced above satisfy the Lyddane-Sachs-Teller[2] relation

$$\frac{\omega_l^2}{\omega_t^2} = \frac{\varepsilon(0)}{\varepsilon(\infty)}. \tag{35.5}$$

This is in fact a phenomenological result which does not depend on the validity of the dipole approximation.

We see from Eq. (35.1) that the frequency of the transverse optic mode depends on a balance between short-range and Coulomb interactions. If the two terms on the right of this equation were equal in magnitude, the crystal would be unstable against this mode of vibration. This is found to coincide with the condition

that is,

$$\left.\begin{aligned} \frac{4\pi}{3}\frac{\alpha(0)}{v} &= 1, \\ \varepsilon(0) &= \infty \end{aligned}\right\} \tag{35.6}$$

so that the descriptions "instability against a transverse optic mode of vibration" and "polarisability catastrophe" (often used to describe a transition to a ferroelectric phase) are in fact synonymous[3].

These results of course apply only in the harmonic approximation, when strictly the frequencies are temperature independent. However if we assume that the expressions are still valid but with the frequencies and force constant temperature dependent, the empirical relation (34.1) implies that in the paraelectric phase

$$\omega_t^2 \propto (T - T_c), \tag{35.7}$$

where ω_t is the frequency of the transverse optic mode, as measured by neutron scattering or infra-red techniques.

36. Normal modes of vibration of the $BaTiO_3$ structure. The unit cell of cubic $BaTiO_3$ contains $r=5$ atoms. When q is parallel to [100] the atomic displacements are constrained by symmetry to be purely longitudinal or transverse, and the determinantal Eq. (5.12) for the frequencies factors into three separate equations, one set giving the frequencies of r longitudinal modes and the remaining two giving the frequencies of r doubly-degenerate transverse modes. It can be shown that for $q \to 0$ with $\boldsymbol{q}$ in a general direction, there are still $r-1=4$ longitudinal optic modes and 4 doubly-degenerate transverse optic modes. Reserving the suffix 1 for acoustic modes, it follows from a theorem due to KUROSAWA[4] that in order of increasing frequency, a transverse optic mode is always followed by a

---

[1] B. SZIGETI: Proc. Roy. Soc. (London) A **204**, 51 (1950).

[2] R. H. LYDDANE, R. G. SACHS, and E. TELLER: Phys. Rev. **59**, 673 (1951).

[3] W. COCHRAN: Phys. Rev. Letters **3**, 412 (1959).

[4] T. KUROSAWA: J. Phys. Soc. Japan **16**, 1298 (1961).

longitudinal optic mode, thus (writing $\omega_{t_2}(q\to 0)=t_2$)

$$t_2 \leqq l_2 \leqq t_3 \leqq \dots l_5 .$$

It can be shown that one transverse optic mode and one longitudinal optic mode of the $BaTiO_3$ structure are degenerate by symmetry for $q\to 0$, experimentally they are found to be l_3 and t_4. This is indicated schematically in Fig. 49.

The Lyddane-Sachs-Teller equation (35.5) can be extended to apply to crystals of general symmetry[1], for this structure the appropriate result is

$$\prod_{j=2}^{5} \frac{(\omega_l)_j^2}{(\omega_t)_j^2} = \frac{\varepsilon(0)}{\varepsilon(\infty)} . \tag{36.1}$$

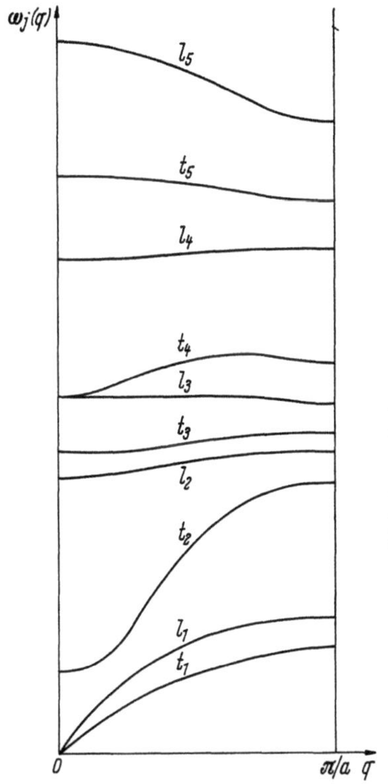

Fig. 49. Schematic representation of $\omega_j(\boldsymbol{q})$ for cubic $BaTiO_3$, with $\boldsymbol{q}$ in the [100] direction.

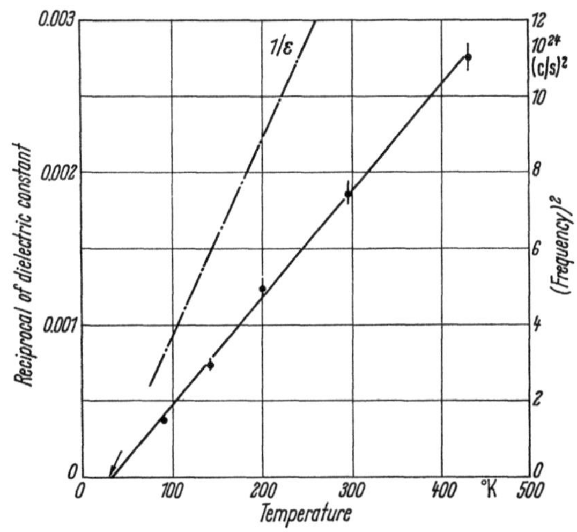

Fig. 50. A graph of $\omega_{t2}^2(0)$ against temperature for $SrTiO_3$. The reciprocal of the static dielectric constant is also shown as a function of temperature.

The fact that the static dielectric constant of $BaTiO_3$ or of $SrTiO_3$ follows a Curie-Weiss law suggests that for these materials[2]

$$(\omega_t)_2^2 \propto (T - T_c), \tag{36.2}$$

if the frequencies of the other optic modes are largely independent of temperature.

Measurements of the frequency of the transverse optic mode of lowest frequency as a function of temperature have been made by infra-red absorption[3] and by neutron scattering[4] techniques. The temperature dependence measured by the latter techniques is shown in Fig. 50 for strontium titanate. The results strikingly confirm Eq. (36.2). Similar conclusions have been reached for both strontium and barium titanates by investigators using infra-red techniques[3,5].

[1] W. Cochran, and R. A. Cowley: J. Phys. Chem. Solids **23**, 447 (1962).
[2] W. Cochran: Advances in Phys. **9**, 387 (1960). — P. W. Anderson, in "Fizika Dielektrikov", Ed. by G. I. Skanavi. Acad. Nauk. S. S. R. Moscow (1960).
[3] A. S. Barker, and M. Tinkham: Phys. Rev. **125**, 1527 (1962).
[4] R. A. Cowley: Phys. Rev. Letters **9**, 159 (1962); — Phys. Rev. **134**, A 981 (1964).
[5] W. G. Spitzer, R. C. Miller, D. A. Kleinman, and L. E. Howarth: Phys. Rev. **126**, 1710 (1962).

The temperature dependence of the other optic modes for which $q \to 0$ have also been studied by optical methods and their frequencies found to be largely independent of temperature[1]. Some of the modes propagating along the [100] direction have been studied at 90 °K and 296 °K for strontium titanate, and only the transverse optic branch of lowest frequency is appreciably temperature dependent[2]. The change in shape of this branch with temperature is shown in Fig. 51. The parameters of a dipole-approximation model (Sect. 22) have been found from these measurements, and the model is consistent with the dynamical properties of these materials[3].

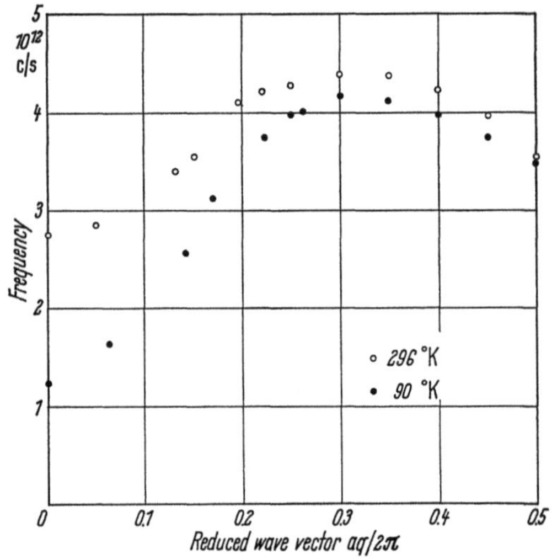

Fig. 51. The shape of the transverse optic branch of lowest frequency with q in the [100] direction for $SrTiO_3$ at two different temperatures.

37. The origin of ferroelectricity. In the preceding sections it was suggested that ferroelectricity in crystals is associated with one branch of the dispersion curve near $q=0$ being anomalously temperature dependent. The origin of this temperature dependence was left open however. Several possible reasons for the temperature dependence have been put forward. One of these is that the interactions between the atoms change with temperature, presumably because the electronic band structure is changing. On this theory, which is implies the breakdown of the adiabatic approximation (Sect. 11), it is difficult to understand why only a few of the modes are strongly temperature dependent. The band gaps of perovskite ferroelectrics are quite large and independent of temperature[4] so that this explanation appears unlikely.

Another suggestion has been that the temperature dependence arises from the change in volume of the crystal with thermal expansion. A change in volume of the crystal can also be produced by applying a hydrostatic pressure. The pressure dependence of the dielectric constant of $BaTiO_3$ has been measured[5] and the dielectric constant is found to decrease with increasing pressure or decreasing volume. As the volume decreases on cooling this effect would lead to a decreasing dielectric constant, contrary to what is found.

[1] A. S. BARKER: Phys. Rev. **145**, 391 (1966).
[2] See footnote 4, p. 139.
[3] R. A. COWLEY: Phil. Mag. **11**, 673 (1965).
[4] H. L. SUCHAN, A. S. BALCHAN, and H. G. DRICKAMER: J. Phys. Chem. Solids **10**, 343 (1959). — H. W. GANDY: Phys. Rev. **113**, 795 (1959).
[5] W. J. MERZ: Phys. Rev. **78**, 52 (1950).

The most likely explanation appears to be that the temperature dependence arises from the anharmonic interactions between the normal modes, as described in section F. In the situation where there is only one transverse optic mode the one-phonon part of the dielectric susceptibility is given by Eq. (31.5). The expression for the frequency shift Δ [Eq. (30.12)] is proportional to temperature if $k_B T \approx \hbar\omega_j(\boldsymbol{q})$ for all the normal modes, so that the static dielectric constant will follow a Curie Law (34.1), where

$$\mathscr{C} = T\left|M_\alpha\binom{0}{j}\right|^2 \Big/ 2\Delta(0jj\,|\,0)\,\omega_j(0). \tag{37.1}$$

and

$$T_c = -T\omega_j^2(0)/2\omega_j(0)\,\Delta(0jj\,|\,0). \tag{37.2}$$

Since $2\omega_j(0)\,\Delta(0jj\,|\,0)$ is independent of $\omega_j(0)$, a material is a ferroelectric if $\omega_j^2(0)$ is negative and $2\omega_j(0)\,\Delta(0jj\,|\,0)$ increases sufficiently with temperature. The crystal is then unstable in the harmonic approximation and only stabilised by the anharmonic interactions.

In Sect. 30 we emphasised that the shift in frequency was frequency dependent. Infra-red absorption and neutron scattering measurements of the frequency yield not $\omega_j^2(0)+2\omega_j(0)\,\Delta(0jj\,|\,0)$, but the solution of $\Omega^2=\omega_j^2(0)+2\omega_j(0)\,\Delta(0jj\,|\,\Omega)$, (Sect. 31 and 32). However since the shift is symmetric in Ω [Eq. (A.33)] and the frequency of the mode concerned is much smaller than for most of the others in the crystal, the difference between the two shifts is quite small. The results of the simple theory given in the preceding section are therefore substantiated by this discussion.

The derivation of the Curie-Weiss Law given above is however unsatisfactory for several reasons. Firstly for both $BaTiO_3$ and $SrTiO_3$ the Curie Law is valid when $k_B T$ is less than $\hbar\omega_j(\boldsymbol{q})$ for many modes, and secondly there is more than one optic mode belonging to the same irreducible representation and the interaction between these modes should be considered.

The calculation of the temperature dependence of the pseudoharmonic frequencies is very difficult at present because little is known of the anharmonic interactions in these crystals. However calculations based on a very simple model[1] suggest that these interactions can explain the observed temperature dependence in $SrTiO_3$, with not unreasonable parameters for the interatomic forces.

These interactions give much larger effects than in alkali halides, for example. The reason for this is not that the shift in frequency is much larger for the unstable mode than for other optic modes. The calculated value of $2\omega_j(0)\,\Delta(0jj\,|\,\Omega)$ for the branch of highest frequency is in fact larger than for the one of lowest frequency in the case of $SrTiO_3$[1]. The reason is the small negative value of $\omega_j^2(0)$ for $j=2$, in agreement with the conclusion that ferroelectricity arises when the forces in the harmonic approximation almost exactly cancel for certain normal modes.

38. The ferroelectric transition. The theory of the transition in ferroelectrics is very satisfactorily described by the phenomenological theory of DEVONSHIRE[2]. It is assumed that the free energy of a stress-free crystal can be expanded in

[1] See footnote 3, p. 140.

[2] A. F. DEVONSHIRE: Phil. Mag. **40**, 1040 (1949); **42**, 1065 (1951); — Advances in Phys. **3**, 85 (1954).

powers of the polarisation of the crystal:

$$\left.\begin{aligned} G &= \frac{1}{2}\frac{T-T_c}{\mathscr{C}}\left(P_x^2+P_y^2+P_z^2\right)+\frac{1}{4}\xi'\left(P_x^4+P_y^4+P_z^4\right)+ \\ &+\frac{1}{6}\zeta'\left(P_x^6+P_y^6+P_z^6\right)+\frac{1}{2}\lambda'\left(P_y^2P_z^2+P_x^2P_y^2+P_x^2P_z^2\right)+\cdots. \end{aligned}\right\} \qquad (38.1)$$

For the cubic to tetragonal transition in $BaTiO_3$ the polarisation is taken to be along the z axis, when

$$G=\frac{1}{2}\left(\frac{T-T_c}{\mathscr{C}}\right)P^2+\frac{1}{4}\xi' P^4+\frac{1}{6}\zeta' P^6 \ldots. \qquad (38.2)$$

If the term ξ' is positive then the spontaneous polarisation below the transition is given by:

$$P_s^2=\frac{T_c-T}{\mathscr{C}\xi'}. \qquad (38.3)$$

The transition is then a second-order transition, and the dielectric constants can be obtained below the transition with the help of Eqs. (38.1) and (38.2).

The behaviour is somewhat different if ξ' is negative. The transition is of first order and occurs when

$$T-T_c=\frac{3\mathscr{C}(\xi')^2}{16\zeta'} \qquad (38.4)$$

giving a difference between the transition temperature and the Curie temperature. The magnitude of the spontaneous polarization just below the transition is:

$$P_s^2=-\tfrac{3}{4}\xi'/\zeta'. \qquad (38.5)$$

The theory outlined here is incomplete since explicit account is not taken of the effects of external strains on the transition. Since the shape of the crystal changes during the transition these effects are important. DEVONSHIRE[1] introduced terms into the free energy to represent the effects of external strains and their coupling to the polarisation. The theory is then able to correlate the temperature dependences of the various elastic and dielectric properties of $BaTiO_3$ in its three phases very successfully[1].

A microscopic theory should lead to expressions for the various coefficients in DEVONSHIRE's free energy expansion in terms of the interatomic forces. Suppose there is a static displacement of the atoms which is described by the normal mode $\binom{0}{j}$. The quadratic change in the free energy is[2]

$$\eta^2\left(\omega_j^2(0)+2\omega_j(0)\,\Delta(0jj\,|\,0)\right) \qquad (38.6)$$

where η is the displacement of the normal mode. This relation shows that the stability of the crystal is determined by precisely the same parameters as the static dielectric susceptibility. The higher terms in the expansion (38.1) can also be obtained by substituting the displacement η into the potential energy expansion [Eq. (29.3)]. The quartic terms for example involve

$$V\begin{pmatrix}0\,0\,0\,0\\ j\,j\,j\,j\end{pmatrix}.$$

The theory is then similar to DEVONSHIRE's theory except that the parameter is the displacement of a normal mode instead of the polarisation. Microscopic expressions for the coefficients of (38.1) are obtained by a comparison of this series with the corresponding expansion of the microscopic theory. An additional

[1] See footnote 2, p. 141.
[2] See footnote 3, p. 140.

feature which is found from the microscopic theory is that the normal mode may interact with others of the same symmetry, so that the eigenvectors change with temperature, and during the transition.

A defect of the thermodynamic theory is its inability to describe the behaviour of the specific heat near the transition[1]. As the temperature is decreased towards the transition the frequencies of some normal modes decrease and give an increase in the specific heat for temperatures above T_c. This is not included in the thermodynamic theory. In principle at least this can be treated by a microscopic theory; however as the quasiharmonic approximation is not valid for describing thermodynamic properties, no very satisfactory way of doing this has yet been proposed.

The microscopic theory leads to essentially the same equations as the thermodynamic theory, and enables expressions to be obtained for the parameters of that theory. Unfortunately at present the parameters of the thermodynamic theory can only be rewritten in terms of the (largely unknown) parameters specifying the atomic interactions. The microscopic theory is however essential for a discussion of the dynamical properties of ferroelectrics. Since the anharmonic effects are so large for some modes the full frequency-dependent shift and lifetime of the normal modes must be used to describe the optical and scattering properties of ferroelectrics.

The theory described here is directly applicable only to the perovskite ferroelectrics. Other ferroelectrics have more complicated crystal structures, nevertheless their dielectric behaviour is probably associated with the instability of a normal mode. However the temperature dependence of the mode may not arise entirely from anharmonic interactions, but involve the effect of the ordering of atomic positions (for example of hydrogen in KH_2PO_4). A theory of antiferroelectrics can be developed in a similar way. The unstable mode must have $\boldsymbol{q}$ at a Brillouin zone boundary rather than at $\boldsymbol{q}\to 0$. There is evidence that the distorted structures of perovskite anti-ferroelectrics can be obtained in terms of displacements contributed by various zone-boundary modes[2]. $NaNO_2$ has a structure in a small temperature range which is probably associated with an unstable normal mode for which $\boldsymbol{q}$ is neither zero nor at a Brillouin zone boundary[3].

The theory of transitions described here is limited in its validity to those in which the displacements of the atoms are small fractions of the lattice spacing. Reconstructive transitions, such as the CsCl to NaCl transition in alkali halides, involve large displacements of the atoms. The potential energy of the atoms at their displaced positions cannot then be obtained from the derivatives of the potential energy about their original positions. These latter transitions frequently have long time constants. There is then no mode which is unstable, but the crystal reaches its new structure through a large fluctuation about the undistorted structure. Displacive transitions are almost instantaneous and fluctuations play a correspondingly less important role[4].

H. Thermodynamic properties

39. Thermodynamic properties in the harmonic approximation. In the harmonic approximation all of the normal modes of vibration are independent of one another, so that the energy of a particular state is the sum of the energies as-

---

[1] A. P. Levanyuk: Fiz. Tverdogo Tela **5**, 1776 (1963).

[2] See footnote 4, p. 139.

[3] S. Tanisaki: J. Phys. Soc. Japan **18**, 1181 (1963).

[4] V. L. Ginzburg: Fiz. Tverdogo Tela **2**, 2031 (1961).

sociated with the different normal modes. The total energy of a state specified by the occupation numbers $\overline{n}\binom{\boldsymbol{q}}{j}$ is then

$$E_{\overline{\boldsymbol{n}}} = \sum_{\boldsymbol{q}j} \left(\overline{n}\binom{\boldsymbol{q}}{j} + \frac{1}{2} \right) \hbar \omega_j(\boldsymbol{q}). \tag{39.1}$$

The partition function [1, 4] is given by

$$\begin{aligned} Z &= \sum \exp(-\beta E_{\overline{\boldsymbol{n}}}) \\ &= \prod_{\boldsymbol{q}j} \frac{\exp(-\frac{1}{2}\beta \hbar \omega_j(\boldsymbol{q}))}{1 - \exp(-\beta \hbar \omega_j(\boldsymbol{q}))} \end{aligned}$$

where $\beta = 1/k_B T$.

The Helmholtz free energy is then given directly from

as

$$\left.\begin{aligned} F &= k_B T \ln Z \\ F &= k_B T \sum_{\boldsymbol{q}j} \ln\left(2 \sinh\left(\tfrac{1}{2} \beta \hbar \omega_j(\boldsymbol{q})\right)\right). \end{aligned}\right\} \tag{39.2}$$

Since this is a sum over the normal modes it can be rewritten in terms of the frequency distribution (Sect. 10) to give

$$F = \frac{1}{\beta} \int_0^\infty \ln\left(2 \sinh\left(\frac{1}{2} \beta \hbar \omega\right)\right) g(\omega)\, d\omega. \tag{39.3}$$

The entropy is given by

$$S = -\left(\frac{\partial F}{\partial T}\right)_V = k_B \int_0^\infty \left(\frac{1}{2} \beta \hbar \omega \coth \frac{1}{2} \beta \hbar \omega - \ln\left(2 \sinh \frac{1}{2} \beta \hbar \omega\right)\right) g(\omega)\, d\omega, \tag{39.4}$$

while the specific heat is given by

$$C_v = T \left(\frac{\partial S}{\partial T}\right)_V = k_B \int_0^\infty \left(\frac{1}{2} \beta \hbar \omega\right)^2 \operatorname{cosech}^2\left(\frac{1}{2} \beta \hbar \omega\right) g(\omega)\, d\omega. \tag{39.5}$$

The thermodynamic properties of a harmonic crystal can therefore be evaluated when the frequency distribution function $g(\omega)$ is known.

The specific heat is directly measured by experiment. Since its temperature dependence is very similar from one material to another, it is usual to represent the results in terms of the Debye Temperature [9]. For the Debye model

$$\begin{aligned} g(\omega) &= 9r N \omega^2/\omega_D^3 \qquad && \omega < \omega_D \\ &= 0 && \omega > \omega_D. \end{aligned}$$

The specific heat of this model is then

$$C_v = 9rN k_B \int_0^{\omega_D} \left(\frac{1}{2} \beta \hbar \omega^2\right)^2 \left(\frac{1}{\omega_D}\right)^3 \operatorname{cosech}^2\left(\frac{1}{2} \beta \hbar \omega\right) d\omega.$$

This integral depends parametrically on ω_D and if C_v is known for a particular temperature (β) the Debye Temperature $\Theta_D = \hbar \omega_D / k_B$ can be obtained uniquely. Experimental results for specific heat are then presented in terms of an effective Debye Temperature for each temperature. Most materials yield a characteristically similar curve for Θ_D against temperature, as shown in Figs. 52 and 53 where experimental measurements of Θ_D are compared with calculations based on lattice dynamical models for potassium bromide[1], and nickel[2]. In each case the models were obtained from inelastic neutron scattering measurements (Sects. 22, and 25), and the agreement between calculation and experiment is fairly typical. It is perhaps worth commenting that the Debye temperature obtained from

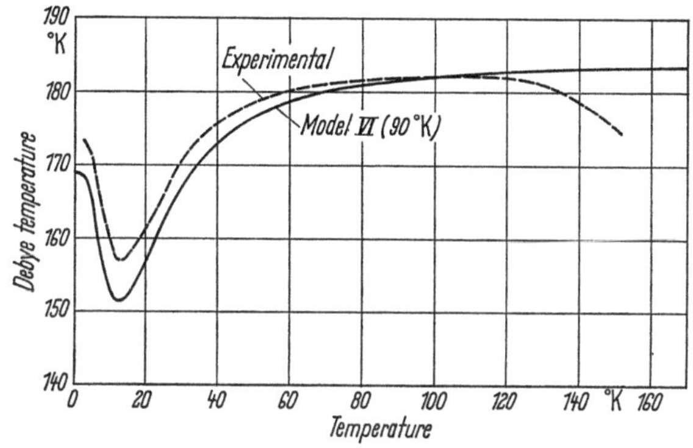

Fig. 52. The Debye temperature $\Theta_D(T)$ for KBr. The dotted line is derived directly from specific heat measurements, the full line from a frequency distribution based on a "shell model" of the crystal. The harmonic approximation is assumed.

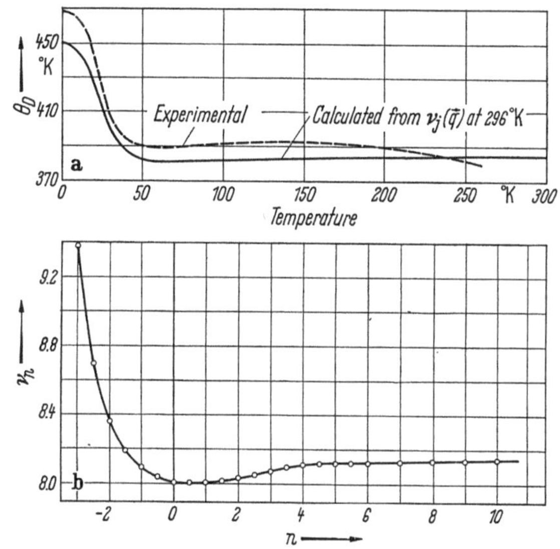

Fig. 53. a Comparison of $\Theta_D(T)$ for nickel derived from specific heat measurements and by calculation from a lattice dynamical model. b A similar comparison of the moments of the frequency distribution.

specific heat measurements is not necessarily the same, and frequently may be quite different, from that obtained by other experimental techniques[3].

The variation of the Debye temperature with temperature is connected with the departure of the frequency distribution from a Debye spectrum. BARRON et al.[4] for alkali halides and SALTER[5] for germanium have deduced various

[1] R. A. COWLEY, W. COCHRAN, N. B. BROCKHOUSE, and A. D. B. WOODS: Phys. Rev. **131**, 1030 (1963).

[2] R. J. BIRGENEAU, J. CORDES, G. DOLLING, and A. D. B. WOODS: Phys. Rev. **136**, A 1359 (1964).

[3] F. H. HERBSTEIN: Advances in Phys. **10**, 313 (1961).

[4] T. H. K. BARRON, and J. A. MORRISON: Proc. Roy. Soc. (London) A **256**, 427 (1960).

[5] L. S. SALTER: Advances in Phys. **14**, 1 (1965).

moments of the frequency distribution from measurements of the specific heat. The technique gives particularly accurate results for the negative moments, which cannot readily be obtained from other experiments. The higher moments are given less accurately however as they are obtained from the measurements at high temperatures. The comparison of these latter moments with those obtained from models based on inelastic neutron scattering results is frequently unsatisfactory, and this presumably arises because the specific heat results involve a range of temperature, while the neutron measurements are usually made at only one temperature. Anharmonicity will clearly affect these two experimental techniques differently.

40. Anharmonic effects and thermodynamic properties. The influence of the anharmonic interactions on the thermodynamic properties have been obtained both by the use of ordinary perturbation theory [6] and with the aid of many-body techniques[1], described in Sect. 30 and in the Appendix. By use of the latter techniques the Helmholtz free energy is found to be given by

$$F = F_0 - \frac{1}{\beta} \Sigma_c \tag{40.1}$$

where F_0 is that part of the free energy which arises from the modes treated as independent of one another, and Σ_c is the sum of all the connected diagrams which have no external lines. These diagrams may be evaluated by using the rules given in the appendix.

The thermal expansion of the crystal gives rise to a shift in the frequency of each mode, $\Delta^T(\boldsymbol{q}jj)$, given by the first term of Eq. (30.12), and since each mode is independent so far as this term is concerned it may be incorporated into F_0. If the shift is small the effect of the thermal expansion can be obtained by expanding (39.2) to give

$$\Delta F^T = \hbar \sum_{\boldsymbol{q}j} \left(\bar{n}\begin{pmatrix}\boldsymbol{q}\\ j\end{pmatrix} + \frac{1}{2} \right) \Delta^T(\boldsymbol{q}jj) . \tag{40.2}$$

The Gibbs free energy has a different dependence on the volume, and since the shifts $\Delta^T(\boldsymbol{q}jj)$ arise from the thermal strains, on these shifts. The result is

$$\Delta G^T = \frac{\hbar}{2} \sum_{\boldsymbol{q}j} \left(\bar{n}\begin{pmatrix}\boldsymbol{q}\\ j\end{pmatrix} + \frac{1}{2} \right) \Delta^T(\boldsymbol{q}jj) . \tag{40.3}$$

Although this simple treatment of the free energy in terms of the thermal shifts can be used to discuss the effects of the thermal expansion, the whole anharmonic contribution to the free energy cannot be obtained by expanding Eq. (39.2) and using the whole shifts of Eq. (30.12). The remainder of the contribution must be obtained from the diagrams, of which the two of lowest order are shown in Fig. 54. Neither of these diagrams depend on the volume of the crystal and so contribute equally to the Gibbs and Helmholtz free energies. The results are[2] [6]:

$$\left.\begin{aligned} \Delta F = {} & 3 \sum_{\substack{\boldsymbol{q}_1 \boldsymbol{q}_2 \\ j_1 j_2}} V\begin{pmatrix}\boldsymbol{q}_1 & -\boldsymbol{q}_1 & \boldsymbol{q}_2 & -\boldsymbol{q}_2\\ j_1 & j_1 & j_2 & j_2\end{pmatrix} (2\bar{n}_1+1)(2\bar{n}_2+1) - \\ & - \frac{6}{\hbar} \sum_{\substack{\boldsymbol{q}_1 \boldsymbol{q}_2 \boldsymbol{q}_3 \\ j_1 j_2 j_3}} \left| V\begin{pmatrix}\boldsymbol{q}_1 & \boldsymbol{q}_2 & \boldsymbol{q}_3\\ j_1 & j_2 & j_3\end{pmatrix}\right|^2 \left(\frac{(\bar{n}_1+1)(\bar{n}_2+\bar{n}_3+1)+\bar{n}_2\bar{n}_3}{\omega_1+\omega_2+\omega_3} \right. + \\ & \left. + 3\,\frac{\bar{n}_1\bar{n}_2+\bar{n}_1\bar{n}_3-\bar{n}_2\bar{n}_3+\bar{n}_1}{\omega_2+\omega_3-\omega_1} \right) . \end{aligned}\right\} \tag{40.4}$$

---

[1] L. van Hove: Interactions of Elastic Waves in Solids. Technical Report 11 of Massachusetts Institute of Technology Solid-State and Molecular theory Group.

[2] R. A. Cowley: Advances in Phys. 12, 421 (1963).

The contributions to the entropy and specific heat of a material can then be calculated by differentiating these expressions with respect to the temperature.

Eq. (40.4) can be simplified considerably if we assume a simplified form (33.3) for the anharmonic coefficients. The free energy is then

$$F_0 + N r^2 k_B^2 T^2 (12 K_2 - 24 K_1 r)$$

where K_2 and K_1 are the proportionality factors for fourth and third order coefficients respectively. Calculations based on this expression have been shown to give approximate agreement with the specific heat of both alkali halides [5] and other materials[1]. However more realistic calculations have been made only for a face centred cubic crystal with forces between nearest neighbour atoms[2], and for alkali halides[3]. As the calculations involve a double sum over the Brillouin zone they are extremely lengthy and the results are in qualitative agreement with experiment only.

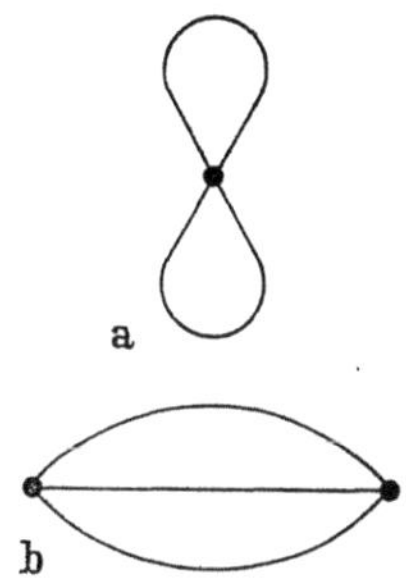

Fig. 54 a and b. Diagrams of lowest-order for anharmonic contributions to the free energy.

41. The quasi-harmonic approximation. In Sects. 31, 32 it was shown that the one-phonon properties of crystals are quite accurately described by the quasi-harmonic frequencies defined by Eq. (32.9). It is of interest to see to what extent these quasi-harmonic frequencies can also describe the thermodynamic properties. As already remarked in the preceding section, the free energy is not given correctly by use of the quasi-harmonic frequencies. However BARRON[4] has shown that to lowest order of perturbation theory the entropy is given correctly, provided that the anharmonic shift in frequency is much smaller than the harmonic frequency. The anharmonic contribution to the entropy is then given by:

$$\varDelta S = \hbar \sum_{\boldsymbol{q} j} \frac{\partial \bar{n}\begin{pmatrix}\boldsymbol{q}\\ j\end{pmatrix}}{\partial T} \varDelta\left(\boldsymbol{q} j j, \omega_T\begin{pmatrix}\boldsymbol{q}\\ j\end{pmatrix}\right). \tag{41.1}$$

The anharmonic part of the specific heat, C_p is then

$$\varDelta C_P = \hbar T \sum_{\boldsymbol{q} j}\left[\frac{\partial^2 \bar{n}\begin{pmatrix}\boldsymbol{q}\\ j\end{pmatrix}}{\partial T^2} \varDelta\left(\boldsymbol{q} j j, \omega_T\begin{pmatrix}\boldsymbol{q}\\ j\end{pmatrix}\right) + \frac{\partial \bar{n}\begin{pmatrix}\boldsymbol{q}\\ j\end{pmatrix}}{\partial T} \frac{\partial \varDelta\left(\boldsymbol{q} j j, \omega_T\begin{pmatrix}\boldsymbol{q}\\ j\end{pmatrix}\right)}{\partial T}\right]. \tag{41.2}$$

This expression shows that the change in specific heat due to anharmonicity is influenced both by the shift in frequency and its rate of change with temperature. At low temperatures $\frac{\partial \varDelta}{\partial T}$ is small and the whole effect arises from the first term, whereas at high temperatures $\frac{\partial^2 \bar{n}}{\partial T^2}$ becomes small and the effect of the rate of change of shift $\varDelta$ with temperature is most important. Eq. (41.2) also provides a direct confirmation of the result of BARRON and KLEIN[5] and of FELDMAN[6], that the specific heat at low temperature is given precisely by the quasi-harmonic approximation.

[1] J. M. KELLER, and D. C. WALLACE: Phys. Rev. **126**, 1275 (1962).
[2] A. A. MARADUDIN, P. A. FLINN, and R. A. COLDWELL-HORSFALL: Ann. Phys. **15**, 337, 360 (1961).
[3] E. R. COWLEY, and R. A. COWLEY: Proc. Roy. Soc. (London) A **292**, 209 (1966).
[4] T. H. K. BARRON, p. 247 of Ref. [11].
[5] T. H. K. BARRON, and M. L. KLEIN: Phys. Rev. **127**, 1997 (1962).
[6] J. L. FELDMAN: Proc. Phys. Soc. (London) **84**, 361 (1964).

Although these results connect the quasi-harmonic frequencies and the thermodynamic properties, their proof necessarily involves the use of perturbation theory. The thermal shifts in frequency were assumed to be small and calculated only in the lowest order of perturbation theory. Unfortunately no such result has been obtained when the shifts are large compared with the harmonic frequencies, as is the case for some normal modes of ferroelectric crystals (Sect. 37). In Sect. 37 it was suggested that $\omega_j^2(\boldsymbol{q})$ may even be negative, and it is then difficult to proceed with the calculation of the thermodynamic properties in the harmonic approximation.

Progress in the direction of expressing the thermodynamic properties in terms of the exact Green's function instead of the harmonic frequencies has been made by Sham[1]. He obtains two expressions for the free energy in terms of the exact Green's function and its proper self-energy. Unfortunately his results involve an integration over all frequencies of the exact Green's function and self-energy, and this detailed information cannot be obtained directly from experiment as yet. It does not seem possible to express his results directly in terms of the quasi-harmonic frequencies in general. However it is comparatively easy to show that if we restrict attention to those free energy diagrams which involve only one vertex, but with any number of lines, the entropy is given exactly by the quasi-harmonic expression. It is tempting to speculate that this result is always exact, but at present this has not been proved and the treatment of the thermodynamic properties of ferroelectrics remains unsatisfactory.

Appendix: Many-body techniques for anharmonic crystals.

In the theory of anharmonic crystals we are interested in predicting the dielectric and elastic constants, the scattering of particles by the crystal, and the thermodynamic properties of the crystal. These cannot be determined from the exact many-body wave functions except for very simple models of crystals. The approach usually adopted is to work with quantities other than the wave functions, which though less detailed than the wave functions are more directly related to experiment. Such quantities are the Green's functions of quantum field theory. Many of the physical properties of crystals are directly related to these Green's functions. The development described below is the use of the thermodynamic Green's function as described by Matsubara[2]. An excellent account of these techniques for both fermions and bosons, with several different applications, is given by Abrikosov et al. [*5*].

The thermodynamic Green's function of two boson operators O_1 and O_2 is defined by

$$G(O_1\,O_2\,\tau) = \langle P\{O_1(\tau)\,O_2(0)\}\rangle \tag{A.1}$$

where $\langle\ldots\rangle$ represents the thermal expectation operator and the operators are written in the Heisenberg representation with τ an imaginary time:

$$O(\tau) = \exp(H\tau/\hbar)\,O\,\exp(-H\tau/\hbar)\,.$$

The operator P is a time ordering operator such that

$$\begin{aligned} G(O_1\,O_2,\,\tau) &= \langle O_1(\tau)\,O_2(0)\rangle, \qquad \tau>0\\ &= \langle O_2(0)\,O_1(\tau)\rangle, \qquad \tau<0. \end{aligned} \tag{A.2}$$

[1] L. J. Sham: Phys. Rev. **139**, A 1189 (1965).
[2] T. Matsubara: Progr. Theoret. Phys. **14**, 351 (1955).

Suppose we write these two expressions out in more detail,

$$\left.\begin{aligned} G(O_1 O_2, \tau) &= \frac{1}{Z}\operatorname{Tr}[\exp(-\beta H + \tau H/\hbar) O_1 \exp(-\tau H/\hbar) O_2], \qquad \tau > 0 \\ &= \frac{1}{Z}\operatorname{Tr}[\exp(-\beta H) O_2 \exp(\tau H/\hbar) O_1 \exp(-\tau H/\hbar)], \quad \tau < 0, \end{aligned}\right\} \tag{A.3}$$

where Z is the partition function $\operatorname{Tr}[\exp(-\beta H)]$, and $\beta = 1/k_B T$.

Since the trace of an expression is invariant under cyclic permutations of the operators, it follows from Eq. (A.3) that:

$$G(O_1 O_2, \tau + \beta\hbar) = G(O_1 O_2, \tau) \qquad O > \tau > -\beta\hbar. \tag{A.4}$$

Within the range $-\beta\hbar > \tau > \beta\hbar$ these GREEN's functions are periodic in the complex time direction, and may be expanded as a Fourier series;

$$G(O_1 O_2, \tau) = \sum_{n=-\infty}^{\infty} G(O_1 O_2, i\omega_n) \exp(i\omega_n \tau), \tag{A.5}$$

where

$$\omega_n = 2\pi n/\beta\hbar \tag{A.6}$$

and

$$G(O_1 O_2, i\omega_n) = \frac{1}{2\beta\hbar} \int_{-\beta\hbar}^{\beta\hbar} G(O_1 O_2, \tau) \exp(-i\omega_n \tau)\, d\tau. \tag{A.7}$$

Now let us introduce complete sets of exact eigenstates of the Hamiltonian into Eq. (A.3)

$$\left.\begin{aligned} G(O_1 O_2, \tau) &= \frac{1}{Z}\sum_{n m} (n|\exp(-\beta H + \tau H/\hbar) O_1 \exp(-\tau H/\hbar)|m)(m|O_2|n), \qquad \tau > 0 \\ &= \frac{1}{Z}\sum_{n m} (n|\exp(-\beta H) O_2|m)(m|\exp(\tau H/\hbar) O_1 \exp(-\tau H/\hbar)|n), \quad \tau < 0. \end{aligned}\right\} \tag{A.8}$$

The energy difference between states m and n is $\hbar\omega_{mn}$, so that

$$(n|\exp(-\beta H + \tau H/\hbar) O_1 \exp(-\tau H/\hbar)|m) = \exp(-\omega_{mn}\tau - \beta E_n)(n|O_1|m).$$

The GREEN's functions (A.8) then become

$$\left.\begin{aligned} G(O_1 O_2, \tau) &= \frac{1}{Z}\sum_{n m} \exp(-\beta E_n - \omega_{mn}\tau)(n|O_1|m)(m|O_2|n), \quad \tau > 0, \\ &= \frac{1}{Z}\sum_{n m} \exp(-\beta E_n - \omega_{nm}\tau)(n|O_2|m)(m|O_1|n), \quad \tau < 0. \end{aligned}\right\} \tag{A.9}$$

The spectral representation is defined by

$$\varrho(O_1 O_2, \omega) = \frac{1}{Z}\sum_{n m} \exp(-\beta E_n)(n|O_1|m)(m|O_2|n)\,\delta(\omega - \omega_{mn}). \tag{A.10}$$

The GREEN's function (A.9) is then

$$G(O_1 O_2, \tau) = \int_{-\infty}^{\infty} \varrho(O_1 O_2, \omega) \exp(-\omega\tau)\, d\omega, \quad \tau > 0. \tag{A.11}$$

For $\tau<0$ a similar expression is obtained by interchanging the labels m and n in Eq. (A.9) to give

$$G(O_1 O_2, \tau) = \int_{-\infty}^{\infty} \varrho(O_1 O_2, \omega) \exp(-\omega\,\tau - \beta\hbar\omega)\, d\omega, \quad \tau<0. \tag{A.12}$$

When these expressions are substituted into Eq. (A.7) the Fourier coefficients of the series (A.5) are

$$G(O_1 O_2, i\omega_n) = \frac{1}{\beta\hbar} \int_{-\infty}^{\infty} \left(1-\exp(-\beta\hbar\omega)\right) \varrho(O_1 O_2, \omega) \frac{d\omega}{\omega + i\omega_n}. \tag{A.13}$$

Now let us relate these GREEN's functions to the physical properties of a crystal. Suppose an external perturbation, H_E, is applied to the material adiabatically ($H_E=0$, $t=-\infty$). This perturbation can be written as the product of an operator O_1 and a force $F\exp(-i\Omega t+\varepsilon t)$, ($\varepsilon>0$, $\varepsilon\to 0$). In the case of the dielectric properties the force is the applied electric field and the operator is minus the dipole moment of the crystal. The response of the crystal is determined by the expectation value of operator O_2, $\langle O_2\rangle$ where $\langle\ldots\rangle$ represents the thermal average with respect to the Hamiltonian $H+H_E$. The density matrix of the system obeys the equation of motion

$$i\hbar \frac{\partial\varrho}{\partial t} = [H+H_E, \varrho],$$

where [ ] represents the commutator.

The equation of motion can be solved if we write $\varrho=\varrho_0+\Delta\varrho$ when

$$i\hbar \frac{\partial\Delta\varrho}{\partial t} = [H, \Delta\varrho] + [H_E, \varrho_0], \tag{A.14}$$

gives the linear response of the system. The boundary condition, $\Delta\varrho=0$ as $t\to-\infty$ enables Eq. (A.14) to be integrated. If $\Delta\varrho$ is written in the interaction representation

$$\Delta\tilde{\varrho}(t) = \exp(iHt/\hbar)\,\Delta\varrho \exp(-iHt/\hbar),$$

then

$$\Delta\tilde{\varrho}(t) = \frac{1}{i\hbar} \int_{-\infty}^{t} \exp(iHt'/\hbar)\,[H_E, \varrho_0] \exp(-iHt'/\hbar)\, dt'.$$

The linear response to the applied force is then

$$\langle O_2\rangle = \frac{1}{i\hbar} \int_{-\infty}^{t} \langle[\tilde{O}_2(t), \tilde{H}_E(t')]\rangle\, dt'.$$

The susceptibility χ is then obtained as

$$\chi(\Omega) = \frac{i}{\hbar} \int_{-\infty}^{t} \langle[\tilde{O}_2(t), \tilde{O}_1(t')]\rangle \exp\left((-i\Omega+\varepsilon)(t'-t)\right) dt'.$$

Since t is arbitrary we may choose $t=0$ and relabel t' as t to give

$$\chi(\Omega) = \frac{i}{\hbar} \int_{-\infty}^{0} \langle[\tilde{O}_2(0), \tilde{O}_1(t)]\rangle \exp\left((-i\Omega+\varepsilon)t\right) dt. \tag{A.15}$$

This expression is the Fourier transform of the retarded GREEN's function

$$\left.\begin{aligned} G_R(O_1 O_2, t) &= 0, & t>0 \\ &= \langle \tilde{O}_1(t)\tilde{O}_2(0) - \tilde{O}_2(0)\tilde{O}_1(t)\rangle, & t<0. \end{aligned}\right\} \tag{A.16}$$

These functions may be rewritten in terms of the spectal representation (A.11) as

$$\begin{aligned} G_R(O_1 O_2, t) &= 0, & t>0 \\ &= \int_{-\infty}^{\infty} \varrho(O_1 O_2, \omega)\left(1-\exp(-\beta\hbar\omega)\right)\exp(-i\omega t)\, d\omega, & t<0. \end{aligned}$$

The Fourier transform of the retarded GREEN's function is defined so as to make the limit at $t=-\infty$ tend to zero, namely by

$$G_R(O_1 O_2, \Omega+i\varepsilon) = \int_{-\infty}^{\infty} G_R(O_1 O_2, t)\exp\left((-i\Omega+\varepsilon)t\right) dt.$$

Hence

$$G_R(O_1 O_2, \Omega+i\varepsilon) = i\int_{-\infty}^{\infty} \varrho(O_1 O_2, \omega)\left(1-\exp(-\beta\hbar\omega)\right)\frac{d\omega}{\omega+\Omega+i\varepsilon}. \tag{A.17}$$

Hence the susceptibility (A.15) is given directly by the retarded GREEN's function as

$$\chi(\Omega) = \frac{i}{\hbar} G_R(O_1 O_2, \Omega+i\varepsilon). \tag{A.18}$$

For some purposes it is convenient to define an advanced GREEN's function

$$\left.\begin{aligned} G_A(O_1 O_2, t) &= \langle \tilde{O}_1(t)\tilde{O}_2(0) - \tilde{O}_2(0)\tilde{O}_1(t)\rangle, & t>0, \\ &= 0, & t<0. \end{aligned}\right\} \tag{A.19}$$

The Fourier transform is defined to make the limit for $t\to\infty$ well behaved; namely

$$G_A(O_1 O_2, \Omega-i\varepsilon) = \int_{-\infty}^{\infty} G_A(O_1 O_2, t)\exp\left((-i\Omega-\varepsilon)t\right) dt.$$

Hence

$$G_A(O_1 O_2, \Omega-i\varepsilon) = -i\int_{-\infty}^{\infty} \varrho(O_1 O_2, \omega)\left(1-\exp(-\beta\hbar\omega)\right)\frac{d\omega}{\omega+\Omega-i\varepsilon}. \tag{A.20}$$

It is of interest to compare these relations with the corresponding equation for the thermodynamic GREEN's functions (A.13). Clearly the Fourier transform of the retarded GREEN's function (A.17) is analytic over the upper half of the frequency Ω plane. Hence for all $\omega_n>0$ we obtain

$$i\, G(O_1 O_2, i\omega_n) = \beta\hbar\, G_R(O_1 O_2, \Omega), \tag{A.21}$$

where $\Omega = i\omega_n$.

Hence from a knowledge of $G_R(O_1 O_2, \Omega)$, we can construct the thermodynamic GREEN's function for all frequencies $\omega_n>0$. However the inverse problem is both more difficult and more useful, since we shall later that it is comparatively easy to construct the $G(i\omega_n)$ from perturbation theory. The problem of analytically continuing from the infinite discrete set of points (A6) to the complex plane, in general has no unique solution. However under certain conditions, which nearly always occur in practice, this continuation can be performed[1]. The susceptibility is then given by

$$\chi(\Omega) = -\beta G(O_1 O_2, \Omega + i\varepsilon). \tag{A.22}$$

The advanced GREEN's functions are analytic in the lower half of the complex Ω plane, and can similarly be obtained by analytic continuation from the thermodynamic GREEN's functions, in that part of the complex plane.

In the first Born approximation the inelastic scattering cross-section of a crystal can be written as proportional to

$$\langle O_1 O_2 \rangle \delta(E_I - E_F - \hbar\Omega), \tag{A.23}$$

where O_1 and O_2 are two operators which depend on the particles scattered and the specimen, while the delta function expresses energy conservation. E_I is the energy of the initial and E_F of the final states of the crystal and $\hbar\Omega$ the energy transfer to the scattered particle. Let us introduce two complete sets of states into the expression (A.23), to obtain

$$\frac{1}{Z}\sum_{nm} \exp(-\beta E_n)\,(n|O_1|m)\,(m|O_2|n)\,\delta(\hbar\,\omega_{mn} - \hbar\Omega).$$

Comparing this equation with the spectral representation (A.10) it is seen that the cross-section is proportional to the spectral representation, $\varrho(O_1 O_2, \Omega)$. The spectral representation can also be obtained by analytically continuing the coefficients $G(O_1 O_2, i\omega_n)$ over the whole frequency plane.

$$G(O_1 O_2, \Omega + i\varepsilon) - G(O_1 O_2, \Omega - i\varepsilon) = \frac{1}{\beta\hbar}\int_{-\infty}^{\infty} (1 - \exp(-\beta\hbar\omega)) \times$$

$$\times\,\varrho(O_1 O_2, \omega)\left(\frac{1}{\omega + \Omega + i\varepsilon} - \frac{1}{\omega + \Omega - i\varepsilon}\right) d\omega.$$

Now taking the limit as $\varepsilon \to 0^+$ gives

$$\left.\begin{aligned}\varrho(O_1 O_2, \Omega) = \frac{i\beta\hbar}{2\pi}\,[1 - \exp(-\beta\hbar\Omega)]^{-1}\,(G(O_1 O_2, \Omega + i\varepsilon) - \\ - G(O_1 O_2, \Omega - i\varepsilon)).\end{aligned}\right\} \tag{A.24}$$

The spectral representation is therefore related to the discontinuity in the GREEN's function across the real frequency axis. This equation then enables us to obtain the scattering properties of crystals from the same functions as the susceptibility.

The main advantage in the use of thermodynamic GREEN's functions is that they are readily calculated by perturbation theory. Two techniques have been developed. The equation of motion method gives a hierarchy of equations connecting successively higher-order GREEN's functions. This series is terminated by making some form of approximation as to the nature of some of the higher-order GREEN's functions[2]. The other approach is to make use of diagrammatic techniques, and this is the one which appears more useful to the present authors. The technique is to draw diagrams representing the different terms in a perturbation series, and the expressions are evaluated directly from the diagrams by a simple set of rules. The derivation of the rules is given in detail by other authors [5] and only the results will be quoted here.

---

[1] G. BAYM, and N. D. MERMIN: J. Math. and Phys. 2, 232 (1961).
[2] D. N. ZUBAREV: Uspekhi Fiz. Nauk 71, 71 (1960).

A contribution to $G(O_1 O_2, i\omega_n)$ with m anharmonic interactions is given by a diagram with m vertices representing the appropriate interactions. The operators O_1 and O_2 are expanded in terms of phonon coordinates, the phonons associated with O_1 entering the diagram on the left, and those associated with O_2 leaving on the right. In between the lines are joined up with the anharmonic interactions.

The contributions from the lines are given by the harmonic one phonon GREEN's function

$$g(\boldsymbol{q}j\,\boldsymbol{q}'j', \tau) = \left\langle P\left\{A\begin{pmatrix}\boldsymbol{q}\\ j\end{pmatrix}\middle|\,\tau\right) A^+\begin{pmatrix}\boldsymbol{q}'\\ j'\end{pmatrix}\middle|\,0\Big)\right\}\right\rangle.$$

In a harmonic crystal the expectation value is zero unless $\boldsymbol{q}=\boldsymbol{q}'$ and $j=j'$, when contracting the notation yields,

$$g(\boldsymbol{q}j, \tau) = \bar{n}\begin{pmatrix}\boldsymbol{q}\\ j\end{pmatrix}\exp\left(|\tau|\,\omega_j(\boldsymbol{q})\right) + \left(\bar{n}\begin{pmatrix}\boldsymbol{q}\\ j\end{pmatrix}+1\right)\exp\left(-|\tau|\,\omega_j(\boldsymbol{q})\right).$$

The Fourier transform (A.7) then gives

$$g(\boldsymbol{q}j, i\omega_n) = \frac{1}{\beta\hbar}\,\frac{2\omega_j(\boldsymbol{q})}{\omega_j^2(\boldsymbol{q}) + \omega_n^2}\,. \tag{A.25}$$

The rules for evaluating the perturbation-theory diagrams with m vertices for $G(O_1 O_2, i\omega_n)$ are[1,2]:

1. Draw all topologically distinct and different connected diagrams in which there are m vertices, joined to one another by the appropriate phonon lines. Phonon lines corresponding to O_1 entering on the left and those corresponding to O_2 leaving the right.

2. Associate momenta, $\boldsymbol{q}$, and frequencies, $i\omega_n$, with the lines of the diagram in such a way that the external lines have the appropriate total momentum and frequency, while satisfying conservation of crystal momentum and $\Sigma\omega_n=0$ at each vertex.

3. Sum over the independent momenta, $\boldsymbol{q}$, and branch labels, j, and frequencies, $i\omega_n$.

4. With each phonon line associate $g(\boldsymbol{q}j, i\omega_n)$, the harmonic GREEN's function.

5. Multiply by the appropriate expansion coefficients for the operators O_1 and O_2, and insert the coefficients for each anharmonic interaction or vertex.

6. Multiply by $(-\beta)^m$.

7. Multiply by the number of ways of pairing up the phonons in the diagram.

It is worth commenting on the restriction to connected diagrams (rule 1). This arises because the disconnected diagrams cancel with the expansion of the $1/Z$ factor in the GREEN's functions. It also ensure that the properties have the correct dependence on the volume of the crystal — a feature which it is difficult to demonstrate with other approaches. If the diagrams have no external lines, as occur in the thermodynamic properties, rule 6 is modified to multiply by $(-\beta)^m/m!$, and the diagrams must be drawn for all possible time orderings of the vertices as well as for every different topology.

These results can be applied to the one-phonon GREEN's function. This is defined by

$$G(\boldsymbol{q}j\,\boldsymbol{q}'j', \tau) = \left\langle P\left\{A\begin{pmatrix}\boldsymbol{q}\\ j\end{pmatrix}\middle|\,\tau\right) A^+\begin{pmatrix}\boldsymbol{q}'\\ j'\end{pmatrix}\middle|\,0\Big)\right\}\right\rangle,$$

[1] A. A. MARADUDIN, and A. E. FEIN: Phys. Rev. **128**, 2589 (1962).
[2] R. A. COWLEY: Advances in Phys. **12**, 421 (1963).

but since there is crystal-momentum conservation at every vertex $\boldsymbol{q}=\boldsymbol{q}'$, the function can be reduced to

$$G(\boldsymbol{q}jj',\tau)=\left\langle P\left\{A\left(\begin{smallmatrix}\boldsymbol{q}\\ j\end{smallmatrix}\middle|t\right)A^+\left(\begin{smallmatrix}\boldsymbol{q}\\ j'\end{smallmatrix}\middle|0\right)\right\}\right\rangle. \tag{A.26}$$

Some typical diagrams are shown in Fig. 55, and these may be divided into two classes. Proper diagrams (a), (b) cannot be broken into two parts by cutting a single phonon line, except an external line, while improper diagrams (c) and (d) can be broken in this way.

Fig. 56 shows the contribution to a one-phonon diagram from all proper diagrams. The bubble is taken to represent the sum of all proper diagrams con-

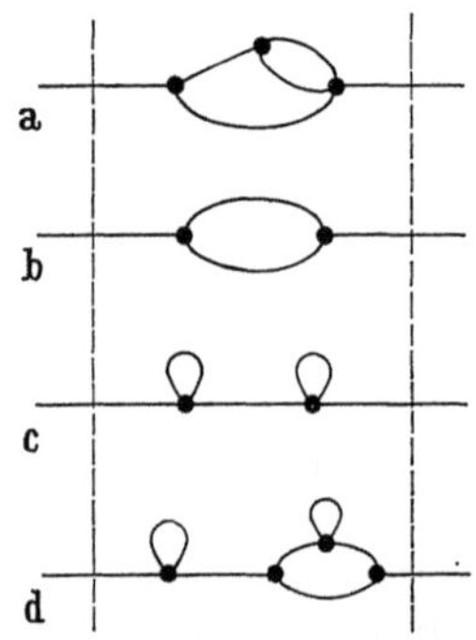

Fig. 55a—d. a and b are proper one-phonon diagrams, c and d are improper diagrams.

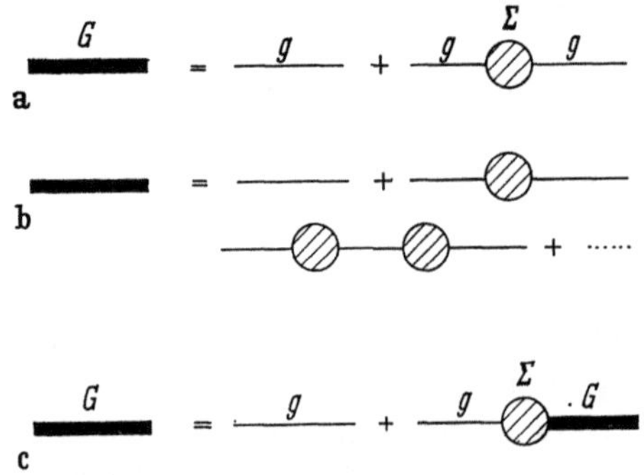

Fig. 56a—c. This illustrates the Dyson equation for the one-phonon diagrams. a Shows the sum of all proper diagrams while b and c show the sum of all proper and improper diagrams.

taining at least one vertex. If $\hbar\Sigma(\boldsymbol{q}jj'|i\omega_n)$ is the contribution from the bubble then the Green's function is given by

$$G(\boldsymbol{q}jj',i\omega_n)=\delta_{jj'}\,g(\boldsymbol{q}j,i\omega_n)-\beta\hbar g(\boldsymbol{q}j,i\omega_n)\sum(\boldsymbol{q}jj'|i\omega_n)\,g(\boldsymbol{q}j',i\omega_n).$$

However because of crystal momentum conservation at each vertex the wave vector of the single phonon in an improper diagram is the same as that of the external lines. The sum of all improper diagrams is the, (b) and (c) of Fig. 56,

$$G(\boldsymbol{q}jj',i\omega_n)=\delta_{jj'}g(\boldsymbol{q}j,i\omega_n)-\sum_{j''}\beta\hbar g(\boldsymbol{q}j,i\omega_n)\sum(\boldsymbol{q}jj''|i\omega_n)\,G(\boldsymbol{q}j''j',i\omega_n)$$

which can be written with the help of (A.25) as

$$\sum_{j''}\left[(\omega_j^2(\boldsymbol{q})+\omega_n^2)\,\delta_{jj''}+2\omega_j(\boldsymbol{q})\sum(\boldsymbol{q}jj''|i\omega_n)\right]G(\boldsymbol{q}j''j',i\omega_n)=\delta_{jj'}\,2\omega_j(\boldsymbol{q})/\beta\hbar. \tag{A.27}$$

This is the basic equation for describing one-phonon effects in an anharmonic crystal.

It is frequently useful to make use of the symmetry properties of the Green's functions. These can be deduced with the aid of the spectral representation (A.10). The definition immediately leads to the relations

$$\varrho^+(O_1\,O_2,\omega)=\varrho(O_2^+\,O_1^+,\omega)$$

and

$$\varrho(O_1\,O_2,-\omega)=\exp(-\beta\hbar\omega)\,\varrho(O_1^+\,O_2^+,\omega).$$

The latter of these equations shows that if

$$F(\omega)=\left(1-\exp(-\beta\hbar\omega)\,\varrho(O_1\,O_2,\omega)\right)$$

then

$$F(-\omega)=\left(\exp(-\beta\hbar\omega)-1\right)\varrho(O_2^+O_1^+,\omega).$$

Eq. (A.13) then leads to the conditions on the GREEN'S functions

$$G(O_1O_2,-i\omega_n)=G(O_2^+O_1^+,i\omega_n),$$

and

$$G^+(O_1O_2,i\omega_n)=G\left(O_2^+O_1^+,(i\omega_n)^*\right).$$

Applying these relations to the one phonon GREEN'S function (A.26), and using the fact that $G(\boldsymbol{q})=G(-\boldsymbol{q})$ gives

$$G(\boldsymbol{q}jj',-i\omega_n)=G(\boldsymbol{q}jj',i\omega_n),$$

and

$$G^+(\boldsymbol{q}jj',i\omega_n)=G\left(\boldsymbol{q}j'j,(i\omega_n)^*\right).$$

The physical properties are determined by the analytic continuation of these GREEN'S functions to the real frequency axis. Suppose $i\omega_n\to\Omega+i\varepsilon$, then Eq. (A.13) becomes, for the case of the one phonon GREEN'S function,

$$\left.\begin{aligned}\lim_{\varepsilon\to0^+}G(\boldsymbol{q}jj',\Omega+i\varepsilon)&=\frac{1}{\beta\hbar}\int_{-\infty}^{\infty}\left(1-\exp(-\beta\hbar\omega)\right)\varrho(\boldsymbol{q}jj',\omega)\frac{d\omega}{(\omega+\Omega)_p}-\\&-\frac{i\pi}{\beta\hbar}\left(1-\exp(-\beta\hbar\Omega)\right)\varrho(\boldsymbol{q}jj',\Omega).\end{aligned}\right\}\tag{A.28}$$

With the aid of Eqs. (A.28) the GREEN'S function matrix is seen to split into Hermitian and anti-Hermitian parts. The Hermitian part is

$$G_H(\boldsymbol{q}jj',\Omega)=\frac{1}{\beta\hbar}\int_{-\infty}^{\infty}\left(1-\exp(-\beta\hbar\Omega)\,\varrho(\boldsymbol{q}jj',\Omega)\omega\right)\frac{d\omega}{(\omega+\Omega)_p}\tag{A.29}$$

and the anti-Hermitian part is

$$G_{AH}(\boldsymbol{q}jj',\Omega)=-\frac{i\pi}{\beta\hbar}\left(1-\exp(-\beta\Omega)\right)\varrho(\boldsymbol{q}jj',\Omega).\tag{A.30}$$

It follows from these two equations that the Hermitian and anti-Hermitian parts obey a dispersion relation

$$G_H(\boldsymbol{q}jj',\Omega)=\frac{i}{\pi}\int_{-\infty}^{\infty}\frac{G_{AH}(\omega)}{(\omega+\Omega)}\,d\omega.\tag{A.31}$$

Eqs. (A.28) also show that

$$G_H(\boldsymbol{q}jj',-\Omega)=G_H(\boldsymbol{q}jj',\Omega)$$

and

$$G_{AH}(\boldsymbol{q}jj',-\Omega)=-G_{AH}(\boldsymbol{q}jj',\Omega).$$

A similar discussion can be given for the self energy $\sum(\boldsymbol{q}jj'|i\omega_n)$. A spectral representation can be introduced for those diagrams for which there is an intermediate state. The diagrams which involve an instantaneous interaction with the background give rise to a constant contribution, C. Then by analogy with Eq. (A.13) the self energy can be written as

$$\sum(\boldsymbol{q}jj'|i\omega_n)=C+\frac{1}{\beta\hbar}\int_{-\infty}^{\infty}\left(1-\exp(-\beta\hbar\omega)\right)\lambda(\boldsymbol{q}jj',\omega)\frac{d\omega}{\omega+i\omega_n}$$

where λ is the spectral representation of $\sum$.

Properties can then be deduced for the self-energy which are essentially similar to those already described for the Green's function. In particular when $i\omega_n \to \Omega + i\varepsilon$, the matrix can be separated into a Hermitian part, $\Delta(\boldsymbol{q}jj'|\Omega)$ and an anti-Hermitian part, $i\Gamma(\boldsymbol{q}jj'|\Omega)$. Eq. (A.27) then becomes

$$\left.\begin{aligned}\sum_{j''}\left[(\omega_j^2(\boldsymbol{q})^2-\Omega^2)\,\delta_{jj''}+2\omega_j(\boldsymbol{q})\,(\Delta(\boldsymbol{q}jj''|\Omega)-i\Gamma(\boldsymbol{q}jj''|\Omega))\right]\\ G(\boldsymbol{q}j''j',\Omega+i\varepsilon)=\delta_{jj'}2\omega_j(\boldsymbol{q})/\beta\hbar.\end{aligned}\right\}\qquad(\text{A.32})$$

The Hermitian and anti-Hermitian parts give rise to temperature-dependent frequencies and lifetime effects respectively. They are also connected by the dispersion relation

$$\Delta(\boldsymbol{q}jj'|\Omega)=C-\frac{1}{\pi}\int_{-\infty}^{\infty}\frac{\Gamma(\boldsymbol{q}jj'/\omega)\,d\omega}{(\omega+\Omega)_p}$$

and satisfy:

$$\Delta(\boldsymbol{q}jj'|\Omega)=\Delta(\boldsymbol{q}jj'|-\Omega),$$

and

$$\Gamma(\boldsymbol{q}jj'|\Omega)=-\Gamma(\boldsymbol{q}jj'|-\Omega).\qquad(\text{A.33})$$

These relations can be of use in calculating the self-energy[1].

References.

[1] Born, M., and K. Huang: Dynamical Theory of Crystal Lattices. Oxford: Oxford University Press 1954.
[2] Bilz, H., u. R. Wehner: Handbuch der Physik, Bd. XXV/2b. Berlin-Göttingen-Heidelberg: Springer.
[3] Birman, J.: Handbuch der Physik, Bd. XXV/2b, Berlin-Göttingen-Heidelberg: Springer.
[4] Maradudin, A. A., E. W. Montroll, and G. H. Weiss: Theory of Lattice Dynamics in the Harmonic Approximation. Solid State Physics, Suppl. **5** (1963).
[5] Abrikosov, A. A., L. P. Gorkov, and J. E. Dzyaloshinski: Methods of Quantum Field Theory in Statistical Physics. London: Prentice Hall 1963.
[6] Leibfried, G., and W. Ludwig: Solid State Physics **12** (1961).
[7] Brillouin, L.: Wave propagation in periodic structures. New York: Dover Publications 1953.
[8] Bak, T. A. (Editor): Phonons and phonon interactions. New York: Benjamin, Inc. 1964. [Proceedings of a Summer School held at Aarhus in 1963.]
[9] Blackman, M.: The Specific Heat of Solids, p. 325, Handbuch der Physik, Bd. VII/1. Berlin-Göttingen-Heidelberg: Springer 1955.
[10] Leibfried, G.: Gittertheorie der mechanischen und thermischen Eigenschaften der Kristalle, p. 104, Handbuch der Physik, Bd. VII/1. Berlin-Göttingen-Heidelberg: Springer 1955.
[11] Wallis, R. F. (Editor): Lattice Dynamics. New York: Pergamon 1965. (Proceedings of an International Conference held at Copenhagen in 1963.)
[12] Kellermann, E. W.: Phil. Trans. Roy. Soc. London **238**, 513 (1940).
[13] Lowdin, P. O.: A theoretical investigation into some properties of ionic crystals. Uppsala: Almquist and Wikseles 1948.

[1] A. A. Maradudin, A. E. Fein, and G. H. Vineyard: Phys. Stat. Solidi **2**, 1479 (1962)

Photon-Electron Interaction, Crystals Without Fields.

By

H. Y. FAN.

With 38 Figures.

A. General theory.

I. Introduction.

1. In the Maxwell equations, the property of matter is represented by a dielectric constant and a permeability. We are interested in materials for which only the dielectric constant, ε, needs to be considered. The dielectric constant is a frequency dependent tensor quantity for crystalline solids. It reduces to a scalar quantity for crystals of cubic symmetry and for polycrystalline materials which may be regarded as isotropic. The dielectric constant depends also on the wave vector, $\bar{K}$, in addition to the frequency and the polarization of the radiation field: $\tilde{\varepsilon}(\bar{K}\,\omega)$. However, for optical radiation K is negligibly small on the scale of the Brillouin zone of the crystal, and the relevant quantity is $\tilde{\varepsilon}(0,\omega)$.

The dielectric constant is a complex quantity, the imaginary part of which corresponds to the conductivity $\tilde{\sigma}$:

$$\tilde{\varepsilon}(\omega)=\tilde{\varepsilon}_1(\omega)+i\,\tilde{\varepsilon}_2(\omega)=\tilde{\varepsilon}_1(\omega)+i\,4\pi\,\tilde{\sigma}/\omega. \tag{1.1}$$

The imaginary part has the plus sign if the field is represented by $\exp(i\,\bar{K}\cdot\bar{r}-i\,\omega t)$. A complex conductivity is sometimes used instead of the complex dielectric constant.

The propagation of plane waves in a crystal can be expressed by three principal indices of refraction. For crystals of sufficiently high symmetries, the principal axes of the tensors $\tilde{\varepsilon}_1$ and $\tilde{\varepsilon}_2$ coincide, and the principal indices of refraction are given by [1]:

$$N_j=(\varepsilon_{1j}+i\,\varepsilon_{2j})^{\frac{1}{2}}, \tag{1.2}$$

where j stands for any one of the principal axes of $\tilde{\varepsilon}$. Each complex quantity N_j may be written:

$$N=n+i\,k, \tag{1.3}$$

where n is the ordinary, real refractive index, and k is the extinction coefficient. The intensity of the wave attenuates exponentially, $\exp(-\alpha\,x)$, in its propagation. The absorption coefficient, α, is:

$$\alpha=2k\,\omega/c=4\pi\,\sigma/c\,n. \tag{1.4}$$

The two parts, $\tilde{\varepsilon}_1(\omega)$ and $\tilde{\varepsilon}_2(\omega)$, of the complex dielectric constant are related to each other by the Kramers-Kronig relations [2]:

$$\tilde{\varepsilon}_1(\omega)-1=(2/\pi)\;P\int_0^\infty \omega'\,\tilde{\varepsilon}_2(\omega')\,(\omega'^2-\omega^2)^{-1}\,d\omega', \tag{1.5}$$

$$\tilde{\varepsilon}_2(\omega)-4\pi\,\tilde{\sigma}_0/\omega=(2\omega/\pi)\;P\int_0^\infty \tilde{\varepsilon}_1(\omega')\,(\omega^2-\omega'^2)^{-1}\,d\omega', \tag{1.6}$$

where $\tilde{\sigma}_0$ is the *dc* conductivity, and P stands for the principal value. These important relations are a direct consequence of causality, i.e. there can be no response before the field is turned on. They are subject only to the general conditions that the response to field is linear and that the quantities on the left hand side are bounded. Similarly, the two parts, $n(\omega)$ and $k(\omega)$, of the complex index of refraction are related:

$$n(\omega)-1=(2/\pi)\,P\int_0^\infty \omega'\,k(\omega')\,(\omega'^2-\omega^2)^{-1}\,d\omega'. \tag{1.7}$$

We discuss now some general considerations concerning the theory of optical properties of crystalline solids. The Hamiltonian of a solid may be written:

$$H_0=H_e+H_L+H_{eL} \tag{1.8}$$

where the first term involves the electronic coordinates, the second term involves the lattice coordinates, and the last term represents the electron-lattice interaction involving both electronic and ionic coordinates. With the presence of a radiation field, there are in addition a term, H_R, involving field coordinates and terms, H_{eR} and H_{LR}, representing interactions of the field with the electrons and the lattice. We shall not be concerned with the effects of the lattice-field interaction. The Hamiltonian of the problem is then:

$$H=H_0+H_R+H_{eR}. \tag{1.9}$$

A proper quantum-mechanical treatment should deal with the solid and the field as a whole. Theoretical works of this nature will be mentioned in Sects. 2 and 7. Usually, semiclassical treatment is used [*3*]. The solid is treated quantum-mechanically with H_0 as the unperturbed Hamiltonian and H_{eR} as perturbation due to a given field. The solution of this problem gives the response of the solid to a given field but does not give the properties of the radiation field in the solid. The properties of the field are obtained from the classical Maxwell equations. In other words, ε is calculated by quantum-mechanical perturbation treatment of the solid, and the index of refraction, N, is obtained from (1.2) given by the Maxwell equations.

The electron-field interaction, H_{eR}, is given by the operator:

$$H_{eR}=\sum_i\left[\frac{e}{mc}\bar{p}_i\cdot\bar{A}(\bar{r}_i)+\frac{ie\hbar}{2mc}\nabla\cdot\bar{A}+\frac{e^2}{2mc^2}A^2\right] \tag{1.10}$$

where $\bar{A}(\bar{r})$ is the vector potential. The second term is zero for transverse fields. In the semiclassical treatment, we may neglect the term in A^2 as a small quantity of higher order. The matrix elements between two stationary states of the solid have then the following form:

$$\langle\psi_m|H_{eR}|\psi_0\rangle=\frac{e}{i\omega}\bar{E}_0\cdot(\bar{j}^{\,+}_{m0}\,e^{-i\omega t}-\bar{j}_{m0}\,e^{i\omega t}) \tag{1.11}$$

$$\sim\frac{e}{im\omega}\bar{E}_0\cdot(\bar{p}_{m0}\,e^{-i\omega t}-\bar{p}_{m0}\,e^{i\omega t}), \tag{1.12}$$

where $2\bar{E}_0$ is the amplitude of the electric field,

$$\bar{p}=\sum_i\bar{p}_i,\qquad \bar{j}=\sum_i(\bar{p}_i/m+\hbar\bar{K}/2m)\,e^{-i\bar{K}\cdot\bar{r}_i}. \tag{1.13}$$

For transverse fields, the term $\hbar\bar{K}/2m$ gives no contribution. Usually, the one-electron approximation is used, in which the solid is represented by a determinant of single-electron wave functions. The operator H_{eR} has then non-vanishing matrix elements only between states which differ in one of the single-electron wave functions, in short, H_{eR} induces only one-electron transitions. The matrix elements $\bar{j}_{mn}$ and $\bar{p}_{mn}$ reduce to matrix elements between the single-electron wave functions.

We are interested primarily in crystalline solids for which the band model is a good approximation. The Bloch single-electron wave function in this model has the form:

$$\psi_{nk}(\bar{r}) = e^{i\bar{k}\cdot\bar{r}}\, U_{nk}(\bar{r}), \tag{1.14}$$

where n is the band index and $U(\bar{r})$ has the periodicity of the lattice. The form of the Bloch function follows from the translational symmetry of the crystal, and the matrix elements between Bloch states are subject to the condition of wave-vector conservation:

$$\left.\begin{aligned} \bar{k}' &= \bar{k}, &&\text{for } \langle n'\bar{k}'|\bar{p}|n\bar{k}\rangle \neq 0, \\ \bar{k}' &= \bar{k}+\bar{K}, &&\text{for } \langle n'\bar{k}'|\bar{j}|n\bar{k}\rangle \neq 0, \\ \bar{k}' &= \bar{k}-\bar{K}, &&\text{for } \langle n'\bar{k}'|\bar{j}^{+}|n\bar{k}\rangle \neq 0. \end{aligned}\right\} \tag{1.15}$$

The absorption and induced emission can be calculated in the usual way using the time-dependent perturbation theory. The transition rate from an initial state i to a final state f is given by:

$$P_{fi} = \frac{2\pi}{\hbar}\, |H_{eR}|^2_{fi}\, \delta(E_f - E_i), \tag{1.16}$$

Here E denotes the energy of the total system, the solid with radiation field. For one-electron transitions:

$$E_f - E_i = \varepsilon_f - \varepsilon_i \mp \hbar\omega \tag{1.17}$$

where ε is the electron energy, the minus sign is for photon absorption, and the plus sign is for emission. Using (1.12), we can get from (1.16):

$$W_{fi} = \hbar\omega\, P_{fi} = E_0^2\, \frac{2\pi e^2}{m^2\omega}\, |\bar{\eta}\cdot\bar{p}_{fi}|^2\, \delta(\varepsilon_f - \varepsilon_i \mp \hbar\omega) \tag{1.18}$$

where $\bar{\eta}$ is a unit vector parallel to $\bar{E}_0$.

We understand that W_{fi} refers to unit volume of the solid. This is accomplished by normalizing the Bloch wave functions for unit volume. For transitions involving imperfection centers, the localized wave functions are normalized, and the result is multiplied by the concentration of centers. Also, for localized electrons it may be necessary to replace the macroscopic field, E_0, with an effective local field, E_{eff}.

The rate of energy absorption per unit volume produced by a given transition gives directly the contribution of the transition to the conductivity of the solid, according to:

$$\bar{\eta}\cdot\tilde{\sigma}\cdot\bar{\eta} = W/2E_0^2. \tag{1.19}$$

Consider transitions between Bloch states. Condition (1.15) requires $\bar{k}' \sim \bar{k}$; therefore, energy conservation (1.17) can only be satisfied by transitions between states in different energy bands, interband transitions. Intraband transitions cannot take place. The absorption at frequency ω due to all possible transitions between two energy bands, n and n', is obtained by integrating (1.18):

$$W_{n'n}(\omega) = E_0^2\, \frac{2\pi e^2}{m^2\omega} \int\limits_S \frac{2\,ds}{(2\pi)^3}\, f(nk)\, [1 - f(n'k)]\, |\bar{\eta}\cdot\bar{p}_{n'\bar{k},\,n\bar{k}}|^2\, [\nabla_{\bar{k}}(\varepsilon_{n'\bar{k}} - \varepsilon_{n\bar{k}})]^{-1}, \tag{1.20}$$

where S is the surface in $\bar{k}$ space given by:

$$\varepsilon_{n'\bar{k}} - \varepsilon_{n\bar{k}} = \hbar\omega. \tag{1.21}$$

$f(n\bar{k})$ is the electron distribution function. Consider the simple case of transitions between the conduction and valence bands of a semiconductor with

$$\varepsilon_{c\bar{k}} = E_G + \hbar^2 k^2/2m_c, \quad \varepsilon_{v\bar{k}} = -\hbar^2 k^2/2m_v. \tag{1.22}$$

E_G is the energy gap between the two bands, and m_c and m_v are the effective masses of the two bands. For a semiconductor, $f(c\bar{k}) \sim 0$ and $f(v\bar{k}) \sim 1$. Putting:

$$|\bar{\eta}\cdot\bar{p}_{c\bar{k},v\bar{k}}| \sim |\bar{\eta}\cdot\bar{p}_{c0,v0}|, \tag{1.23}$$

we get the simple result for absorption near the threshold:

$$\sigma_{cv} = W_{cv}/2E_0^2 = \frac{e^2}{6\pi m^2}\,\frac{(2\mu)^{\frac{3}{2}}}{\hbar^2}\,|\bar{p}_{cv}|^2\,\frac{(\hbar\omega - E_G)^{\frac{1}{2}}}{\hbar\omega}, \tag{1.24}$$

where μ is the joint density-of-state mass for the two bands

$$\mu = m_c\, m_v/(m_c + m_v). \tag{1.25}$$

In case $p_{c\bar{k},v\bar{k}}$ vanishes at $\bar{k}=0$, we take it to be proportional to k:

$$|\bar{p}_{c\bar{k},v\bar{k}}|^2 \sim C\,\hbar^2 k^2 \tag{1.26}$$

thereby getting:

$$\sigma_{cv} = \frac{e^2}{6\pi m^2}\,\frac{(2\mu)^{\frac{5}{2}}}{\hbar^2}\,C\,\frac{(\hbar\omega - E_G)^{\frac{3}{2}}}{\hbar\omega}. \tag{1.27}$$

II. Dielectric constant of solids.

2. The semi-classical treatment of optical properties of solids has been discussed in various standard textbooks [*3*], [*4*], in the usual one-electron approximation. Recently, a great deal of theoretical work has been devoted to the study of the dielectric response of solids as a many-electron problem. The following expression was obtained by Nozieres and Pines [*5*]:

$$\left.\begin{aligned}\tilde{\varepsilon}(\bar{K},\omega) &= \left(1 - \frac{4\pi e^2 N}{m\omega^2}\right)\tilde{1} + \frac{4\pi e^2}{\hbar\omega^2}\times \\ &\times \left\{\sum_\nu \left[\frac{\bar{j}^+_{0\nu}(\bar{K})\,\bar{j}_{\nu 0}(\bar{K})}{\omega - \omega_{\nu 0} + iS} - \frac{\bar{j}_{0\nu}(\bar{K})\,\bar{j}^+_{\nu 0}(\bar{K})}{\omega + \omega_{\nu 0} + iS}\right]\right\}_{S\to 0}\end{aligned}\right\} \tag{2.1}$$

where the opertor $\bar{j}$ is defined in (1.13). The term involving N, the electron density, corresponds to the diamagnetic current. The subscript 0 denotes the unperturbed state, and the subscript ν refers to the various other states of the many-electron solid. The quantity S is introduced by the addition of a factor e^{st} to the vector potential $\bar{A}$. This procedure insures that the field is switched on adiabatically so that the response of the solid to the field will be causal. Since

$$\lim_{S\to 0} \frac{1}{x + iS} = P\frac{1}{x} - i\pi\,\delta(x) \tag{2.2}$$

where P denotes the principal value, the introduction of S gives automatically an imaginary part of the dielectric constant in accordance with the Kramers-Kronig relation. The imaginary part corresponds to the conductivity which can also be calculated separately as shown in the previous section.

In case the dielectric constant is isotropic, the expression becomes

$$\varepsilon(\overline{K},\omega)=1-\frac{4\pi e N}{m\omega^2}+\frac{4\pi e^2}{\hbar\omega^2}\sum_{\nu}|\bar{\eta}\cdot\bar{j}_{\nu 0}(\overline{K})|^2\left\{P\frac{2\omega_{\nu 0}}{\omega_{\nu 0}^2-\omega^2}-i\pi\delta(\omega_{\nu 0}-\omega)\right\}. \tag{2.3}$$

In this case, we have purely transverse radiation fields. The dielectric constant involved in the optical properties is often referred to as the transverse dielectric constant, $\varepsilon_\perp$. In the treatments of many-electron systems, the random phase approximation is often used, which neglects in the electron-field and electron-electron interactions, the coupling between interactions of different momentum transfers. In the optical properties of isotropic materials, the random phase approximation is equivalent to using Bloch wave states in place of the exact states of the solid [6]. The expression reduces then to the usual form. We give in the following the expression based on Bloch states for the general case, noting that its justification has been studied only for the isotropic case.

As pointed out in the previous section, in the one-electron approximation non-vanishing matrix elements of $\bar{j}$ exist only between states which differ by one Bloch function. By neglecting $\overline{K}$, the following expression is obtained [7]:

$$\left.\begin{aligned}\tilde{\varepsilon}(0,\omega)=\left(1-\frac{4\pi e^2 N}{m\omega^2}\right)\tilde{1}+\frac{4\pi e^2}{m^2\omega^2\hbar}\times\\ \times\left\{\sum_{nn'}\frac{(n\bar{k}|\bar{p}|n'\bar{k})(n'\bar{k}|\bar{p}|n\bar{k})}{\omega-\omega_{n'\bar{k},n\bar{k}}+iS}\left[f(\varepsilon_{n'\bar{k}})-f(\varepsilon_{n\bar{k}})\right]\right\}_{S\to 0}\end{aligned}\right\} \tag{2.4}$$

where f is the electron distribution function in the unperturbed solid. This expression can be put in a more convenient form by introducing oscillator strengths and effective mass of electrons. The reciprocal effective mass tensor is given by:

$$\widetilde{\left(\frac{1}{m^*}\right)}_{n\bar{k}}=\frac{1}{\hbar^2}\nabla_{\bar{k}}\nabla_{\bar{k}}\,\varepsilon_n(\bar{k}). \tag{2.5}$$

We shall define tensor oscillator strength $\tilde{f}$ by:

$$\tilde{f}_{\bar{k},n'n}=\frac{2}{m}\,\frac{\bar{p}_{\bar{k},n'n}\bar{p}_{\bar{k},n'n}}{\hbar\omega_{\bar{k},n'n}} \tag{2.6}$$

where $\hbar\,\omega_{\bar{k},n'n}=\varepsilon_{n'}(\bar{k})-\varepsilon_n(\bar{k})$. By using the relations [4] satisfied by the matrix elements of $\bar{p}$ it can be shown:

$$m\widetilde{\left(\frac{1}{m^*}\right)}_{n,k}=\tilde{1}-\frac{1}{2}\sum_{n'\neq n}(\tilde{f}_{\bar{k},n'n}-\tilde{f}_{\bar{k},nn'}) \tag{2.7}$$

which may be regarded as the f-sum rule. We can now rewrite the expression for the dielectric constant:

$$\left.\begin{aligned}\tilde{\varepsilon}_1(0,\omega)&=\tilde{1}-\frac{4\pi e^2}{\omega^2}\sum_{n\bar{k}}f(\varepsilon_{n\bar{k}})\widetilde{\left(\frac{1}{m^*}\right)}_{n\bar{k}}+\\ &\quad+\frac{4\pi e^2}{2m\omega}\sum_{\bar{k}nn'}{}'\left[f(\varepsilon_{n\bar{k}})-f(\varepsilon_{n'\bar{k}})\right]\frac{\tilde{f}_{\bar{k},n'n}}{\omega_{\bar{k},n'n}-\omega},\\ &=\tilde{1}-\frac{4\pi e^2}{\omega^2}\sum_{nk}f(\varepsilon_{n\bar{k}})\widetilde{\left(\frac{1}{m^*}\right)}_{n\bar{k}}+\\ &\quad+\frac{4\pi e^2}{2m\omega}\sum_{\bar{k}nn'}{}'f(\varepsilon_{n\bar{k}})\left[\frac{\tilde{f}_{\bar{k},n'n}}{\omega_{\bar{k},n'n}-\omega}-\frac{\tilde{f}_{\bar{k},nn'}}{\omega_{\bar{k},nn'}-\omega}\right],\end{aligned}\right\} \tag{2.8}$$

$$\tilde{\varepsilon}_2(0,\omega)=4\pi\frac{\pi e^2}{2m\omega}\sum_{\bar{k}nn'}{}'\left[f(\varepsilon_{n\bar{k}})-f(\varepsilon_{n'\bar{k}})\right]\tilde{f}_{\bar{k},n'n}\,\delta(\omega_{\bar{k},n'n}-\omega). \tag{2.9}$$

The prime over the summation sign indicates that $n' \neq n$. The last term of $\tilde{\varepsilon}_1$ represents interband contributions. It depends on the oscillator strengths connecting different bands. The second term gives contributions of individual energy bands and may be called the intraband effect. An energy band fully occupied by electrons does not contribute to intraband effect, since the summation of $1/m^*$ over a complete band gives zero. Thus, the second term may also be called the free carrier effect. While electrons in a completely filled energy band contribute only to the interband effect; carriers in an incompletely filled band provide intraband as well as interband effects. There is no intraband term in ε_2. There is no energy dissipation associated with the intraband effect in a perfect, rigid lattice. Like the resistance, intraband contribution to ε_2 arises from electron scattering by lattice vibration or lattice imperfection. This problem will be discussed later as free carrier absorption.

For an isotropic solid, ε_1 and ε_2 are scalar quantities. With

$$\tilde{f}_{\bar{k},n'n} = -f_{\bar{k},nn'} = \frac{2}{m} |(\bar{\eta}\cdot\bar{p})_{\bar{k},n'n}|^2/\hbar\omega_{\bar{k},n'n} = \frac{2}{3m} |\bar{p}_{\bar{k},n'n}|^2/\hbar\omega_{\bar{k},n'n}, \tag{2.10}$$

the expressions for the dielectric constant become:

$$\varepsilon_1 = 1 - \frac{4\pi e^2}{\omega^2} \sum_{n\bar{k}} f(\varepsilon_{n\bar{k}}) \left(\frac{1}{m^*}\right)_{n\bar{k}} + \frac{4\pi e^2}{m} \sum_{\bar{k}nn'}{}' f(\varepsilon_{n\bar{k}}) \frac{f_{\bar{k},n'n}}{\omega^2_{\bar{k},n'n} - \omega^2}, \tag{2.11}$$

$$\varepsilon_2 = \frac{4\pi}{\omega} \frac{\pi e^2}{2m} \sum_{\bar{k},nn'}{}' [f(\varepsilon_{n\bar{k}}) - f(\varepsilon_{n'\bar{k}})] f_{\bar{k},n'n}\, \delta(\omega_{\bar{k},n'n} - \omega). \tag{2.12}$$

It is of interest to consider a different dielectric constant which is used in problems concerning the response of the many-electron solid to longitudinal fields produced by an applied charge distribution. Plasma oscillations and electron excitations produced by the passage of fast, charged particles are problems of this category. In these problems, the property of the material is characterized by a longitudinal dielectric constant. Let $\varrho_{\text{ext}}(\bar{K}, \omega)$ be the Fourier transform of wave vector $\bar{K}$ and angular frequency ω of the applied charge distribution. The longitudinal dielectric constant, $\varepsilon_{\|}(\bar{K}, \omega)$, is defined by [8]:

$$\frac{1}{\varepsilon_{\|}(\bar{K}, \omega)} - 1 = \frac{\varrho\langle(\bar{K}, \omega)\rangle}{\varrho_{\text{ext}}(\bar{K}, \omega)} \tag{2.13}$$

where $\langle\varrho(\bar{K}, \omega)\rangle$ is the expectation value of the electron charge fluctuation induced by the applied charge. The rate at which a fast charged particle transfers momentum $\hbar\bar{K}$ and energy $\hbar\omega$ to the electron gas is given by:

$$W(\bar{K}, \omega) = -\frac{8\pi (Ze)^2}{\hbar^2 K^2} \operatorname{Im} \frac{1}{\varepsilon_{\|}(\bar{K}, \omega)}. \tag{2.14}$$

The dispersion relation of plasma oscillations, collective oscillations of the electrons, is determined by the condition:

$$\varepsilon_{\|}(\bar{K}, \omega) = 0. \tag{2.15}$$

For isotropic solids, theory shows that $\varepsilon(\bar{K}, \omega) = \varepsilon_{\|}(\bar{K}, \omega)$ in the random phase approximation, and that $\varepsilon(0, \omega) = \varepsilon_{\|}(0, \omega)$ is valid in general. Thus, optical properties can be related to the phenomena involving $\varepsilon_{\|}$.

The semiclassical treatment is a make-shift approach which lacks rigorous justification. It does not give the behavior of the field directly. In a material medium of low density where H_{eR} in (1.9) is small compared with H_R as well as

H_0, it can be legitimately treated as a perturbation in the overall problem. The attenuation and dispersion of the radiation field can be obtained by using perturbation theory to calculate the absorption and the scattering of photons [9]. However, this procedure is not applicable in dealing with radiation field in solids since H_{eR}, though small compared to H_0, is not small in comparison with H_R, which is evident from the strong influence of the material on the dispersion. NOZIERES and PINES [10] treated the problem quantum mechanically for the case of transverse field and isotropic solid. Short-range electron-electron Coulomb interaction is included in H_0, and long-range Coulomb interaction is neglected on the ground that it could only have a very small effect on the transverse photons. By using a canonical transformation, the Hamiltonian is reduced to the form:

$$H = H_0 + \tfrac{1}{2}\,(P^+ P + \omega^2\, Q^+\, Q) + H_m \tag{2.16}$$

where H_m represents higher order effects including non-linear electron-field interactions and electron interaction by the exchange of virtual photons. The electromagnetic field is represented essentially by the second term with Q and P being the canonical variables of the field. The frequency ω of the field is determined by the canonical transformation used:

$$\omega^2 = c^2 K^2 + \frac{4\pi e^2 N}{m} - \frac{4\pi e^2}{\hbar} \sum_{\nu} |\bar{\eta} \cdot \bar{j}_{\nu 0}(\bar{K})|^2 \left[P \frac{2\omega_{0\nu}}{\omega_{\nu 0}^2 - \omega^2} + i\pi\delta(\omega_{\nu 0} - \omega) \right]. \tag{2.17}$$

The expression agrees with (2.3), in view of the relation: $\varepsilon = c^2 K^2/\omega^2$.

III. Electron-lattice interaction.

3. Perturbation theory. The electron-lattice interaction, H_{eL}, of the crystal Hamiltonian (1.8) involves the nuclear coordinates, R, as well as the electronic coordinates, r. Since the nuclear displacements are small, it is customary to use the approximation which assumes that H_{eL} is linearly dependent on R. H_{eL} is then a linear function of the phonon creation and destruction operators, a^+ and a:

$$H_{eL} = f(\bar{r}, \bar{q})\, a_{\bar{q}}\, e^{i\bar{q}\cdot\bar{r}} + f^*(\bar{r}, \bar{q})\, a_{\bar{q}}^+ e^{i\bar{q}\cdot\bar{r}}\,, \tag{3.1}$$

where $\bar{q}$ is the phonon wave vector.

If the electron-lattice interaction, H_{eL}, is small, we may treat it together with the radiation field interaction, H_{eR}, as perturbations. The wave function of the solid in the zero-order approximation has the form:

$$\psi(\bar{r}, \bar{R}) = \varphi(\bar{r})\, \chi(\bar{R}) \tag{3.2}$$

where $\chi(\bar{R})$ represents the lattice vibration and may be considered as a product of harmonic oscillator wave functions. The existence of H_{eL} gives rise to transitions between ψ's which are not produced by H_{eR} alone. The probability of such transitions between an initial state ψ_0 and a final state ψ_f is:

$$P_{f0} = \frac{2\pi}{\hbar} \left| \sum_i \frac{(H_{eL} + H_{eR})_{fi}\, (H_{eL} + H_{eR})_{i0}}{E_0 - E_i} \right|^2 \varrho\,(E_0 - E_i)_{E_0 = E_f} \tag{3.3}$$

where i is an index of the intermediate state. Such transitions are generally referred to as indirect transitions in contrast to the direct transitions produced by H_{eR} alone. Under the linear approximation (3.1), the indirect transition given by (3.3) involves the absorption or emission of a single phonon of some mode.

Sometimes the perturbation by H_{eL} is considered first, separately. This procedure would seem to be more appropriate when the effect of H_{eL} is large as compared to the perturbation by the radiation field. The diagonal matrix element of H_{eL} gives a change of the energy level in first order. For an electronic state belonging to an energy band of the solid, $\varphi(\bar{r})$ is represented by a Bloch wave function, and the diagonal matrix element of H_{eL} given by (3.1) vanishes, which is to be expected from the requirements of wave vector and energy conservations. A change of the energy levels may then be calculated from the non-diagonal matrix elements in the second order perturbation:

$$\Delta \varepsilon_n = \sum_m \frac{|(H_{eL})_{nm}|^2}{E_n - E_m}, \tag{3.4}$$

which arises from the emission and reabsorption of virtual phonons.

The non-diagonal matrix elements of H_{eL} have also the effect of broadening the energy levels by producing transitions among the states with emissions or absorptions of phonons. The broadening of a level is inversely proportional to its lifetime, τ:

$$\Gamma_n \equiv \frac{1}{\tau_n} = \frac{2\pi}{\hbar} \sum_m |(H_{eL})_{nm}|^2 \, \delta(E_m - E_n), \tag{3.5}$$

and the frequency breadth of a transition between two states m and n is given by [*11*]:

$$\Gamma = |\Gamma_m - \Gamma_n|. \tag{3.6}$$

For strong electron-lattice interactions, the usual perturbation treatments discussed above are not adequate. It is then necessary to treat the interaction to higher order. Transitions with emission or absorption of multiple phonons will appear even within the linear approximation (3.1). Such calculations will be discussed in connection with excitions and imperfection centers. We discuss briefly here the polaron problem which concerns the strong interaction of electrons with the longitudinal optical phonons in an ionic crystal. The strong interaction arises from the polarization field associated with the lattice vibration. The eigenstates of the system correspond to the electron accompanied by a cloud of virtual phonons which is called a polaron. The polaron problem has been a subject of extensive theoretical studies which have been summarized by severals reviews [*12*], [*13*], [*14*]. The desirable procedure in the calculation of optical transitions would be to use polaron wave functions to take care of the strong interaction from the start. Other electron-phonon interactions can be then treated by the perturbation method.

The solution of the polaron problem depends on the strength of the coupling defined by:

$$\alpha = \frac{e^2}{\hbar} \left(\frac{m^*}{2\hbar\,\omega_l}\right)^{\frac{1}{2}} \left(\frac{1}{\varepsilon_\infty} - \frac{1}{\varepsilon_0}\right), \tag{3.7}$$

where m^* is the effective mass of the electron in the absence of the polaron effect, $\hbar\,\omega_l$ is the energy of a longitudinal optical phonon, ε_0 and ε_∞ are respectively the static and infrared dielectric constants which differ by the lattice polarizability. For weak coupling, $\alpha < 1$, perturbation can be used. Up to $\alpha \sim 6$, the so-called intermediate coupling theory is applicable giving polaron wave functions of the form [*15*]:

$$\psi = \exp\left\{i\left(\bar{k} - \sum_{\bar{q}} a_{\bar{q}}^+ a_{\bar{q}}\, \bar{q}\right)\cdot \bar{r}\right\} \exp\left\{\sum_{\bar{q}} \left[a_{\bar{q}}^+ f(\bar{q}, \bar{k}) - a_{\bar{q}}\, f^*(\bar{q}, \bar{k})\right]\right\} \Phi \tag{3.8}$$

where Φ is the free vacuum state and

$$f(\bar{q},\bar{k})=\frac{i\hbar\omega}{q}\left(\frac{\hbar}{2m^*\omega_l}\right)^{\frac{1}{4}}\left(\frac{4\pi\alpha}{V}\right)^{\frac{1}{2}}\left(\hbar\omega_l+\frac{\hbar^2 q^2}{2m^*}+\frac{\hbar^2\bar{q}\cdot\bar{k}}{m^*}\right)^{-1}. \tag{3.9}$$

The approximation for the energy is:

$$E(\bar{k})=-\alpha\hbar\omega_l+\frac{\hbar^2k^2}{2m^*}\left(1-\frac{\alpha}{6}\right)+O(k^4). \tag{3.10}$$

The range of intermediate coupling covers most materials of interest. The solution given is applicable, however, only for polaron states of low energies and is not proper for calculations involving electronic states deep inside the energy bands. The problem for strong coupling or large α has been treated by PEKAR, BOGOLIUKOV and TIABLIKOV, and others. The problem is very complex and will not be discussed here.

4. Indirect interband transitions. Studies of interband transitions near the transition threshold is of particular interest for semiconductors and insulators. On account of the requirement of wave-number conservation (1.15), the threshold photon energy for transition between two energy bands may be larger than the energy gap, E_G, between the two bands. The two are equal, $\hbar\omega_0=E_G$, only when the extrema of the two bands, the minimum of the conduction band and the maximum of the valence band, lie at the same $\bar{k}$. Indirect transitions can, however, begin at about E_G in any case, and may therefore have a lower threshold than the direct interband transitions. We are interested specifically in indirect transitions due to lattice vibrations. It should be pointed out that the interaction of the excited electron with other electrons significantly influences the excited state of the crystal. This is the excition effect which is especially important near the threshold of excitation. The treatment of transitions which takes into account the exciton effect is discussed in Sect. 9. We consider the problem here as interband transitions in the individual-electron model for the purpose of providing a background. The theory for this case was first given by HALL, BARDEEN and BLATT [*16*].

Consider, for example, the energy diagram shown in Fig. 1. The electron energy as a function of the wave number $\bar{k}$ is shown for the conduction band, $\varepsilon_c(\bar{k})$, and the valence band $\varepsilon_v(\bar{k})$. Consider absorption processes. The arrows T_d and T_i indicate respectively a direct and an indirect transition. For the indirect transition, the initial state of the crystal may be denoted by $|v\bar{k}_a, v\bar{k}_a, n_{\bar{q}j}\rangle$ or $|v\bar{k}_b, v\bar{k}_b, n_{\bar{q}j}\rangle$. There are possible final states: $|c\bar{k}_b, v\bar{k}_a, n_{\bar{q}j}+1\rangle$ for the transition process with phonon emission and $|c\bar{k}_b, v\bar{k}_a, n_{\bar{q}j}-1\rangle$ corresponding to phonon absorption, where $n, \bar{q}, j$ are respectively the quantum number, the wave vector, and the mode of the phonon involved. Since we have in the final state an electron in the conduction band and a hole in the valence band, it is useful to specify the wave vector of the hole as well as that of the electron. The initial state may be visualized as having a hole and an extra electron in the valence band, both having the same, arbitrary wave vector. Each of the two final states can be arrived at via two intermediate states as indicated below:

$$\begin{array}{ccc} \text{Initial state} & \text{Intermediate states} & \text{Final states} \\ |v\bar{k}_a, v\bar{k}_a, n_{\bar{q}j}\rangle \rightarrow & |c\bar{k}_a, v\bar{k}_a, n_{\bar{q}j}\rangle & \nearrow |c\bar{k}_b, v\bar{k}_a, n_{\bar{q}j}+1\rangle \\ |v\bar{k}_b, v\bar{k}_b, n_{\bar{q}j}\rangle \rightarrow & |c\bar{k}_b, v\bar{k}_b, n_{\bar{q}j}\rangle & \searrow |c\bar{k}_b, v\bar{k}_a, n_{\bar{q}j}-1\rangle \end{array} \tag{4.1}$$

The two intermediate states are indicated by the points i_1 and i_2 in Fig. 1. Transition via either intermediate state has a "vertical" step, $(a-i_1)$ or (i_2-b). The matrix element of electron-radiation coupling H_{eR} is responsible for this step; this step is vertical since the wave vector of radiation field is negligible. The non-vertical step, $(i_1 - b)$ or $(a - i_2)$, depends on the matrix element of electron-lattice coupling H_{eL}. For example, the intermediate state $|c\,\bar{k}_a, v\,\bar{k}_a, n_{\bar{q}j}\rangle$ contributes to (3.3) a term with the following product of matrix elements:

$$\langle 0, n_{\bar{q}j}|\,H_{eR}\,|c\,\bar{k}_a, v\,\bar{k}_a, n_{\bar{q}j}\rangle\,\langle c\,\bar{k}_a, v\,\bar{k}_a, n_{\bar{q}j}|\,H_{eL}\,|c\,\bar{k}_b, v\,\bar{k}_a, n_{\bar{q}j} \pm 1\rangle, \qquad (4.2)$$

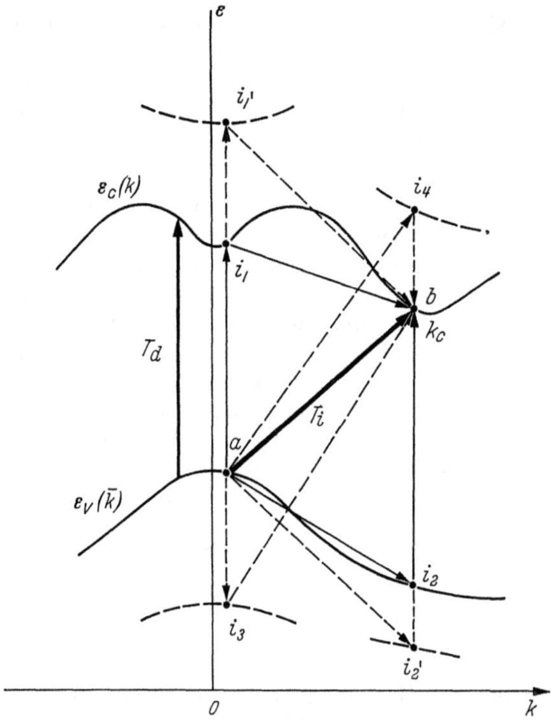

Fig. 1. Schematic energy diagram with the conduction band and the valence band given by $\varepsilon_c(\bar{k})$ and $\varepsilon_v(\bar{k})$, respectively. Arrow T_d indicates a direct transition. Arrow T_i indicates an indirect transition with intermediate states, i_1, i_2, etc.

where 0 denotes the ground state of the electronic system. The plus sign is for transition with phonon emission, and the minus sign is for transition with phonon absorption. The change in wave vector of the electron or hole is made up by the wave vector $\bar{q}$ of the phonon:

$$\bar{k}_a = \bar{k}_b \pm \bar{q}. \qquad (4.3)$$

Energy bands lying above the conduction band and below the valence band may contribute to the indirect transition. These bands are shown by the dashed curves in the regions around $\bar{k} = 0$ and $\bar{k}_c$. We may classify the intermediate states into four types, i_1, i_2, i_3, and i_4. There will be several of each type, depending upon the number of additional energy bands we wish to take into account.

The energy difference between the initial and the final states of the system, crystal and radiation field, is

$$E_f - E_0 = \hbar\,\omega_{f0} - \hbar\,\omega \qquad (4.4)$$

which should be equal to zero for energy conservation. $\hbar\,\omega_{f0}$ is the change in the energy of the crystal, given by:

$$\hbar\,\omega_{f0}(\bar{k}_a, \bar{q}) = [\varepsilon_c(\bar{k}_a \mp \bar{q}) - \varepsilon_v(\bar{k}_a) \pm \hbar\,\omega_{\bar{q}j}]. \qquad (4.5)$$

With a given $\bar{k}_a$, the density-of-states function ϱ in (3.3) can be written:

$$\varrho(E_f - E_0)_{\hbar\omega=\hbar\omega_{f0}} = \left| \frac{2}{(2\pi)^3} \int\limits_S \frac{ds}{|\nabla_{\bar{q}}\, \hbar\omega_{f0}|} \right|_{\hbar\omega_{f0}=\hbar\omega} \tag{4.6}$$

where S denotes a surface of constant $\hbar\,\omega_{f0}$ in the $\bar{q}$ space.

The above discussion outlines the essential considerations involved in the problem. The transition probability can now be calculated by using (3.3). An approximate expression for the absorption coefficient near the threshold can be written in the following form [17]:

$$\left.\begin{aligned} \alpha(\omega) = \frac{A}{\hbar\omega} \sum_j \frac{1}{\hbar\omega_{\bar{k}_c j}} \{B_j^+ (n_{\bar{k}_c j}+1)\, [\hbar\,\omega - (E_G + \hbar\,\omega_{\bar{k}_c j})]^2 + \\ + B_j^-\, n_{\bar{k}_c j}\, [\hbar\,\omega - (E_G - \hbar\,\omega_{\bar{k}_c j})]^2\}. \end{aligned}\right\} \tag{4.7}$$

The constant A involves the effective masses of the electrons and holes. The coefficients B^+ and B^- contain optical matrix elements at the extrema of the two bands as well as matrix elements of H_{eL} between $\bar{k}=0$ and $\bar{k}_c$ states of each energy band. $n_{\bar{k}_c j}$ is the average quantum number for phonons of wave vector $\bar{q}=\bar{k}_c$:

$$n_{\bar{k}_c j} = (e^{\hbar\omega_{\bar{k}_c j}/kT} - 1)^{-1}. \tag{4.8}$$

Each phonon branch, j, effective for indirect transitions, gives two absorption thresholds $(E_G + \hbar\,\omega_{\bar{k}_c j})$ and $E_G - \hbar\,\omega_{\bar{k}_c j})$. The absorption curve is continuous with steps at the thresholds given by each phonon branch. The phonon absorption processes with thresholds $(E_G - \hbar\,\omega_{\bar{k}_c j})$ should become negligible as $n_{\bar{k}_c j}$ approaches zero with decreasing temperature.

The quantities B^+ and B^- actually contain matrix elements of H_{eL} and H_{eR} between states of various $\bar{k}$. B^+ and B^- become constants when for all the matrix elements each state is taken to be either $\bar{k}=0$ or $\bar{k}_c$. For example, (4.2) is approximately replaced by:

$$\langle 0, n_{\bar{q}j}|\, H_{eR}\, |co, vo, n_{\bar{q}j}\rangle\, \langle co, vo, n_{\bar{q}j}|\, H_{eL}\, |c\bar{k}_c, vo, n_{\bar{q}j} \pm 1\rangle. \tag{4.9}$$

This approximation is not appropriate when a matrix element vanishes at the band extrema, in other words when the transition given by the matrix element is forbidden at the band extrema. The value of the matrix element is assumed then to be proportional to k or $|\bar{k} - \bar{k}_c|$ as the case may be. When forbidden transitions are important, the frequency dependence may be expressable by $[\hbar\,\omega - (E_G \pm \hbar\,\omega_{\bar{k}_c j})]^3$ instead of the quantity squared.

5. Free carrier absorption. By free carrier effect we mean the contribution of intraband term to ε_1 and ε_2. It is pointed out in connection with (2.8) and (2.9) that ε_2 has no intraband term. This is due to the fact that real transitions can not occur between Bloch states within the same energy band since wave vector conservation and energy conservation cannot be satisfied simultaneously. The missing intraband effect in ε_2 corresponds at zero frequency to the *dc* resistivity which arises from electron collisions with the lattice. In the Drude-Kronig theory for free carriers [3, 4], an ε_2 or conductivity, σ, is obtained by introducing the electron-lattice collision frequency, $1/\tau$, giving:

$$\sigma_{\text{complex}} = \sigma_0 \frac{1}{1 - i\,\omega\,\tau}, \tag{5.1}$$

or

$$\varepsilon_1 = -4\pi\sigma_0 \frac{\tau}{1+(\omega\tau)^2} = -4\pi \frac{N e^2 \tau^2}{m^*} \frac{1}{1+(\omega\tau)^2}, \tag{5.2}$$

$$\sigma = \sigma_0 \frac{1}{1+(\omega\tau)^2} = \frac{N e^2 \tau}{m^*} \frac{1}{1+(\omega\tau)^2}. \tag{5.3}$$

With vanishing collision frequency, $1/\tau \to 0$, the ε_1 contribution reduces to the second term in (2.8), and the σ contribution vanishes. These expressions have long been used for the interpretation of optical properties of conduction electrons in metals. LINDHARD [*18*] treated the Boltzmann equation of free electrons quantum mechanically and obtained expressions for longitudinal as well as transverse dielectric constant, $\varepsilon(K, \omega)$, of a Fermi gas. For $K \sim 0$, the result is the same as the expression given above.

Using a constant parameter, $1/\tau$, for the effect of electron collisions with lattice is a rough approximation, particularly for semiconductors in which the average energy of free carriers is normally small, equal to the thermal energy. The energy of a carrier is raised considerably in the absorption of a photon, depending on the frequency of radiation. The fact that electron scattering is energy dependent affects the frequency dependence of absorption. The absorption arises from intraband indirect transitions and may be calculated by perturbation theory using (3.3) [*19*—*21*]. This method was first used by FRÖHLICH to treat the scattering by accoustical vibrations. Imperfection centers in the lattice scatter carriers and may also produce absorption. Scattering centers of which impurity atoms are the kind of common interest can often be represented by point charges. Their effect can also be calculated by the same method using the Born approximation for electron interaction with a screened Coulomb center [*20*], [*22*]. However, the absorption in this case is essentially the inverse process of bremstrahlung, and a better approach is to make use of SOMMERFELD's treatment of the problem regarding the absorption as direct transitions between electronic states modified by the point charge [*20*], [*23*], *24*].

Using the deformation potential approximation for H_{eL}, one obtains the contribution to conductivity due to accoustical vibrations [*21*]:

$$\sigma = N \frac{e^2 (2m^*)^{\frac{1}{2}}}{3\pi\hbar^2} \frac{E^2 k T}{d c_l^2} (\hbar\omega)^{-\frac{3}{2}} f(x) \tag{5.4}$$

where N is the carrier concentration, d is the density of the crystal, c_l is the sound velocity, E is the deformation potential, and [*22*]

$$f(x) = \left(\frac{2}{\pi}\right)^{\frac{1}{2}} x^{\frac{1}{2}} e^x (1 - e^{-2x}) K_2(x), \tag{5.5}$$

$$x = \hbar\omega/2kT. \tag{5.6}$$

$K_2(x)$ is the modified Bessel function. The expression is calculated for a simple energy band characterized by a scalar effective mass m^*. For a many-valley energy band with a tensor effective mass, the expression may be written in the same from, with E and m^* standing for appropriate combinations of deformation potential components and mass tensor components, respectively [*22*], [*24*]. The function $f(\hbar\omega/2kT)$ approaches unity for $\hbar\omega \gg 2kT$, giving a $\omega^{-1.5}$ dependence for σ.

Carriers in semiconductors have normally small wave vectors corresponding to the thermal energy. Their scattering involves phonons of small wave vectors owing to the requirement of wave vector conservation. In the case of a many-valley energy band, inter-valley scattering of carriers may involve phonons of large wave vectors and correspondingly higher energy. Also, optical phonons have a nearly constant, large energy for all wave vectors. The contribution of deformation potential scattering by energetic phonons has been shown to have the following frequency dependence [22]:

$$\sigma \propto N(z/\sinh z)\,[x\,K_2(x)]^{-1}\,[(x+z)^2 K_2(x+z)-(x-z)^2 K_2(x-z)] \tag{5.7}$$

where

$$z=\hbar\,\omega_1/2kT, \tag{5.8}$$

$\hbar\,\omega_1$ being the mean energy of the phonons involved. The magnitude of absorption depends on the electron-phonon coupling strength which requires detailed information about the material.

The contribution of polar mode scattering, scattering due to the polarization field of optical phonons, is given by the following expression [25]:

$$\sigma=N\frac{\sqrt{2}}{3}\frac{e^4\hbar\,\omega_l}{m^*}\left(\varepsilon_\infty^{-1}-\varepsilon_0^{-1}\right)\frac{e^{2z}+1}{e^{2z}-1}(\hbar\,\omega)^{-2.5}\left(1-\frac{11}{8x}-\frac{4}{(2\pi\,x)^{\frac{1}{2}}}\cdots\right) \tag{5.9}$$

under the limiting condition:

$$x-z=(\hbar\,\omega-\hbar\,\omega_l)/2kT\gg 1. \tag{5.10}$$

Here $\hbar\omega_l$ is the optical phonon energy.

For scattering by centers of charge Ze, the calculated contribution is [20], [24]:

$$\sigma=NN_i\frac{8\pi}{3}\left(\frac{Z\,e^2}{\varepsilon}\right)^2\frac{e^2\hbar^2}{m^*}(2\pi\,m^*\,kT)^{-\frac{1}{2}}(\hbar\omega)^{-3}(1-e^{-2x})\,e^x\,K_0(x), \tag{5.11}$$

where N_1 is the concentration of scattering centers and $K_0(x)$ is the modified Bessel function of order zero. The expression is an approximation valid for classical distribution of carriers with a thermal energy large compared to the electron binding energy of the charge center. Using the matrix elements of H_{eL} known from the bremsstrahlung problem, it is straight forward to obtain approximations appropriate for other conditions.

The expressions involve $1/\omega$ to some power. They are evidently not applicable in the low frequency limit. In the calculations, the broadening of electron energy levels due to all kinds of collisions has not been taken into account. Therefore, the expressions are valid only when $\hbar\,\omega$ is large compared to the broadening, $\hbar/\tau$, of the energy levels involved.

6. Temperature dependence of energy gap [26], [21]. The threshold of transitions between two energy bands, direct or indirect transitions as the case may be, is determined by the energy gap, E_G, between the two bands. The absorption edge near threshold varies in general with temperature. The temperature dependence may be caused by two effects, lattice dilatation and lattice vibration. Lattice dilation afffects the energy levels of the solid, thus shifting the absorption edge. The coefficient of this shift, $\partial E_G/\partial V$, is the difference between the deformation potentials of the bands. It is related to the coefficient of pressure shift $\partial E_G/\partial P$, which gives a shift of absorption edge under pressure. The coefficient for the temperature shift due to lattice dilatation is:

$$\beta\left(\frac{\partial E_G}{\partial V}\right)_T=-\frac{\beta}{\chi}\left(\frac{\partial E_G}{\partial P}\right)_T \tag{6.1}$$

where β is the thermal expansion coefficient and χ is the compressibility.

Lattice vibration may produce both a broadening and a shift of the energy levels, as pointed out in Sect. 3. RADKOWSKY [27] calculated the broadening and considered the combined half width of the two band edges as a shift of the absorption edge. Actually, broadening should affect the shape rather than the position of the absorption edge. In many semiconductors including some ionic crystals, a shift in position is the major effect, especially when the temperature is not too high. Owing to lattice vibrations, a transition producing a electron-hole pair requires an additional energy equal to the lattice interaction energies of the electron and the hole. The interaction energies can be calculated according to (3.4). For simple energy bands, the calculated shift due to acoustical vibrations can be written approximately [26]:

$$\Delta E_G = -\frac{1}{(2\pi)^3}\frac{1}{\hbar\, d c_l}\left(m_2^* E_c^2 + m_v^* E_v^2\right)\left(\frac{kT}{\hbar c_l}\, 8\pi q_{\max} + 2\pi q_{\max}^2\right) \qquad (6.2)$$

for temperatures comparable or higher than the Debye temperature. The temperature dependence diminishes at lower temperatures. E_c and E_v are the deformation potentials for the conduction and valence bands, respectively. $q_{\max}$ is the maximum reduced wave number of the lattice. A similar expression was also obtained by MUTO and OYAMA [28]. The expression gives results which are in reasonable agreement with the experimental observations. The use of m_c and m_v to characterize the two energy bands is a rough approximation. When the expression is used with m_c and m_v values characteristic of only the band edges, the effect is likely to be underestimated. In comparison, the broadening calculated according to RADKOWSKY is smaller by orders of magnitude.

The calculated shift due to polar interaction with optical modes is given by [26]:

$$\Delta E_G = -\frac{1}{2}\, e^2 (\hbar\omega_l)^{\frac{1}{2}}\left(\frac{1}{\varepsilon_\infty} - \frac{1}{\varepsilon_0}\right)\left[\left(\frac{2m_c}{\hbar^2}\right)^{\frac{1}{2}} + \left(\frac{2m_v}{\hbar^2}\right)^{2}\right](n_l+1), \qquad (6.3)$$

where the temperature dependence enters through $n_l = [\exp(\hbar\omega_l/kT) - 1]^{-1}$. For this case, the broadening given by the calculation of RADKOWSKY has essentially the same expression with n_l in place of (n_l+1). The important difference between the energy shift and level broadening calculations is that the former deals with virtual transitions whereas broadening involves real transitions. For real transitions, energy as well as wave vector has to be conserved while only wave vector conservation is required for virtual transitions. This is particularly important for acoustical vibrations since the matrix element for deformation potential interaction is proportional to the phonon wave vector. For electronic states near the band edge, only phonons of small wave vectors are effective for real-transitions on account of both energy and wave vector conservations. Virtual-transitions are not limited to phonons of small wave vectors; therefore the energy shift is much larger than the broadening. In the case of polar mode, the interaction matrix element is inversely proportional to the phonon wave vector. Phonons of large wave vector are inherently less effective, hence the energy shift and the broadening turn out to be the same.

The polar mode interaction may give relatively large effects in ionic crystals of the order of 10^{-3} eV/°K for the energy shift or broadening. Even for compound semiconductors with small effective masses and small $(\varepsilon_\infty^{-1} - \varepsilon_0^{-1})$, the effect of polar interactions may be appreciable, and it can be largely responsible for the temperature broadening of absorption edge in such materials.

IV. Excitons.

7. Theories of exciton [29]. Exciton phenomena are clearly a problem which is beyond the one-electron approximation. In the band model of one-electron approximation, an excitation corresponds to simply replacing an electron wave function of one band with one of another band, and transitions between two bands lead to a continuous spectrum of excitation. In simple terms, exciton effect arises from the interaction between the excited electron and the hole left behind. Thus, the effect concerns electron-electron interaction which is not properly taken into account in the one-electron approximation. In particular, the interaction may lead to discrete excited states of the crystal which may be visualized as the electron and the hole held together in bound states. Even when the electron and the hole are dissociated from each other and move separately in their respective bands, there will be modification of their wave functions and energy spectra, due to the interaction. These consequences will be called generally the exciton effect, although the term exciton is commonly used to refer specifically to the discrete excitation states of the crystal.

It is sufficient for practical purposes to consider an insulating crystal, covering cases where the concentration of free carriers is sufficiently low to be of no significant consequence. It appears then that the exciton effect can be treated as a two particle problem with one excited electron and one hole to consider. Originally, FRENKEL who introduced the term exciton used the Heitler-London rather than the band model. With the ground state of the crystal represented by a Slater determinant of atomic ground state wave functions, the excition state is obtained by using an excited state wave function for one of the atoms. WANNIER [*30*] gave the connection between the atomic and band approaches and showed that in the band approximation the problem reduces to the solution of a wave equation for two particles with effective masses.

The use of atomic wave functions seems to be advantageous for describing excitations of localized character. Since there must be some overlapping, the atomic functions are not orthogonal to each other, hence they are not really appropriate to use. Orthogonal wave functions centered arround individual atoms were introduced by WANNIER. These functions, $a_{n\bar{R}}$, bear the following relationship to the Bloch functions, $\psi_{n\bar{k}}$, in the band model:

$$a_{n\bar{R}}(\bar{r}) = N^{-\frac{1}{2}} \sum_{\bar{k}} e^{-i\bar{k}\cdot\bar{R}} \psi_{n\bar{k}}(\bar{r}), \tag{7.1}$$

where $\bar{R}$ is the coordinate of the atom and N is the number of lattice cells. Each band, designated by the index n, is characterized by its own Wannier functions which reduce to atomic wave functions centered around the atoms in case of negligible overlapping. Determinantal wave functions constructed from the localized Wannier functions can be shown to be equivalent to the determinants of Bloch functions.

For an insulator, the ground state of the crystal is non-degenerate, and its wave function, Ψ_0, is a normalized determinant of $\psi_{v\bar{k}}$ or $a_{v\bar{k}}$ of the valence band. If only the conduction band is taken into consideration, then the wave function for an excited state may be written:

$$\Psi_{\text{ex}} = \sum_{\bar{k}\bar{k}'} A_{c\bar{k},v\bar{k}'}\, \Phi_{c\bar{k},v\bar{k}'} \tag{7.2}$$

where $\Phi_{c\bar{k}',v\bar{k}}$ denotes a determinant in which a Bloch function $\psi_{v\bar{k}'}$ of the valence band is replaced by $\psi_{c\bar{k}}$ of the conduction band. Without the electron-hole interaction, each $\Phi_{c\bar{k},v\bar{k}'}$ would give an excited state of the crystal. The problem then is to determine the coefficients $A_{c\bar{k},v\bar{k}'}$ so as to diagonalize the Hamiltonian with the interaction. Let

$$\bar{K} = \bar{k} - \bar{k}'. \tag{7.3}$$

Translational symmetry requires that Ψ_{ex} should only be multiplied by a constant, under translation by a lattice vector, $\bar{\beta}$. On the other hand, each $\Phi_{c\bar{k},v\bar{k}'}$ is multiplied by a factor $\exp[i(\bar{k}-\bar{k}')\cdot\bar{\beta}]$. Thus, Ψ_{ex} should correspond to a definite K:

$$\Psi_K=\sum_{\bar{k}} A_{c\bar{k},v(\bar{k}-\bar{K})}\,\Phi_{c\bar{k},v(\bar{k}-\bar{K})}=\sum_{\bar{k}} N^{-\frac{1}{2}}\sum_{\bar{\beta}} U_{\bar{K}}(\bar{\beta})\,e^{-i\bar{k}\cdot\bar{\beta}}\,\Phi_{c\bar{k},v(\bar{k}-\bar{K})}, \tag{7.4}$$

where

$$U_{\bar{K}}(\bar{\beta})=N^{-\frac{1}{2}}\sum_{\bar{k}} A_{\bar{k},\bar{k}-\bar{K}}\,e^{i\bar{k}\cdot\bar{\beta}}$$

is the Fourier transform of A. In order to show the equivalence of Bloch and Wannier functions, it is convenient to introduce the following function:

$$\left.\begin{aligned}\Phi_{cv,\bar{K}\bar{\beta}}&=N^{-\frac{1}{2}}\sum_{\bar{k}} e^{-i\bar{k}\cdot\bar{\beta}}\,\Phi_{c\bar{k},v(\bar{k}-\bar{K})}\\&=N^{-\frac{1}{2}}\sum_{\bar{R}} e^{i\bar{K}\cdot\bar{R}}\Phi_{c(\bar{R}+\bar{\beta}),v\bar{R}}\end{aligned}\right\} \tag{7.5}$$

where $\Phi_{c(\bar{R}+\bar{\beta}),v\bar{R}}$ denotes the determinant of Wannier functions $a_{n\bar{R}}$. In terms of these functions, we have:

$$\Psi_{\bar{K}}=\sum_{\bar{\beta}} U_{\bar{K}}(\bar{\beta})\,\Phi_{cv,\bar{K}\bar{\beta}}\equiv\sum_{\bar{\beta}} e^{i\bar{K}\cdot\bar{\beta}/2}\,U'_{\bar{K}}(\bar{\beta})\Phi_{cv,\bar{K}\bar{\beta}}. \tag{7.6}$$

The Wannier model: WANNIER derived the following equation for determining the coefficients $U'_{\bar{K}}(\bar{\beta})$ which diagonalize the crystal Hamiltonian approximately:

$$[\varepsilon_c(-i\nabla+\tfrac{1}{2}\bar{K})-\varepsilon_v(-i\nabla-\tfrac{1}{2}\bar{K})-V(\bar{\beta})]\,U'_{\bar{K}}(\bar{\beta})=E\,U'_{\bar{K}}(\bar{\beta}). \tag{7.7}$$

In this equation, $\bar{\beta}$ is to be regarded as a continuous variable. The operator $\varepsilon_c(-i\nabla+\frac{1}{2}\bar{K})$ is derived from the single electron energy $\varepsilon_c(\bar{k})$ of the band model by replacing $\bar{k}$ with $(-i\nabla+\frac{1}{2}\bar{K})$. It should be noted that $U'_{\bar{K}}(\bar{\beta})$ in (7.6) are coefficients for discrete lattice vectors $\bar{\beta}$. The solution of (7.7) normalized as a continuous function of $\bar{\beta}$ should be multiplied by $\Omega^{\frac{1}{2}}$ to give the coefficients for discrete $\bar{\beta}$, Ω being the volume of the unit cell.

In the derivation of (7.7), certain terms in the Hamiltonian have been neglected. RASHBA [31] showed that a part of the neglected terms, known as transfer interaction, has the effect of splitting some states which are degenerate according to the solution of (7.7). This effect has been shown by KNOX [29] to be analogous to the splitting of longitudinal and transverse excitons discovered by HELLER and MARCUS [32] in their studies using atomic wave functions. Consider the atomic wave function to be s-like in the ground state and p-like in the excited state. There exists a matrix element of dipole moment connecting the two states, the direction of which is determined by the orientation of the p-type function, whether it is longitudinal or transverse to the exciton wave vector $\bar{K}$. The longitudinal-transverse effect is quite general. HOPFIELD and THOMAS [33] discussed the effect for cubic and uniaxial crystals from the point of view of symmetry.

Electron-hole interaction potential: Eq. (7.7) is a kind of two particle wave equation with a potential energy $V(\bar{\beta})$, which consists of matrix elements of $e^2/(\bar{r}_i-\bar{r}_j)$ between various Bloch or Wannier functions. For values of β large compared to the dimension of the Wannier function, $V(\bar{\beta})$ approaches a simple Coulomb potential, $-e^2/\bar{\beta}$, β being related to the separation between the electron and the hole. Intuitively, one expects the electron-hole interaction to represent a Coulomb potential screened by the dielectric constant ε of the crystal

$$V(\beta)=-\frac{e^2}{\varepsilon\beta}. \tag{7.8}$$

However, the dielectric constant did not enter the Wannier equation. In the first place, proper treatment should not be limited by considering only the conduction and valence bands, as is done in (7.2). A complete orthonormal set of functions should be used for the representation. The main problem, however, is that an approximation of higher order than that used to derive (7.7) is necessary in order to bring out the screening. The problem belongs to the realm of many-body theory, and a complete treatment is not yet available. TOYOZAWA [34] considered the system of one electron in an electronically polarizable crystal and described the polarization associated with the extra electron in terms of a field of Frenkel excitons. Following this line, HAKEN and SCHOTTKY [35] treated the exciton as an excess electron and an excess hole, each with its own polarization field, and considered the direct electron-hole Coulomb interaction as well as the interaction between the two fields. Under certain simplifying conditions, they obtained

$$-\frac{e^2}{\beta}\left\{1-\left(1-\frac{1}{\varepsilon_\infty}\right)\left[1-\frac{1}{2}\left(e^{-\beta u_e}+e^{-\beta u_h}\right)\right]\right\} \tag{7.9}$$

where ε_∞ is the high frequency dielectric constant,

$$u \equiv \left(\frac{2m^*}{\hbar^2}\,\Delta\varepsilon\right)^{\frac{1}{2}}$$

m^* is the effective mass of the electron or the hole, and $\Delta\varepsilon$ is the excitation energy of a localized electron-hole pair (Frenkel exciton). In addition to the electronic polarization, there is also the polarization associated with ion displacement. This part was treated by HAKEN and SCHOTTKY following the polaron theory. The following result is obtained by adding the effects of electronic and ionic polorizations:

$$\left.\begin{aligned} V(\beta) = &-\frac{e^2}{\beta}\left\{1-\left(1-\frac{1}{\varepsilon_\infty}\right)\left[1-\frac{1}{2}\left(\varepsilon^{-\beta u_e}+e^{-\beta u_h}\right)\right]\right\}+\\ &+\frac{e^2}{\beta}\left(\frac{1}{\varepsilon_0}-\frac{1}{\varepsilon_\infty}\right)\left[1-\frac{1}{2}\left(e^{-\beta v_e}+e^{-\beta v_h}\right)\right]\end{aligned}\right\} \tag{7.10}$$

where ε_0 is the static dielectric constant, and

$$v = \left(\frac{2m^*}{\hbar^2}\,\hbar\,\omega_l\right)^{\frac{1}{2}}, \tag{7.11}$$

ω_l being the frequency of longitudinal optical phonon. For large β corresponding to weak electron-hole binding, we get:

$$V(\beta) \sim -\frac{e^2}{\varepsilon_0\,\beta}, \quad \text{if} \quad \beta \gg 1/u \quad \text{and} \quad \beta \gg 1/v; \tag{7.12}$$

$$V(\beta) \sim -\frac{e}{\varepsilon_\infty\,\beta}, \quad \text{if} \quad 1/v \gg \beta \gg 1/u. \tag{7.13}$$

Although the theory involves assumptions and is based on a simplified model, it is satisfying that the results, (7.12) and (7.13), are obtained which are expected intuitively.

Various other models: The Wannier equation is effective for the treatment of excitons with weak binding. Not only $V(\bar{\beta})$ can be approximated by a simple potential. Also, only a small portion of each energy band near the extremum is of importance, since small values of $\bar{k}$ are involved when the electron-hole separation is large. Thus the energy bands need to be specified only by a few parameters pertaining to the neighborhood of the extrema. Various models have been employed to treat cases of stronger binding, using to some extent the atomic point of view [29]. The so-called excitation models treat the problem of an electron in the potential field of a fixed hole located on a particular ion. Wave functions of the ion and of its neighbors are made use of in parts of the calculation. The electron-transfer models consider the transfer of an electron from one ion to another ion. In more refined versions, the transferred electron is regarded as shared by the neighboring ions. Wave functions of various excited states of the ions are combined to form the wave function of the electron. Models of both types with various degrees of refinement have been applied to excitons in alkali halides which are the excitons studied earliest. Recent works along the line of the exitation model are those of DEXTER [36] and MUTO and co-workers [37]. The work of OVERHAUSER [38] is representative of modern treatments based on the electron-transfer concept. The Frenkel exciton is a model for the extreme case of tight binding. The exciton is pictured as a local excitation confined essentially to one atom. The model has been applied to excitons in molecular crystals and rare gas solids [29].

Interaction with radiation field: It has been pointed out in Sect. 1 that properly the solid and the radiation field should be treated together as one system. The semiclassical treatment needs justification especially for regions of high absorption, and exciton states give rise to strong absorption. In addition, a consideration peculiar to sharp line exciton transitions was pointed out by HOPFIELD [39]. For sharp-line transitions in isolated atoms, resonance fluorescence is equivalent to absorption since photons are scattered out of the initial state in the process. In the case of excitons, however, resonance fluorence does not represent a scattering process. Because the intermediate-state exciton is subject to the wave vector conservation rule, the final-state photon has the same wave vector as the incident photon. Therefore, it is evident that the excitons and the radiation field constitute one system, and there can be no absorption due to internal interaction. Absorption occurs only as a result of interactions with other elements: phonons, lattice imperfections, and the surface of the solid. In the approximation of semiclassical treatment, an oscillator strength may be calculated, but external interaction must be considered in order to obtain an absorption.

The quantum treatment mentioned in connection with (2.20) is based on the model of electron plasma. FANO [40] used an assembly of oscillators of various frequencies as a model for the solid. By diagonalizing the Hamiltonian with a unitary transformation, it was shown that the normal modes of excitation are mixed electronic and field oscillations. This approach has been applied to excitons [39], [41]. Each exciton band may be represented by an oscillator which in coupling with the field gives two normal modes for any wave vector. Away from the resonance, the two normal modes represent approximately photons and excitons respectively. They are mixtures of "clothed" excitons and photons near the resonance. HOPFIELD [39] showed that excitons are formally equivalent to quantized polarization fields in a dielectric and introduced the term polariton for particles of the polarization field. The term has since been used for particles of the mixed exciton-photon field. The polariton problem has yet to be fully investigated.

8. Effective mass theory of exciton states. The Wannier equation (7.7) provides the basis for treating excitons of weak binding. As mentioned, in this case the energy bands need to be specified only by parameters describing the band edges. In the simplest case, the bands have their extrema at $\bar{k}=0$ and are parabolic for small values of $\bar{k}$. The bands are given then by effective masses, m_e and m_h:

$$\varepsilon_c(\bar{k}) = E_G + \hbar^2 k^2/2m_e; \qquad \varepsilon_v(\bar{k}) = -\hbar^2 k^2/2m_h. \tag{8.1}$$

Introducing:

$$F_{\bar{K}}(\bar{\beta}) = e^{-i\alpha \bar{K}\cdot\bar{\beta}} U'_{\bar{K}}(\bar{\beta}) \tag{8.2}$$

with

$$\alpha = \tfrac{1}{2}(m_e - m_h)/(m_e + m_h),$$

(7.7) can be reduced to a hydrogen-like equation:

$$\left(-\frac{\hbar^2}{2\mu}\nabla^2-\frac{e^2}{\varepsilon\beta}\right)F(\bar\beta)=\left[E-E_G-\frac{\hbar^2 K^2}{2(m_e+m_h)}\right]F(\bar\beta), \tag{8.3}$$

μ being the reduced mass:

$$1/\mu=1/m_e+1/m_h. \tag{8.4}$$

The energy eigenvalues of the exciton is given by:

$$E_{\text{ex}}=E_G-\frac{e^4\mu}{2\hbar^2}\frac{1}{n^2}+\frac{\hbar^2 K^2}{2(m_e+m_h)} \tag{8.5}$$

where n is an integer quantum number. The last term corresponds to the kinetic energy of the hole-electron pair and leads to an energy band for each discrete value of n. The coefficients of the exciton wave function in (7.2) are given by:

$$A^{(n,l)}_{\bar k,\bar k-\bar K}=(\Omega/N)^{\frac{1}{2}}\sum_{\bar\beta} e^{-i(\bar k-\alpha'\bar K)\cdot\bar\beta}F_{n,l}(\bar\beta) \tag{8.6}$$

where l is the orbital quantum number, and

$$\alpha'=\alpha+\tfrac{1}{2}=m_e/(m_e+m_n).$$

The wave function is a linear combination of excitations with the electron in various states, $\bar k$, and the hole in corresponding states $(\bar k-\bar K)$. Often, a wave vector $\bar k_h$ is used for the hole which is the negative of the wave vector of the electronic state. In terms of this notation, (7.3) becomes:

$$\bar K=\bar k_e+\bar k_h. \tag{8.7}$$

The application of Wannier equation to more complicated energy band structures was considered by DRESSELHAUS [42]. Let the conduction and valence bands have the more general form.

$$\left.\begin{aligned}\varepsilon_c(\bar k)&=E_G+\Delta\varepsilon_c(\bar k_e-\bar k_{e0}),\\ \varepsilon_v(\bar k)&=\Delta\varepsilon_v(\bar k_h-\bar k_{h0})\end{aligned}\right\} \tag{8.8}$$

where $\Delta\varepsilon_c$ and $\Delta\varepsilon_e$ are quadratic functions. The departure of the band extrema from $k=0$ does not cause significant complication. Introducing

$$U''_{\bar K}(\bar\beta)=\exp\left[-i(\bar k_{e0}-\bar k_{h0})\cdot\bar\beta/2\right]U'_{\bar K}(\bar\beta), \tag{8.9}$$

(4.7) can be readily transformed to:

$$\left[\Delta\varepsilon_c(-i\nabla+\tfrac{1}{2}\bar\varkappa)+\Delta\varepsilon_v(-i\nabla-\tfrac{1}{2}\bar\varkappa)+V(\beta)\right]U''_{\bar K}(\beta)=(E-E_G)\,U''_{\bar K}(\bar\beta), \tag{8.10}$$

$$\bar\varkappa=\bar K-(\bar k_{e0}+\bar k_{h0}). \tag{8.11}$$

The equation has the same form as (7.7) in the simpler case. However, if the $\Delta\varepsilon$'s are more general quadratic functions of $(\bar k-\bar k_0)$ instead of parabolic functions, then cross products of differential operators $(\partial/\partial\beta_\mu)(\partial/\partial\beta_\nu)$ are introduced. The equation will not have the simple hydrogenic form but becomes that of a particle with a tensor effective mass. The situation is more complicated if one or both bands are degenerate at the extrema. More than one conduction band and/or more than one valence band have to be taken into account. The expression (7.2) for the exciton wave function will include summations over the degenerate conduction bands, j, and degenerate valence bands, j'. We get a set of simultaneous equations for $U_{jj'\bar K}(\bar\beta)$. The problem is analogous to the impurity state problem treated by KOHN and LUTTINGER [43]. There are no simple solutions which can be discussed here.

Fig. 2a shows schematically a usual individual-electron diagram, and Fig. 2b shows the corresponding exciton energy diagram. Confine first our attention to the part of the diagrams shown by the solid curves near $\bar{k}=0$ and $\bar{K}=0$. The ground state is represented by the point G. The curves in Fig. 2b represent the exciton bands for the bound states with a continuum of dissociated excitons above them. The Wannier exciton states are derived from individual-electron states near the extrema of the energy bands which for the present are assumed to be at $\bar{k}=0$ in Fig. 2a. In this case, the effective mass equation is applicable to exciton states with $\bar{K}$ in the neighborhood of 0. Each exciton state such as point B, Fig. 2b, is composed of electron-hole pair states, ck and $v(\bar{k}-\bar{K})$, connec-

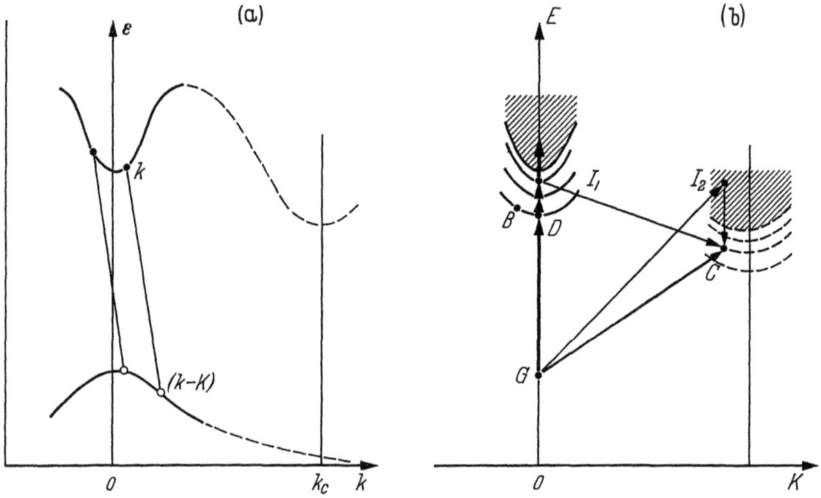

Fig. 2. Electronic energy diagram (a) and exciton energy diagram (b). In (a), $\varepsilon_c(\bar{k})$ and $\varepsilon_v(\bar{k})$ are the conduction and valence bands, respectively. In (b), the point G represents the ground state of the crystal. The curves represent bound state exciton bands. The continuum of dissociated excitons is shaded.

ted by lines in Fig. 1a. The pair states enter the exciton state with the weighting factor $A_{\bar{k},\bar{k}-\bar{K}}$. If the orbit of electron-hole relative motion as measured by β is large compared to the atomic distance, then the coefficients A have large values only for small values of $\bar{k}$ and $\bar{K}$. With increasing energy in the continuum, exciton states approach single pair states.

Optical transitions [*44*].

The usual semiclassical treatment will be used. The matrix element:

$$\int \Phi_{c\bar{k},v\bar{k}'}\Big(\sum_i e^{i\bar{k}_\lambda\cdot\bar{r}_i}\bar{\eta}\cdot\bar{p}_i\Big)\Phi_0=\int\psi^*_{c\bar{k}}(\bar{r})\,(e^{i\bar{k}_\lambda\cdot\bar{r}}\,\bar{\eta}\cdot\bar{p})\,\psi_{v\bar{k}'}(\bar{r})\,dr. \tag{8.12}$$

is the usual one-electron matrix element for interband transition which requires:

$$\bar{k}-\bar{k}'=\bar{k}_\lambda\sim 0. \tag{8.13}$$

As usual, the wave vector $\bar{k}_\lambda$ of the radiation field will be neglected. The matrix element for transition to an exciton state (7.2) is given by:

$$M=\int\Psi_{\bar{K}}\Big(\sum_i\bar{\eta}\cdot\bar{p}_i\Big)\Phi_0=\sum_k A_{\bar{k},\bar{k}-\bar{K}}\langle c\bar{k}|\bar{\eta}\cdot\bar{p}|v,\bar{k}-\bar{K}\rangle. \tag{8.14}$$

It follows from (8.13) that optical transition requires:

$$\bar{k}-(\bar{k}-\bar{K})=\bar{K}=\bar{k}_\lambda\sim 0. \tag{8.15}$$

Therefore, an exciton band gives a sharp line transition at $\bar{K}=0$.

Consider the case of simple energy bands represented by (8.1). Since the coefficients A are large only for a small range of $\bar{k}$, the expression (8.14) may be simplified by taking: $\langle c\bar{k}|\,\bar{\eta}\cdot\bar{p}\,|v\bar{k}\rangle\sim\langle c o|\,\bar{\eta}\cdot\bar{p}\,|v o\rangle$. Thus, we get

$$M=\langle c o|\,\bar{\eta}\cdot\bar{p}\,|v o\rangle\sum_{\bar{k}}A_{\bar{k},\bar{k}}=\langle c o|\,\eta\cdot p\,|v o\rangle N^{\frac{1}{2}}U_0(0) \tag{8.16}$$

according to (7.4). For the case under consideration, the coefficients $A_{\bar{k},\bar{k}-\bar{K}}$ are given by (8.6). The oscillator strength of the exciton line corresponding to the quantum number n is:

$$f_{n0}=\frac{2}{m\hbar\omega}|M_{n0}|^2=\frac{2N\Omega}{m\hbar\omega}|\langle c o|\,\bar{\eta}\cdot\bar{p}\,|v o\rangle|^2|F_{n,0}(0)|^2. \tag{8.17}$$

Only s-type solutions, $F_{n,0}$, have non-vanishing magnitude at the origin with:

$$|F_{n,0}(0)|^2=\frac{1}{\pi a^3 n^3}, \tag{8.18}$$

a being the effective Bohr radius. There will be a series of sharp lines at energies:

$$\hbar\omega=E_G-R_{\text{ex}}/n^2, \tag{8.19}$$

R_{ex} being the effective Rydberg unit. The line intensity is proportional to $1/n^3$.

The continuum of exciton ionized states corresponds to ordinary interband transitions. However, the electron-hole interaction modifies the Bloch states of the electron and the hole. The effect corresponds to the departure of the wave function of ionized hydrogen from the plane wave. For simple bands, the absorption coefficient in the continuum of interband transtions becomes:

$$\left.\begin{aligned}\alpha(\omega)&=\alpha_0(\omega)\,|F(0)|^2=\alpha_0(\omega)\,\frac{\pi\Delta\,e^{\pi\Delta}}{\operatorname{Sinh}\pi\Delta},\\ \Delta&=[(\hbar\omega-E_G)/R_{\text{ex}}]^{-\frac{1}{2}}.\end{aligned}\right\} \tag{8.20}$$

$\alpha_0(\omega)$ is the absorption coefficient in the band model without electron-hole interaction.

If interband transition is forbidden at the extrema, the approximation: $\langle ck|\bar{\eta}\cdot\bar{p}|vk\rangle\sim\langle co|\bar{\eta}\cdot\bar{p}|vo\rangle$ is not applicable. Perturbation calculation gives:

$$\langle c\bar{k}|\bar{\eta}\cdot\bar{p}\,|v\bar{k}\rangle=\langle co|\,\bar{\eta}\cdot\bar{p}\,|vo\rangle+(\hbar/m)\,\bar{k}\cdot\langle c|\,\bar{P}\,|v\rangle, \tag{8.21}$$

$$\langle c|\bar{P}|v\rangle=\sum_m\left[\frac{\langle co|\bar{p}\,|mo\rangle\langle mo|\bar{p}\cdot\bar{\eta}\,|vo\rangle}{E_c-E_m}+\frac{\langle co|\bar{p}\cdot\bar{\eta}\,|mo\rangle\langle mo|\,\bar{p}\,|vo\rangle}{E_v-E_m}\right]. \tag{8.22}$$

Using this approximation, we get:

$$M_{n0}=\frac{\hbar}{m}\langle c|\bar{P}|v\rangle\cdot\sum_{\bar{k}}A_{k,k}\bar{k}=(N\Omega)^{\frac{1}{2}}\frac{\hbar}{im}\langle c|\bar{P}|v\rangle\cdot\nabla F_{n,1}(0). \tag{8.23}$$

The last expression is obtained by using (8.6) with the fact that only p-type, $l=1$, functions have non-vanishing gradients at the origin. The series of lines with energies given by (8.19) begin with $n=2$. The oscillator strength of the lines is given by:

$$f_{n0}=\frac{2N\Omega}{m\hbar\omega}\,\frac{\hbar^2}{m^2}\,|\langle c|\bar{P}|v\rangle|^2\,\frac{n^2-1}{\pi a^5 n^5}, \tag{8.24}$$

and the absorption coefficient in the continuum takes the form:

$$\alpha(\omega) = \alpha_0(\omega) \frac{\pi \Delta (1+\Delta^2) e^{\pi\Delta}}{\operatorname{Sinh} \pi\Delta}. \tag{8.25}$$

The exciton effect stands out when the sharp lines can be observed. However, the transitions in the continuum are also strongly influenced. Consider (8.20). In the absence of exciton effect, $\alpha_0(\omega)$ varies as Δ^{-1} according to (1.24), being zero at $\hbar\omega = E_G$. With the exciton effect, we have $\alpha(\omega) \to 2\pi \Delta \alpha_0(\omega)$ with a finite value at $\hbar\omega = E_G$. Even at frequencies corresponding to $(\hbar\omega - E_G) \sim 100 R_{ex}$, α is still about 35 per cent higher than α_0.

9. Exciton-phonon interaction.

Indirect exciton transition.

The restriction (8.15) for optical transitions can be relaxed by lattice imperfections or lattice vibration. We consider the latter mechanism, using perturbation theory. The effect of indirect transitions in the case where the extrema of the electronic energy bands do not occur at the same $\bar{k}$ has been discussed in Sect. 4 without the exciton effect. The problem has been treated with the exciton effect by ELLIOTT [44].

The diagram in Fig. 2a, as extended by the dashed curve, shows the minimum of the conduction band at $\bar{k}_c$ whereas the valence band is shown with maximum at $\bar{k}=0$. Wannier equation in the form (8.10) should be used for this case. In the effective mass approximation, we again get (8.3) and (8.5) with $\varkappa$ of (8.11) substituting for $\bar{K}$. The exciton states obtained in this way are valid for values of $\bar{k}$ close to $\bar{k}_c$ corresponding to the minimum of the conduction band. They are represented by the dashed curves in Fig. 2b. An exciton state represented by the point C will be given by $\nu\bar{K}$ where ν is index of the exciton band. The state of the crystal will be specified by the exciton state and a set of quantum numbers, $n_{\bar{q}j}$, for the lattice modes. An indirect transition $G \to C$ corresponds to $|0, n_{\bar{q}j}\rangle \to |\nu\bar{K}, n'_{\bar{q}j}\rangle$. Possible intermediate states for the transition involve $\bar{K}=0$ exciton states in various bands, indicated in Fig. 2b by the tips of arrows along the vertical axis. The products of matrix elements in the expression (3.3) of perturbation theory are of the following type:

$$\langle 0, n_{\bar{q}j} | H_{eR} | \nu_0\, 0, n_{\bar{q}j}\rangle \langle \nu_0\, 0, n_{\bar{q}j} | H_{eL} | \nu\bar{K}, n_{\bar{q}j} \pm 1\rangle, \tag{9.1}$$

ν_0 being the index of exciton bands with $\bar{K}$ near zero. Another type of intermediate states is $|\nu'\bar{K}, n_{\bar{q}j} \pm 1\rangle$ which is represented by point I_2 in Fig. 2b. These intermediate states contribute terms of the type:

$$\langle 0, n_{\bar{q}j} | H_{eL} | \nu'\bar{K}, n_{\bar{q}j} \pm 1\rangle \langle \nu'\bar{K}, n_{\bar{q}j} \pm 1 | H_{eR} | \nu\bar{K}, n_{\bar{q}j} \pm 1\rangle. \tag{9.2}$$

An exciton state being represented by a linear combination of determinants, $\Phi_{c\bar{k}, v(\bar{k}-\bar{K})}$, each matrix element in (9.1) or (9.2) consists of matrix elements of the operator between two Φ's. The electron-lattice interaction H_{eL} is a sum of one-electron operators just as the electron-radiation interaction H_{eR}. In such cases, a matrix element between two Φ's reduces to that between two single-electron wave functions as shown by (8.12). Therefore, the exciton wave function may be written in the simple form

$$\Psi_{\bar{K}} = \sum_{\bar{k}} A_{c\bar{k}, v(\bar{k}-\bar{K})} \psi_c(\bar{k})\, \psi_v(\bar{k}-\bar{K}) \tag{9.3}$$

instead of (7.4), so long as the problem deals only with single-electron operators. Thus, (9.1) becomes:

$$\left.\begin{aligned}\sum_{\bar{k}'} |A^{\nu_0 0}_{\bar{k}',\bar{k}'}|^2\, A^{\nu \bar{K}}_{\bar{k},\bar{k}-\bar{K}}\, \langle 0, n_{\bar{q}j}|\, H_{eR}\,|\, c\bar{k}', v\bar{k}', n_{\bar{q}j}\rangle \times \\ \times \langle c\bar{k}', v\bar{k}', n_{\bar{q}j}\,|H_{eL}|\, c\bar{k}, v(\bar{k}-\bar{K}) \times n_{\bar{q}j} \pm 1 \rangle.\end{aligned}\right\} \tag{9.4}$$

The product of matrix elements has the same form as (4.2) for the case without exciton effect. The exciton effect brings in the coefficients A characterizing the final, $\nu\bar{K}$, and the intermediate, $\nu_0 0$, exciton states.

Details of the calculation are involved, and the exact solution will be given by very complicated expressions of little use. Approximate expressions of absorption coefficient have been given by ELLIOTT for simple energy bands which are parabolic and non-degenerate at the extrema. For a bound exciton band, m, the absorption coefficient has the form:

$$\alpha_m \propto (m_e+m_h)^{\frac{3}{2}} |F_m(0)|^2 \sum_j \{(n_{\bar{k}_c j}+1)\,[\hbar\omega-(E_{\text{ex}}+\hbar\omega_{\bar{k}_c j})]^{\frac{1}{2}} + n_{\bar{k}_c j}\,[\hbar\omega-\hbar\omega_{\bar{k}_c j}]^{\frac{1}{2}}\}. \tag{9.5}$$

The exciton energy, E_{ex}, is given by (8.5) with $(\bar{K}-\bar{k}_c)$ replacing $\bar{K}$. Each bound exciton band gives a continuous absorption owing to the fact that indirect transitions do not require the final exciton state to have a specific $\bar{K}$. Each phonon branch j gives two transition thresholds, for phonon emission and absorption processes, respectively.

As pointed out in connection with (4.7), a higher frequency dependence occurs when transitions are forbidden at the band extrema. Under certain conditions, an expression of similar form may apply with $\frac{3}{2}$ instead of $\frac{1}{2}$ for the power of the square brackets.

An approximate expression for the absorption coefficient has also been given for the continuum of dissociated excitons. It approaches (4.7) of the individual-particle model at $(\hbar\omega - E_G) \gg R_{\text{ex}}$.

Exciton line shape.

Broadening: The shape of an exciton line can be treated on the basis of indirect transitions. Referring to Fig. 2b, indirect transitions to various exciton states, such as B, determine the spectrum in the neighborhood of the exciton line $(G \to D)$. The problem may be treated approximately in terms of lifetime broadening. From this point of view, an optically produced exciton is scattered to other exciton states by lattice interaction. The lifetime against scattering and the corresponding line-width can be calculated by perturbation theory according to (3.5). In addition, a shift of exciton energy is produced by lattice interaction, which can be estimated according to (3.4).

A scattering process is called intraband or interband depending on whether the exciton is scattered to a state in the same or in another band, e.g. transition $(D \to B)$ in Fig. 2b is an intraband scattering. Intraband scattering may dominate for the lowest exciton band at low temperatures, particularly when the next band lies considerably higher in terms of the phonon energies. Calculations of lifetime due to intraband scatterings have been made for the lowest, 1 s, band of the effective mass model [*45*], [*46*]. For acoustic mode deformation potential scattering, the reciprocal lifetime for states of small $\bar{K}$ is [*46*]:

$$\frac{1}{\tau} = \frac{8\,(m_e+m_h)^2\, k\,T\,(E_{1e}-E_{1h})^2}{9\pi\,\hbar^4\,\varrho\, c_l}\,, \quad \text{for} \quad K < (m_e+m_h)\, c_l/\hbar. \tag{9.6}$$

The expression has the form of relaxation time for individual electrons or holes, with the sum of electron and hole mass replacing the effective mass of individual carriers and with the difference of deformation potentials replacing the deformation potential of an individual energy band. The expression for polar optical mode scattering is [47]:

$$\frac{1}{\tau} = \omega_0 e^2 \left(\frac{1}{\varepsilon_\infty} - \frac{1}{\varepsilon_0}\right) \frac{1}{2} \left(\frac{m_e - m_h}{m_e + m_h}\right)^2 \left(\frac{m_e + m_h}{\hbar^2}\right) a^4 \left[\frac{2(m_e + m_h)\,\omega_0}{\hbar}\right]^{\frac{3}{2}} n_0 \tag{9.7}$$

for $K < 2(m_e + m_h)\,\omega_0/\hbar$, where a is the exciton Bohr radius. ω_0 and n_0 are the frequency and the statistical quantum number of the optical phonons. Calculations taking into account interband as well as intraband scatterings have been made by GENKIN [48]. Interband scattering may be important particularly when the exciton band is not the lowest.

Polar optical mode scattering is generally weaker for excitons than for individual electrons or holes. Physically, this is expected on the ground that exciton is electrically neutral as an entirely; therefore, the scattering effectiveness of polar modes is reduced, especially for modes with wavelengths longer than the exciton radius. Optical modes may, however, give rise to an interesting phenomenon. Since all the modes have about the same frequency, $\omega_0 \neq 0$, indirect transitions involving optical phonons may give peaks separated from the main line by ω_0 and its multiples.

The lifetime consideration cannot be expected to give more than a rough indication for the width of an exciton line. For example, from the point of view of lifetime broadening, a Lorentzian line shape of the form:

$$\frac{1}{(\hbar\omega - E_{\text{ex}}) + i\,2\pi/\tau}, \tag{9.8}$$

would be expected. It is easily seen that this is not correct, at least for the lowest exciton state. Since there are no lower lying exciton states, there will be more indirect transitions on the high energy side of the main transition. As a result, the overall line shape will be asymmetric instead of the symmetric Lorentzian shape.

Formalism for accurate treatment [49], [50], [51]: The indirect transition or lifetime approach is based on the ordinary perturbation theory in which the lattice interaction is treated to the first order of approximation. A formalism has been developed by TOYOZAWA which can in principle treat the lattice interaction accurately, while limiting to first order with regard to photon interaction. It is shown that the probability, $W(\omega)$, for a photon of frequency ω to be absorbed may be written in the form [51]:

$$W(\omega) = \frac{1}{2\pi i} M^+ \left[\frac{1}{\hbar\,\omega + H_0^e - \Delta_0(\omega) - i\,\hbar\,\Gamma_0(\omega)} - \frac{1}{\hbar\,\omega - H_0^e - \Delta_0(\omega) + i\,\hbar\,\Gamma_0(\omega)}\right] M \tag{9.9}$$

where M, H_0^e, $\Delta_0(\omega)$ and $\Gamma_0(\omega)$, are matrices. M is a column matrix with elements $M_{\lambda g}$'s which are matrix elements of photon interaction connecting the ground state of the crystal with various exciton bands, λ. M^+ is the Hermitian conjugate matrix of M. H_0^e is a diagonalized square matrix with exciton energies E_λ as diagonal elements, the subscript 0 referring to $\overline{K} = 0$. The matrices $\Gamma_0(\omega)$ and $\Delta_0(\omega)$ correspond to the damping and energy shift, respectively, having elements of the type $\Delta_{\lambda'\lambda'',0}$ and $\Gamma_{\lambda'\lambda'',0}$.

The theory gives $\Delta_{\lambda'\lambda''}$'s and $\Gamma_{\lambda'\lambda''}$'s each as a series of increasing powers of lattice interaction. In the limit of weak lattice interaction, the first term of each series reduces to:

$$\left.\begin{aligned}\hbar\Gamma_{\lambda'\lambda'',0}&=\pi\sum_{\lambda,\overline{K},j,\pm} V^{(j,\pm)*}_{\lambda\overline{K},\lambda'0}\, V^{(j,\pm)}_{\lambda\overline{K},\lambda''0}\,\delta[\hbar\omega-(E_{\lambda\overline{K}}\pm\hbar\omega_{\overline{K},j})],\\ \Delta_{\lambda'\lambda'',0}&=\sum_{\lambda\overline{K},j,\pm} V^{(j,\pm)*}_{\lambda\overline{K},\lambda'0}\, V^{(j,\pm)}_{\lambda\overline{K},\lambda''0}\, P[\hbar\omega-(E_{\lambda\overline{K}}\pm\hbar\omega_{\overline{K},j})]^{-1}.\end{aligned}\right\}\tag{9.10}$$

$V^{j,\pm}_{\lambda\overline{K},\lambda'0}$ denotes the matrix element of lattice interaction for the scattering of an exciton from $\lambda'0$ to $\lambda\overline{K}$ with the emission (+) or absorption (−) of a phonon of $(\overline{q}=\overline{K},j)$. P denotes the principal part. The result in this limit reduces to that given by the perturbation treatment of indirect transitions, if Δ and $\hbar\Gamma$ are neglected in the denominators in comparison with $(\hbar\omega-E_{\lambda 0})$.

The expression for $W(\omega)$ can also be written in the form of a sum of terms:

$$W(\omega)=\frac{1}{\pi}\sum_l F_l(\omega)\,\frac{[\Gamma_{l0}(\omega)+2A_l(\omega)]\,[\hbar\omega-E_{l0}(\omega)]}{[\hbar\omega-E_{l0}(\omega)]^2+[\hbar\Gamma_{l0}(\omega)]^2},\tag{9.11}$$

by a transformation. If the exciton phonon coupling is sufficiently weak, the quantities $F_l(\omega)$, $E_{l0}(\omega)$, $\Gamma_{l0}(\omega)$ and $A_l(\omega)$ can be taken to be approximately constant within the energy range $|\hbar\omega-E_{l0}|\lesssim\Gamma_{l0}(E_{l0})$. In this case, the expression shows that the exciton spectrum is decomposed into a number of components each of which has an asymmetric Lorentzian shape, with A_l giving the degree of symmetry. The reader is referred to the original articles for details of the theory and its results.

V. Imperfection centers.

10. Effective mass theory. Imperfections of the lattice, impurity atoms or physical defects, perturb the electronic energy states of the crystal and may give rise to orbitals bound around the imperfection centers. The bound states have energy levels in the forbidden gaps of the energy band spectrum. In insulators and semiconductors, optical transitions involving bound states give absorption, emission and photoconductivity at photon energies less than the threshold for intrinsic absorption. The optical effects of some impurity ions can be related to the properties of the free ions, and the influence of the surrounding crystal on the so-called internal excitations can be treated as a perturbation. Rare earth ions with shielded incomplete $4f$ shells and transition metal ions incorporated in certain crystals belong typically to this category. This is a special subject outside of the scope of the present article. The color centers in alkali halides and similar ionic crystals constitute another special subject on which there is a large amount of literature. We shall touch on the subject only in connection with the effect of phonons on optical transitions.

We are concerned with localized electronic states which are closely related to the energy bands of the crystals. For bound states of this kind, the electron orbitals must be large in dimension as compared to the interatomic distance. Such states may be treated as the result of a perturbation on the periodic potential in the crystal. For an electrically charged center, the long-range electrostatic interaction is the important effect, and the problem is simplified if the center can be approximately represented by a point charge. We differentiate between two kinds of imperfection centers: donors and acceptors. A simple donor center can bind an electron in a localized orbital, and it is electrically neutral when an electron is bound to it. An acceptor center can bind a hole with which it becomes electrically neutral; speaking in terms of electrons, an acceptor center becomes negatively charged when the localized orbital is occupied by an electron (the hole is removed). Some centers can bind more than one electron and, depending on the number of electrons bound, may be either positive or negative electrically. Impurities of this kind in semiconductors are called amphoteric.

We consider the so-called shallow levels which lie close to the edge of one band [52]. The wave functions associated with such levels can be approximately expressed in terms of Bloch functions, $\psi_{\bar{k}}(\bar{r})$, of that particular band:

$$\psi(\bar{r}) = \sum_{\bar{k}} A_{\bar{k}}\, \psi_{\bar{k}}(\bar{r}). \tag{10.1}$$

Let $F(\bar{r})$ be the Fourier transform of the expansion coefficients $A_{\bar{k}}$:

$$F(\bar{r}) = V^{-\frac{1}{2}} \sum_{\bar{k}} A_{\bar{k}}\, e^{i\bar{k}\cdot\bar{r}}. \tag{10.2}$$

It can be shown that $F(\bar{r})$ is a solution of equation:

$$[\varepsilon_{\bar{k}}(-i\nabla) + V(\bar{r})]\, F(\bar{r}) = E\, F(\bar{r}), \tag{10.3}$$

which is analogous to the Eq. (8.3) for the relative motion of a bound electron-hole pair. The operator $\varepsilon^{\bar{k}}(-i\nabla)$, is obtained by replacing $\bar{k}$ with $-i\nabla$ in the electron energy $\varepsilon(\bar{k})$. $V(\bar{r})$ is the perturbation potential. We have discussed at some length the problem of dielectric screening for the exciton case. Theoretical studies [53] show that the interaction between an electron and a fixed point charge, Ze, can be represented by:

$$V(\bar{r}) = -Z\, e^2/r\, \varepsilon_0 \tag{10.4}$$

at large distances. This expression can therefore be used for shallow levels with large orbits. With a Coulomb-type of potential, the solution of (10.3) is covered by the discussion in Sect. 8 e.g. $F(\bar{r})$ has the form of hydrogen wave functions in the case of simple energy bands. Solutions have been obtained for multivalley conduction bands [52] and for degenerate valence bands [52], [54] of germanium and silicon.

The optical transitions depend mainly on the envelope function $F(\bar{r})$. Consider the transition between a bound i state to a state j which may be another bound state or a state in the energy band. The transition matrix element for the case of a non-degenerate band is:

$$\int \psi_j^* \bar{p}\, \psi_i\, d\bar{r} = \sum_{\bar{k},\bar{k}'} A_{\bar{k}'j}^* A_{\bar{k}i} \int \psi_{\bar{k}'}^* \bar{p}\, \psi_{\bar{k}}\, d\bar{r} = \sum_{\bar{k}} A_{\bar{k}j}^* A_{\bar{k}i} \frac{m}{\hbar} \nabla_{\bar{k}}\, \varepsilon. \tag{10.5}$$

The component along an axis, α of the effective mass tensor is:

$$\int \psi_j^*\, p_\alpha\, \psi_i\, d\bar{r} = \frac{m}{m_\alpha} \sum_{\bar{k}} A_{\bar{k}j}^* A_{\bar{k}i}\, \hbar k_\alpha = \frac{m}{m_\alpha} \int F_j(\bar{r})\, p_\alpha F_i(\bar{r})\, d\bar{r}. \tag{10.6}$$

Thus, the transitions are determined by the matrix elements of $F(\bar{r})$. A simple energy band with a scalar effective mass gives a hydrogen-like spectrum.

It appears from recent studies that transitions from shallow impurity states near one energy band to the energy band on the opposite side of the energy gap are often observed in various semiconductors. In this case, Bloch functions and envelope functions for two different bands are involved. Consider for example the simple case where both the conduction and valence bands are non-degenerate and the two band edges correspond to $k=0$. The transition matrix element is:

$$\int \psi_{cj}^* \bar{p}\, \psi_{vi}\, dr = \sum_{\bar{k}} A_{\bar{k},cj}^* A_{\bar{k},vi} \int \psi_{c\bar{k}}^* \bar{p}\, \psi_{v\bar{k}}\, dr \sim \bar{p}_{cv} \sum_{\bar{k}} A_{\bar{k},cj}^* A_{\bar{k},vi} = \bar{p}_{cv} \int F_{cj}^*(\bar{r})\, F_{vi}(\bar{r})\, dr. \tag{10.7}$$

With the approximation of neglecting the k dependence of $\bar{p}_{cv}$, the matrix elements are determined by the overlap of the two F functions. We note that a perturbing potential which is attractive for one type of carriers is repulsive for the other type of carriers. Hence, we would deal with transitions from bound states near one band to continuum states in the other band.

11. Effect of lattice vibration.

General theory [55].

We consider the effect of lattice vibration on the optical transition between two discrete, bound states of an imperfection center. Most theoretical treatments are based on the approximation that the wave function of the solid can be represented by a product of electronic and lattice functions:

$$\psi(r, R) = \varphi_{a\overline{R}}(\overline{r})\, \chi_{a\alpha}(\overline{R}) \tag{11.1}$$

where $\overline{r}$ and $\overline{R}$ are the electronic and nuclear coordinates, respectively. In the commonly adopted adiabatic approximation the electronic function is assumed to depend on $\overline{R}$ as parameters. The lattice function is a solution of the equation:

$$[T(\overline{R}) + V(\overline{R}) + \varepsilon_a(\overline{R})]\, \chi_{a\alpha}(\overline{R}) = E_{a\alpha}\, \chi_{a\alpha}(\overline{R}) \tag{11.2}$$

where T represents the kinetic energy of the ions, and $V(R)$ is the potential energy without the effect of the electron under consideration. $\varepsilon_a(\overline{R})$ is the contribution of the electron to the potential energy. It represents the electron-lattice interaction. Due to the effect of this contribution, the equilibrium positions as well as the vibration frequencies of the ions are influenced by the state, a, of the electron.

With the approximation (11.1) the matrix element for optical transition between states $a\alpha$ and $b\beta$ becomes:

$$\langle a\alpha | \overline{r} | b\beta \rangle = \int d\overline{R}\, \chi_{a\alpha}^{*}\, \overline{r}_{ab}(\overline{R})\, \chi_{b\beta} = \overline{r}_{ab} \int d\overline{R}\, \chi_{a\alpha}^{*}\, \chi_{b\beta}, \tag{11.3}$$

where

$$\overline{r}_{ab}(\overline{R}) = \int d\overline{r}\, \varphi_{a\overline{R}}^{*}(\overline{r})\, \overline{r}\, \varphi_{b\overline{R}}(\overline{r}), \tag{11.4}$$

and $\overline{r}_{ab}$ is the mean value of $\overline{r}_{ab}(\overline{R})$. In principle, $\overline{r}_{ab}$ depends on α and β. Lacking accurate knowledge of the wave functions, it is customary to use a single parameter $\overline{r}_{ab}$ irrespective of α and β, a procedure known as the Condon approximation. According to (11.3), the transition depends on the overlap of the lattice functions. The dependence of χ on the electronic state makes it possible to have non-vanishing overlap intergral for $\alpha \neq \beta$. The effect of electron-lattice interaction brought in through its effect on χ can be calculated more accurately than in the ordinary perturbation treatments such as outlined in Sect. 3. However, the electron-lattice interaction should affect the state of the electron. Thus, in the perturbation treatment various electronic states other than the final state are involved. In the adiabatic approximation, this effect is included in a way through the dependence of $\varphi_{a\overline{R}}$ on $\overline{R}$, but this is lost with the introduction of the Condon approximation. Basically, the present approach has merit for transitions between states which are well separated in energy from each other and from other states of the system. This condition is essential for the adiabatic approximation to be applicable as well as for the effect of other electronic states to be small. Assuming the approach is good for the calculation of optical transitions, actual departure from the adiabatic approximation has important consequences in other respects, e.g. it is responsible for radiationless transitions [55].

Let us write $\varepsilon(\overline{R})$ in the form:

$$\left.\begin{aligned} \varepsilon_a(\overline{R}) &= \varepsilon_a(\overline{R}_0) + \Delta\, \varepsilon_a(\overline{R} - \overline{R}_0), \\ \varepsilon_b(\overline{R}) &= \varepsilon_b(\overline{R}_0) + \Delta\, \varepsilon_b(\overline{R} - \overline{R}_0). \end{aligned}\right\} \tag{11.5}$$

$\overline{R}_0$ denotes the equilibrium ion positions given by (11.2) without $\varepsilon(\overline{R})$. The linear approximation is usually used, taking $\Delta\varepsilon$ to be proportional to $(\overline{R}-\overline{R}_0)$. Introducing the normal coordinates q_j of various vibration modes j, the linear approximation can be expressed by:

$$\left.\begin{aligned}\Delta\varepsilon_a &= N^{-\frac{1}{2}}\sum_{j=1}^{N} A_j^a\, q_j,\\ \Delta\varepsilon_b &= N^{-\frac{1}{2}}\sum_{j=1}^{N} A_j^b\, q_j.\end{aligned}\right\} \tag{11.6}$$

Using these expressions, we get:

$$\hbar\,\omega_{ba} = \varepsilon_b(\overline{R}_b) - \varepsilon_a(\overline{R}_a) = \varepsilon_b(\overline{R}_0) - \varepsilon_a(\overline{R}_0) - \frac{1}{2N}\sum_{j=1}^{N}\left[\left(\frac{A_j^b}{\omega_j^b}\right)^2 - \left(\frac{A_j^a}{\omega_j^a}\right)^2\right]. \tag{11.7}$$

The last terms represent the effect of the shift of equilibrium coordinates from R_0 to R_a and R_b, respectively for the two electronic states.

The transition probability between a state with φ_a to a state with φ_b is determined by:

$$I_{ba}(\hbar\,\omega) = |\overline{r}_{ba}|^2\, A v_\alpha \sum_\beta \left|\int d\overline{R}\,\chi_{b\beta}^*\,\chi_{a\alpha}\right|^2 \delta(E_{b\beta} - E_{a\alpha} - \hbar\,\omega). \tag{11.8}$$

The spectral distribution is given by:

$$G_{ba}(\hbar\,\omega) = I_{ba}(\hbar\,\omega)/|\overline{r}_{ba}|^2. \tag{11.9}$$

Using the linear approximation (11.6), O'ROURKE [56] has derived the following exact expression:

$$\left.\begin{aligned}G_{ba}(\hbar\,\omega) = \frac{I_{ba}(\hbar\,\omega)}{|\overline{r}_{ba}|^2} &= \frac{1}{2\pi\hbar}\int_{-\infty}^{\infty} dt \exp\Bigg[i(\omega_{ba}-\omega)t + \\ &\quad - \frac{it}{2}\sum_{j=1}^{N}(\omega_j^b - \omega_j^a)\coth(\hbar\,\omega_j^a/2kT)\Bigg] \times \\ \times \exp\Bigg\{-\sum_{j=1}^{N} 2S_j &\left[\coth(\hbar\,\omega_j^a/2kT + i\omega_j^a t/2) + \coth(-i\omega_j^b t)\right]^{-1}\Bigg\},\end{aligned}\right\} \tag{11.10}$$

where

$$2S_j = \frac{\omega_j^a}{N\hbar}\left[\frac{A_j^b}{(\omega_j^b)^2} - \frac{A_j^a}{(\omega_j^a)^2}\right]^2. \tag{11.11}$$

The expression shows the dependence of $I_{ba}(\hbar\,\omega)$ on the vibrational frequencies, ω_j^a and ω_j^b, and the coefficients, A_j^a and A_j^b, of electron coupling with various vibration modes. If the approximation $\omega_j^a = \omega_j^b$ is made, the expression reduces to:

$$\left.\begin{aligned}G_{ba}(\hbar\,\omega) &= \frac{1}{2\pi\hbar}\int_{-\infty} dt\, e^{i(\omega_{ba}-\omega)t} \exp\left\{\sum_{j=1}^{\infty} S_j\left[i\sin\omega_j t - (2n_j+1)(1-\cos\omega_j t)\right]\right\},\\ &= \frac{1}{2\pi\hbar}\int_{-\infty} dt\, e^{i(\omega_{ba}-\omega)t}\, e^{f(t)}.\end{aligned}\right\} \tag{11.12}$$

This expression was derived earlier by LAX [57].

The overall absorption band may be described by its various moments [57] defined by:

$$\langle(\hbar\,\omega)^m\rangle = \int d(\hbar\,\omega)\,(\hbar\,\omega)^m\, G(\hbar\,\omega), \tag{11.13}$$

and it is often more convenient to use $(\hbar\omega - \langle\hbar\omega\rangle)^m$ instead of $(\hbar\omega)^m$ for the higher moments, $m>1$. The following results are obtained by using (11.12):

$$\langle\hbar\omega\rangle = \hbar\omega_{ba} + \sum_j S_j\hbar\omega_j \tag{11.14}$$

and

$$\langle(\hbar\omega - \langle\hbar\omega\rangle)^2\rangle = \sum_j (S_j\hbar\omega_j)^2 (2n_j+1). \tag{11.15}$$

The first moment gives the center of gravity of the band; the second term which takes the minus sign for emission is the STOKE's shift of the average frequency from $\hbar\omega_{ba}$. The second moment provides an indication of the broadening of the band. We note that the expression for the first moment is independent of temperature. However, this is a consequence of using (11.12) which neglects the differences between ω_j^a's and ω_j^b's. We shall return to this point presently.

Effect of polar optical modes.

The problem may be greatly simplified by special conditions. In the pioneering works of PEKAR [*58*] and HUANG and RHYS [*59*] on F-center absorption, the electrostatic interaction of the electron with the polarization produced by long wavelength optical modes was treated. These modes may be assumed to have approximately the same frequency. The results of the treatments based on this model can be obtained from (11.12) by using one frequency ω_0 for all the modes [*59*]: $\omega_0 = \omega_j = \omega_j^a = \omega_j^b$. Thus, we get:

$$\left.\begin{aligned} G_{ba}(\hbar\omega) &= \frac{1}{\hbar\omega_0}\exp\left[-S(2n+1)\right]\left(\frac{n+1}{n}\right)^{p/2}\times \\ &\times I_p\{2S\,[n(n+1)]^{\frac{1}{2}}\}\left\{\sum_{k=-\infty}^{\infty}\delta(k-p)\right\}, \end{aligned}\right\} \tag{11.16}$$

where $n = [\exp(\hbar\omega_0/kT)-1]^{-1}$ is the average quantum number of each vibrational mode, I_p is the modified Bessel function of order p,

$$\left.\begin{aligned} p &= \frac{\omega-\omega_{ba}}{\omega_0} \\ \text{and}\qquad S &= \frac{1}{2\hbar\omega_0^3}\,\frac{1}{N}\sum_j (A_j^b - A_j^a)^2. \end{aligned}\right\} \tag{11.17}$$

The quantity p corresponds to the number of phonons involved in the transition at frequency ω. The maximum of I_{ba} occurs at $\sim S$, i.e. the number of phonons emitted in an electronic transition at the peak of the absorption band is close to S. The δ functions ensure that p has only integer values. The expression gives a spectrum of a series of sharp lines differing by $\hbar\omega_0$ in energy. Beside the frequency, ω_0, the expression involves only one parameter S and is therefore easily applicable to the interpretation of experimental data.

The sharp lines given by the δ-functions may not be resolvable under experimental conditions. The observed smooth absorption band is described then by:

$$\left.\begin{aligned} \overline{G}_{ba}(\hbar\omega) &= \frac{1}{\hbar\omega_0}\int\limits_{\omega-\frac{1}{2}\omega_0}^{\omega+\frac{1}{2}\omega_0} G_{ba}(\hbar\omega)\,d\hbar\omega = \frac{1}{\hbar\omega_0}\left[\frac{n+1}{n}\right]^{p/2}\times \\ &\times\exp\left[-S(2n+1)\right] I_p\{2S\,[n(n+1)]^{\frac{1}{2}}\}. \end{aligned}\right\} \tag{11.18}$$

In the range:

$$4S\,[n(n+1)]^{\frac{1}{2}} > |p| > 1,$$

the shape of the spectrum is approximately Gaussian:

$$\overline{G}_{ba}(\hbar\,\omega) \propto \exp\left[-\frac{1}{4S\,[n(n+1)]^{\frac{1}{2}}}\left(\frac{\omega-\omega_{\max}}{\omega_0}\right)^2\right], \tag{11.19}$$

where $\omega_{\max}$ is the frequency at the peak. At high temperatures where n is large, this range may cover a considerable part of the spectrum. On the other hand, for

$$p \gg S^2\,n\,(n+1),$$

we get

$$\overline{G}_{ba}(\hbar\,\omega) \sim S^p/p. \tag{11.20}$$

At low temperatures where $n\to 0$, this expression becomes appropriate.

As pointed out earlier, the first moment as given by (11.14) shows no temperature dependence. Experimentally, absorption bands often show frequency shifts with increasing temperature. A classical example is F-center absorption in alkali halides. It was pointed out by HUANG and RHYS that a shift can be obtained from differences between the vibrational frequencies for the two electronic states. We should then use (11.10) rather than (11.12). Consider that the frequencies ω_j^a and ω_j^b depart slightly from a medium frequency $\omega_0=\omega_j$. For small departures, (11.10) gives the following approximate expression for the first moment [*56*]:

$$\langle\hbar\,\omega\rangle = \hbar\,\omega_{ba} + S\,\hbar\,\omega_0 + \left[\frac{1}{2}\sum_j \frac{\omega_j^a-\omega_j^b}{\omega_j^a}\right](2n+1)\,\hbar\,\omega_0. \tag{11.21}$$

The last term gives a temperature dependence of $\langle\hbar\,\omega\rangle$.

Model of configurational coordinates.

The problem is naturally simplified if there are only a few vibrational modes which play a role in the transitions. The familiar model of configurational coordinates which has been extensively used to treat color centers is based on this assumption. In this model, the lattice potential energy is regarded as a function of one or a few parameters describing the positions of the surrounding ions. Such an approximation could be applicable for centers with tightly bound electrons. In such cases, the equilibrium positions of the neighboring ions may be strongly influenced by the electronic state, and the motion of these ions may give rise to a few local modes with frequencies distinct from those of the normal modes of the crystal. The existence of local modes has been shown theoretically [*60*], and experimental studies of local modes in various types of crystals are one of the subjects of current interest. The problem becomes very simple if only one mode needs to be considered or if the frequencies of the important modes may be taken approximately as the same .The problem reduces then to the approximation of one vibrational frequency discussed above.

The Franck-Condon principle has been frequently used in conjuction with the model of configurational coordinates. In this approximation, the nuclei are regarded as standing still while the electron makes the transition. Thus, (11.8) is replaced by:

$$I_{ba}(\hbar\,\omega) = |\overline{r}_{ba}|^2\,A\,v_\alpha \int d\overline{R}\,|\chi_{a\alpha}(\overline{R})|^2\,\delta[E_b(\overline{R}) - E_{a\alpha} - \hbar\,\omega]. \tag{11.22}$$

Since the quantization of the final vibrational state is neglected, sharp lines in the spectrum would be smeared out. The Franck-Condon principle puts the problem in classical, more familiar terms. The transitions can be simply visualized in a diagram of potential energy versus the configurational coordinate. It has been shown that under certain conditions the use of the additional approximation does not affect the final results [*61*].

Zero-phonon line.

Let us consider the zero-phonon transition, i.e. transition without changes of the vibrational state. TRIFONOV [62] noted that:

$$\lim_{t\to\pm\infty}[e^{f(t)}] = -\sum_j S_j(2n_j+1) \neq 0 \tag{11.23}$$

and therefore the integral of (11.12) may diverge. The integral can be separated into two parts:

$$G(\hbar\omega) = \frac{1}{2\pi\hbar}\int_{-\infty}^{\infty} dt\, e^{i(\omega_{ba}-\omega)t}\left[e^{f(t)} - e^{f(\infty)}\right] + \frac{1}{2\pi\hbar}\int_{-\infty}^{\infty} dt\, e^{i(\omega_{ba}-\omega)t} e^{f(\infty)}. \tag{11.24}$$

The first integral remains finite at $\omega = \omega_{ba}$ whereas the second integral gives

$$\delta(\hbar\omega - \hbar\omega_{ba}) \exp\left[-\sum_j S_j(2n_j+1)\right]. \tag{11.25}$$

which causes divergence at $\omega = \omega_{ba}$ and corresponds evidently to zero-phonon transition. The integrated zero-phonon transition is given by the exponential factor which decreases with increasing temperature. Since, the distribution function G normalizes the absorption of the band, we have

$$\frac{\text{integrated zero-phonon absorption}}{\text{integrated absorption of the band}} = \exp\left[-\sum_j S_j(2n_j+1)\right]. \tag{11.26}$$

The same result was obtained also by KANE [63] and HOPFIELD [64] in different ways. The zero-phonon transition is predominant for weak electron-phonon couplings and becomes unobservable for very strong couplings. Some intermediate coupling is favorable for the observation of a sharp zero-phonon line above an absorption band. HOPFIELD calculated the shape of emission band in CdS due to excitons bound to acceptor impurities. Piezoelectric coupling with acoustical phonons is assumed to be the dominant mechanism. The calculated shape is in general agreement with the observed emission, consisting of a sharp, outstanding zero-phonon line with a low-energy wing of phonon emisson at very low temperature and a high energy wing of phonon absorption growing with increasing temperature.

B. Experimental observations.

VI. Absorption edge and edge emission.

12. Intrinsic absorption edge. In insulators and semiconductors, long wavelength radiation is absorbed due to the effect of free carriers, imperfection centers, and lattice vibration. Absorption due to lattice vibration occurs in the infrared, in the range 10 to 100 microns, and is strong only in ionic crystals; it is not a subject matter of this article. The absorption due to electronic excitation is not very strong unless the concentrations of free carriers and imperfection centers are unusually large. Strong absorption due to interband transitions of the valence electrons begins when the photon energy exceeds the energy gap. The fast rise of absorption associated with the onset of interband transition is called the intrinsic absorption edge. Study of the absorption edge gives information about the energy gap and the band structure near the extrema of conduction and valence bands.

The curves shown in Fig. 3 for InSb represent absorption edges corresponding to direct transitions. The curve for liquid nitrogen temperature, curve 2, has been satisfactorily fitted by using (1.20) with energy, $\varepsilon(\bar{k})$, and matrix element,

$p_{c\bar{k},v\bar{k}}$, obtained from perturbation calculation of the energy bands [65]. Usually, the edge shifts toward smaller photon energies with increasing temperature as shown by curves 1 and 2. In most semiconductors, the shift is of the order of magnitude of -10^{-4} eV/° K near room temperature and becomes very small at low temperatures. Referring to (6.2), the effect of lattice vibration, $(\partial E_G/\partial T)_v$, gives always a negative temperature shift. The effect of lattice dilation, $(\partial E_G/\partial V)_T$ or $-(\beta/\chi)\,(\partial E_G/\partial P)_T$, may give, however, a shift of the same or opposite sign. Calculated $(\partial E_G/\partial T)_v$ combined with measured $(\partial E_G/\partial P)_v$ give in general estimates which are in fair agreement with the observed temperature shift [66].

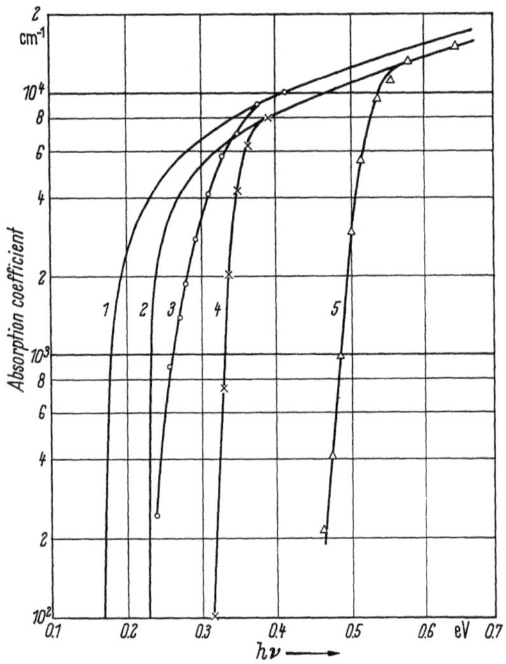

Fig. 3. Absorption edge of *n*-type InSb. Curves *1* and *2* are for pure samples at room temperature and liquid nitrogen temperature, respectively. Curves *3* and *4* show the absorption at the two temperatures for a sample having $6.1\times10^{17}\,\mathrm{cm}^{-3}$ carriers. Curve *5* is for a sample of $2.3\times10^{18}\,\mathrm{cm}^{-3}$ carriers, at 5° K. [After G. W. GOBELI, and H. Y. FAN, Phys. Rev. **119**, 613 (1960).]

In some semiconductors, e.g. Ge and InSb, the absorption edge shifts under pressure to shorter wavelengths, and in others, e.g. Si and Te, it shifts in the opposite direction. Pressure shift has been studied for a number of semiconductors [67]. It can be useful for the investigation of energy band structure. Consider the group of semiconductors consisting of the group IV elements and the group III—V compounds. These materials all have the diamond or zincblende type of structure, and their energy bands may be expected to have similar characteristics. The maximum of the valence band is near $\bar{k}=0$ in all cases. The overlapping bands constituting the conduction band have valleys at $\bar{k}=0$, along ⟨III⟩ directions, and along ⟨100⟩ direction. The different sets of valleys are not far apart in energy. One or another type of valleys has the lowest energy, depending on the material. Studies of the edge shift and pressure dependence of electrical properties indicate that the pressure coefficients, $(\partial E/\partial P)_T$, of the different valleys are similar in the various materials: $(\partial E_{000}/\partial P)_T > (\partial E_{111}/\partial P)_T > 0$ and $(\partial E_{100}/\partial P) < 0$. Thus, Ge with ⟨111⟩ minima and InSb with (000) minimum have values of $(\partial E_G/\partial P)_T$ about 5×10^{-6} eV/kg cm^{-2} and 15×10^{-6} eV/kg cm^{-2}, respectively. The edge of conduction band is given by the $\langle 100 \rangle$ valleys in Si which has $(\partial E_G/\partial P) \sim -1.5 \times 10^{-6}$ eV/kg cm^{-2}. In GaAs[68] and GaSb[69], the (000) valley is the lowest, and the pressure coefficient, normally positive, has been found to reverse its sign at sufficiently high pressures, indicating that the $\langle 100 \rangle$ valleys had become the lowest. Germanium-silicon alloys present another interesting case. The pressure coefficient is positive as in Ge for Si content less than 15%,

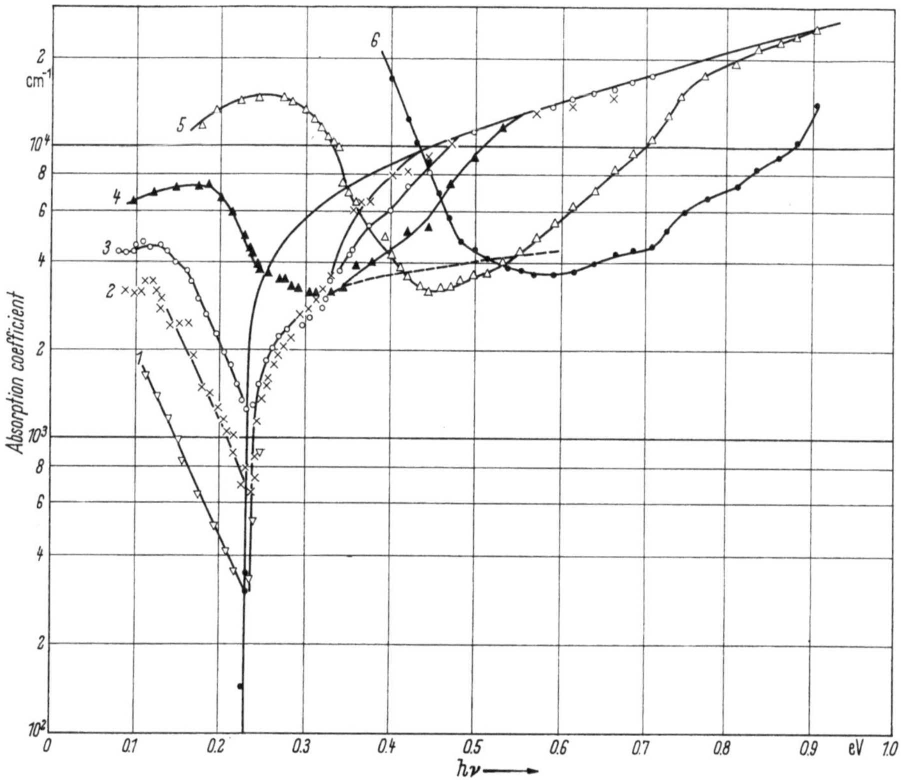

Fig. 4. Absorption edge of degenerate p-type InSb samples at 5° K. The hole concentrations in cm^{-3} are: *1* 5.5×10^{17}; *2* 9.0×10^{17}; *3* 1.56×10^{18}; *4* 2.6×10^{18}; *5* 9.4×10^{18}; *6* 2.0×10^{19}. The smooth curve is for pure samples. [After G. W. Gobeli, and H. Y. Fan, Phys. Rev. **119**, 613 (1960).]

above which it becomes negative [*67*]. Again, the reversal indicates the cross over of the $\langle 100 \rangle$ and $\langle 111 \rangle$ valleys. Similar correlations have been found for other groups of materials such as the II—VI and I—VII compounds.

The absorption edge of a semiconductor is significantly affected by the presence of free carriers when the latter have a sufficiently large concentration such that the Fermi level is inside the energy band. The states between the Fermi level and the band edge become ineffective for interband transitions being almost fully occupied by carriers. Therefore, the absorption edge will be shifted to higher energies. This effect can be seen in Fig. 3 by comparing curve 1 and 2 for samples of low carrier concentrations with curves 3 and 4. The shift provides a means for determining the Fermi level as a function of carrier concentration or the density-of-states of the energy band. If both the conduction and valence bands are non-degenerate with extrema at the same $\bar{k}$, then the absorption edge at low temperatures should be very sharp when it is determined by the Fermi level of degenerate carriers. Curve 5 shows that actually this is not the case. This is due to a more complicated structure of the valence band the effect of which can be seen more clearly in Fig. 4 for *p*-type InSb. The curves for samples with various concentrations of holes are far from sharp absorption edges. These results are understood on the basis that the valence band of InSb is degenerate near the maximum having two sub-bands of quite different curvatures. Only a small part of the holes were in the band of large curvature, the light hole band, and transitions from this branch

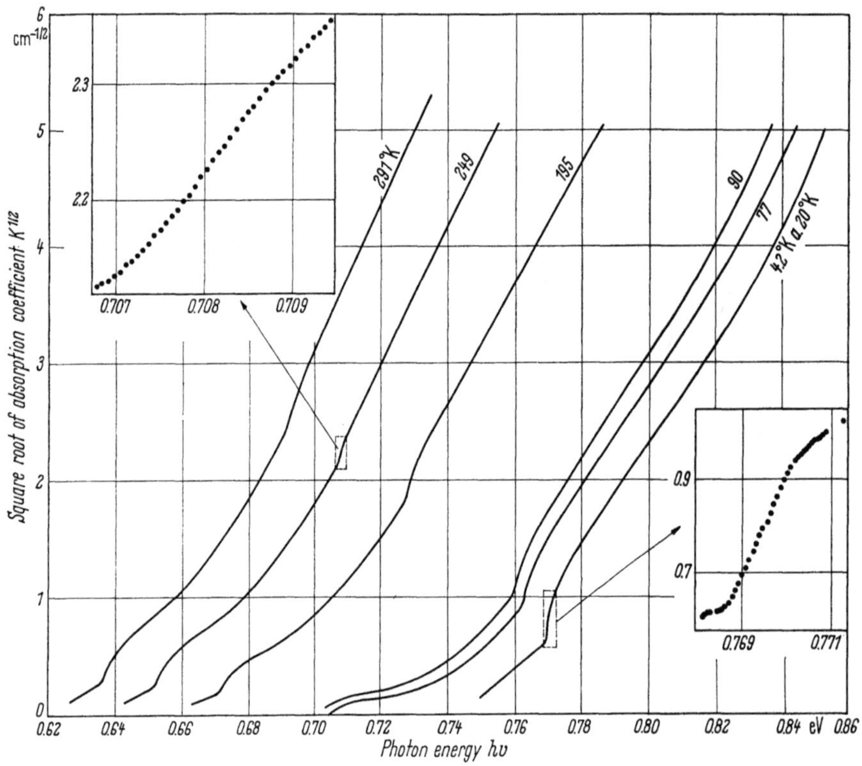

Fig. 5. Absorption edge in germanium at various temperatures. (After G. G. MACFARLANE, T. P. MCLEAN, J. E. QUARRINGTON, and V. ROBERTS.)

to the conduction band were not appreciably affected. Due to these transitions, the absorption below a certain level nearly coincided with the edge of pure samples. The fact that the absorption did not show a sharp edge at high levels is understandable on the ground that the heavy hole band does not have spherical surfaces of constant energy. The shape of the curves was used to deduce the anisotropy of the heavy hole band.

Indirect transitions were discovered first in germanium and silicon [*70*]. One of the characteristics revealing the indirect nature of the transitions is the temperature dependence of the absorption edge. The edge sharpens with decreasing temperature as transitions with phonon absorption vanishes and transition with phonon emission reduces to a low level. The more important distinguishing characteristic is the presence of steps in the absorption edge which mark the onsets of transitions involving different types of phonons [*71*], [*72*]. Fig. 5 shows the absorption edge of germanium at different temperatures [*71*]. The minima of the conduction band in germanium lie along $\langle 111 \rangle$ directions at the boundary of the Brillouin zone while the top of the value band is at $\bar{k}=0$. Therfore, the phonons involved have wave vectors near the zone boundary along ⟨111⟩, and there are four phonon branches: TO, LO, LA, TA. Analysis of the curves for different temperatures identified the transitions with absorption of each type of phonons and transitions with emisson of either type of acoustical phonons. The spectral shape of the absorption were analyzed using the theoretical expressions given in Sect. 9, and the presence of exciton transitions was deduced. In germanium, an absorption peak corresponding to direct exciton transitions at $\bar{k}=0$ has also been observed at a higher energy. Details about the absorption edges in germanium and silicon can be found in the review article by McLean [*17*]. Indirect transition absorption has been found now in many materials. In a number of cases, e.g. GaP [*73*], AgCl [*74*], SiC [*75*], characteristic steps in the absorption edge have been clearly resolved.

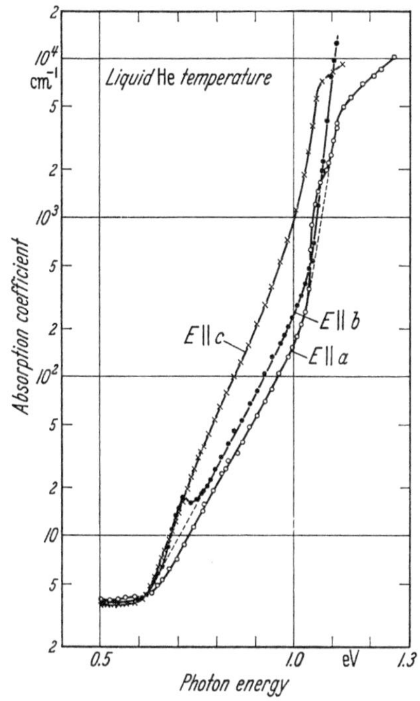

Fig. 6. Absorption edge of ZnSb measured with the electric vector, E, of radiation along each of the three axes, a, b, and c. The departure of the points from the dashed line in the lower part of $E \| b$ curve is due to some surface effect and can be eliminated by careful polishing procedure. (After H. Komiya, K. Masumoto, and H. Y. Fan.)

The absorption edge of non-cubic crystals is likely to depend on the polarization of radiation. Such dependence has been investigated for several materials including Te [*76*], CdS [*77*], CdSb [*78*], and ZnSb [*79*]. It provides additional information on the energy bands. The data for ZnSb are shown in Fig. 6 as an example. The crystal has a structure of orthorhombic symmetry with lattice parameters: $a=6.218$ Å, $b=7.741$ Å, $c=8.165$ Å along the three axes. The curves are steep in the upper part (above $\sim 10^3\,cm^{-1}$ for $E\|c$, above $\sim 300\,cm^{-1}$ for $E\|b$ and $E\|a$) which is interpreted as given by direct transitions. The curves are shifted relative to each other. The lower part of the curves corresponds to indirect transitions. The common threshold at 0.61 eV corresponds approximately to the energy gap. The result shows that one of the energy bands has three sub-bands, separate in energy at the extremum of the band. Direct transition from each branch to the opposite band is allowed only for a particular polarization, at the threshold of transition.

13. Exciton absorption. Structures in the absorption edge of the alkali halides and a few other insulating crystals have long been known and interpreted as exciton absorption [*80*]. As mentioned in Sect. 7, theoretically exciton in alkali halides is a difficult case for which neither atomic or band approximation is satisfactory. Exciton absorption in the form of a hydrogen-like series of sharp lines was discovered in Cu_2O by Gross's group [*81*] in 1952. This discovery and the development of semiconductor physics stimulated great activity in the study of excitons. Exciton lines often correspond to high absorption coefficients of the order of $10^5\,cm^{-1}$. Very thin samples have to be used for transmission measurements. Strong lines can be studied with reflection instead of transmission measurements as is done in some investigations. A variety of materials have been studied, and the results have been reviewed in several papers [82], [83], [84].

An exciton spectrum of the Wannier type may be discussed on the basis of the following considerations. 1. Sharp lines forming a series arise from direct transitions to various exciton bands, although often only one of the lines is observed. In the simple case of non-degenerate bands, the line spacings follow the hydrogenic expression (8.19), beginning with $n=1$ if interband transition is allowed at the band extrema or with $n=2$ in case the transition is forbidden. Cu_2O provides an outstanding example of the latter case. 2. There may be more than one series of lines owing to the fact that one or both of the energy bands have several subbands with small energy separations, e.g., there are four series of lines in Cu_2O [84] and three in CdS [85] and ZnO [86]. Polarization effect may be expected in noncubic crystals. For example, both CdS and ZnO show polarization effects. In

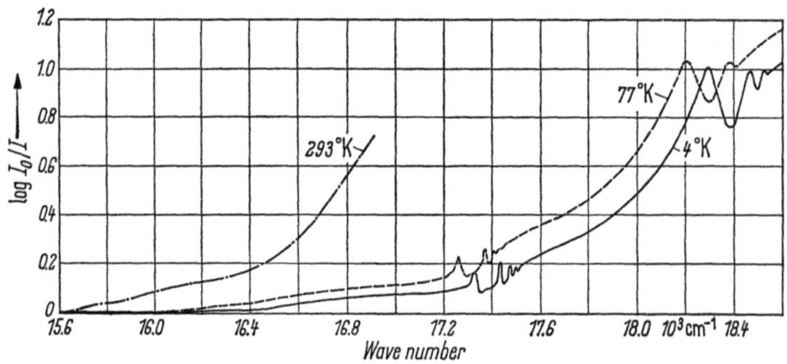

Fig. 7. Absorption spectra of Cu_2O at three temperatures. (After S. NIKITINE.)

ZnO two of the series are observed mainly with electric vector, E, perpendicular to the hexagonal c-axis whereas the third series shows up for $E \parallel c$. 3. The absorption may begin with indirect exciton transitions, giving a continuous background beginning at lower energies on which sharp lines are superimposed. We mentioned that many substances show indirect transitions. In some of them, e.g. Ge [87] and AgCl [74], exciton lines have also been observed.

In addition to the intrinsic exciton lines characteristic of the material, lines associated with impurities may be found in the neighborhood of the absorption edge. Such lines should be separated out from the spectra. They will be discussed separately in Sects. 21—23.

The absorption curves for Cu_2O shown in Fig. 7 [83] serve the purpose of demonstrating the points mentioned above. The absorption begins in the red region showing no line structure until $\sim 17{,}200\ \text{cm}^{-1}$, at low temperatures. This part of the absorption with its temperature dependence can be satisfactorily explained on the basis of indirect transitions. Two series of lines are seen in the low temperature curves, so called the yellow and the green series. At 4.2° K, they are represented by:

$$\text{yellow series: } \tilde{\nu} = 17525 - 790/n^2\ (\text{cm}^{-1}),$$
$$\text{green series: } \tilde{\nu} = 18598 - 1242/n^2\ (\text{cm}^{-1}),$$

beginning with $n=2$. The green series has no $n=1$ line. The $n=1$ line of the yellow series lies at $\sim 16395\ \text{cm}^{-1}$. The line is, however, relatively very weak and does not appear in Fig. 7. The interpretation is that both exciton series are associated with energy bands for which transitions are forbidden at the extrema.

Two additional series at shorter wavelengths have been found by PASTRNYAK [88] in reflection measurements. Two lines of each series were observed, given by:

$$\text{blue-green series: } \tilde{\nu} = 21220 - 368/n^2 \ (\text{cm}^{-1}),$$

$$\text{blue series: } \tilde{\nu} = 22302 - 415/n^2 \ (\text{cm}^{-1}),$$

with $n=1$, 2. On the basis of the theoretical treatments of ELLIOTT [89] and ZHILICH [90], GROSS [84] attributes the four series of excitons to the effect of two valence bands, Γ_7^+ and Γ_8^+, and two conduction bands, Γ_6^+ and Γ_{12}^-. The lower conduction band, Γ_6^+, gives with Γ_7^+ and Γ_8^+ the yellow and green series, respectively, and the upper conduction band Γ_{12}^- gives with the same valence bands the blue-green and the blue series.

In semiconductors with large dielectric constants and small carrier effective masses, the exciton rydberg, R_{ex}, can be very small, of the order of 10^{-2} or 10^{-3} eV. It is difficult to observe directly a series of resolved lines. However, single exciton lines have been observed in a number of such materials, for example: in Ge [87] with $R_{\text{ex}}=0.0012$ eV, in GaSb [91] with $R_{\text{ex}}=0.0028$ eV, in GaAs [92] with $R_{\text{ex}}=0.0034$ eV, and InP [93] with $R_{\text{ex}}=0.0040$ eV. The absorption on the high energy side of the line can be fitted with (8.20).

Several special points of interest are discussed in the following.

Quadrupole transition: The $n=1$ line of the yellow series in Cu_2O is unusual in that it has been shown to be a quadrupole transition. GROSS and KAPLYANSKII [94] found that, in the cubic crystal, this absorption line is anisotropic. For light propagating along an axis of four-fold symmetry, $\bar{k}_\lambda \| C_4$, or along an axis of three-fold symmetry, $\bar{k}_\lambda \| C_3$, the absorption is independent of polarization, but it is stronger. for $\bar{k}_\lambda \| C_4$ than for $\bar{k}_\lambda \| C_3$. More striking is the case of $\bar{k}_\lambda \| C_2$, in which case the line is completely polarized with electric vector perpendicular to the cube plane containing the C_2 axis. The quadrupole transition has been treated theoretically by ELLIOTT [89].

Use of uniaxial stress: Application of elastic deformation is in general a useful means for optical studies of solids since degenerate energy levels or degenerate bands may split by the deformation. Uniaxial compression was used by GROSS and KAPLYANSKII to study Cu_2O [95]. The yellow series and the steps in the red absorption were studied. Splittings and polarization effects were observed with compression applied along various directions. Particular attention was paid to the $n=1$ line and one of the steps. From the results, it was deduced that both are triply degenerate, indicating that the red absorption corresponds to indirect transitions to the yellow $n=1$ exciton band which is the lowest exciton band.

Longitudinal and transverse excitons: The nature of longitudinal and transverse excitons has been explained in Sect. 7. In cubic crystals, excitons with $\bar{K} \sim 0$ are purely transverse or purely longidinal. Radiation field, being a transverse wave, interacts only with transverse excitons and does not produce longitudinal excitons in cubic crystals. In non-cubic crystals, exciton of $\bar{K} \sim 0$ may have a mixed character, and nearly longitudinal excitons can be produced by light through a small transverse component of polarization. Longitudinal exciton lines have been observed in the hexagonal crystals ZnO [33] and CdS [96]. HOPFIELD and THOMAS gave a phenomenological theory of the problem. According to the theory, the exciton has two modes, a pure transverse and a mixed mode. The transverse mode has a polarization perpendicular to both the wave vector, $\bar{K}$, and the hexagonal axis, C. The polarization vector of the mixed mode lies along the projection of the wave vector in the plane normal to the hexagonal axis. Thus, the mixed mode becomes purely transverse for $\bar{K} \| C$ and purely longitudinal for $\bar{K} \perp C$. Using light polarized in the plane of $\bar{k}_\lambda$ and c-acis, only the mixed mode will be observable, and with $\bar{k}_\lambda$ nearly normal to the c-axis, short of 90° by a small angle Θ, excitons which are almost longitudinal can be observed. In this way, two lines were observed in ZnO, on the high energy side of each of two known transverse exciton lines. These lines increased in intensity with increasing Θ as is expected from the increase of the longitudinal component of mixed modes. The energies of those lines in relation to the energies of the corresponding transverse exciton lines are as expected of longitudinal excitons.

Spatial dispersion, anomalous waves.

For most optical properties, the wavelength of the radiation may be considered as infinite corresponding to a zero wave vector $\bar{k}_\lambda$. The effect of a small but finite $\bar{k}_\lambda$ is called spatial dispersion. It is known that spatial dispersion leads to optical activity and may give rise to double refraction in crystals normally not expected to be birefringent [*97*]. The effects are usually small. It was first shown theoretically by PEKAR [*98*] that spatial dispersion may have a large effect at frequencies close to the exciton absorption band, as a result of which it is possible to have two waves of the same polarization propagating in the same direction with different phase velocities. The phenomenon known as anomalous waves has since been investigated extensively, particularly by scientists in the Soviet Union [*99*]. Experimental observations of such waves in several materials have been reported.

The essential physics of the phenomenon may be brought out by the following simple consideration. In the absence of damping, the refractive index n is related to the real dielectric constant by:

$$n^2 \equiv \frac{c^2 k_\lambda^2}{\omega^2} = \varepsilon(k_\lambda, \omega). \tag{13.1}$$

When the spatial dispersion is neglected, $\varepsilon(0, \omega)$ is a function only of ω. Each frequency corresponds to a single value for the magnitude of $\bar{k}_\lambda$ and a single value of n. With ε dependent on $\bar{k}_\lambda$, the equation may be of higher order in k_λ^2 giving more than one solution for the magnitude of $\bar{k}_\lambda$, hence anomalous waves are obtained. Classically as well as quantum mechanically, the expression of the dielectric constant is equivalent to that of oscillators of various frequencies. It is sufficient for our purpose to consider the oscillators of a frequency ω_0 close to the frequency of interest. In this case:

$$\varepsilon = \varepsilon' + \frac{A}{\omega_0^2 - \omega^2} \sim \varepsilon' + \frac{A}{2\omega(\omega_0 - \omega)} \tag{13.2}$$

where ε' and A are constants, i.e.; except for the oscillators under consideration the material would have a constant $\varepsilon = \varepsilon'$ near ω_0. We get a dependence of ε on the wave vector if ω_0 varies with k. Such a variation is not difficult to expect, e.g. a coupling between the oscillators would produce the effect [*29*], [*100*]. Specifically in the case of excitons, photons of wave vector $\bar{k}_\lambda$ produce, according to (8.5) and (8.15), excitons of energy:

$$\hbar\omega_0 = \hbar\omega_0(0) + (\hbar^2/2M)\, k_\lambda^2. \tag{13.3}$$

Using (13.3) and (13.2), we get for (13.1) two solutions of k_λ^2 which correspond to:

$$\left.\begin{aligned} n_\pm^2 &= \tfrac{1}{2}(\mu + \varepsilon') \pm [\tfrac{1}{3}(\mu - \varepsilon')^2 + b]^{\frac{1}{2}}, \\ \text{where}\quad \mu &= \frac{2Mc^2}{\hbar\omega}\left(1 - \frac{\omega_0}{\omega}\right), \qquad b = \frac{Mc^2}{\hbar\omega^3} A \sim \frac{Mc^2}{\hbar\omega_0^3} A. \end{aligned}\right\} \tag{13.4}$$

The general behavior of n_+ and n_- are illustrated by the curves in Fig. 8a [98]. The curves approach asymptotically a sloped line representing μ, giving a n_+ and n_- for each frequency. In contrast, the usual Lorentz model of oscillators

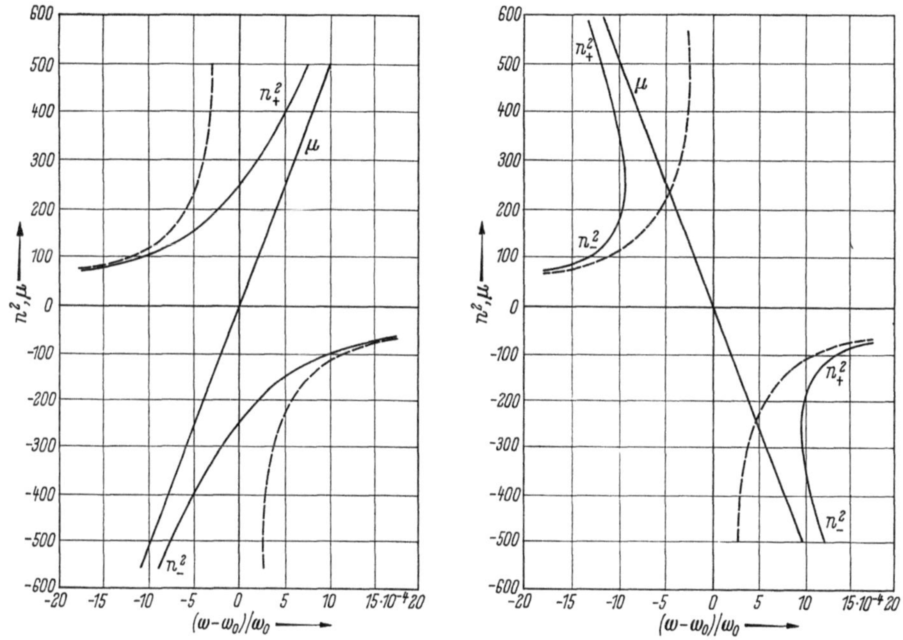

a b

Fig. 8a and b. Refractive indices squared, n_+^2 and n_-^2, of anomalous waves at frequencies ω close to the oscillator frequency ω_0. The curves are calculated with $|M|=m$, $\hbar\omega_0=2$ eV, and $b=58{,}400$. The dashed curves are for constant ω_0. a) For $M>0$. b) For $M<0$. (After S. I. PEKAR.)

with fixed ω_0, gives the dashed curves which approach asymtotically the vertical line at $\omega=\omega_0$, giving a single value of n for each ω. Fig. 8b shows the case where the dependence of $\hbar\omega_0$ on k^2 corresponds to a negative effective mass M.

Experimentally, detection of anomalous waves have been reported for Cu_2O, anthracene, CdS and ZnTe. Since the two waves have the same polarization, interference effect should occur and transmission through the material would oscillate with variation of the sample thickness, in the region of the exciton band. A pronounced oscillatory effect has been observed in Cu_2O by GORBAN' and TIMOFEEV [101] and in anthracene by BRODIN and PEKAR [102]. Fig. 9 shows the results obtained for anthracene by using a large number of samples with various thicknesses. From the period of oscillations, the difference between the two refractive indices was estimated to be $n_+ - n_- = 6.9$.

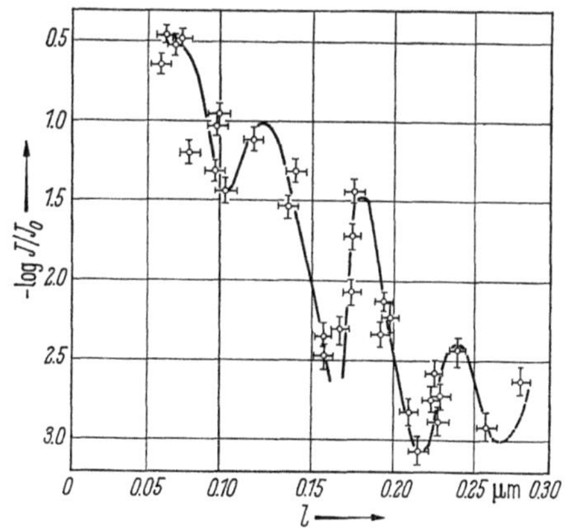

Fig. 9. Variation of transmission, I/I_0, with sample thickness, l (in microns), for anthracene at 20° K. Measurements were made at frequency of 25,108 cm^{-1}. (After M. S. BRODIN and S. I. PEKAR.)

From transmission measurements on thin films of CdS, Brodin and Strashnikova [*103*] found that the dispersive and absorptive parts of the refractive index as obtained from classical analysis of the data are incompatible, and they attributed the incompatibility to spatial dispersion. Hopfield and Thomas [*100*] made reflectivity measurements on CdS and ZnTe crystals. Some of the results on CdS are shown in Fig. 10. The curve in (b), calculated according to the classical oscillator model, shows that a line width of about 10^{-3} eV is required in order to reproduce the width of the observed reflectivity variation. On the other hand, the width of the exciton line found in transmission measurements is less than 10^{-4} eV. This discrepancy is understandable on the basis of spatial dispersion with which we get the composite effect of two waves, e.g. there is always one

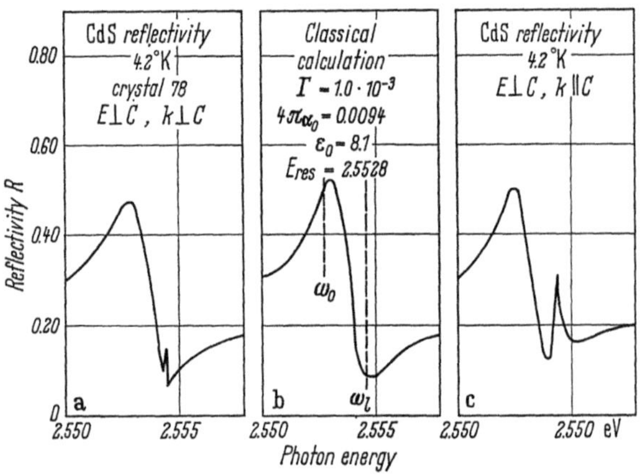

Fig. 10a–c. Reflectivity of CdS at normal incidence, in the vicinity of the first exciton peak. (a) and (c): curves measured with wave vector $\boldsymbol{k}$ perpendicular and parallel to the *C*-axis, respectively. (b): curve calculated according to the classical theory. ω_0 and ω_l are respectively the transverse and longitudinal frequencies of the exciton. [After J. J. Hopfield, and D. G. Thomas.]

wave that can propagate into the crystal, and the classical total reflection region is absent. It is clear then that the reflectivity curve is not directly indicative of the linewidth. In the case of ZnTe, the measured reflectivity curves have little resemblance to the shape given by the classical theory. There was a pronounced minimum, but the maximum was almost completely suppressed.

In Fig. 10 (a) and (c), the sharp peak near 2.555 eV is a striking anomaly. It is situated at ω_l corresponding to the minimum of reflectivity in the classical picture (b). The explanation given for this peak requires, in addition to the presence of anomalous waves, the existence of a potential barrier for excitons at the sample surface. One of the suggested causes for a potential barrier is the surface image force which can be expressed in terms of a potential energy:

$$V(x) = \frac{1}{2}\left(\frac{\varepsilon-1}{\varepsilon+1}\right) E_B \left(\frac{a}{x}\right)^3,$$

where E_B is the exciton binding energy, a is the exciton Bohr radius, and ε is the static dielectric constant of the crystal. With $\varepsilon > 1$, the force is repulsive leading to a depletion of excitons near the surface. Thus, there is a surface layer having a different effective index of refraction. The reflectivity calculated on this model shows in fact a peak at ω_l which is more prominant for larger thicknesses of the depletion layer. With thicknesses of the order of 10^2 Å, curves are obtained which resemble the observed reflectivity of CdS as well as that of

ZnTe. It is clear that anomalous dispersion and surface boundary conditions are important in experimental studies of exciton reflection.

14. Intrinsic edge emission. Radiation may be emitted from a solid due to electron transitions under a non-equilibrium condition. The crystal can be excited by irradiation with electromagnetic waves or high energy charged particles. Emission may be produced also by electrically injecting carriers into a semiconductor from a contact or *p-n* junction. Radiation emitted with frequencies close to that of the absorption edge is called edge emission. Such emission may come from the recombination of free electrons with free holes or from the decay of excitons in a bound state. There may be also recombination emission involving localized states of imperfection centers. The emission given by such processes occurs in general at longer wavelengths, but it may still be close enough to the absorption edge to be classified as edge emission. It is the cause for the common complication that edge emission of a given material varies with the specimen. It will be discussed separately in connection with impurity effects.

The kinetics of excitation and emission is a complicated problem which is outside of the scope of the present article. We shall limit our discussion to the nature of intrinsic emission resulting from excitations that produce interband transitions.

The luminescence spectra of a number of substances [*83*], [*84*], e.g. CdS, CuCl, HgI_2, etc., contain sharp lines which correspond closely to the exciton absorption lines, with slight shifts to longer wavelengths. These lines have been attributed to the decay of intrinsic excitons. Consider the case of CdS which has been extensively studied by different investigators. The crystal gives strong green emission with peaks at wavelengths longer than 5100 Å, and a number of sharp lines, so called blue emission, at shorter wavelengths. According to THOMAS and HOPFIELD [*85*], only two of the observed lines correspond to intrinsic excitons. The others are close in frequency to the absorption lines found to be associated with impurities. CdS has three exciton series, each involving the conduction band and one of three sub-bands in the valence band. The $n=1$ exciton of the series given by the top valence band is identified by an absorption line at 4873 Å (at 77° K), designated as the A line. The $n=1$ exciton of the series formed with the next valence band gives an absorption line at 4844 Å designated as the B line. The two fluorescence lines are close in frequency to the A and B lines and have correspondingly the same polarization properties. The line close to A is polarized with $E \perp C$, and the line close to B has both polarizations. The intensity ratio of the two fluorescent lines corresponds approximately to Boltzmann distribution of holes between the two valence bands.

It is interesting to note that the two fluorescent lines of CdS which are identified with intrinsic excitons are weak compared to the fluorescence lines associated with impurities. This observation was explained by the authors on the basis of the theoretical conclusion of HOPFIELD [*39*] that the excitons decay only when they encounter some crystal imperfection such as an impurity or a surface, as discussed in connection with polaritons (see Sect. 7).

In semiconductors, exciton emission as well as free electron hole recombination emission has been observed. Fig. 11a shows the photoluminescence curves of *n*-type InAs samples of various electron concentrations [*104*]. The emission band becomes broader with increasing electron concentration showing the effect of electron distribution in the conduction band on the electron-hole recombination. The solid curve is calculated according to the electron distribution with an appropriate effective mass for the $1.8\times10^{17}\ \mathrm{cm}^{-3}$ sample. The spectra measured under an applied magnetic field are shown in Fig. 11b. The dashed curve shows a peak at about 0.41 eV, with a shoulder extending to higher energies. Apparently, the peak and the shoulder correspond to exciton and electron-hole

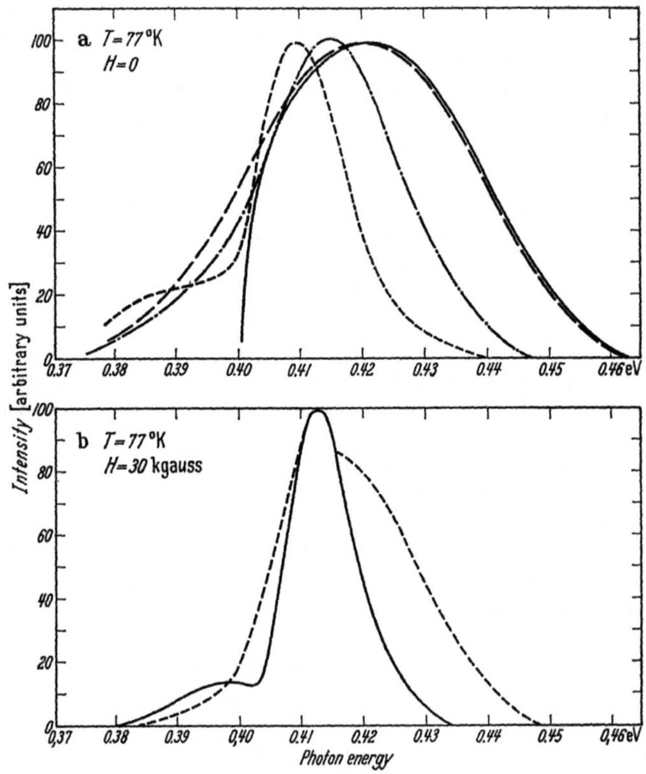

Fig. 11a and b. Photoluminescence of n-type InAs. a) Spectra of samples with various carrier concentrations: short dashed curve 2.3×10^{16} cm^{-3}, dot-dash curve 9×10^{16} cm^{-3}, dashed curve 1.8×10^{17} cm^{-3}. The solid curve is calculated for an electron concentration of 1.8×10^{17} cm^{-3}. b) Spectra of two samples: solid curve for 2.3×10^{16} cm^{-3} electron concentration and dashed curve for 9×10^{16} cm^{-3} electron concentration, measured under a magnetic field of 30 kgauss. (After A. Mooradian, and H. Y. Fan.)

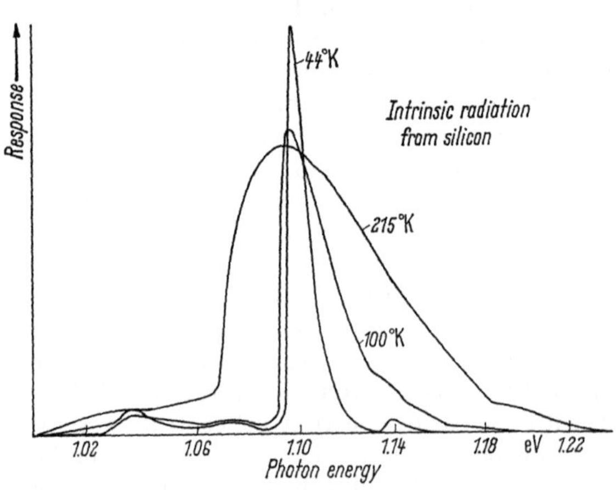

Fig. 12. Recombination radiation from silicon obtained at three different temperatures. (After J. R. Haynes, M. Lax, and F. Flood.)

recombinations, respectively. Exciton emission probably dominates at low carrier concentrations as indicated by the solid curve.

In a semiconductor with indirect transition edge, emission from exciton or free carrier recombination can occur with phonon cooperation. Emission accompanied by phonon creation occurs at longer wavelengths and can be detected without the interference of internal absorption in the crystal. Such emission is

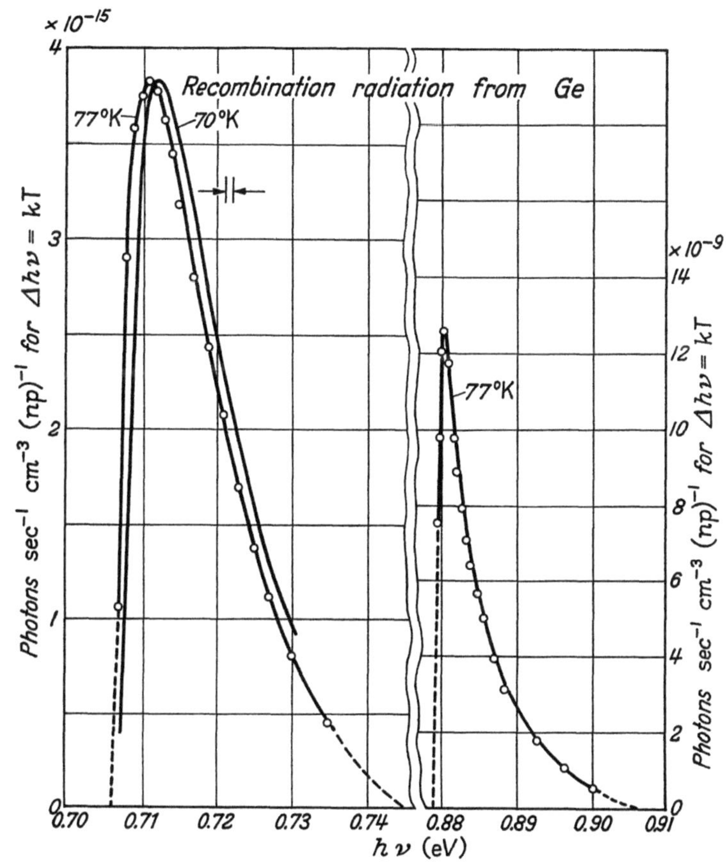

Fig. 13. Recombination radiation from germanium. The efficiency is plotted. The curves are obtained from measurement. The points are calculated from absorption using the principle of detailed balance. The right part of the figure shows the direct transition without reabsorption. The left part shows indirect transition involving longitudinal acoustic phonons. (After J. R. Haynes, and N. G. Nilsson.)

shown in Fig. 12 for silicon [*105*]. The curve for the lowest temperature shows clearly four peaks. They are identified as emissions involving acoustic transverse phonon, acoustic longitudinal phonon, optical transverse phonon, and optical longitudinal phonon, respectively in order of decreasing energy of the peak. In the case of germanium, emission of both direct transition and indirect transition excitons have been observed as shown in Fig. 13 [*106*]. The indirect transition is interpreted as primarily due to exciton recombination. Its width is attributed to the thermal energy distribution of the excitons together with their scattering by the lattice. In the case of the direct transition emission, the peak region is attributed to recombination via excitons and the part above 0.8825 eV is attributed to the recombination of free electrons and holes.

VII. Free carrier effects in semiconductors.

15. Absorption. In semiconductors, effects of free carriers can be observed at long wavelengths below the threshold of interband transitions. Studies were made first on n-type germanium [*19*], and the theory of free carrier absorption, discussed in Sect. 5, developed in connection with these studies [*20*]. According to the usual Drude-Kronig theory (5.3), the absorption is simply related to the *dc* conductivity and should vary as $1/\omega^2$ when $\omega\tau\gg 1$. The inadequacy of this simple theory was shown clearly by the measurements made at different temperatures on samples of various carrier concentrations. Fig. 14 shows the absorption per unit carrier concentration or the absorption cross section of carriers [*20*]. It is seen that the frequency dependence is distinctly different for the two temperatures. Scattering by lattice vibration dominates at the higher temperature, and the samples of various carrier concentrations gave the same curve of absorption cross section. At 78° K, the absorption cross section was found to be higher in samples or larger carrier concentrations. This is due to the fact that the scattering by charged impurities becomes important at low temperatures, and a higher carrier concentration is associated with a larger concentration of impurities. The curves were calculated by using (5.4) and (5.11) with an average effective mass for the conduction band. The curves are seen to fit the data reasonably well. Improved calculations have been made [*22*], [*107*] using detailed information about the many-valley conduction band and taking into account also scattering by energetic modes (5.7).

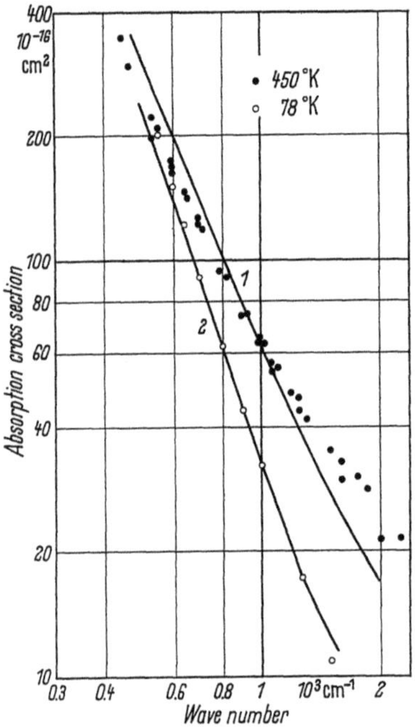

Fig. 14. Absorption cross section of conduction electrons in n-type germanium. The solid points for 450° K represent combined data for samples with carrier concentrations $n=(4.1-75)\times 10^{16}$ cm^{-3}. The open circles gives the 78° K data for a sample of $n=10^{18}$ cm^{-3}. The calculated curves take into account scatterings by acoustical vibrations and charged impurities. (After H. Y. FAN, W. SPITZER, and R. J. COLLINS.)

Free carrier absorption observed in various semiconductors [*108*] generally has a frequency dependence varying from $\omega^{-1.5}$ to $\omega^{-3.5}$. In some compounds, e.g. InP [*25*] and InAs [*109*], the polar optical mode scattering appears to be important. Free carrier absorption is potentially an effective means for investigating the carrier scattering processes. For this purpose, measurements over wide frequency and temperature ranges should be made on samples of known electrical properties, and the data have to be carefully analyzed. If only the order of magnitude of absorption is of interest, even the simple Drude-Kronig theory may give a reasonable estimate under certain conditions. Thus, for $\hbar\omega > kT$, we can write approximately:

$$\sigma \sim \frac{\sigma_0}{\omega^2\tau^2}\left[\frac{4}{9\pi^{\frac{1}{2}}}\left(\frac{\hbar\omega}{kT}\right)^{\frac{1}{2}}\right] \tag{15.1}$$

for (5.4), and

$$\sigma \sim \frac{\sigma_0}{\omega^2\tau^2}\left[\frac{64}{3\sqrt{\pi}}\left(\frac{kT}{\hbar\omega}\right)^{\frac{3}{2}}\right] \tag{15.2}$$

for (5.11). Often, absorption measurements are conveniently made at room temperature with wavelengths of several microns. Under such conditions, the quantity in the square brackets does not affect the order of magnitude of the estimate.

In non-cubic crystals, the free carrier absorption may be dependent on the polarization of radiation. In tellurium [*110*] the ratio of free hole absorptions for electric vector perpendicular and parallel to the crystal *C*-axis was found to be: $\alpha_\perp/\alpha_\parallel \sim 0.8$ at 300 °K and $\alpha_\perp/\alpha_\parallel \sim 2.2$ at 77° K. In ZnSb [*79*], the free hole absorption for electric vector parallel to the *C*-axis is higher than for electric

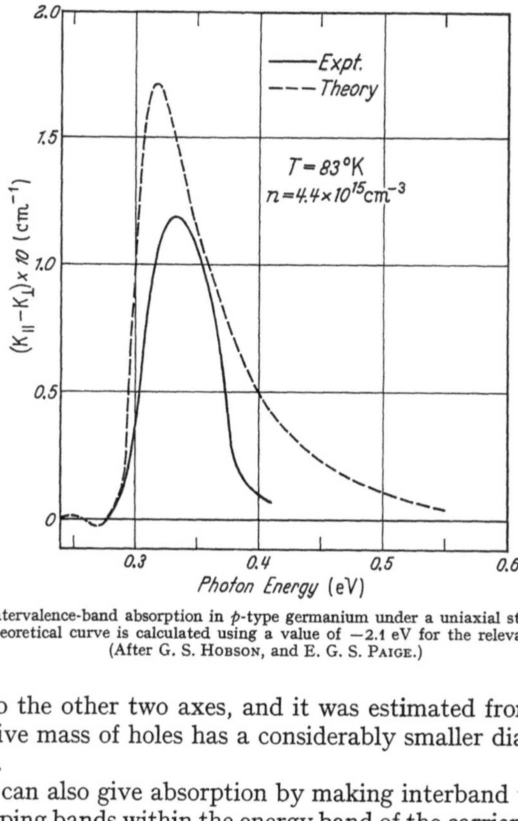

Fig. 15. Anisotropy of intervalence-band absorption in *p*-type germanium under a uniaxial stress of −3 kg/mm² in the ⟨100⟩ direction. The theoretical curve is calculated using a value of −2.1 eV for the relevant deformation potential. (After G. S. Hobson, and E. G. S. Paige.)

vector parallel to the other two axes, and it was estimated from this result that the tensor effective mass of holes has a considerably smaller diagonal component along the *C*-axis.

Free carriers can also give absorption by making interband transitions. When there are overlapping bands within the energy band of the carriers, such transitions may occur at lower frequencies than the threshold of the valence band to conduction band transitions, and the associated absorption is observable on the long wavelength side of the intrinsic absorption edge. Free carrier interband absorption in semiconductors was first observed in *p*-type germanium [*111*] with three absorption bands corresponding to hole transitions between each two of the three valence bands. Theoretical calculation of the absorption has been made with valence bands obtained from perturbation treatment [*112*]. Anologous absorption has been observed by various workers in most of the III—V compounds [*108*] which have a similar kind of valence band. Transitions of free electrons to other minima inside the conduction band has also been observed in silicon and some III—V compounds.

Intervalence-band absorption of holes has been also observed in the non-cubic crystals Te [*110*] and ZnSb [*79*]. The absorption band is highly polarization dependent in both cases. Recently, HOBSON and PAIGE [*113*] studied the intervalence-band absorption of germanium under uniaxial stress. A polarization dependence was produced by the stress. The effect can be calculated according to the theory of strain effect on the valence band. The comparison of observed and calculated differences in absorption is shown in Fig. 15. Anisotropy of the intervalence-band absorption can also be produced by applying a high electric field which causes an anisotropic distribution of the holes [*114*].

16. Effect on reflection [*115*]. The intraband effect of free carriers contributes to the dielectric constant ε_1. The contribution is represented by the second terms of (2.8) and can be written in the form:

$$\tilde{\varepsilon}_1^{f} = -\frac{4\pi e^2}{(\hbar\omega)^2} \int \frac{2}{(2\pi)^3}\, d\bar{k}\, f(\bar{k})\, \nabla_{\bar{k}} \nabla_{\bar{k}}\, \varepsilon(\bar{k}) \tag{16.1}$$

for carriers within one energy band. The contribution corresponds to the out-of-phase current of free carriers in an electric field. The conductivity which corresponds to an in-phase current owes its existence to the scattering of the carriers. For the out-of-phase current, however, the effect of scattering can be neglected when the photon frequency is large in comparison with the collision frequency $1/\tau$. Thus, (5.2) of the Drude theory reduces, for $(\omega\tau)^2 \gg 1$, to (16.1) for the case of a scalar effective mass: $\hbar^{-2} \nabla_{\bar{k}} \nabla_{\bar{k}} \varepsilon = (1/m^*)\tilde{1}$.

At infrared frequencies, the condition $(\omega\tau)^2 \gg 1$ is usually satisfied. Free from the complications of scattering, the carrier contribution to electric susceptibility gives directly information about the energy band. We define an effective mass for susceptibility $\tilde{m}_s$:

$$\tilde{\varepsilon}_1^{f} = -\frac{4\pi e^2 N}{\omega^2} \widetilde{\left(\frac{1}{m_s}\right)}. \tag{16.2}$$

Comparison of (16.1) and (16.2) shows that $\tilde{m}_s$ depends in general on the temperature through the distribution function $f(\bar{k})$ of carriers. However, when $\varepsilon(\bar{k})$ can be approximated by a quadratic function, m_s is independent of the carrier distribution:

$$\widetilde{\left(\frac{1}{m_s}\right)} = \hbar^2 \nabla_{\bar{k}} \nabla_{\bar{k}} \nabla(\bar{k}) = \widetilde{\left(\frac{1}{m^*}\right)}. \tag{16.3}$$

For a multivalley energy band in a cubic crystal, we get

$$1/m_s = (\tfrac{1}{3})\,(1/m_\alpha + 1/m_\beta + 1/m_\gamma)\,, \tag{16.4}$$

where $m_\alpha, m_\beta, m_\gamma$ are the principal components of the tensor effective mass of an individual valley. When there are two isotropic bands which are degenerate at the band extrema, the expression is:

$$1/m_s = (m_1^{\frac{1}{2}} + m_2^{\frac{1}{2}})/(m_1^{\frac{3}{2}} + m_2^{\frac{3}{2}})\,. \tag{16.5}$$

If $\varepsilon(\bar{k})$ cannot be represented by a quadratic function, m_s will vary with carrier distribution. In case $\varepsilon(\bar{k})$ is isotropic, we can write

$$\varepsilon(\bar{k}) = \hbar^2 k^2/2m^*(\varepsilon)\,, \tag{16.6}$$

then

$$m_s = m^*(\zeta) \tag{16.7}$$

for a completely degenerate distribution with Fermi energy ζ.

The effect of ε_1^f shows up clearly in the reflectivity, R, which is related to the optical constants by:

$$R = \frac{(1-n)^2 + k^2}{(1+n)^2 + k^2}. \tag{16.8}$$

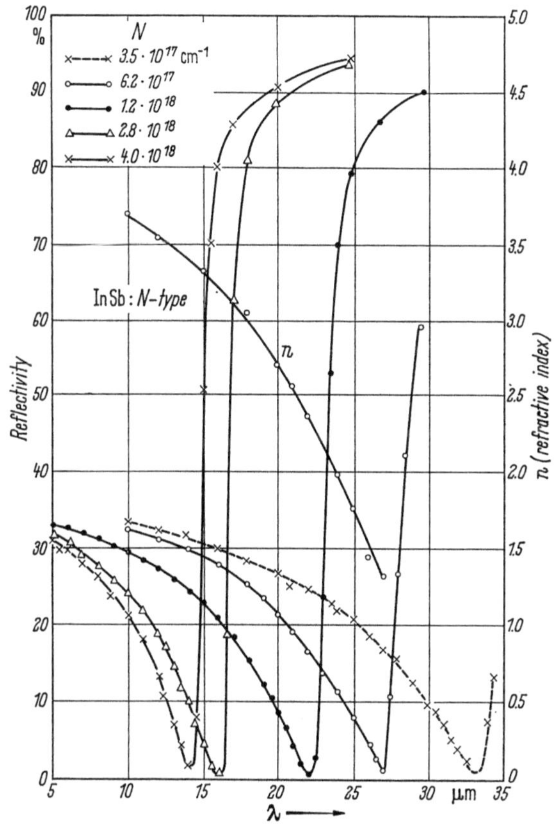

Fig. 16. Reflectivity as a function of wavelength for n-type samples of InSb. The curve of refractive index, n, is for the sample $N = 6.2 \times 10^{17}$ cm^{-3}. (After W. G. SPITZER, and H. Y. FAN.)

With a sufficient carrier concentration, it is possible for ε_1^f to become appreciable in the infrared while the extinction coefficient k is still small. Then

$$R \sim \frac{(1-n)^2}{(1+n)^2} \sim \frac{(1-\varepsilon_1^{\frac{1}{2}})^2}{(1+\varepsilon_1^{\frac{1}{2}})^2}, \tag{16.9}$$

$$\varepsilon_1 = \varepsilon_1^0 + \varepsilon_1^f = \varepsilon_1^0 - 4\pi\, e^2 N/\omega^2\, m_s, \tag{16.10}$$

where ε_1^0 is the dielectric constant in the absence of free carriers. It is seen that a minimum in reflectivity occurs near the frequency where $\varepsilon_1 = 1$. How close the minimum approaches to zero depends on the magnitude of the extinction coefficient in this region. The smaller the scattering frequency, $1/\tau$, in the material, the smaller will be the extinction coefficient and the more pronounced will be the minimum of reflectivity. At still higher frequencies, $\varepsilon_1 \leqq 0$ and high reflectivity is obtained. The frequency at which $\varepsilon_1 = 0$ corresponds to the plasma frequency, according to (2.15).

Fig. 16 shows as an example the experimental reflectivity data for several InSb samples of various carrier concentrations. In order to obtain both optical constants for the determination of ε_1, reflectivity measurement should be supplemented by absorption measurement. In this particular case, the extinction coefficient was very low near the region of the reflectivity minimum, and the minimum of each curve was very close to zero. Under such conditions, (16.9) and (16.10) are good approximations, and estimate of ε_1^f can be obtained from reflectivity measurement alone. This conclusion was actually born out by absorption measurement made on one of the samples. The ε_1^f obtained was proportional to $1/\omega^2$, as expected, and the results for the degenerate samples with various Fermi energies gave $m^*(\varepsilon)$ approximately, according to (16.7). For other materials, e.g. Ge and Si, absorption correction was not negligible.

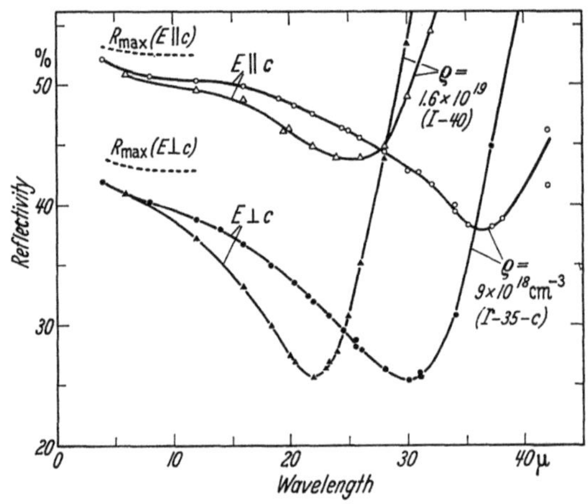

Fig. 17. Reflectivity as a function of wavelength measured with polarized radiations, for two samples of tellurium with hole concentrations of 9×10^{18} cm^{-3} and 1.7×10^{19} cm^{-3}, respectively. (After R. S. CALDWELL, and H. Y. FAN.)

Studies of the carrier effect on reflectivity have been made by different workers on various semiconductors [*115*], [*108*]. When interband absorption of free carriers is present, the contribution of interband transitions to ε_1 can be estimated from the absorption by using the Kramers-Kronig relation. This contribution should be subtracted from the total carrier effect obtained from the measurements, in order to obtain the intraband effect alone. Another complication was encountered in the case of GaSb [*116*] where carriers were present in the lowest as well as in a set of higher valleys of the conduction band.

Fig. 17 shows the case of an anisotropic energy band [*110*] .The measurements with polarizations parallel and perpendicular to the crystal C-axis gave an estimate of the components, $m_\perp$ and $m_\parallel$, of the anisotropic effective mass. The results were $m_\perp \sim m_\parallel \sim 0.45$ m at 300° K, and $m_\perp \sim 0.30$ m, $m_\parallel \sim 0.45$ m at 100° K.

VIII. Reflection spectra of solids.

Interband transitions give rise to high absorption corresponding to absorption coefficients of the order 10^4 cm^{-1} or higher. The threshold for the onset of interband transitions ranges about 1 eV or less for semiconductors and is of the order of a few electron-volts for insulators and metals. At frequencies above the threshold, it is difficult to make transmission measurements on the bulk material. For metals, the free electron or intraband absorption is so high that transmission measurements on bulk material is usually impracticable at all frequencies. Under such conditions, optical studies have to be made on thin films or be limited to reflection measurements. In reflection studies, various methods have been used to obtain two independent measurements in order to determine ε and σ. These methods include measurements of reflected intensity for different angles of incidence, measurements of intensity and phase, and measurements with polarized radiations. The basic problem associated with high absorption is that the properties of a thin layer is measured owing to the small depth of light penetration. Thus, the

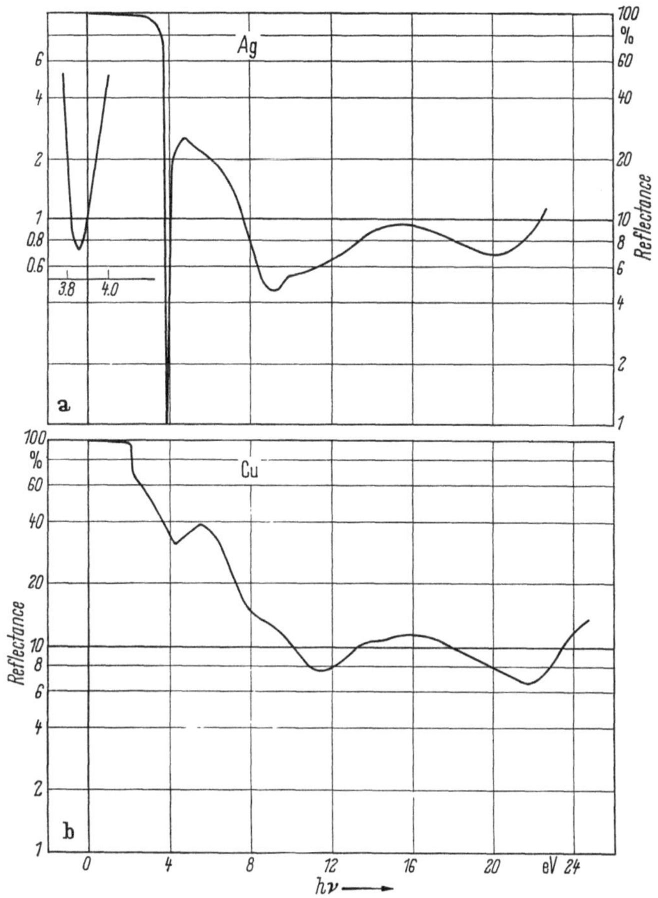

Fig. 18a and b. Spectra of reflectance for Ag and Cu. (After H. EHRENREICH, and H. R. PHILIPP.)

surface preparation is important. Electro-polishing is generally used. Many studies have been carried out on films deposited on a chosen substrate by cathode sputtering or vacuum evaporation. In any case, it is difficult to be sure that the results obtained correspond truly to the property of the bulk material, and there are considerable scatters among the data accumulated in the literature. Early works have been discussed in various texts [*3*], [*4*] and review articles [*117*].

There has been basically little change in the interpretation of the optical properties of metals. The expressions (2.11) and (2.12) have been used for a long time, except that recent treatments of the many-electron problem have justified them on a more sophisticated basis and established the connection between the transverse and longitudinal dielectric constants. A relatively recent concept is the anomalous skin effect which is important at low temperatures and long wavelengths. The effect is now well known, and we shall not be concerned with it here. Strong interest in the reflection spectra over a wide spectral range is renewed recently, due mainly to the better knowledge of the energy bands of the materials. Theoretical calculations and various experimental studies have provided for some materials sufficient information with which the spectra can be interpreted concretely in terms of the structure of the energy bands. Studies of reflection spectra have been made for a number of metals [*118*], semiconductors [*119*], and insulators [*120*]. Examples of such spectra are given in Fig. 18 [*121*]

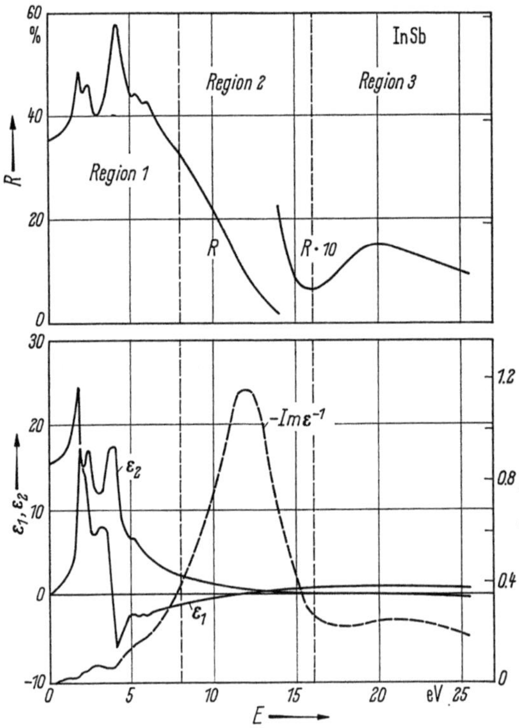

Fig. 19. The spectral dependence of the reflectance R, the real and imaginary parts of the dielectric constant, ε_1 and ε_2, and the energy loss function, $-\operatorname{Im}\varepsilon^{-1}$, for InSb. (After H. R. PHILIPP, and H. EHRENREICH.)

and Fig. 19 [*119*]. The emphasis has been to cover a wide energy range in order to get a broad picture of the energy bands. In such studies the procedure of deducing both ε_1 and ε_2 from a reflection spectrum with the help of the Kramers-Kronig relation has been widely used.

17. Effect of conduction electrons in metals. Let us write:

$$\varepsilon_1(\omega) = 1 + \varepsilon_1^f(\omega) + \varepsilon_1^i(\omega), \tag{17.1}$$

$$\varepsilon_2(\omega) = 1 + \varepsilon_2^f(\omega) + \varepsilon_2^i(\omega), \tag{17.2}$$

using superscript f for the contribution of the intraband effect of free carriers and superscript i for the interband effect of all electrons. According to (5.2) and (5.3):

$$\varepsilon_1^f(\omega) = -4\pi \frac{Ne^2}{m^*} \frac{1}{m^2 + (1/\tau)^2}, \tag{17.3}$$

$$\varepsilon_2^f(\omega) = \frac{4\pi}{\omega} \frac{Ne^2}{m^*} \frac{1/\tau}{\omega^2 + (1/\tau)^2}. \tag{17.4}$$

The effective mass m^* is the same as m_s defined by (16.2). The onset of interband transitions shows up in an increase in ε_2. This can be seen at about 4 eV in the curve for Ag shown in Fig. 20 [*121*]. Consider the low frequency region below the threshold, ω_i, of interband transitions. In this region, $\varepsilon_2 = \varepsilon_2^f$, and ε_1 is also dominated by ε_1^f and is consequently negative. An estimate of m^* can be obtained from ε_1^f which is not sensitive to $1/\tau$ at frequencies $\omega > 1/\tau$. In order to obtain ε_1^f, we have to subtract ε_1^i from the total ε_1 given by the analysis of the experimental data. The interband contribution ε_1^i may be estimated from the Kramers-Kronig

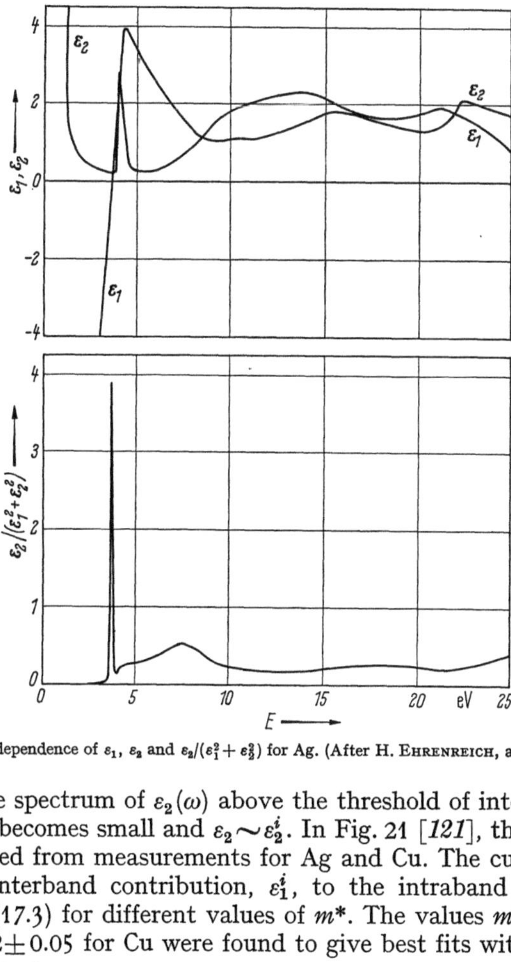

Fig. 20. Spectral dependence of ε_1, ε_2 and $\varepsilon_2/(\varepsilon_1^2+\varepsilon_2^2)$ for Ag. (After H. EHRENREICH, and H. R. PHILIPP.)

relation using the spectrum of $\varepsilon_2(\omega)$ above the threshold of interband transition, where ε_2^f usually becomes small and $\varepsilon_2 \sim \varepsilon_2^i$. In Fig. 21 [*121*], the points show the total ε_1 determined from measurements for Ag and Cu. The curves are obtained by adding the interband contribution, ε_1^i, to the intraband contribution, ε_1^f, calculated from (17.3) for different values of m^*. The values $m^* = 1.03 \pm 0.06$ for Ag and $m^* = 1.42 \pm 0.05$ for Cu were found to give best fits with the data points over the energy range.

As mentioned, ε_1^f is not sensitive to $1/\tau$. The curves for Ag in Fig. 21 were calculated by using the value of $1/\tau$ estimated from the *dc* conductivity, and $1/\tau$ was neglected in the calculation for Cu. The values of m^* obtained from the analysis of ε_1 were used to calculate τ from the data on ε_2. The calculated values varied with frequency; from a low frequency value of 3.7×10^{-14} sec to 1.6×10^{-14} sec at 2 eV in the case of Cu. A variation is to be expected from the fact that the expression (17.4) is a rough approximation, as discussed in Sect. 5.

According to (2.15), the condition for the existence of long wavelength plasma oscillations is $\varepsilon(\omega) = 0$. For real materials, oscillations are always damped, and the condition leads to a complex frequency:

$$\omega = \omega_p + i\Gamma. \tag{17.5}$$

The real part, ω_p, may be regarded as the plasma frequency while Γ corresponds to the damping of the oscillation. When $\varepsilon_2(\omega_p) \ll 1$, the damping is small, and the frequency is approximately given by:

$$\varepsilon_1(\omega_p) = 0 = 1 + \varepsilon_1^f(\omega_p) + \varepsilon_1^i(\omega_p). \tag{17.6}$$

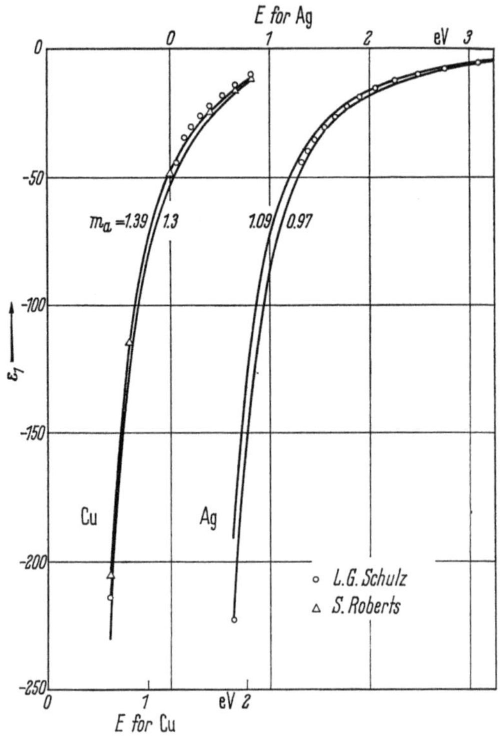

Fig. 21. Comparison of experimental points of ε_1 for Ag and Cu with curves calculated for various values of effective mass for susceptibility. (After H. R. EHRENREICH, and H. R. PHILIPP.)

The intraband contribution ε_1^f is negative in sign and decreases in amplitude steadily with increasing frequency. The interband contribution, ε_1^i, is positive below the threshold, ω_i, of interband transitions and goes through a peak near ω_i. Above ω_i, ε_1^i varies depending on the structure of energy band; a given transition gives a positive or negative contribution, depending on whether the frequency is higher or lower than the transition frequency. Neglecting $1/\tau$, we get from (17.3) and (17.6):

$$\omega_p^2 = (4\pi N e^2/m^*)/(1+\varepsilon_1^i) = \omega_{p0}^2/(1+\varepsilon_1^i), \tag{17.7}$$

ω_{p0} being the plasma frequency of the conduction electrons if there were no interband contribution ε_1^i. The curve of ε_1 in Fig. 20 shows that ω_p is close to 4 eV for Ag.

According to (2.14), the quantity:

$$-\mathrm{Im}\,\frac{1}{\varepsilon(\omega)} = \frac{\varepsilon^2}{\varepsilon_1^2 + \varepsilon_2^2} \tag{17.8}$$

determines the energy loss of a fast charged particle to the electrons of the solid. The quantity, called the energy loss function, should show a maximum close to ω_p. This is seen clearly in the $-\mathrm{Im}\,\varepsilon^{-1}$ curve of Ag shown in Fig. 20 which exhibits a sharp peak at a frequency corresponding to $\varepsilon_1=0$. The peak is very pronounced in the case of Ag where ε_2 is small at ω_p.

18. Effect of interband transitions. We consider cubic crystals. According to (2.11), the interband contribution to ε_1 may be written:

$$\varepsilon_1^i(\omega)=\frac{4\pi e^2}{m}\sum_{n,n'}\int\frac{2}{(2\pi)^3}\,d\bar{k}\,f(\varepsilon_{n\bar{k}})\,\frac{f_{n'n}(\bar{k})}{\omega_{n'n}(\bar{k})-\omega^2}\,, \tag{18.1}$$

where the summation over $\bar{k}$ is replaced by integration. For the correlation of the optical data with the structure of the energy bands, it is more profitable to consider ε_2^i which is determined by the band structure at the transition frequency whereas ε_1^i at any ω depends on transitions of all frequencies. The expression (2.12) for ε_2^i may be written:

$$\left.\begin{aligned}\varepsilon_2^i(\omega)=\frac{4\pi^2 e^2}{3m^2\omega^2\hbar}\sum_{n,n'}\int\frac{2}{(2\pi)^3}\,d\bar{k}\,[f(\varepsilon_{n\bar{k}})-{}&\\ -f(\varepsilon_{n'\bar{k}})]\,|\bar{p}_{nn'}(\bar{k})|^2\,\delta(\omega_{n'n}(\bar{k})-\omega)\,.&\end{aligned}\right\} \tag{18.2}$$

This expression can be transformed into integration over a surface S_ω of constant transition frequency, $\omega_{nn'}(\bar{k})-\omega$:

$$\varepsilon_2^i(\omega)=\frac{4\pi^2 e^2}{3m^2\omega^2\hbar}\,\frac{2}{(2\pi^3)}\sum_{n,n'}\int_{S_\omega}\frac{dS_\omega}{|\nabla_{\bar{k}}\,\omega_{n'n}(\bar{k})|}\,[f(\varepsilon_{n\bar{k}})-f(\varepsilon_{n'\bar{k}})]\,|\bar{p}_{nn'}(\bar{k})|^2. \tag{18.3}$$

In semiconductors and insulators, we deal with transitions between nearly filled bands to nearly unoccupied bands; therefore the square bracket in the integrand may be replaced with unity. The same applies to metals except when the S_ω intersects the Fermi surface. The consequence of the latter case will be discussed later. As to the matrix element $p_{nn'}(\bar{k})$, it is not expected to vary drastically with $\bar{k}$. Thus, structures in the spectrum of $\varepsilon_2^i(\omega)$ reflect the joint density of states of the energy band:

$$\varrho_{n'n}(\hbar\omega)=\frac{2}{(2\pi)^3}\int_{S_\omega}\frac{dS_\omega}{|\nabla_{\bar{k}}(\hbar\,\omega_{n'n})|} \tag{18.4}$$

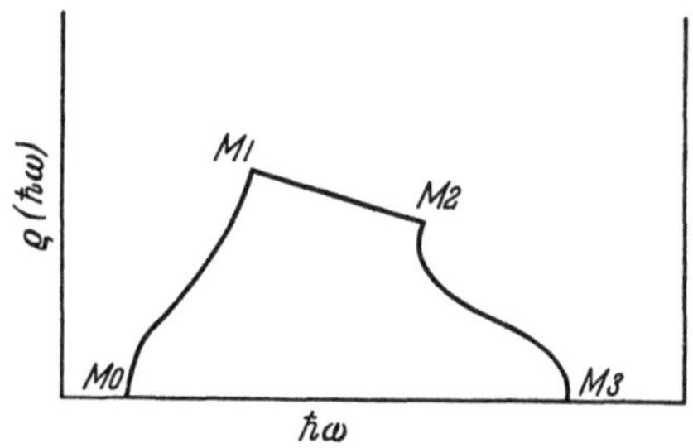

Fig. 22. Schematic diagram of density of states, $\varrho(\hbar\omega)$, showing four types of critical points.

which is the number of states in either band per unit transition energy. Pronounced structures in $\varepsilon_2^i(\omega)$ occur at critical points of density of states which are points in $\bar{k}$ space where $\nabla_{\bar{k}}\,\omega_{nn'}(\bar{k})=0$. Such points were discussed first by VAN HOVE in connection with lattice vibration spectrum [*122*]. They are known as van Hove singularities. The sketch of $\varrho(\hbar\omega)$ in Fig. 22 shows four kinds of critical points: M_0 corresponds to a minimum of $\omega_{nn'}(\bar{k})$, M_3 corresponds to a maximum, and M_1 and M_2 correspond to two types of saddle points. At a critical point, there is a break, a discontinuity in slope, in the density of states with an infinite slope on one side, on the high frequency side of a M_0 or M_2 point and on the low frequency side in the case of a M_3 or M_1 point. Symmetry points of the Brillouin zone which are specially favorable for energy band calculations are also necessarily critical points of density of states [*123*].

Fig. 23 shows the comparison of experimental $\varepsilon_2(\omega)$ obtained from reflectance data with the curve given by theoretical energy band calculations. The calculations took into account the transitions between the valence band and two lowest conduction bands. The calculated transition matrix elements for each pair of bands did not vary greatly. The dashed lines were calculated by taking a suitable constant matrix element for each pair of bands. The dotted curve for silicon was calculated without this approximation. The calculated curve for germanium show clearly the breaks associated with the critical points. Two critical points, a M_1 and a M_2, closely spaced in energy produce a peak. Such is the case of the calculated peak at ~ 3.7 eV. The calculated curves reproduce quite well the structure in the experimental data. The examples show clearly the usefulness of the optical data for the study of energy bands.

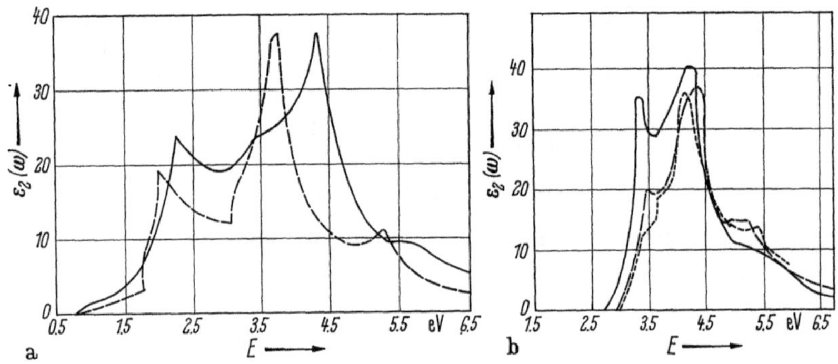

Fig. 23a and b. Comparison of experimentally determined (solid line) and theoretically calculated curves of $\varepsilon_2(\omega)$. a) For germanium. b) For silicon; the dashed and dotted lines give two different calculations (see text). [After D. BRUST, Phys. Rev. **134**, A 1337 (1964).]

In the absence of detailed energy band calculations, various experimental observations and some theoretical considerations may be utilized to aid the interpretation of the reflectance spectrum. A particularly useful consideration is the splitting of degenerate energy bands due to spin-orbit interaction. Such splitting can be deduced from symmetry considerations and helps to identify the energy bands involved. For example, the two peaks at ~ 2 eV in Fig. 19 are attributed to conduction band — valence band transitions, $L_3' \leftarrow L$, at the point L of the Brillouin zone. The presence of two closely spaced peaks is consistent with the expected splitting of the valence band at L_3', thus providing support to the identification of the transition. The study of various materials of similar band structures is obviously advantageous. An example is the series of III—V compound semiconductors. There is a general similarity in the reflectance spectra of these materials. Experimental informations on the band structure of the various compounds are commonly useful for the interpretation of the spectra of these materials.

In addition to critical points of density of states, the sharp change of electron distribution at the surface of the Fermi level may give rise to abrupt changes in $\varepsilon_2(\omega)$. Transitions from a lower band to the conduction band take place over the part of S_ω which is above the Fermi surface. If an appreciable part of a S_ω surface nearly coincides with the Fermi surface, then there will be an abrupt increase of possible transitions with a corresponding rise in $\varepsilon_2(\omega)$, at the frequency ω where S_ω crosses the Fermi surface. Such effect has been found in reflectance of Au [*118*].

19. Reflectance in the ultraviolet. Transitions of frequencies smaller than the radiation frequency ω give negative contributions to $\varepsilon_1(\omega)$. Fig. 19 shows that

ε_1 of InSb drops below zero above ~ 4 eV due to the effect of transitions of lower energies. In the range marked region 2, ω is large compared to the frequencies of the important transitions involving the valence band as shown by the fact that ε_2 has dropped to low values in this region. Under this condition the negative contribution of the valence electrons to ε_1 decreases with increasing frequency much as the free carrier effect. As a result, ε_1 increases, becoming positive again above 11 eV. The curve of energy loss function, $-\operatorname{Im}\varepsilon^{-1}$, shows a strong peak, the maximum of which corresponds to 12 eV. The difference between the two energies, at $\varepsilon_1 = 0$ and at the maximum of $-\operatorname{Im}\varepsilon^{-1}$, is due to the fact that $\varepsilon_2 \neq 0$. The energy at the maximum of $-\operatorname{Im}\varepsilon^{-1}$ corresponds more closely to ω_p as defined by (17.5).

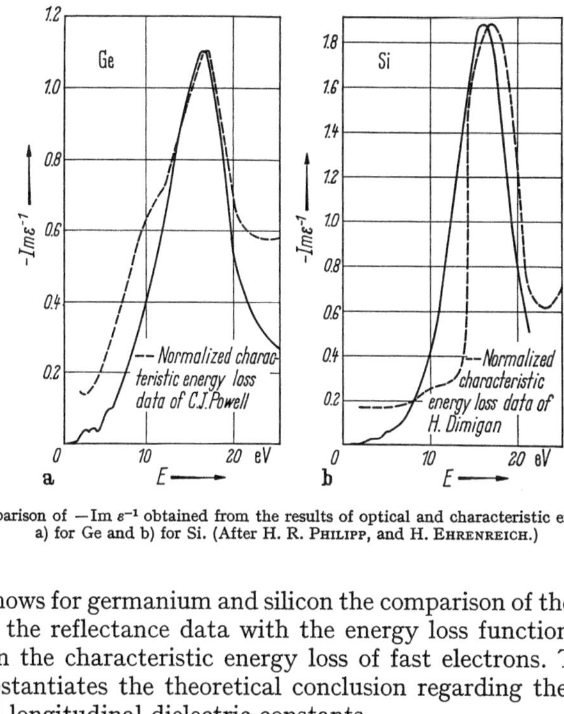

Fig. 24a and b. Comparison of $-\operatorname{Im}\varepsilon^{-1}$ obtained from the results of optical and characteristic energy loss experiments; a) for Ge and b) for Si. (After H. R. PHILIPP, and H. EHRENREICH.)

Fig. 24 [*119*] shows for germanium and silicon the comparison of the $-\operatorname{Im}\varepsilon^{-1}$ curve obtained from the reflectance data with the energy loss function obtained from experiments on the characteristic energy loss of fast electrons. The satisfactory agreement substantiates the theoretical conclusion regarding the equivalence of transverse and longitudinal dielectric constants.

The case of InSb discussed above represents the situation where the transition frequencies for a group of filled energy bands (transitions between the valence band, v, and the higher bands, h, in the case of InSb) are widely separated from the transition frequencies involving lower filled bands (the next filled band in InSb is the d band of In). There is thus a range of frequency between the two groups of transition frequencies such as region 2 for InSb where $\omega_{hv}(\bar{k}) \ll \omega \ll \omega_{hd}(\bar{k})$. In this frequency range, the effect of the low frequency transitions resembles that of free carriers whereas the high frequency transitions give a constant positive contribution to ε_1 with no contribution to ε_2. Thus, for region 2 and lower frequencies, the effect of the d band and deeper lying bands of InSb may be described by a constant dielectric susceptibility. This situation simplifies the analysis, and it applies for most of the III—V compound semiconductors as well as for germanium and silicon. In Fig. 19 for InSb, the rise of reflectance and energy loss function in region 3 shows that the frequency in this region overlaps the transition frequencies of the d-band.

IX. Impurity effects.

20. Absorption and emission. The simple model of effective mass and Coulomb potential (see Sect. 10) has been quite successful for certain impurity states in semiconductors. The orbit of the bound state will be large if effective mass is small and if the material has a large dielectric constant. The perturbation may be expected to be small when an impurity atom from a neighboring column of the periodic table substitutes for a regular atom. The effective donor and acceptor

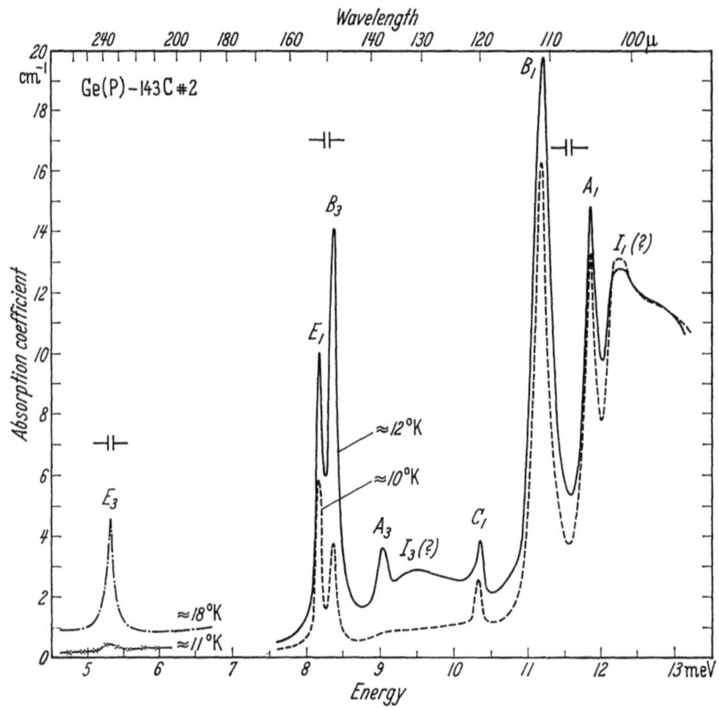

Fig. 25. Absorption spectrum of phosphorus impurity in germanium at different sample temperatures. Impurity concentration $\sim 8\times 10^{14}\,cm^{-3}$. (After REUSZER and FISHER.)

impurities for a semiconductor are often of this kind which has the conditions favorable for the applicability of the effective mass model. Typical examples are the group III acceptor and group V donor impurities in germanium and silicon, for which considerable optical studies have been made. Earlier works on the donor and acceptor impurities in silicon and on donor impurities in germanium have been reviewed by HROSTOWSKI [*124*]. Studies on the acceptor impurities in germanium have been reported by FISHER and FAN [*125*]. Recent results on the donor impurities are given by REUSZER and FISHER [*126*] for germanium and by AGGARWAL and RAMDAS [*127*] for silicon.

Figs. 25—27 show for illustration the low temperature absorption spectra of a donor impurity, phosphorus, and an acceptor impurity, boron, in germanium and silicon. The absorption lines correspond to excitations from the ground state to various excited states. The spectra of the various donors, Bi, P, As, Sb, are similar, so are the spectra of the various acceptors. The energy separations between the various lines are nearly the same for different donors and for different acceptors. This is consistent with the simple model, according to which the energy levels would be independent of the nature of the impurity. In fact, the observed energy separations are in excellent agreement with the calculated energy separations between the various excited states of donors [*128*] and acceptors [*54*],

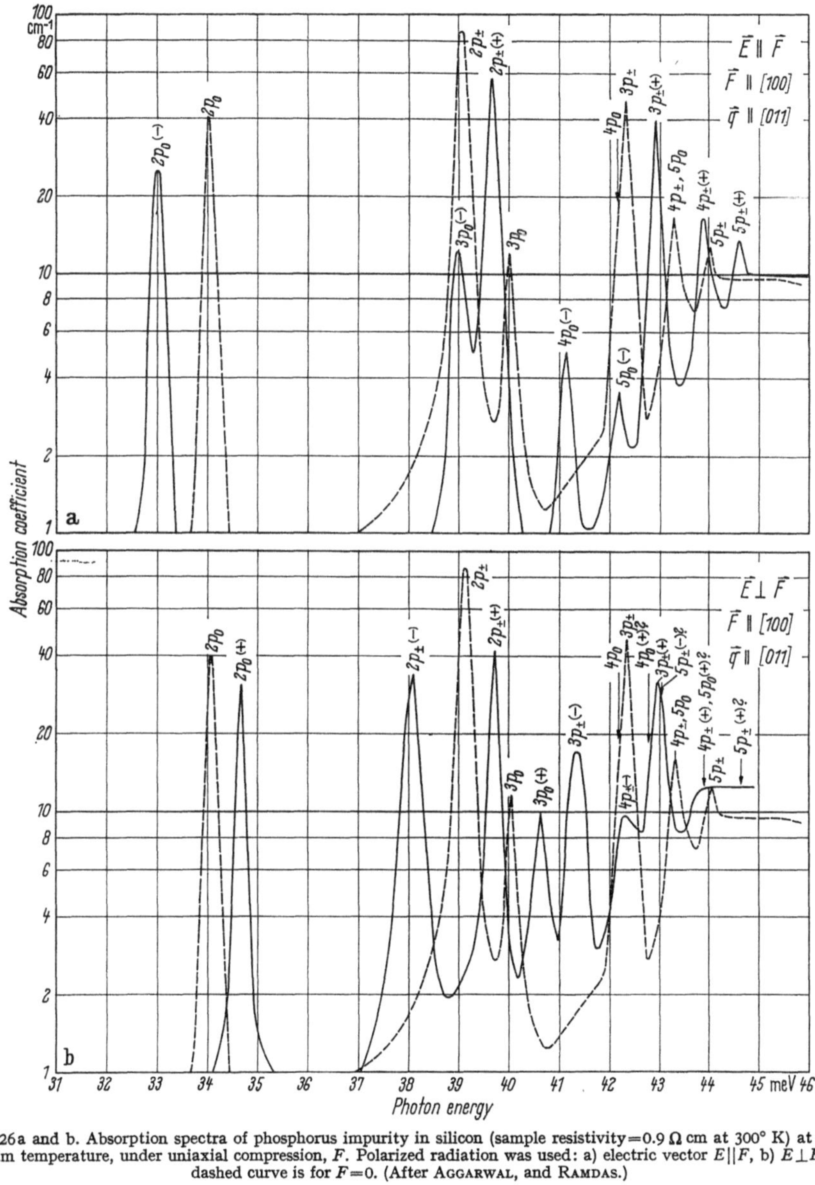

Fig. 26a and b. Absorption spectra of phosphorus impurity in silicon (sample resistivity=0.9 Ω cm at 300° K) at liquid helium temperature, under uniaxial compression, F. Polarized radiation was used: a) electric vector $E \| F$, b) $E \perp F$. The dashed curve is for $F=0$. (After AGGARWAL, and RAMDAS.)

respectively. On the other hand, it is well known that the different donors and acceptors do not have the same ionization energies. The observed spectra for different impurities of each kind are shifted in energy accordingly. Thus, the simple model is inadequate for the ground state but gives good results for the excited states. This is understandable since the ground state is s in character whereas the excited states observed are generally of p character and have, in addition, larger orbits. Therefore, the wave function of the ground state is more concentrated in the neighborhood of the impurity ion where the effect of the nature of the impurity is important. Theoretical treatments have been made in the attempt to correct the simple model calculations [*128*], [*129*]. The problem is not yet satisfactorily solved.

Transitions from levels close to the ground state may be observed at a temperature where the levels have sufficient occupation probabilities. The interpretation of the measurements shown in Fig. 25 for different temperatures is that the lines denoted with subscript 3 are given by excitations from a higher ground state to the same excited states as the corresponding lines with subscript 1 given by transitions from the lowest ground state. The conduction band in germanium has valleys on the ⟨111⟩ axes. The ground state of the effective

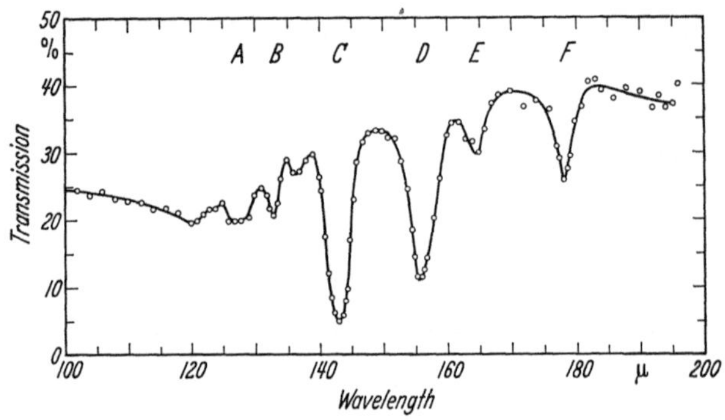

Fig. 27. Transmission spectrum of boron impurity in germanium at ~5° K. Sample resistivity = 14 Ωcm at room temperature. (After FISHER, and FAN.)

mass Hamiltonian is therefore 4-fold degenerate. The symmetry around an atom is tetrahedral. Insofar as the Coulomb potential approximation is not rigorous, the levels calculated should split according to the irreducible representations of the tetrahedral symmetry. Thus, the 4-fold degenerate ground state splits into a non-degenerate and a triply degenerate level. The energy difference between the corresponding lines provides a direct determination of the energy splitting of the ground state. The ground state splitting for donors in silicon have been obtained by AGGARWAL [*130*] from similar measurements.

The application of an elastic stress may shift the energy and also split the degenerate levels. In the effective mass approximation the effect of strain on the bound states can be derived from its effect on the energy band [*128*], [*131*]; the perturbation introduced by the strain can be expressed in terms of the deformation potential. For donor impurities, the various valleys of the conduction band shift by different energies depending upon the orientation of the valley axis relative to the strain. The degeneracy of the bound states due to the equivalence of the valleys can then be lifted in accordance with the reduced symmetry under the strain. The effect of uniaxial stress on the absorption spectra of donor impurity was studied first by WEINREICH and WHITE [*132*] for germanium. More detailed information can be obtained by using polarized light as illustrated by the data shown in Fig. 26. Transitions involving excited states split by the strain depend on the direction of polarization as expected. The results are consistent with the effective mass theory providing a means for estimating the deformation potential.

The observed spectrum of the continuous absorption corresponding to the ionization of the bound carriers is roughly similar to that of the hydrogen atom. The nature of the impurity through its effect on the ground state may influence the spectrum to some extent. Also, the structure of the energy band should affect the spectrum. These aspects have not yet received special attention.

Deep-level impurities: Impurities with ionization energies much larger than that given by the effective mass theory have shallow excited states according to the prediction of the theory. An example is In in silicon [*124*] which has an ionization energy of 0.16 eV, about three times larger than the value given by the effective mass theory. The excitation spectrum is, however, similar to those of other group III acceptors B, Al, Ga, which have the normal ionization energy.

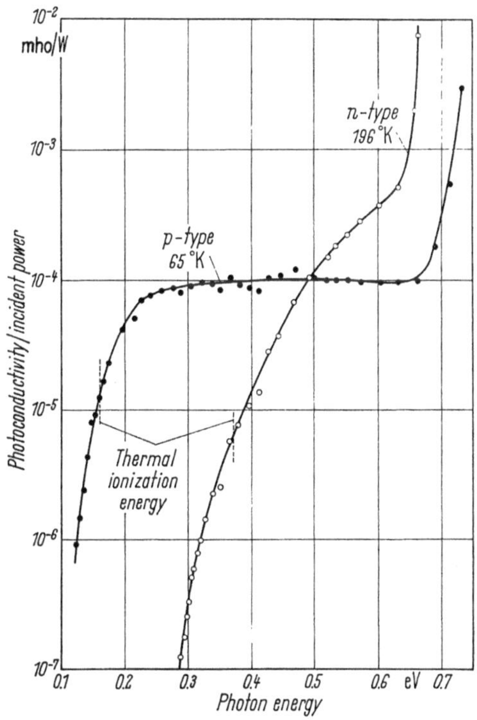

Fig. 28. Photoconductivity spectra of *n*- and *p*-type Mn-doped germanium. (After R. NEWMAN, H. H. WOODBURY, and W. W. TYLER.)

Other examples are Cu and Zn in germanium [*133*] which can trap three and two holes respectively. The first ionization energy of Cu is 0.04 eV. The first and second ionization energies of Zn are 0.03 and 0.09 eV. The excitation lines observed for Cu, Zn and Zn^- show that the excited states involved conform approximately to the effective mass theory. Even the excitation spectrum of a defect produced by neutron bombardment in silicon shows three excitation lines near 0.35 eV which correspond apparently to the $2p$ and $3p$ states of a doubly charged center according to the effective mass theory [*134*]. Such phenomena are in agreement with the general expectation that excited states, particularly those of *p*-character, are less sensitive to the nature of the center.

Little can be said in general about the optical spectra involving levels deep inside the energy gap. The spectra of photoconductivity associated with the ionization of a number of impurities in germanium show an interesting characteristic. The spectra in *p*-type germanium are generally rather flat with a well-defined low energy threshold, whereas the spectra in *n*-type germanium decrease gradually toward low energies [*135*]. This is illustrated by the spectra of Mn in n- and p-type germanium, shown in Fig. 28. The impurities in question have more than one acceptor level, i.e. more than one hole can be trapped at the impurity. In p-type samples, the photoconductivity is produced by the ionization of holes from an acceptor level to the valence band and the ionized center has an attractive potential for the hole. On the other hand, the removal of an electron from one of the acceptor levels to the conduction band leaves the multilevel acceptor still negatively charged which exerts a repulsive potential on the electron. The repulsive potential tends to supress the electron wave function near the impurity thereby reduces

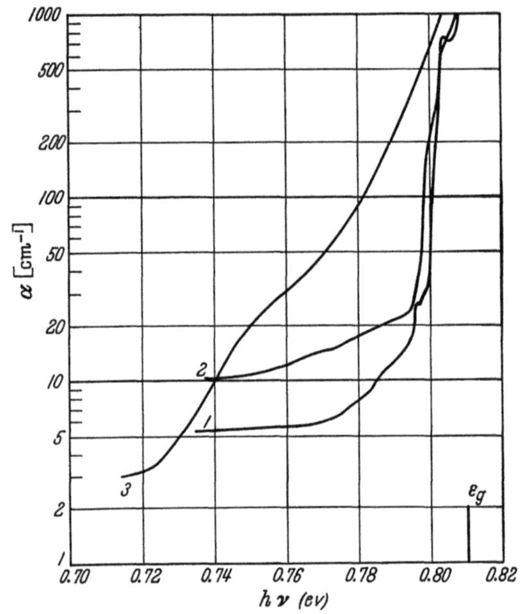

Fig. 29. Absorption edge of GaSb. *1* Undoped p-type sample with a hole concentration of $p=1.4\times10^{17}\,\mathrm{cm}^{-3}$ and a hole mobility of $\mu=2800\,\mathrm{cm}^2/\mathrm{volt\text{-}sec}$ at 80° K. *2* Undoped p-type sample, $p=2.5\times10^{17}$ and $\mu=1520$. *3* p-type sample compensated with Te, $p=0.39\times10^{17}$ and $\mu=93$. (After E. J. JOHNSON and H. Y. FAN.)

the transition probability. The effect becomes less effective with increasing electron energy; therefore, the ionization probability continues to increase with increasing photon energy.

Impurity effect near the intrinsic edge: Transitions from impurity levels near one band edge to the opposite energy band can give absorption in the neighborhood of the intrinsic edge of a semiconductor, and may sometimes produce an apparently sloping absorption edge. Such effect appears to be quite common in the III—V compound semiconductors. Fig. 29 shows the absorption edges observed in three samples of GaSb [*136*]. Sample 1 shows an absorption tail rising from the background at $h\nu\sim0.77$ eV, due to the excitation of electrons to the conduction band from an impurity level at 0.034 eV above the valence band. Samples 2 and 3 show the effect of levels lying deeper in the energy gap, with the absorption extending to smaller $h\nu$. Normally, shallow impurity levels near one band would not be expected to be associated with shallow levels near the other band. Thus, the absorption under discussion should have only a continuous spectrum determined by (10.7). Calculations of such absorption for the case of simple nondegenerate bands have been made by EAGLES [*137*]. It should be pointed out that should there be bound state levels at both band edges, excitation of the impurity between the levels may be considered as an exciton impurity complex discussed in Sect. 23. Thus, electron excitation from an acceptor level to a bound state level near the conduction band may be regarded as the formation of an exciton attached to an ionized acceptor.

Transitions of free carriers from one energy band to impurity states near the opposite energy band give emission bands situated close to the energy gap. The well-known green emission from CdS was found to have such origin [*138*], and the edge emission of other II—VI compounds also contains this type of impurity bands [*139*]. We shall say more about them in the next section, in connection with the phonon effect. Emission bands have been observed in various III—V compounds, e.g. GaAs [*140*], GaSb [*91*], InAS [*141*], [*104*], InSb [*142*], which are associated with impurities and are apparently due to free carierr-impurity transitions. This type of emission involving various acceptor and donor impurities have been observed also in silicon [*143*] which has an indirect transition energy gap.

21. Phonon effects. *Color centers of alkali halides.* Phonon effects have been extensively studied in connection with color center absorptions in alkali halides [*144*], and a great deal of theoretical work has been done with the color centers in mind. The theories of PEKAR and of HUANG and RHYS based on polar inter-

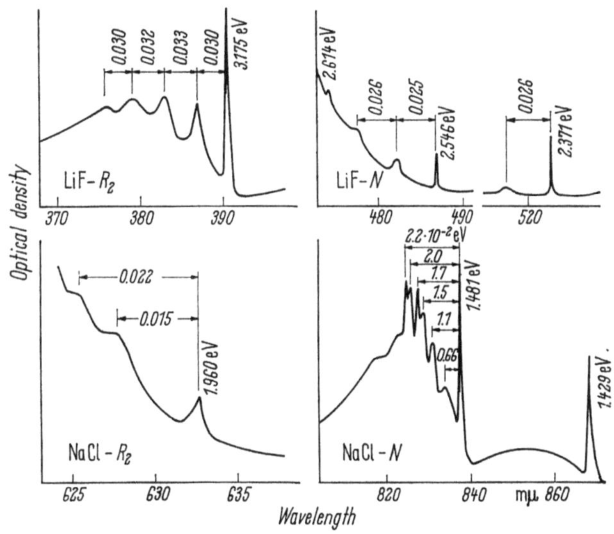

Fig. 30. Absorption spectrum of R_2 and N bands of LiF and NaCl, showing details of fine structure due to phonon emission peaks at 10° K. (After C. B. PIERCE.)

action with optical phonons, can account for the shape and the temperature dependence of the absorption of F-centers by taking a suitable value for the coupling parameter S. We pointed out in Sect. 11 that the model of local modes, in which short range electron-ion interaction is important, leads to similar results. There is some experimental evidence that the frequency of the important modes is significantly different from the frequency of the optical modes. This is an indication in favor of the local modes. However, the basic question about the nature of the important modes and their interaction with the electron is not settled.

An interesting development is the observation of sharp lines associated with some color center absorption bands. Sharp lines associated with several centers were first observed in LiF. Recently, sharp line structures associated with the R_2 and N bands have been observed and studied systematically in a number of alkali halides [*145*]. The R_2 and N bands arise from complex electron-excess centers. Fig. 30 shows some of the spectra. The strong line at the high energy end was identified by FITCHEN et al. as the zero-phonon absorption. Pierce found that the energy separations between the various peaks and the zero-phonon line can be correlated with multiples of various phonons at the boundary of the Brillouin zone. Thus, the peaks of R_2 band in LiF are separated from the zero-phonon line at 3.175 eV by approximately multiples of the calculated energy, 3.1×10^{-2} eV, of transverse optical phonon at the boundary of the Brillouin zone along $\langle 111\rangle$ directions, TO $\langle 111\rangle$. In the case of the N band in NaCl, the peaks 2, 3, 4, 5, 6 are separated from the zero-phonon line by energies close to: 0.0103 eV (TA $\langle 100\rangle$ phonon), 0.015 eV (TO $\langle 111\rangle$), 0.018 eV (LA $\langle 100\rangle$), 0.020 eV (LA $\langle 111\rangle$), 0.023 eV (TO $\langle 100\rangle$), respectively. PIERCE concludes that normal lattice phonons near the zone boundary are excited with electron transitions in these centers. This is understandable on the ground that phonons have high densities of states near the zone boundary. It is also reasonable that for normal lattice phonons those of short wavelengths have strong coupling on account of the wavelength being comparable to the dimensions of the defect centers.

Impurity line width of semiconductors.

The absorption lines associated with the excitation of group III acceptor and group V donor impurities in silicon and germanium are very narrow. The observed line widths at low temperatures were of the order of 10^{-3} eV in silicon [*146*] and $\gtrsim 10^{-4}$ eV in germanium [*133*]. Recently, studies on B and P in silicon were reported by PAJOT [*147*] in which the instrumental resolution was carefully taken into account in the analysis. The results show that even at a boron concentration as low as $1.2\times10^{15}\,\mathrm{cm}^{-3}$, the observed width ($2.5\times10^{-4}$ eV) of a prominent line was considerably larger than the width (0.76×10^{-4} eV) found in a sample of still lower ($0.84\times10^{14}\,\mathrm{cm}^{-3}$) boron concentration. For a sample $4.4\times10^{14}\,\mathrm{cm}^{-3}$ phosphorus concentration, the "$1s\leftrightarrow 2p_{\pm1}$" line had a width of $\sim 0.6\times10^{-4}$ eV at 4.2° K and increased to $\sim 2.5\times10^{-4}$ eV at 63° K. LAX and BURSTEIN [*146*] calculated the broadening due to phonon effect using the deformation potential to estimate the electron-phonon coupling. A value of about 3×10^{-3} eV was obtained for the low temperature width of acceptor absorption lines in silicon. In the light of PAJOT's results, the value is more than an order of magnitude too high.

It was pointed out by KANE [*63*] that the observed absorption lines should correspond to zero-phonon transitions since the coupling factor, $\sum_j S_j(2n_j+1)$ in (11.26), is very small, 0.15 for silicon and 0.05 for germanium, as estimated for accoustical phonons from the deformation potential. The author treated the width of the lines as produced by the lifetime-broadening of the excited states. Consider as initial approximation the wave function $\varphi_{R_0}(r)\,\chi_0(R)$ where the electronic part φ corresponds to the equilibrium nuclear coordinates R_0, and the vibrational part $\chi_0(R)$ is independent of φ. Electron-lattice interaction is a perturbation. Its matrix elements connecting $\varphi\chi_0$ with states $\varphi\chi_0'$ lead to a dependence of the vibrational state on φ, which corresponds to $\Delta\varepsilon(R)$ of (11.5) and leads to optical transitions accompanied by phonon emission or absorption. The matrix elements connecting $\varphi\chi_0$ with $\varphi'\chi_0'$ produce transitions between states of different electronic wave functions. The effect is a broadening of the level $\varphi\chi_0$. At low temperatures, the broadening of a level depends on the presence of levels lying below it. For group III acceptors in silicon, KANE estimated that an excited level having a level at 4×10^{-3} eV below it, would have a line width of about 3×10^{-5} eV. The value seems to be of the right order of magnitude. However, detailed studies have yet to be made. Impurities with larger ionization energies, e.g. Cu and Zn^- in germanium, seem to give absorption lines of much larger widths [*133*].

Absorption from the excitation of various defect centers of radiation damage have been observed in silicon [*134*]. Some centers give rather broad bands which may be caused by multi-phonon effect as in the case of *F*-center absorption.

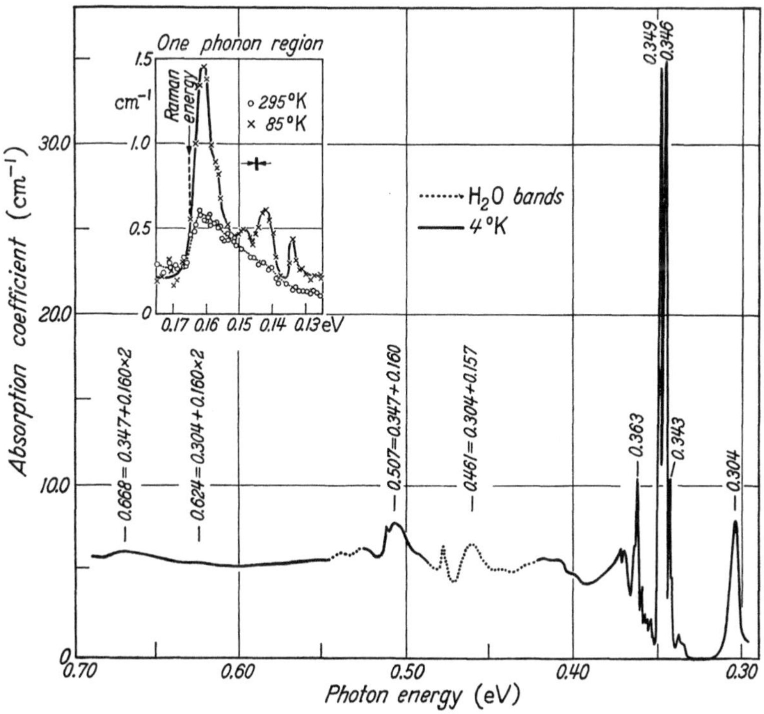

Fig. 31. Absorption spectrum of semiconducting diamond. Broken line indicates spectrum obscured by atmospheric absorption. (After J. R. HARDY, S. D. SMITH, and W. TAYLOR.)

Phonon peaks of impurity excitation in semiconductors.

Absorption peaks produced by phonon effect such as discussed for the R_2 and *N* bands of alkali halides, are observed also in semiconductors. Fig. 31 shows the absorption spectrum of the acceptor in semiconducting diamond observed by HARDY et al. [*148*]. The lines at energies $\lesseqgtr 0.363$ eV correspond to transitions to various excited states. The peak at 0.461, 0.507, 0.624 and 0.668 eV are identified as excitations with phonon emission. The peaks form two pairs (0.461, 0.507) and (0.624, 0.668). The energy separation between the components of each pair is about the same as the separation, 0.043 eV, between the two excitation lines at 0.347 and 0.304 eV. The first pair of peaks is shifted from the excitation lines by 0.159 eV, and the second pair is shifted by twice the energy. Within the experimental error, the energy 0.159 eV is equal to the energy of transverse optical phonons near the zone boundary. The phonon energy is larger than the difference between the ionization energy and the energy of the excited states; therefore, the phonon emission peaks are superimposed on the continuum of ionization absorption. The condition favoring the observation of phonon-emission peak is that there be a large number of modes with a small frequency spread and adequate coupling. HARDY et al. estimated the coupling, $\sum_j S_j$, for the optical phonons to be 0.2, 0.03, 0.002 respectively for diamond, silicon and germanium, with corresponding values: 0.2, 0.15, 0.05 for acoustical phonons. In comparison to silicon and germanium, diamond has a larger combined coupling and a much larger coupling for optical phonons. These estimates indicate that phonon peaks would be difficult to observe in germanium and silicon.

Structure due to phonon effects was observed also in the absorption of acceptors in ZnTe [*149*] where the absorption was found to rise stepwise with increasing energy. The steps were attributed to ionizations with the emission of

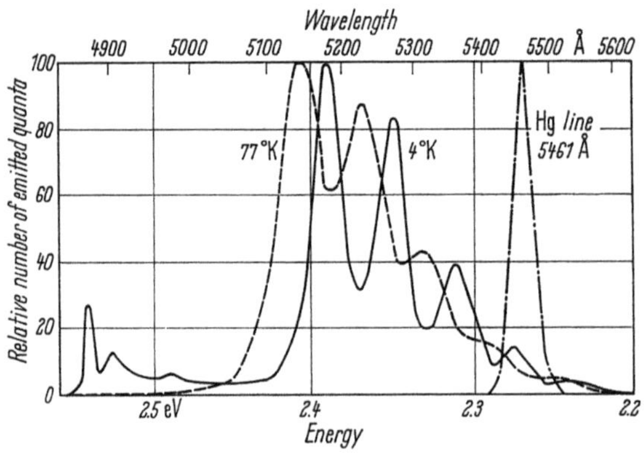

Fig. 32. Emission spectra of CdS showing the series of lines in the region 5100—5600 Å. (After C. C. KLICK.)

successively larger number of phonons. PEAKS were observed in the steps which were presumably due to phonon-assisted excitations.

Emission. It has long been known that the luminescence in each of the crystals: CdS, ZnS and ZnO, under irradiation in the fundamental absorption band contains a series of equally spaced, narrow bands on the long wavelength side of the absorption edge. In the literature, the term edge emission is sometimes used to refer to such lines specifically. The lines decrease in intensity with increasing wavelength, as shown in Fig. 32 for CdS [*150*]. The spacing between the lines is close to the energy of longitudinal, optical phonon [*151*], [*152*]. It was thought by some authors that these bands are associated with excitons, on the ground that emission associated with impurity centers would involve energies different from the lattice phonon. However, it has become now established from studies on photoconductivity, decay time of emission, etc. [*138*] that the green series in CdS originates from the recombination of free carriers with the opposite type of carriers trapped at some centers.

In the Huang and Rhys model, the effect of optical phonons on transitions between localized states is given by expression (11.16). For emission, the exponent of $[(n+1)/n]^{p/2}$ should be taken with a negative sign, and emission processes accompanied by photon creation have $p<0$. For temperatures where kT is small compared to the phonon energy, $n \sim 0$ and $(n+1) \sim 1$, and the intensity of emission with the creation of $|p|$ phonons is proportional to:

$$S^{|p|}/|p|!. \tag{21.1}$$

HOPFIELD [*152*] considered transitions of carriers from an energy band to a localized state using basically the same model but a somewhat different approach, making use of polaron wave function. The same dependence on $|p|$ is obtained. Assuming a Gaussian charge distribution for the trapped carrier,

$$\varrho(r) = (\pi^{\frac{1}{2}} a)^{-3} \exp(-r^2/a^2), \qquad (21.2)$$

HOPFIELD gets for S, the mean number of phonons emitted, the following expression:

$$S = \left(\frac{e^2}{a}\right) \frac{1}{\hbar\,\omega_0} \frac{1}{\sqrt{2\pi}} \left(\frac{1}{\varepsilon} - \frac{1}{\varepsilon_0}\right). \qquad (21.3)$$

It was shown that the intensity variation between the emission lines of CdS can be fitted with reasonable values for a and S.

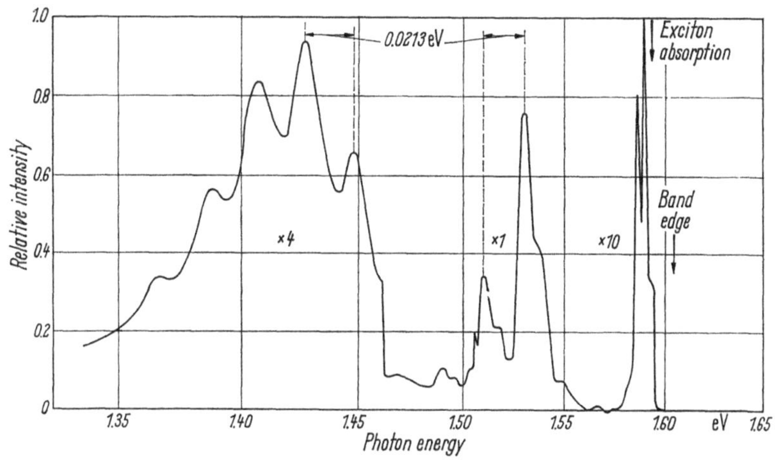

Fig. 33. Emission spectra of CdTe near the band edge, at 20° K. Series of lines separated by $k=0$ longitudinal optical phonon energy, 0.021 (3) eV, are observed. Sharp lines of width $< kT$ occur at ~ 1.51 eV and 1.59 eV. (After R. E. HALSTED, M. R. LORENZ, and B. SEGALL.)

Series of lines with separations equal to the energy of longitudinal optical phonon have been observed in various other II—VI compounds. The spectra for CdTe [*153*] are shown in Fig. 33.

22. Donor-acceptor pair. Optical transitions involving a pair of impurity centers, a donor and an acceptor, was first considered by WILLIAMS [*154*] and co-workers in their studies of luminescence of zinc sulphide. In ZnS containing Cu or Ag acceptors and Ga or In donors, an emission band was observed, the spectrum of which depended on the identities of both the donor and the acceptor. The emission was attributed to transitions between the ground states of the donor and the acceptor. Transitions between excited states of the donor and the ground state of the acceptor were also postulated in connection with another observed emission band. We note that when the electron and hole involved can be characterized by effective masses of the respective bands, the emission may be described as that of an exciton bound to a pair of centers (see Sect. 23). CHOKE, HAMILTON and PATRICK [*155*] observed in the photoluminescence of hexagonal SiC, complex spectra of polarized emission, at photon energies near the energy gap. The spectra consisted of several narrow lines and wide bands. The emission was attributed to donor-acceptor pairs on the basis of its polarization properties.

WILLIAMS treated the problem of donor-acceptor pair by perturbation method. The product $\psi_D(-)\,\psi_A(+)$ is used as the zero order approximate wave function, where $\psi_D(-)$ and $\psi_A(+)$ are respectively the wave functions of electron bound to an isolated donor and hole bound to an isolated acceptor. The binding energy of the pair with bound electron and hole is then:

$$\left.\begin{aligned} E &= E_D + E_A + \frac{e^2}{\varepsilon R} - \\ &\quad - \frac{e^2}{\varepsilon} \int \psi_D^*(-)\, \psi_A^*(+) \left[\frac{1}{r_{A-}} + \frac{1}{r_{D+}} - \frac{1}{r_{+-}}\right] \psi_D(-)\, \psi_A(+)\, dr_+\, dr_- \\ &\equiv E_D + E_A + \frac{e^2}{\varepsilon R} + I, \end{aligned}\right\} \tag{22.1}$$

E_D and E_A are the ionization energies of isolated donor and acceptor, respectively. The term $e^2/\varepsilon R$ is the attraction between the oppositely charged cores of the centers. The integral represents the interaction between the electron and acceptor core, the interaction between the hole and the donor core, and the hole-electron interaction. The energy required to free the electron and hole to their respective bands is:

$$E_i = E - \frac{e^2}{\varepsilon R} = E_D + E_A + I \tag{22.2}$$

since the attraction interaction, $e^2/\varepsilon R$, remains after the hole and electron are removed. Thus, the energy of the photon emitted by the recombination of electron and hole in the pair is:

$$h\nu = E_G - E_i = E_G - (E_D + E_A + I). \tag{22.3}$$

This treatment neglects the correlation energy of the electron and hole; therefore, it becomes inaccurate for small separations between the centers. In the limit where R is large compared to the Bohr radii of $\psi_D(-)$ and $\psi_A(+)$, the correlation energy becomes the van der Waals interaction which has been considered by HOOGENSTRAATEN [*156*].

Recently, striking spectra of photoluminescence consisting of numerous sharp lines near the energy gap have been observed in GaP [*157*]. Fig. 34 illustrates the type of spectrum observed [*158*]. HOPFIELD, THOMAS and GERSHENZON attributed the emission to distant donor-acceptor pairs, the concept of isolated pairs being applicable so long as R is small, compared to the average separation R_0 corresponding to random distribution. The following expression was used for the energy of the photon emitted by a pair:

$$h\nu = E_G - E_D - E_A + \frac{e^2}{\varepsilon R} - \frac{e^2}{\varepsilon}\left(\frac{A}{R}\right)^6. \tag{22.4}$$

The last term represents the van der Waals interaction, A being an effective coefficient. The rest of the terms represent the limit of (22.3) for distant pairs since I approaches $-e^2/\varepsilon R$ for large R.

The possible values of R are discrete according to the geometry of the lattice, and the number of pairs, N_R, having various possible discrete values of $h\nu_R$ can be readily calculated. The assumption was made that the intensities of the fluorescent lines are proportional to N_R multiplied by a slowly varying function of R. The calculated spectrum of N_R was matched with the observed fluorescence spectrum for short range variation of intensity of the lines. The matchings obtained are illustrated by Fig. 35. Two types of spectra, I and II, were identified which correspond to impurities on substitutional sites. In type I pairs, both the donor and the acceptor are on the same type of lattice sites, gallium sites or phosphorus sites, whereas the donor and acceptor in a type II pair are on opposite types of sites. Different spectra of each type were observed which were associated with various impurities.

The van der Waals coefficient, A, is an adjustable parameter which can be determined from the matching of the spectra of N_R with the observed lines. It

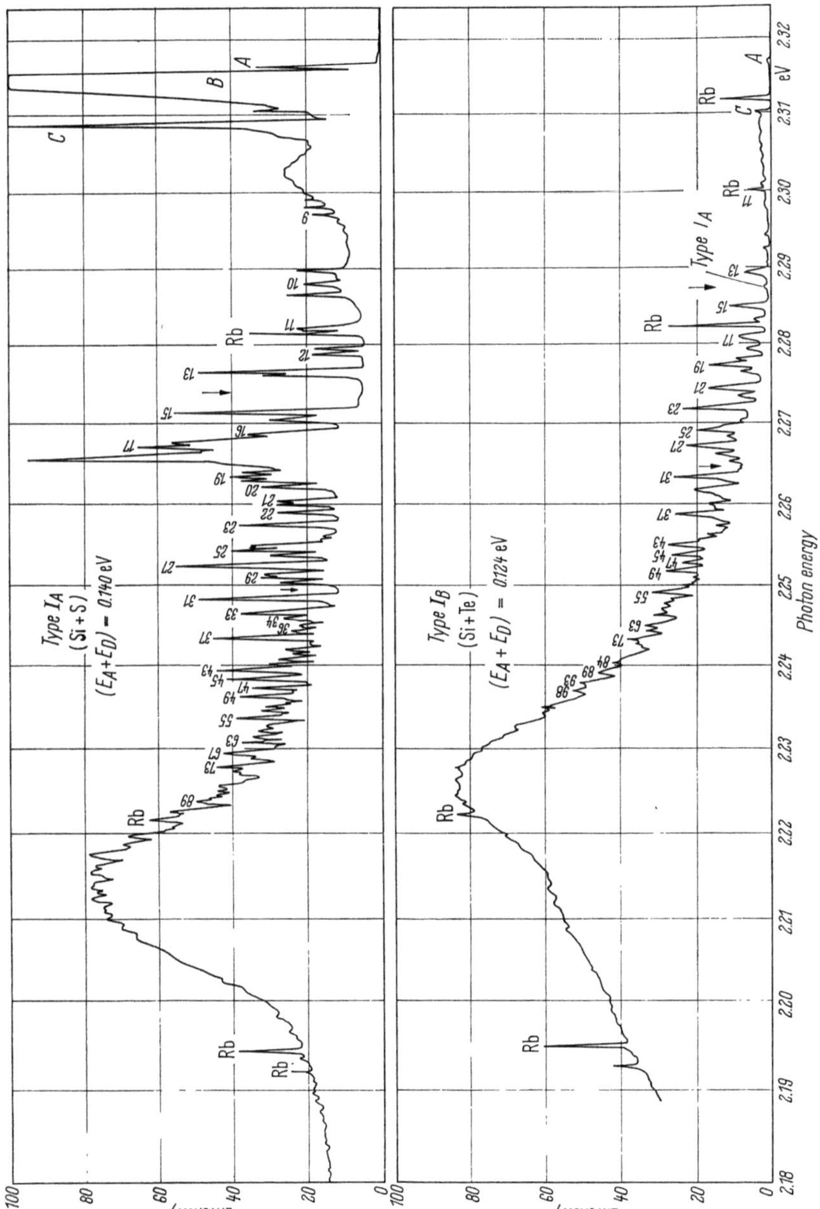

Fig. 34. Photoluminescent spectra of GaP taken at 1.6° K. The type I_A crystals are grown without the addition of impurities; the I_B crystals are produced by adding Te. The lines marked Rb are rubidium calibration lines. A, B, and C are bound exciton lines. The small numbers indicate the order numbers of the discrete values of R, in increasing order. The arrows indicate some positions for which $N_R=0$, and mark a gap in the spectra. (After D. G. Thomas, M. Gershenson, and F. A. Trumbore).

was found, however, that the observed spectra could be fitted without including the van der Waals term at all. It was also found that pairs with R less than about 10 Å did not contribute to the spectra, indicating that electrons and holes are not bound to very close pairs. Finally, it was found that the broad peak seen in Fig. 34 at the low energy end of the line spectrum varied its position for various impurities, and the variation appeared to correspond to the change in $(E_D + E_A)$.

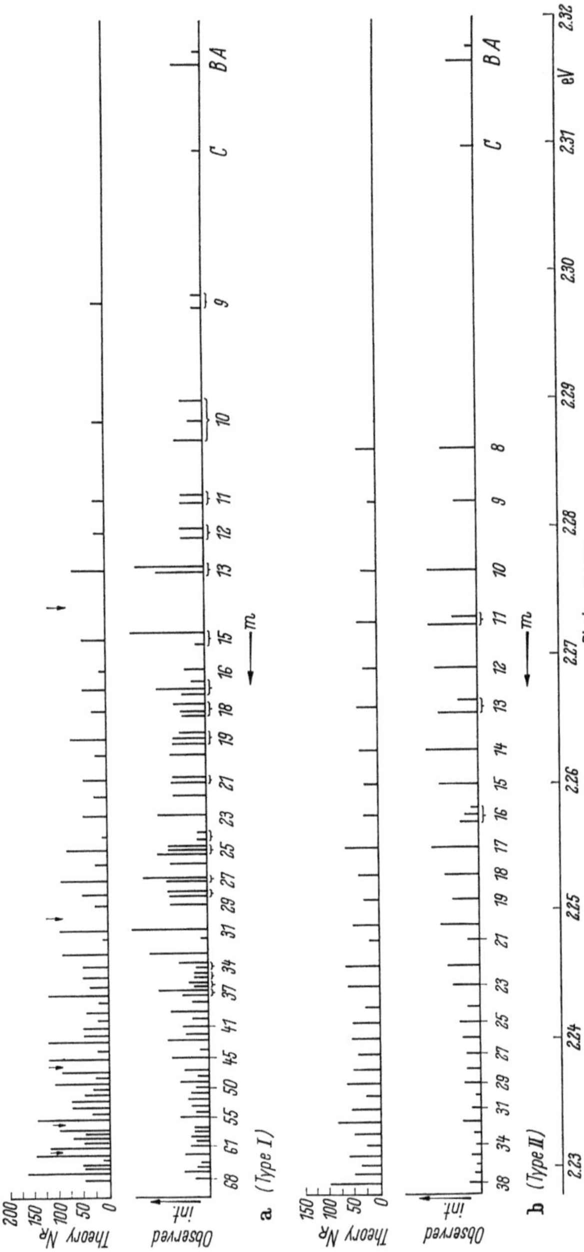

Fig. 35a and b. Calculated spectra of number of pair N_R at distance R and observed spectra of photoluminescence at 1.6° K in GaP, for type I and II pairs. (After J. J. Hopfield, D. G. Thomas, and M. Gershenzon.)

The position of the peak corresponds to the energy of pairs of $R \sim 50$ Å irrespective of the impurities. Hence, it was suggested that the broad peak and the pair lines are intimately related.

23. Exciton impurity complexes. Extending the concept of excitons, LAMPERT [*159*] postulated the possible existence of various aggregates consisting of more charged particles than a hole-electron pair. The aggregates may be of two kinds, the mobile ones consisting of electrons and holes only, and those involving imperfection centers. So far no aggregates of the first kind beside the exciton have been observed. The simplest aggregates of the second kind would be an electron-hole pair attached to a charged center. The addition of a second electron or a second hole gives another aggregate of interest. An electron-hole pair attached to a positively (negatively) charged center may be regarded as an exciton-ionized donor (acceptor) complex, for which a convenient notation is $\oplus - + (\ominus + -)$. The addition of another electron (hole), $\oplus - - + (\ominus + + -)$, gives an aggregate which is equivalent to an exciton-neutral donor (acceptor) complex. Consider the binding of these complexes. Let E_D be the energy of dissociation of the complex into an exciton and an ionized or neutral center. Estimates of E_D can be obtained by analogy with molecular binding, in the limit of extremely different effective masses between the electron and the hole. For $m_h \gg m_e$: the complex $\oplus - +$ is equivalent to H_2^-, the complex $\oplus - - +$ is equivalent to H_2, and the complex $\ominus + + -$ may be considered as a negative particle bound to a positive center, the removal of which leaves $\ominus + +$ analogous to H^-. In this way, we get the estimates of E_D in the Table for the various cases. Calculations made by HOPFIELD [*160*] show that for the complexes: $\oplus - - +$ and $\ominus + + -$, E_D varies monotonically between the limiting values as the ratio m_h/m_e varies from $\gg 1$ to $\ll 1$. The calculations also indicate that $\oplus - +$ is not stable for $m_h/m_e < 1.4$, and, similarly, $\ominus + -$ is not stable for $m_e/m_h < 1.4$.

Table. *Exciton-impurity complex.*

E_D is the dissociation energy. E_e and E_h are the ionization energies for an electron bound to a positive center and a hole bound to a negative center, respectively. E_x is the binding energy of an exciton.

Complex		m_h/m_e	E_D	Molecular analogy
Exciton-ionized donor . .	$\oplus - +$	$\gg 1$	$0.21\, E_e$ [*159*]	H_2^-
Exciton-ionized acceptor .	$\ominus + -$	$\ll 1$	$0.21\, E_h$ [*159*]	H_2^-
Exciton-neutral donor . .	$\oplus - - +$	$\gg 1$	$0.33\, E_e$ [*163*]	H_2
		$\ll 1$	$E_h + 0.055\, E_e - E_x = 0.055\, E_e$ [*163*]	$0.055\, E_e$ analogous to electron affinity of H^- $(\oplus - -)$
Exciton-neutral acceptor .	$\oplus + + -$	$\ll 1$	$0.33\, E_h$ [*163*]	
		$\gg 1$	$0.055\, E_h$ [*163*]	analogous to the case $\oplus - - +$ with $m_h/m_e \ll 1$

Sharp lines observed in the absorption and emission spectra of various semiconductors have been attributed to exciton complexes. These lines are situated close to the intrinsic absorption edge and are very narrow in width, often narrower than kT. The materials for which exciton complexes have been reported include some semiconductors, GaSb [*91*], CdS [*161*], ZnTe [*162*], with direct transition energy gaps, and some semiconductors, Si [*163*], Ge [*164*], SiC [*165*], GaP [*166*], with indirect transition energy gaps. The sharp lines at 1.51 and 1.59 eV in Fig. 33 are apparently of this origin, and such sharp lines have been seen in the emission spectra of various other II—VI compounds. Fig. 36 shows collections of absorption and emission lines observed in selected CdS crystals. The intrinsic exciton lines are marked with asterisks. The A lines given by the top valence band, should be active only for $E \perp c$ whereas the B lines given by the next band B should be active for both polarizations. From the energies and the polarization behavior, it was deduced that the lines I_1, I_2, I_3 are given by exciton A-impurity complexes, while the lines (I_{1B}, I'_{1B}), I_{2B} are derived from exciton B and the same centers involved in I_1 and I_2. Studies of the Zeeman effect of these lines led to the conclusion that center 1 is a neutral acceptor, center 2 is a neutral donor, and center 3 is an ionized donor.

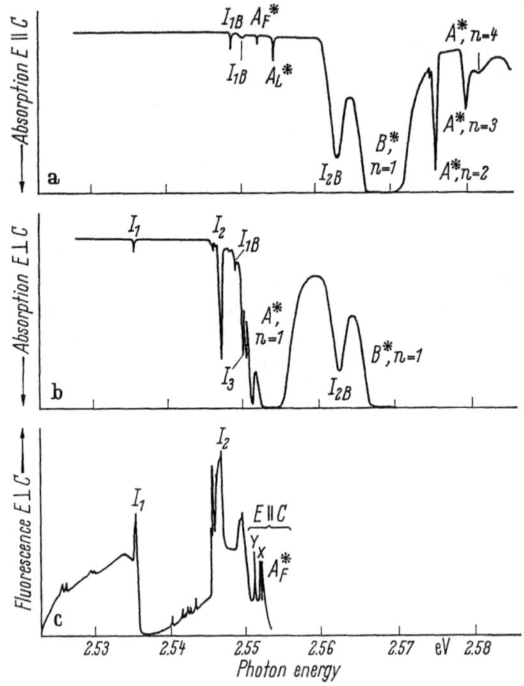

Fig. 36a—c. Composite diagrams of absorption lines and emission lines seen in selected CdS crystals at 1.6° K. The intrinsic excitons are each marked with an asterisk. Polarized light is used with electric vector parallel to the C axis of the crystal, $E \| C$, or perpendicular to the c axis, $E \perp C$. (After D. G. Thomas, and J. J. Hopfield.)

The effect of interaction with phonons can be seen in the fluorescence spectrum of Fig. 36. There is a broad continuum at the low photon energy side of the line I_1 which corresponds to electronic transition at I_1 with acoustic phonon emission. Not seen in Fig. 36 on account of its weakness is a corresponding high energy wing of I_1 in absorption. Phonon effects are clearly displayed by a center in ZnTe [*162*]. Absorption and fluorescence were observed around 1.986 eV. Each showed structures, and at 20° K the two spectra appeared to mirror in energy as shown in Fig. 37. The central line common to both absorption and fluorescence corresponds to zero-phonon transitions. The longitudinal optical phonon energy is 0.026 eV. The apparent repetition of structure at this energy interval is given by transitions accompanied by the emissions of successively larger number of LO phonons. The two broad peaks in each repeated interval are supposed to arise from the transverse and longitudinal acoustic phonons. Transitions with the emission of a transverse optical phonon are marked by the arrows.

Exciton-impurity complexes were first observed in silicon [*163*] in which the extrema of the conduction and valence bands do not coïncide in k space and the intrinsic absorption edge corresponds to indirect transitions. The intrinsic exciton emission shown by the dashed curve in Fig. 38 for a specimen of high purity corresponds to exciton recombination with emission of phonons. The exciton energy is $E_0=1.156$ eV, and the principal line corresponds to recombination

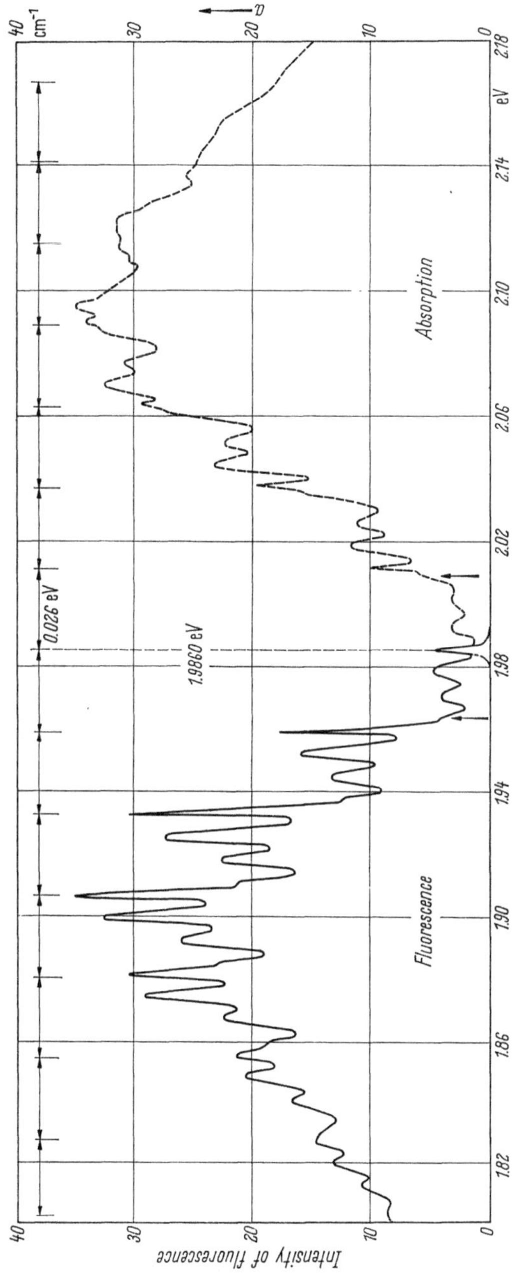

Fig. 37. The fluorescence and absorption of a center in ZnTe, at 20° K. The central line is common to both spectra. (After R. E. Dietz, D. G. Thomas, and J. J. Hopfield.)

accompanied by the emission of transverse optical (TO) phonon. The two additional lines at 1.091 and 1.149 eV in the solid curve for an As-doped specimen are produced by an exciton-As impurity complex. The energy difference between the two lines is equal to the energy of the TO phonon, and the lines are attributed to emissions of the exciton-As complex with zero and one TO phonon emission, respectively. In contrast to the free exciton, the complex can recombine without phonon emission since the conservation of crystal momentum presents no problem for the bound exciton. The energy difference between the 1.091 eV line and the principal free exciton line gives the dissociation energy, E_D, of the complex.

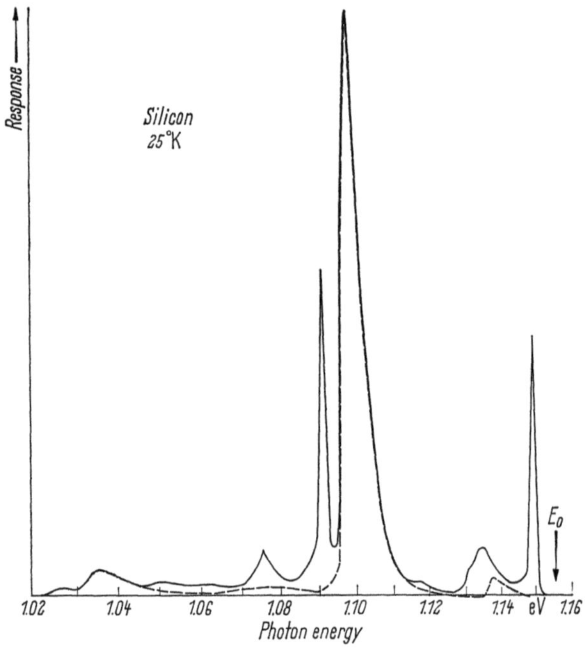

Fig. 38. Photoluminescence spectra for two silicon crystals. The dashed curve is for a specimen containing a negligible amount of impurity. The solid curve is for a specimen containing 8×10^{16} cm^{-3} arsenic atoms. (After J. R. HAYNES.)

Similar complexes were observed in crystals doped with various group III acceptor and group V donor impurities. In all cases, $E_D \sim 0.1\, E_i$ where E_i is the ionization energy of the impurity. The value of E_D falls within the range, 0.055 E_i to 0.33 E_i, indicating that exciton-neutral impurity complexes were observed.

References.

[1] BORN, M.: Optik, p. 267. Berlin: Springer 1933.
[2] See for example F. STERN: In: Solid State Physics, Vol. 15, edit. by F. SEITZ, and D. TURNBULL, p. 300. New York: Academic Press 1963.
[3] SEITZ, F.: Modern Theory of Solids. New York: McGraw Hill Book Co. 1939.
[4] FRÖHLICH, H.: Electronentheoric der Metalle. Berlin: Springer 1936. — WILSON, A. H.: Theory of Metals. Cambridge: Cambridge University Press 1936. — MOTT, N. F., and H. JONES: Theory of the Properties of Metals and Alloys. Oxford: Oxford University Press 1956.
[5] NOZIÈRES, P., and D. PINES: Phys. Rev. **113**, 1254 (1959).
[6] EHRENREICH, H., and M. H. COHEN: Phys. Rev. **715**, 786 (1959), investigated in detail the approximations involved in reducing the problem of many electrons with electromagnetic field to a single-particle self-consistent-field problem.
[7] ADLER, S. L.: Phys. Rev. **126**, 413 (1962).
[8] PINES, D.: Elementary excitations in Solids. New York: W. A. Benjamin, Inc. 1963.

[9] WENTZEL, G.: In: Handbuch der Physik, Vol. 24, part 1. Berlin: Springer 1933. — HEITLER, W.: Quantum Theory of Radiation. Oxford: Oxford University Press 1954.

[10] NOZIÈRES, P., and D. PINES: Phys. Rev. **109**, 762 (1958).

[11] BROUT, R.: Phys. Rev. **107**, 664 (1957).

[12] FRÖHLICH, H.: Advance in Phys. **3**, 325 (1954).

[13] ALLCOCK, G. R.: Advance in Phys. **5**, 412 (1965).

[14] Polarons and Excitons, edit. by C. G. KUPER, and G. D. WHITFIELD. Edinburgh and London: Oliver & Boyd 1963.

[15] LEE, T. D., F. LOW, and D. PINES: Phys. Rev. **90**, 29 (1953).

[16] HALL, L. H., J. BARDEEN, and F. BLATT: Phys. Rev. **95**, 559 (1954). In: Photoconductivity Conference, Atlantic City, 1954, edit. by R. G. BRECKENRIDGE, B. R. RUSSELL, and E. E. HAHN, p. 146. New York: John Wiley & Sons 1956.

[17] McLEAN, T. P.: In: Progress in Semiconductors, Vol. 5, edit. by A. F. GIBSON, R. E. BURGESS, and F. A. KRÖGER, p. 53. New York: John Wiley & Sons 1960.

[18] LINDHARD, J.: Kgl. Danske Videnskab. Selskabs, Mat.-fys. Medd. **28**, No. 8 (1954).

[19] FAN, H. Y., and M. BECKER: In: Semiconducting Materials, Proceedings of the Reading Conference 1950, edit. by H. K. HENISCH, p. 132. London: Butterworths 1951.

[20] FAN, H. Y., W. SPITZER, and R. J. COLLINS: Phys. Rev. **101**, 566 (1956).

[21] FAN, H. Y.: Repts. Progr. in Phys. **19**, 107 (1956).

[22] ROSENBERG, R., and M. LAX: Phys. Rev. **112**, 843 (1958).

[23] WOLFE, R.: Proc. Phys. Soc. (London) A **67**, 74 (1954).

[24] MEYER, H. J. G.: Phys. Rev. **112**, 298 (1958).

[25] VISVANATHAN, S.: Phys. Rev. **120**, 376 (1960).

[26] FAN, H. Y.: Phys. Rev. **82**, 900 (1951).

[27] RADKOWSKY, A.: Phys. Rev. **73**, 749 (1948).

[28] MUTO, T., and S. OYAMA: Progr. Theoret. Phys. (Osaka) **5**, 833 (1950).

[29] KNOX, R. S.: Theory of Excitons, Solid State Physics, Suppl. 5, edit. by F. SEITZ and D. TURNBULL. New York: Academic Press 1963. This book gives a comprehensive review and detailed discussions of the theories.

[30] WANNIER, G. H.: Phys. Rev. **52**, 191 (1937).

[31] RASHBA, E. I.: Zhur. Eksptl. i Teoret. Fiz. **36**, 1703 (1959) [English translation: Soviet Phys. JETP **9**, 1213 (1959)].

[32] HELLER, W. R., and A. MAREUS: Phys. Rev. **91**, 273 (1953).

[33] HOPFIELD, J. J., and D. G. THOMAS: J. Phys. Chem. Solids **12**, 276 (1960).

[34] TOYOZAWA, Y.: Progr. Theoret. Phys. (Kyoto) **12**, 421 (1954).

[35] HAKEN, H., u. W. SCHOTTKY: Z. physik. Chem., N.F. **16**, 218 (1958). — HAKEN, H.: J. Phys. Chem. Solids **8**, 166 (1959). — HAKEN, H.: In: Polarons and Excitons, edit. by C. G. KUPER, and G. D. WHITFIELD, p. 295. Edinburgh and London: Oliver & Boyd 1963.

[36] DEXTER, D. L.: Phys. Rev. **108**, 707 (1957).

[37] MUTO, T., S. OYAMA, and H. OKUNO: Progr. Theoret. Phys. (Kyoto) **20**, 804 (1958). — MUTO, T.: Progr. Theoret. Phys. (Kyoto), Suppl. **12**, 3 (1959).

[38] OVERHAUSER, A. W.: Phys. Rev. **101**, 1702 (1956).

[39] HOPFIELD, J. J.: Phys. Rev. **112**, 1555 (1958).

[40] FANO, U.: Phys. Rev. **103**, 1202 (1956); **118**, 451 (1960).

[41] AGRANOVICH, V. M.: Zhur. Eksptl. i Teor. Fiz. **37**, 430 (1959) [English translation: Soviet Phys. JETP **10**, 307 (1960)].

[42] DRESSELHAUS, G.: J. Phys. Chem. Solids **1**, 14 (1956).

[43] KOHN, W., and J. M. LUTTINGER: Phys. Rev. **98**, 915 (1955).

[44] ELLIOTT, R. J.: Phys. Rev. **108**, 1384 (1957). — ELLIOTT, R. J.: In: Polarons and Excitons, edit. by C. G. KUPER, and G. D. WHITEFIELD, p. 269. Edinburgh and London: Oliver & Boyd 1963.

[45] LEURGANS, P. J., and J. BARDEEN: Phys. Rev. **87**, 200 (1952).

[46] ANSEL'M, A. I., i Y. A. FIRSOV: Zhur. Eksptl. i Teoret. Fiz. **28**, 151 (1955) [English translation: Soviet Phys. JETP **1**, 139 (1956)].

[47] ANSEL'M, A. I., i Y. A. FIRSOV: Zhur. Eksptl. i Teoret. Fiz. **30**, 719 (1956) [English translation: Soviet Phys. JETP **3**, 564 (1956)]. See also Y. TOYOZAWA: Progr. Theoret. Phys. (Kyoto) **20**, 53 (1958).

[48] GENKIN, G. M.: Fiz. Tverd. Tela **3**, 2097 (1961) [English translation: Soviet Phys. — Solid State **3**, 1523 (1962)].

[49] TOYOZAWA, Y.: Progr. Theoret. Phys. (Kyoto) **20**, 53 (1958).
[50] TOYOZAWA, Y.: Progr. Theoret. Phys. (Kyoto) **27**, 89 (1962).
[51] TOYOZAWA, Y.: J. Phys. Chem. Solids **25**, 59 (1964).
[52] LUTTINGER, J. M., and W. KOHN: Phys. Rev. **97**, 869 (1955). — KOHN, W.: In: Solid State Physics, Vol. 5, edit. by F. SEITZ, and D. TURNBULL, p. 258. New York: Academic Press 1957.
[53] KOHN, W.: J. Phys. Chem. Solids **8**, 43 (1959). — ROTH, L. M., and G. W. PRATT: J. Phys. Chem. Solids **8**, 47 (1959).
[54] MENDELSON, K. S., and H. M. JAMES: J. Phys. Chem. Solids **25**, 729 (1964). — SCHECHTER, D.: J. Phys. Chem. Solids **23**, 237 (1962).
[55] DEXTER, D. L.: In: Solid State Physics, Vol. 6, edit. by F. SEITZ, and D. TURNBULL, p. 355. New York: Academic Press 1958. — MARKHAM, J. J.: Rev. Mod. Phys. **31**, 956 (1959).
[56] O'ROURKE, R. C.: Phys. Rev. **91**, 265 (1953).
[57] LAX, M.: J. Chem. Phys. **20**, 1752 (1952).
[58] PEKAR, S. I.: Untersuchungen über die Elektronen-Theorie der Kristalle. Berlin: Akademische Verlagsgesellschaft 1954.
[59] HUANG, K., and A. RHYS: Proc. Roy. Soc. (London) A **204**, 406 (1950).
[60] Reviewed by A. A. MARADUDIN, E. W. MONTROLL, and C. H. WEISS: Solid State Physics, Suppl. 3, edit. by F. SEITZ, and D. TURNBULL, New York: Academic Press 1963.
[61] LAX, M.: In: Photoconductivity Conference, Atlantic City, 1954, edit. by R. G. BRECKENRIDGE, B. R. RUSSELL, and E. E. HAHN, p. 111. New York: John Wiley & Sons 1956.
[62] TRIFONOV, E. D.: Doklady Akad. Nauk. S.S.S.R. **147**, 826 (1962) [English translation: Soviet Phys. — Doklady **7**, 1105 (1963)].
[63] KANE, E. O.: Phys. Rev. **119**, 40 (1960).
[64] HOPFIELD, J. J.: Proceedings of the Internat. Conference on the Physics of Semiconductors, Exeter, 1962, p. 75. London: The Institute of Physics and The Physical Society.
[65] KANE, E. O.: J. Phys. Chem. Solids **1**, 249 (1957).
[66] MOSS, T. S.: Optical Properties of Semiconductors. New York: Academic Press 1949.
[67] See review articles by W. PAUL, and D. M. WARSCHAUER, and by H. G. DRICKAMER: In: Solids Under Pressure, edit. by W. PAUL and D. M. WARSCHAUER. New York: McGraw Hill Book Co. 1963.
[68] EDWARDS, D. F., T. E. SKYLHOUSE, and H. G. DRICKAMER: J. Phys. Chem. Solids **11**, 140 (1959).
[69] EDWARDS, A. J., and H. G. DRICKAMER: Phys. Rev. **122**, 1149 (1961).
[70] FAN, H. Y., M. L. SHEPHERD, and W. SPITZER: Photoconductivity Conference, Atlantic City, edit. by R. G. BRECKENRIDGE, B. R. RUSSELL, and E. E. HAHN, p. 184. New York: John Wiley & Sons 1957.
[71] MACFARLANE, G. G., T. P. MCLEAN, J. E. QUARRINGTON, and V. ROBERTS: Phys. Rev. **108**, 1377 (1957).
[72] MACFARLANE, G. G., T. P. MCLEAN, J. E. QUARRINGTON, and V. ROBERTS: Phys. Rev. **111**, 1245 (1958).
[73] GERSHENZON, M., D. G. THOMAS, and R. E. DIETZ: Proceedings of the Internat. Conference on Semiconductor Physics, Exeter, 1962, p. 752. London: The Institute of Physics and the Physical Society 1962.
[74] BROWN, F. C., T. MASUMI, and H. H. TIPPINS: J. Phys. Chem. Solids **22**, 101 (1961).
[75] CHOYKE, W., and L. PATRICK: Proceedings of the Internat. Conference on Semiconductor Physics, Prague, 1960, p. 432. Prague: Czechoslovak Academy of Sciences 1961.
[76] LOFERSKI, J. J.: Phys. Rev. **93**, 707 (1954).
[77] GIBRECHT, H., u. A. BARTSCHAT: Z. Physik **136**, 224 (1953). — DUTTON, D.: Phys. Rev. **112**, 785 (1958).
[78] STOURAC, L., J. TAUC, and M. ZAVETOVA: Proceedings of the Internat. Conference on Semiconductor Physics, 1960, p. 1091. Prague: Czechoslovak Academy of Sciences 1961. — ABRAHAM, A., V. FREI, J. TAUC, B. VELIČKY, and M. ZAVETOVA: Proceedings of the 7th Internat. Conference on Physics of Semiconductors, 1964, p. 209. Paris: Dunod 1964.
[79] KOMIYA, H., K. MASUMOTO, and H. Y. FAN: Phys. Rev. **133**, A 1679 (1964).

[80] See for example F. N. MOTT, and R. W. GURNY: Elcctronic Processes in Ionic Crystals. Oxford: Oxford University Press 1953.
[81] GROSS, E. F., i N. A. KARRYEV: Doklady Akad. Nauk S.S.S.R. **84**, 471 (1952).
[82] UETA, M.: Progr. Theoret. Phys., Suppl. **12**, 40 (1959).
[83] NIKITINE, S.: In: Progress in Semiconductors, Vol. 6, edit. by A. F. GIBSON, R. E. BURGESS, and F. A. KRÖGER, p. 269. New York: John Wiley & Sons 1962.
[84] GROSS, E. F.: Uspekhi Fiz. Nauk **76**, 433 (1962) [English translation: Soviet Phys. — Uspekhi **5**, 195 (1962)].
[85] THOMAS, D. G., and J. J. HOPFIELD: Phys. Rev. **116**, 573 (1959).
[86] THOMAS, D. G.: J. Phys. Chem. Solids **15**, 86 (1960).
[87] MACFARLANE, G. G., T. P. McLEAN, J. E. QUARRINGTON, and V. ROBERTS: Proc. Phys. Soc. (London) **71**, 863 (1958).
[88] PASTRNYAK, I.: Fiz. Tverd. Tela **3**, 869 (1961) [English translation: Soviet Phys. — Solid State **3**, 633 (1961)].
[89] ELLIOTT, R. J.: Phys. Rev. **124**, 340 (1961).
[90] ZHILICH, A. G.: Vestnik Leningrad State Univ., Phys. Chem. Series No. 2, No. 105 (1960).
[91] JOHNSON, E. J., I. FILINSKI, and H. Y. FAN: Proceedings of the Internat. Conference on the Physics of Semiconductors, Exeter, 1962, p. 375. London: The Institute of Physics and The Physical Society 1962.
[92] STURGE, M. D.: Phys. Rev. **127**, 768 (1962).
[93] TURNER, W. G., W. E. REESE, and G. D. PETTIT: Phys. Rev. **136**, A 1467 (1964).
[94] GROSS, E. F., i. A. A. KAPLYANSKII: Doklady Akad. Nauk S.S.S.R. **139**, 75 (1961) [English translation: Soviet Phys. — Doklady **6**, 592 (1962)].
[95] GROSS, E. F., i A. A. KAPLYANSKII: Fiz. Tverd. Tela **2**, 1676, 2968 (1960) [English translation: Soviet Phys. — Solid State **2**, 1518, 2637 (1960)].
[96] GROSS, E. F., i B. S. RAZBIRIN: Fiz. Tverd. Tela **4**, 207 (1962).
[97] BORN, M., and K. HUANG: Dynamical Theory of Crystal Lattices. Oxford: Oxford University Press 1954.
[98] PEKAR, S. I.: Zhur. Eksptl. Theoret. Fiz. **34**, 1176 (1958) [English translation: Soviet Phys. JETP **7**, 813 (1958)].
[99] PEKAR, S. I.: Proceedings of the Internat. Conference on the Physics of Semiconductors, Exeter, 1962, p. 419. London: The Institute of Physics and the Physical Society 1962. See references in this paper for earlier works.
[100] HOPFIELD, J. J., and D. G. THOMAS: Phys. Rev. **132**, 563 (1963).
[101] GORBAN', I. S., i V. B. TIMOFEEV: Doklady Akad. Nauk S.S.S.R. **140**, 791 (1961) [English translation: Sovient Phys. — Doklady **6**, 878 (1962)].
[102] BRODIN, M. S., i I. S. PEKAR: Zhur. Eksptl. i Teoret. Fiz. **38**, 1910 (1960) [English translation: Soviet Phys. JETP **11**, 1373 (1960)].
[103] BRODIN, M. S., i M. I. STRASHNIKOVA: Fiz. Tverd. Tela **4**, 2454 (1962) [English translation: Soviet Phys. — Solid State **4**, 1798 (1963)].
[104] MOORADIAN, A., and H. Y. FAN: Radiative Recombination in Semiconductors, 7th International Conference on the Physics of Semiconductors, Paris, 1964, p. 39. Paris: Dunod 1964.
[105] HAYNES, J. R., M. LAX, and W. F. FLOOD: J. Phys. Chem. Solids **8**, 392 (1959).
[106] HAYNES, J. R., and N. G. NILSON: Radiative Recombination in Semiconductors, 7th Internat. Conference on the Physics of Semiconductors, Paris, 1964, p. 21. Paris: Dunod 1964.
[107] VEER, S. M. DE, and H. J. G. MEYER: Proceedings of the Internat. Conference on the Physics of Semiconductors, Exeter, 1962, p. 358. London: The Institute of Physics and The Physical Society 1962.
[108] A review of the studies on III—V compounds will be published in a volume of the series "Physics of III—V Compounds", edit. by R. K. WILLARDSON, and A. C. BEER. New York: Academic Press.
[109] HAGA, E., and H. KIMURA: J. Phys. Soc. Japan **19**, 471 (1964).
[110] CALDWELL, R. S., and H. Y. FAN: Phys. Rev. **114**, 664 (1959).
[111] KAISER, W., R. J. COLLINS, and H. Y. FAN: Phys. Rev. **91**, 1380 (1953). — BRIGGS, H. B., and R. C. FLETCHER: Phys. Rev. **91**, 1342 (1953).
[112] KANE, E. O.: J. Phys. Chem. Solids **1**, 82 (1956).
[113] HOBSON, G. S., and E. G. S. PAIGE: Proceedings of the 7th Internat. Conference on the Physics of Semiconductors, Paris, 1964, p. 143. Paris: Dunod 1964.

[114] BRAY, R., and W. E. PINSON: Phys. Rev. Letters **11**, 246 (1963). — BAYNHAM, A. C., and E. G. S. PAIGE: Proceedings of the 7th Internat. Conference on the Physics of Semiconductors, Paris, 1964, p. 149. Paris: Dunod 1964.

[115] SPITZER, W. G., and H. Y. FAN: Phys. Rev. **106**, 882 (1957).

[116] BECKER, W. M., A. K. RAMDAS, and H. Y. FAN: J. Appl. Phys. **32**, 2094 (1961).

[117] SHULTZ, L. G.: Suppl. Phil. Mag. **6**, 102 (1957). — GIVENS, M. P.: Solid State Physics, Vol. 6, edit. by F. SEITZ, and D. TURNBULL, p. 313. New York: Academic Press 1958.

[118] See the following articles with references to other works: COOPER, B. R., H. EHRENREICH, and H. R. PHILIPP: Phys. Rev. **138**, A 494 (1964). — EHRENREICH, H., H. R. PHILIPP, and D. J. OLECHRA: Phys. Rev. **131**, 2469 (1963).

[119] See the following articles with references to other works: CARDONA, M.: Proceedings of the 7th Internat Conference on the Physics of Semiconductors, Paris, 1964, p. 181. Paris: Dunod 1964. — PHILLIPS, J. C.: Phys. Rev. **133**, A 452 (1964). — PHILIPP, H. R., and H. EHRENREICH: Phys. Rev. **129**, 1550 (1963).

[120] See the following articles with references to other works: PHILIPP, H .R., and E. A. TAFT: Phys. Rev. **136**, A 1445 (1965). — PHILLIPS, J. C.: Phys. Rev. **136**, A 1705 (1964).

[121] EHRENREICH, H., and H. R. PHILIPP: Phys. Rev. **128**, 1622 (1962).

[122] HOVE, L. VAN: Phys. Rev. **89**, 1189 (1953).

[123] PHILLIPS, J. C.: Phys. Rev. **104**, 1263 (1956).

[124] HROSTOWSKI, H. J.: In: Semiconductors, edit. by N. B. HANNAY, p. 437. New York: Reinhold Publ. Corp. 1959.

[125] FISHER, P., and H. Y. FAN: Phys. Rev. Letters **2**, 456 (1959).

[126] REUSZER, J. H., and P. FISHER: Phys. Rev. **135**, A 1125 (1964).

[127] AGGARWAL, R. L., and A. K. RAMDAS: Phys. Rev. **137**, A 602 (1965).

[128] KOHN, W.: In: Solid State Physics, Vol. 5, edit. by F. SEITZ, and D. TURNBULL, p. 257. New York: Academic Press 1957.

[129] MORITA, A., M. AZUMA, and H. NARA: J. Phys. Soc. Japan **17**, 1570 (1962). — CSAVINSKY, P. J.: J. Phys. Chem. Solids **24**, 1003 (1963). — MULLER, K.: Solid State Comm. **2**, 205 (1964). — Appel, J.: Phys. Rev. **133**, A 280 (1964).

[130] AGGARWAL, R. L.: Solid State Comm. **2**, 162 (1964).

[131] PRICE, P. J.: Phys. Rev. **104**, 1223 (1956).

[132] WEINREICH, G.: Proceedings of the Internat. Conference on Semiconductor Physics, Prague, 1960, p. 360. Prague: Czechoslovak Academy of Sciences 1961.

[133] FISHER, P., and H. Y. FAN: Phys. Rev. Letters **5**, 195 (1960).

[134] FAN, H. Y., and A. K. RAMDAS: J. Appl. Phys. **30**, 1127 (1959).

[135] NEWMAN, R., and W. W. TYLER: In: Solid State Physics, Vol. 8, edit. by F. SEITZ, and D. TURNBULL, p. 50. New York: Academic Press 1959.

[136] JOHNSON, E. J., and H. Y. FAN: Phys. Rev. **139**, A 1991 (1965).

[137] EAGLES, D. M.: J. Phys. Chem. Solids **16**, 76 (1960).

[138] LAMBE, J. J., C. C. KLICK, and D. L. DEXTER: Phys. Rev. **103**, 1715 (1956).

[139] HALSTED, R. E., M. R. LORENZ, and B. SEGALL: J. Phys. Chem. Solids **22**, 109 (1961).

[140] NATHAN, M. I., and G. BURNS: Phys. Rev. **129**, 125 (1963). For absorption see M. D. STURGE: Phys. Rev. **127**, 768 (1962).

[141] TRIC, C., C. BENOIT A LA GUILLAUME, et J. M. DE LEVER: Compt. rend. **225**, 3152 (1962).

[142] MOORADIAN, A., and H. Y. FAN: Bull. Am. Phys. Soc. **10**, 368 (1965).

[143] HAYNES, J. R., and W. C. WESTPHAL: Phys. Rev. **101**, 1676 (1956).

[144] See J. H. SCHULMAN, and W. D. COMPTON: Color Centers in Solids, Chap. III. New York: Macmillan 1962.

[145] FITCHEN, D. B., R. H. SILSBEE, T. A. FULTON, and E. L. WOLF: Phys. Rev. Letters **11**, 275 (1963), references to earlier works on LiF are given. — VON DER OSTEN, W.: Z. Physik **174**, 423 (1963). — PIERCE, C. B.: Phys. Rev. **135**, 85 (1964).

[146] LAX, M., and E. BURSTEIN: Phys. Rev. **100**, 592 (1955).

[147] PAJOT, B.: J. Phys. Chem. Solids **25**, 613 (1964).

[148] HARDY, J. R., S. D. SMITH, and W. TAYLOR: Proceedings of the Internat. Conference on the Physics of Semiconductors, Exeter, 1962, p. 521. London: The Institute of Physics and The Physical Society.

[149] HAANSTRA, J. H., and C. HAAS: Phys. Letters **2**, 21 (1962).

[150] KLICK, C. C.: Phys. Rev. **89**, 274 (1953).

[151] KRÖGER, F. A., and H. J. G. MEYER: Physica **20**, 1149 (1954).
[152] HOPFIELD, J. J.: J. Phys. Chem. Solids **10**, 109 (1959).
[153] HALSTED, R. E., M. R. LORENZ, and B. SEGALL: J. Phys. Chem. Solids **22**, 109 (1961).
[154] For summary and references see F. E. WILLIAMS: J. Phys. Chem. Solids **12**, 265 (1960).
[155] CHOYKE, W. J., D. R. HAMILTON, and L. PATRICK: Phys. Rev. **117**, 1430 (1960). — PATRICK, L.: Phys. Rev. **117**, 139 (1960).
[156] HOOGENSTRAATEN, W.: Philips Research Repts. **13**, 515 (1958).
[157] GROSS, E. F., i D. S. NEDZVETSKI: Doklady Akad. Nauk. S.S.S.R. **146**, 1047 (1962) [English translation: Soviet Phys. — Doklady **7**, 908 (1963)]. — VINK, A. T., and C. Z. VAN DOORN: Phys. Letters **1**, 332 (1962). — GERSHENZON, M., D. G. THOMAS, and R. E. DIETZ: Proceedings of the Internat. Conference on the Physics of Semiconductors, Exeter, 1962, p. 752. London: The Institute of Physics and The Physical Society.
[158] THOMAS, D. G., M. GERSHENZON, and F. A. TRUMBORE: Phys. Rev. **133**, A 269 (1964). — HOPFIELD, J. J., D. G. THOMAS, and M. GERSHENZON: Phys. Rev. Letters **10**, 162 (1963).
[159] LAMPERT, M.: Phys. Rev. Letters **1**, 450 (1958).
[160] HOPFIELD, J. J.: Proceedings of the Internat. Conference on the Physics of Semiconductors, Paris, 1964, p. 725. Paris: Dunod 1964.
[161] THOMAS, D. G., and J. J. HOPFIELD: Phys. Rev. **128**, 2135 (1962).
[162] DIETZ, R. E., D. G. THOMAS, and J. J. HOPFIELD: Phys. Rev. Letters **8**, 391 (1962).
[163] HAYNES, J. R.: Phys. Rev. Letters **4**, 361 (1960).
[164] BENOIT A LA BUILLAUME, C., and O. PARODI: Proceedings of the Internat. Conference on Semiconductor Physics, Prague, 1960, p. 426. Prague: Czechoslovak Academy of Sciences 1960.
[165] HAMILTON, D. R., W. J. CHOYKE, and L. PATRICK: Phys. Rev. **131**, 127 (1963).
[166] THOMAS, D. G., M. GERSHENZON, and J. J. HOPFIELD: Phys. Rev. **131**, 2397 (1963).

Magneto-Optics in Crystals.

By

S. D. Smith.

With 50 Figures.

I. Introduction.

1. General. The first magneto-optical effect was discovered by MICHAEL FARADAY in 1845 when he observed the rotation in the plane of polarisation of light which was transmitted through glass in a direction parallel to an applied magnetic field[1]. Experiments then spread to other materials, particularly gases, leading up to the discovery of the Zeeman effect in 1896. By the early years of the twentieth century, following the formulation of the electromagnetic theory by MAXWELL and of dispersion theory by LORENTZ, DRUDE, VOIGT and others, the magneto-optical effects were remarkably well understood in terms of the classical dynamics of an electron in a magnetic field[2,3].

The equation of motion for an electron of mass m, bound with a force constant $m\omega_0^2$ driven by an electric field $\boldsymbol{E}$ and a static magnetic field $\boldsymbol{B}$ is, including a damping term, $g\dot{\boldsymbol{r}}$, proportional to velocity,

$$m\ddot{\boldsymbol{r}}+m g\dot{\boldsymbol{r}}+m\omega_0^2\boldsymbol{r}=e\boldsymbol{E}+e\dot{\boldsymbol{r}}\times\boldsymbol{B} \tag{1.1}$$

where e is the charge on the electron and $\boldsymbol{r}$ is the position vector. The solution is an oscillatory function from which the dipole moment and hence a complex dielectric constant may be found as

$$\varepsilon_\pm=(n-i\kappa)^2_\pm=1+\frac{N e^2/m\,\varepsilon_0}{\omega_0^2-\omega^2\pm\omega\omega_c-i\omega g} \tag{1.2}$$

in which the positive and negative subscripts refer to right and left circularly polarised waves[4] arising from

$$\boldsymbol{E}=\boldsymbol{E}_0\cdot e^{i\omega t}=(\boldsymbol{E}_{0x}\pm i\boldsymbol{E}_{0y})\,e^{i\omega t} \tag{1.3}$$

and

$$\omega_c=e\boldsymbol{B}/m. \tag{1.4}$$

ω_c is known as the cyclotron resonance frequency .

The conclusions from Eq. (1.1) were known to the early workers including the relation between the dispersive effects (Faraday effect and magnetic double refraction or Voigt effect) and the absorption effects (the inverse Zeeman effect). In addition, in 1897 Becquerel derived the relation between dispersion and Faraday effect as

$$\Theta=\frac{1}{2}\,\omega\omega_c\frac{dn}{d\omega}\,. \tag{1.5}$$

---

[1] M. FARADAY: Phil. Mag. **29**, 153 (1846).

[2] See, for example, P. ZEEMAN: Researches in Magneto-Optics. London: Macmillan 1913.

[3] W. VOIGT: Magneto- and Electro-Optics. Tübingen, Leipzig 1908.

[4] See also Sect. 3.

A number of measurements of Faraday effect were made during the next fifty years on various crystalline materials and the results compared with Eq. (1.5)[1]. The experiments showed that Eq. (1.5) was inadequate both qualitatively and quantitatively since rotations of opposite signs were observed. These were empirically included by defining a quantity known as "the magneto-optic anomaly". Theory appropriate to describe molecular levels is given by SERBER[2].

Little progress was made until the measurements spread to semiconductors and quantum mechanical interpretation in terms of energy band structure became possible during the 1950's. Magneto-optics as an experimental tool for studying the band structure of crystals dates from 1956 when the first infra-red observations of *resonant* effects were made, but these were preceded by the observation at microwave frequencies of the cyclotron resonance of free carriers in 1954.

Although the classical theory is inadequate, it is qualitatively useful for *free carrier* effects and is useful by analogy for *interband* effects which classically correspond to effects of bound charges. We shall make some use of classical theory in this section to describe a number of different magneto-optical effects that have been observed since 1956. The importance of magneto-optics in semiconductors lies in the fact that very explicit quantitative information about the band structure can be obtained. There are distinct classes of experimental measurements that may be made: firstly, analogous to the Zeeman effect, the absorption of a crystal may be measured with the object of finding the *position* (frequency) of any resonances; secondly, as in the Faraday effect, the state of polarisation before and after passing through the specimen may be examined and the *magnitude* of the rotation or ellipticity used in analysis. These latter effects are interference phenomena of high sensitivity and may be used both on and off resonance.

The simplest and most fundamental effect, cyclotron resonance absorption of free carriers, was also the first resonance effect to be observed. It is classically a very simple phenomenon — the charged particle moves according to Eq. (1.1) but with $\omega_0=0$ and describes circular orbits whose radii increase as energy is absorbed from the applied electric field until, after a time τ a collision takes place and the process begins again. The cyclotron resonance of free carriers was discussed theoretically (and independently) by DORFMAN[3], DINGLE[4] and SHOCKLEY[5] during the early nineteen-fifties and was observed experimentally, with the resonance at 9,000 M cps (~ 3 cm) by DRESSELHAUS, KIP and KITTEL[6]. From the relation $\omega_c = eB/m^*$ extensive information about the shape of the energy surfaces near the band extrema was obtained in experiments on Germanium and Silicon. Extension of cyclotron resonance observations to infra-red frequencies required high magnetic fields but enabled the necessary condition that $\omega_c\tau > 1$ to be more easily satisfied. Such observations, in 1956, by BURSTEIN, PICUS, GEBBIE and BLATT[7] at NRL and KEYES et al. at MIT[8], using pulsed fields, working on InSb, marked the beginning of modern magneto-optics in semiconductors in which "optical" as opposed to microwave techniques were used and quantitative interpretation in terms of the energy band structure of the crystal became possible.

---

[1] C. G. DARWIN, and W. H. WATSON: Proc. Roy. Soc. (London) A **114**, 474 (1937).
[2] R. SERBER: Phys. Rev. **41**, 489 (1932).
[3] J. DORFMAN: Doklady Acad. Sci. (U.S.S.R.) **81**, 765 (1951).
[4] R. B. DINGLE: Proc. Int. Conf. on Very Low Temperatures. Oxford 1951.
[5] W. SHOCKLEY: Phys. Rev. **90**, 491 (1953).
[6] G. DRESSELHAUS, A. F. KIP, and C. KITTEL: Phys. Rev. **92**, 827 (1953).
[7] E. BURSTEIN, G. PICUS, H. A. GEBBIE, and F. BLATT: Phys. Rev. **103**, 825 (1956).
[8] R. J. KEYES, S. ZWERDLING, S. FONER, H. H. KOLM, and B. LAX: Phys. Rev. **104**, 1804 (1956).

A new development in the oldest magneto optic effect — the Faraday effect — was also made during this period. MITCHELL[1], in 1955, suggested that in the dispersion term $dn/d\omega$ in Eq. (1.5) there would be a contribution from free carriers giving rise to a *free carrier* Faraday rotation. This effect, quite distinct from the older observations on bound electrons, was measured by SMITH and MOSS in 1958[2] and proved to be an accurate method for determination of effective mass. It is best understood as the *differential dispersion* of the cyclotron resonance absorption. It has the useful property that it is a dispersive effect and so, by the Kramers Kronig relation related to an integral over *all* the magneto-absorption, it is independent of the scattering processes and hence of the collision time. This

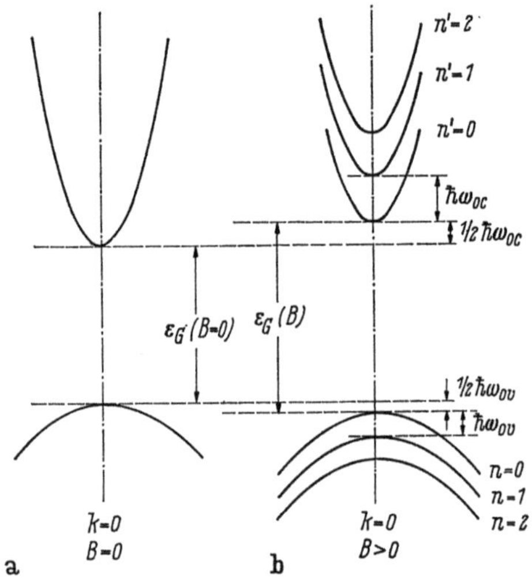

Fig. 1 a and b. Landau levels for simple bands (after BURSTEIN et al., Ref. 4, p. 257).

implies that it can be measured well off resonance and be detected under conditions which preclude the actual observation of cyclotron resonance absorption, i.e., when $\omega_c \tau < 1$. Non-resonant Faraday effect however is unable to detect any anisotropy of effective mass in a cubic crystal so that this technique is complementary to cyclotron resonance absorption which is sensitive to anisotropy, but is only observable when $\omega_c \tau > 1$.

At the same time as the observation of infra-red CR, magneto-optical effects were also found at the absorption edge associated with the transitions of electrons from valence to conduction band[3, 4]. These effects which are the analogy in the solid of the Zeeman effect cannot be usefully described by classical theory but are a consequence of the magnetic quantization of the quasi-continuous levels in the allowed energy bands of crystals. The solution of the Schrödinger equation for a free electron in a magnetic field, given by Landau in 1930 can be adapted to give a simple model for this situation by substituting the respective masses m_c^*

[1] E. W. J. MITCHELL: Proc. Phys. Soc. (London) B **68**, 973 (1955).

[2] S. D. SMITH, and T. S. MOSS: Proc. Brussels Solid State Conference 1958, Vol. II, p. 671. New York: John Wiley & Sons.

[3] E. BURSTEIN, G. S. PICUS, H. A. GEBBIE, and F. BLATT: Phys. Rev. **103**, 826 (1956).

[4] S. ZWERDLING, R. J. KEYES, S. FONER, H. H. KOHN, and B. LAX: Phys. Rev. **104**, 1805 (1956).

and m_v^* of electrons and holes. The allowed energies are then given by (see Sect. 8)

$$E_c = E_{c0} + \frac{\hbar^2 k^2}{2m_c^*} + \hbar\omega_{cc}\left(n' + \frac{1}{2}\right), \tag{1.6}$$

$$E_v = E_{v0} + \frac{\hbar^2 k^2}{2m_v^*} + \hbar\omega_{cv}\left(n + \frac{1}{2}\right), \tag{1.7}$$

where $\omega_{cc} = eB/m_c^*$ $\omega_{cv} = eB/m_v^*$ and n is a positive integer. These levels are known as Landau levels and are shown in Fig. 1.

Cyclotron resonance (CR) transitions are between Landau levels in one band with a change of quantum number, $\Delta n = \pm 1$; transitions between *different* bands, e.g., valence and conduction bands can also occur; in this case n is unchanged for the simplest (direct) transition — such transitions are known as interband

Table 1. *Magneto-optical effects in semiconductors.*

F and V refer to orientation of the Poynting vector of the radiation beam with respect to the magnetic field — Faraday configuration (parallel to B) and Voigt configuration (perpendicular to B) respectively. $n_\pm$ $\kappa_\pm$ refer to the optical constants for right- and left-circularly polarised radiation respectively associated with F and $n_\perp$, $n_\parallel$, $\kappa_\perp$, $\kappa_\parallel$ to parallel and perpendicularly polarised radiation associated with V.

1. Free carrier effects.

Effect	Orientation	n, κ	Resonant or nonresonant	Reference to observation
		a) In transmission		
Cyclotron resonance	F or V	κ_+	Res.	BURSTEIN et al., 1956[1] ZWERDLING et al., 1956[2]
Faraday rotation	F	$n_+ - n_-$	Non res.	SMITH and MOSS, 1958[3]
Resonant Faraday rotation	F	n_+	Res.	PALIK, 1963[4]
Faraday ellipticity	F	$\kappa_+ - \kappa_-$	Non res.	SMITH and PIDGEON, 1960[5]
Voigt effect	V	$n_\perp - n_\parallel$	Non res.	TEITLER and PALIK, 1960[6]
Interference fringe shift	F or V	$n_\pm - n$ $n_\parallel$ or $n_\perp - n$	Non res.	PALIK, 1963[4]
Oscillatory free carrier absorption	F or V	κ_+		PALIK and WALLIS, 1963[7]
		b) In reflection		
Magnetoplasma reflection	F or V	$n_\pm$ $n_\parallel$, $n_\perp$	Shift of resonance	LAX and WRIGHT, 1960[8]
Magnetoplasma rotation (Kerr effect)	F	$\kappa_\pm$	—	LAX, 1960[9], PALIK et al., 1962[10]
Magnetoplasma ellipticity	F	$n_\pm$	—	LAX, 1960[9], PALIK[10] et.al., 1962

[1] See ref. 7, p. 235.
[2] See ref. 4, p. 236.
[3] See ref. 2, p. 236.
[4] See ref. 7, p. 279.
[5] See ref. 2, p. 280.
[6] See ref. 5, p. 280.
[7] See ref. 3, p. 284.
[8] See ref. 2, p. 282.
[9] See ref. 3, p. 286.
[10] See ref. 6, p. 278.

Table 1 (Continued).

2. Interband effects.

Effect	Orientation	n, κ	Resonant or nonresonant	Reference to observation
a) In transmission.				
Interband magneto-absorption (I.M.O.)	F or V	κ_+, κ_- $\kappa_\parallel, \kappa_\perp$	Res.	BURSTEIN et al., 1956[1] ZWERDLING et al., 1956[2]
Faraday rotation	F	$n_+ - n_-$	Non res.	KIMMEL, 1957[3]
Resonant Faraday rotation	F	$n_+ - n_-$	Res.	NISHINA et al., 1962[4] SMITH et al., 1962[5] MITCHELL and WALLIS, 1963[6]
Faraday ellipticity	F	$\kappa_+ - \kappa_-$	Non res.	SMITH and PIDGEON, 1960[7]
Resonant Faraday ellipticity	F	$\kappa_+ - \kappa_-$	Res.	MITCHELL and WALLIS, 1963[6]
Voigt effect	V	$n_\parallel - n_\perp$	Non res.	not reported
Resonant Voigt effect	V	$n_\parallel - n_\perp$	Res.	NISHINA, KOLODZIEJCZAK and LAX, 1962[4]
Zeeman-effect for impurities	F or V	$\kappa_\pm$, $\kappa_\parallel, \kappa_\perp$	Resonance shift	FAN and FISHER, 1958[8] BOYLE, 1958[9]
Magnetoabsorption of the impurity photoionisation spectrum	F or V	κ_+, κ_- $\kappa_\parallel, \kappa_\perp$	Resonance shift	BOYLE, 1958[9]
Zeeman effect of excitons	F or V	κ_+, κ_- $\kappa_\parallel, \kappa_\perp$	Resonance shift	GROSS and ZAHARCENJA, 1957[10]
Cross-field magneto absorption			Res.	VREHEN and LAX, 1964[11]
b) In reflection.				
Interband magneto-reflection	F or V	$n_\pm, \kappa_\pm$ $n_\parallel n_\perp, \kappa_\perp \kappa_\parallel$	Res.	WRIGHT and LAX, 1961[12]
Interband Kerr rotation	F	$\kappa_\pm$		LAX and NISHINA, 1961[13]
Interband Kerr ellipticity	F	$n_\pm$		

[1] See ref. 3, p. 236.
[2] See ref. 4, p. 236.
[3] See ref. 2, p. 302.
[4] See ref. 6, p. 298.
[5] See ref. 7, p. 298.
[6] See ref. 8, p. 298.
[7] See ref. 2, p. 280.
[8] See ref. 4, p. 314.
[9] See ref. 5, p. 314.
[10] See ref. 3, p. 297.
[11] See ref. 4, p. 312.
[12] See ref. 2, p. 311.
[13] See ref. 1, p. 312.

transitions and the effects associated with them as *interband* magneto-optic (IMO) effects (Fig. 1).

We have therefore two classes of magneto-optic effect arising from either free carrier or interband transitions. Further distinctions can be made depending upon whether the effect is dispersive or absorptive, resonant or non-resonant, and upon the relative orientation of the magnetic field to the direction of the radiation beam and its polarised components. There is therefore a variety of effects many of which differ in the type of information which they give concerning the crystal band structure. Most of the possible effects have now been observed and a list of these and some others which may prove to be observable is given below. Classification of the effects by the information which they give concerning band structure is necessarily postponed until the conclusion of this article after examing the detailed explanations.

All the free carrier effects can be expressed by the classical Drude-Zener theory in terms of an effective mass and a single relaxation time yielding a variety of formulae in terms of these parameters and the frequency. If the mass is anisotropic, different combinations of the mass tensor components are found; these can be related by discussing the effects in terms of the conductivity or dielectric constant tensors in the presence of a magnetic field. Thus the anisotropy of various magneto-optical effects can be discussed using macroscopic theory and these results apply whether the origin of the conductivity or permittivity is from free carrier or interband effects. We therefore commence with a review of the macroscopic formulae (see II).

For the low field cases usually appropriate for the free carrier effects, the band parameters may be introduced by applying the Boltzmann transport equation to calculation of $\underline{\sigma}$ or $\underline{\varepsilon}$. In certain cases the simple Boltzmann theory is strikingly successful. For interband effects and high field cases it is necessary to use quantum mechanical methods and since this is fundamental to the understanding of all the effects it is discussed in Sect. 3. The low field Boltzmann theory is deferred to IV, free carrier effects.

II. Macroscopic theory.

2. Introduction. Classically the fundamental motion of the charge carrier is that of cyclotron resonance, but this must be considered as analogous to a linear oscillator in quantum mechanics, with levels $(n+\frac{1}{2})\hbar\omega_c$, to give correct results for magneto-optical effects. Consequently the fundamental mode of the applied electromagnetic radiation is the circularly polarised mode — which drives the cyclotron resonance. This can be represented by

$$\boldsymbol{E}(\omega)=\boldsymbol{E}_0(\omega)\,e^{i(\omega t-\boldsymbol{k}\cdot\boldsymbol{r})} \tag{2.1}$$

where $\boldsymbol{E}_0(\omega)=(E_{0x}\pm i\,E_{0y})$ for a wave propagating in the z-direction, and the $\pm$ signs give the right- and left-circularly polarised modes when viewed *along* the direction of propagation[1].

The electromagnetic wave equation can be written

$$\nabla(\nabla\cdot\boldsymbol{E})-\nabla^2\boldsymbol{E}=-\mu_0\frac{\partial\boldsymbol{j}}{\partial t}-\mu_0\frac{\partial^2\boldsymbol{D}}{\partial t^2} \tag{2.2}$$

in a non-magnetic medium using m.k.s. units and we have in addition the constitutional equations relating $\boldsymbol{j}$ to $\boldsymbol{E}$ and $\boldsymbol{D}$ to $\boldsymbol{E}$ which introduce the conductivity $\underline{\sigma}$ and permittivity $\underline{\varepsilon}$ of the medium from the relations

$$\boldsymbol{j}=\underline{\sigma}\,\boldsymbol{E} \quad\text{and}\quad \boldsymbol{D}=\underline{\varepsilon}\,\boldsymbol{E}$$

where $\underline{\sigma}$ and $\underline{\varepsilon}$ are, in general, tensor quantities. Since both $\boldsymbol{j}$ and $\partial\boldsymbol{D}/\partial t$ are currents proportional to $\boldsymbol{E}$ but 90° out of phase, they may be combined, using the harmonic true dependence of Eq. (2.1) by defining either a *complex* permittivity tensor $\underline{\varepsilon}$

[1] Right-circular-polarised (r.c.p.) and left-circular-polarised (l.c.p.) modes are also defined looking *against* the direction of propagation, e.g. F. A. JENKINS, and H. E. WHITE, Fundamentals of Optics. New York: McGraw Hill Book Co. (1960) then having the opposite sense.

or a *complex* conductivity tensor $\underline{\sigma}$. These alternative descriptions are related by

$$i\omega\underline{\varepsilon}=\underline{\sigma}. \tag{2.3}$$

For a system having at least threefold symmetry about the z axis and subject to a magnetic field along the z axis, the permittivity and conductivity tensors may be shown from symmetry alone to have the forms[1]

$$\underline{\varepsilon}=\begin{pmatrix}\varepsilon_{xx} & \varepsilon_{xy} & 0\\ -\varepsilon_{xy} & \varepsilon_{xx} & 0\\ 0 & 0 & \varepsilon_{zz}\end{pmatrix},\quad \underline{\sigma}=\begin{pmatrix}\sigma_{xx} & \sigma_{xy} & 0\\ -\sigma_{xy} & \sigma_{xx} & 0\\ 0 & 0 & \sigma_{zz}\end{pmatrix}. \tag{2.4}$$

Either of these tensors can represent the response of the system to harmonic disturbance, the real part of the permittivity representing the displacement and the imaginary part the current and vice versa for the conductivity. The complex displacement

$$D(\omega)=\underline{\varepsilon}(\omega)\boldsymbol{E}(\omega) \tag{2.5}$$

and also Eq. (2.2) are to be understood in the sense that if

$$\boldsymbol{E}=\int_{-\infty}^{\infty}\boldsymbol{E}(\omega)\,d\omega$$

then

$$\boldsymbol{D}=\int_{-\infty}^{+\infty}\underline{\varepsilon}(\omega)\boldsymbol{E}(\omega)\,d\omega$$

and similarly for $\boldsymbol{j}$ and $\underline{\sigma}$

Reality of field and displacement then requires that

$$\varepsilon_{ij}(-\omega)=\varepsilon_{ij}^{*}(\omega) \tag{2.6}$$

for all components of the tensor $\underline{\varepsilon}$.

3. The propagation constant. Substituting from Eqs. (2.1) and (2.5) in the wave Eq. (2.2) we obtain the equation defining k, the propagation constant in terms of $\underline{\varepsilon}$ and the frequency ω

$$-\boldsymbol{k}(\boldsymbol{E}_0\cdot\boldsymbol{k})+k^2\boldsymbol{E}_0=\omega^2\mu_0\underline{\varepsilon}E_0$$

and a similar equation in terms of $\underline{\sigma}$ using Eq. (2.3).

Longitudinal propagation — Faraday configuration. For propagation along the Z axis, with $\boldsymbol{B}$ along the same axes we have

$$k^2\boldsymbol{E}_0=\omega^2\mu_0\underline{\varepsilon}\boldsymbol{E}_0$$

from which, using (2.4), we can write the component equations

$$k^2E_{0x}=\omega^2\mu_0(\varepsilon_{xx}E_{0x}+\varepsilon_{xy}E_{0y}),$$
$$k^2E_{0y}=\omega^2\mu_0(-\varepsilon_{xy}E_{0x}+\varepsilon_{xx}E_{0y}).$$

Multiplying the latter equation by $\pm i$ and adding we obtain

$$k^2E_{0\pm}=\omega^2\mu_0\varepsilon_{\pm}E_{0\pm} \tag{3.1}$$

where

$$E_{0\pm}=E_{0x}\pm i\,E_{0y}$$

and

$$\varepsilon_{\pm}=\varepsilon_{xx}\mp i\,\varepsilon_{xy} \tag{3.2}$$

[1] I. M. BOSWARVA, R. E. HOWARD, and A. B. LIDIARD: Proc. Soc. A **269**, 125 (1962).

so that we have formed circularly polarised modes, the upper sign corresponding to the propagation along $+z$ of waves which appear right circularly polarised to an observer looking in the same direction, and vice versa for the lower signs. Thus we are led to define the permittivity $\varepsilon_\pm$ governing the propagation of right- and left-circularly polarised modes in terms of the components of the tensor [Eq. (3.2)].

Similarly we may define

$$\sigma_\pm = \sigma_{xx} \mp i\,\sigma_{xy}. \tag{3.3}$$

Transverse propagation — Voigt configuration. The magnetic field is again chosen along the z-axis but we now consider propagation in the $x-y$ plane. Since $\varepsilon_{xx} = \varepsilon_{yy}$ there is no loss in generality in choosing k to be in the y direction. Once again writing out components we obtain

$$k^2 E_{0x} = \omega^2 \mu_0 (\varepsilon_{xx} E_{0x} + \varepsilon_{xy} E_{0y}) \tag{3.4}$$

as before, but for the y direction

$$-k^2 E_{0y} + k^2 E_{0y} = 0 = \omega^2 \mu_0 (-\varepsilon_{xy} E_{0x} + \varepsilon_{xx} E_{0y}) \tag{3.5}$$

and

$$k^2 E_{0z} = \omega^2 \mu_0 \varepsilon_{zz} E_{0z}. \tag{3.6}$$

We note that by virtue of the polarisation of the medium there is a component of $\boldsymbol{E}$ along the direction of propagation given from (3.5) as

$$E_{0y} = \frac{\varepsilon_{xy} E_{0x}}{\varepsilon_{xx}}. \tag{3.7}$$

There are two cases for transverse propagation — the incident beam may be polarised so that E is either *perpendicular* or *parallel* to B (or a polarised beam in some other direction can be resolved into parallel or perpendicular components). For the *perpendicular* case, $E_{0z} = 0$ and, substituting for E_{0y} from (3.7) in (3.4) we find

$$k^2 E_{0x} = \omega^2 \mu_0 \left(\varepsilon_{xx} + \frac{\varepsilon_{xy}^2}{\varepsilon_{xx}}\right) E_{0x}$$

whence we define

$$\varepsilon_\perp = \varepsilon_{xx} + \frac{\varepsilon_{xy}^2}{\varepsilon_{xx}} \tag{3.8}$$

as the permittivity for "perpendicular" radiation. For the *parallel* case, $E_{0x} = 0$ and the propagation constant is given from Eq. (3.6) so that we may define

$$\varepsilon_\parallel = \varepsilon_{zz} \tag{3.9}$$

and ε_{zz} is just the value of the permittivity in the absence of the magnetic field. This case therefore corresponds to motion of the electron parallel to the magnetic field so that by the Lorentz force expression $\boldsymbol{F} = e\boldsymbol{v} \times \boldsymbol{B}$, no interaction takes place. Conversely the perpendicular component would cause a free particle to undergo cyclotron resonance if driven at the correct frequency, ω_c.

4. The complex refractive index. From the result

$$k^2 \boldsymbol{E} = \omega^2 \mu_0 \underline{\varepsilon} \boldsymbol{E} \tag{4.1}$$

we note that when $\underline{\varepsilon}$ is complex $\boldsymbol{k}$ is complex and the wave is attenuated. The equivalent symmetry of the real and imaginary parts of $\underline{\varepsilon}$ ensures that the directions defined by the real and imaginary parts of $\boldsymbol{k}$ coincide and we may therefore introduce the optical constants by writing

$$\boldsymbol{k}=\frac{\omega}{c}(n-i\kappa)\hat{\boldsymbol{k}} \tag{4.2}$$

where $\hat{\boldsymbol{k}}$ is real unit vector in the direction of $\boldsymbol{k}$. The quantity $(n-i\kappa)$ is the *complex* refractive index. The real part, n, is the refractive index and controls the phase velocity in the medium from the relation

$$v_p=c/n.$$

The imaginary part, κ, we shall call the *absorption index* and this gives the attenuation of the intensity (i.e., time average of the Poynting vector) from the absorption coefficient α, where

$$\alpha=(2\omega/c)\,\kappa \tag{4.3}$$

and has dimensions of metres^{-1} in MKS units. Magneto-optical constants appropriate to longitudinal and transverse propagation may now be defined from the appropriate tensor components.

Longitudinal propagation. From (4.2), (4.1) and (3.2) we have

$$(n_\pm-i\kappa_\pm)^2=\varepsilon_\pm=\varepsilon_{xx}\mp i\varepsilon_{xy}. \tag{4.4}$$

Transverse propagation. Similarly from (4.2), (4.1), (3.8) and (3.9)

$$(n_\perp-i\kappa_\perp)^2=\varepsilon_\perp=\varepsilon_{xx}+\varepsilon_{xy}^2/\varepsilon_{xx} \tag{4.5}$$

and

$$(n_\parallel-i\kappa_\parallel)^2=\varepsilon_\parallel=\varepsilon_{zz}.$$

In terms of conductivity we have the equivalent expressions

$$(n_\pm-i\kappa_\pm)^2=\sigma_\pm/i\omega=-(i\sigma_{xx}\pm\sigma_{xy})/\omega, \tag{4.6}$$

$$(n_\perp-i\kappa_\perp)^2=\sigma_\perp/i\omega=(\sigma_{xx}+\varepsilon\sigma_{xy}^2/\sigma_{xx})/i\omega. \tag{4.7}$$

5. The optical mechanism of the effects and expressions in terms of magneto-optical constants and the tensor components.

(1) *Magneto-absorption.*

(a) *Cyclotron resonance absorption.* For free carriers the resonance is givenby either κ_+ or κ_- depending upon the sign of the charge, which selects which of the circularly polarised modes resonates.

α) *Faraday configuration.* For *electrons* only right-circularly polarised radiation is absorbed so cyclotron resonance absorption is given by

$$\alpha_+=\frac{2\omega}{c}\kappa_+$$

which from (4.4) can be written

$$\alpha_+=-\frac{\omega}{c n_+}\varepsilon_+^I \tag{5.1}$$

where ε_+^I is the imaginary part of ε_+. If the free carrier concentration and magnetic fields are small the magnetic contribution to n will be small, even near resonance, and we may write approximately $n_\pm=n$, the zero field value. We then have

$$\alpha_+\simeq\frac{\omega}{cn}(\varepsilon_{xx}^I+\varepsilon_{xy}^R)$$

or more appropriately for an absorption effect

$$\alpha_+ = (\sigma^R_{xx} + \sigma^I_{xy})/cn \tag{5.2}$$

in terms of conductivity.

β) *Voigt configuration and plane polarised radiation.* The above result is for circularly polarised radiation — in the Voigt configuration we consider plane polarised radiation and obtain

$$\alpha = \frac{\omega}{c}\,\kappa_+,$$

i.e., half the value for circularly polarised radiation, since the plane polarised radiation must be resolved into two circular components, only one of which is absorbed. This result also applies to plane polarised radiation in the Faraday configuration.

(*b*) *I.M.O. absorption.* The above results also apply to interband effects except that absorption takes place for both circular components.

α) *The Faraday effect.* The rotation of the plane of polarisation arises from the different phase velocities c/n_+ and c/n_- of the two circular components making up the plane polarised beam incident on the crystal. At the end of a crystal of length l the two components recombine to give a rotation Θ given by[1]

$$\Theta = \frac{\omega l}{2c}\,(n_- - n_+)\,. \tag{5.3}$$

The sense of rotation is that positive Θ is a clockwise rotation for a wave propagating along the magnetic field for an observer also looking in that direction. This is the situation as it is observed for free electrons.

From Eq. (4.4)

$$\varepsilon^R_\pm = n^2_\pm - \kappa^2_\pm = \varepsilon^R_{xx} \pm \varepsilon^I_{xy} \tag{5.4}$$

and we may re-write (5.3) as

$$\Theta = \frac{-\omega l}{4nc}\,(n^2_+ - n^2_-)\,,$$

where $n = \frac{n_+ + n_-}{2}$ and may be approximated to the refractive index at zero field. Thus we obtain from (5.4), if $\kappa^2 \ll n^2$,

$$\Theta = \frac{-\omega l}{4nc}\,(\varepsilon^R_+ - \varepsilon^R_-)$$

$$= \frac{\omega l}{2nc}\,\varepsilon^I_{xy}\,, \tag{5.5}$$

$$= \frac{\sigma^R_{xy}\, l}{2nc}\,. \tag{5.6}$$

The non-resonant Faraday effect described by (5.5) or (5.6) is an isotropic effect in a cubic crystal.

β) *Faraday ellipticity.* In addition to the differential magneto-dispersion responsible for rotation there is the corresponding absorption effect depending upon $\kappa_\pm$. This attenuates one of the circular modes more than the other and makes the emergent beam elliptically polarised. The attenuation of amplitude is given by $E = E_0\, e^{-\frac{\omega \kappa_\pm z}{c}}$ and for small absorption this can be expanded to

[1] The formulae for both Θ and Δ given here assume no effects from multiple reflections. These have been included in a discussion by B. Donovan and T. Medcalf, Brit. J. Appl. Phys. **15**, 1139 (1964).

$(1-\omega\kappa z/c)E_0$. In this approximation the ellipticity Δ, the amplitude ratio of the axes of the ellipse is given by[1]

$$\Delta = \tfrac{1}{2}(\alpha_+ - \alpha_-)l = \frac{\omega l}{2c}(\kappa_- - \kappa_+), \tag{5.7}$$

$$\simeq \frac{\omega l}{2cn}(n_-\kappa_- - n_+\kappa_+)$$

$$= \frac{\omega l}{cn}\varepsilon^R_{xy}, \tag{5.8}$$

$$= \frac{l}{cn}\sigma^I_{xy}. \tag{5.9}$$

γ) *Voigt effect*. The Voigt[2] effect, which could also have been called the Cotton-Mouton effect, is magnetic birefringence and arises from the difference between $n_\parallel$ and $n_\perp$ in the transverse orientation. It is usually observed by inclining incident plane polarised radiation with the electric vector at 45° to the direction of B. The components resolved parallel and perpendicular to B then have different phase velocities, $c/n_\parallel$ and $c/n_\perp$, and recombine at the end of the crystal to give emergent radiation which is elliptically polarised. Measurement of this ellipticity then determines the Voigt effect.

The Voigt phase shift δ, is given by.

$$\delta = \frac{\omega l}{c}(n_\parallel - n_\perp) \tag{5.10}$$

In the absence of any absorption, this phase shift is related to the resultant ellipticity Δ by

$$\Delta^2 = \frac{1-\cos\delta}{1+\cos\delta} = \tan^2\frac{\delta}{2} \tag{5.11}$$

and the ellipse is oriented at 45° to B.

In the presence of absorption the ellipse is *rotated* by differential attenuation caused by $\kappa_\parallel$ and $\kappa_\perp$ and the angle of orientation is given from

$$\tan 2\varphi = \frac{2f\cos\delta}{f^2-1}$$

where $f = e^{(\kappa_\parallel - \kappa_\perp)l}$. φ becomes 45° when $f=0$. Re-writing (5.10) as

$$\delta = \frac{\omega l}{2cn}(n^2_\parallel - n^2_\perp)$$

we can find δ in terms of the tensor components from (4.5) when $\kappa \ll n$, as

$$\delta = \frac{\omega l}{2cn}(\varepsilon_{zz} - \varepsilon_{xy} - \varepsilon^2_{xy}/\varepsilon_{xx}). \tag{5.12}$$

From symmetry relations of the tensor components, it can be shown from (5.12) that Voigt effect is second order in the magnetic field and, unlike the non-resonant Faraday effect described by (5.5) or (5.6) it can be *anisotropic* in a cubic crystal (see. Sect. 17).

δ) *Magneto-reflection*. The direct effects in reflectivity can be derived by simply substituting the magneto-optical constants $n_\pm\,\kappa_\pm$, $n_\parallel\,n_\perp$, $\kappa_\parallel\,\kappa_\perp$ in the

---

[1] See footnote 1, p. 243.

[2] Originally observed in gases.

Fresnel formulae for the reflection coefficient, i.e.,

$$R_{\pm}=\frac{(n_{\pm}-1)^2+\kappa_{\pm}^2}{(n_{\pm}+1)^2+\kappa_{\pm}^2} \tag{5.13}$$

in the Faraday configuration and

$$R_{\parallel,\perp}=\frac{(n_{\parallel,\perp}-1)^2-\kappa_{\parallel,\perp}^2}{(n_{\parallel,\perp}+1)^2-\kappa_{\parallel,\perp}^2} \tag{5.14}$$

for the Voigt configuration. Usually $\kappa \ll n$ so that the absorption terms can be neglected.

The rotation effect in reflection arises from the differential phase shift, δ, given by

$$\tan\delta_{\pm}=2\kappa_{\pm}/(1-n_{\pm}^2-\kappa_{\pm})^2 \tag{5.15}$$

for the Faraday configuration, which leads to a rotation

$$\Theta=\tfrac{1}{2}(\delta_{-}-\delta_{+}). \tag{5.16}$$

The differential reflectivity, on the other hand, gives ellipticity:

$$\Delta=\frac{[|r_+|-|r_-|]}{[|r_+|+|r_-|]} \tag{5.17}$$

where $r_{\pm}$ are Fresnel amplitude coefficients.

$$r_{\pm}=\frac{(1-n_{\pm}-i\kappa_{\pm})}{(1+n_{\pm}-i\kappa_{\pm})}. \tag{5.18}$$

The role of absorption and dispersion is thus reversed compared with the Faraday rotation and ellipticity in transmission.

Corresponding effects also exist for the Voigt configuration, but the separation of causes is not so complete.

Using the methods of the previous section and Eqs. (2.5) and (2.6) the magneto-reflection effects can be expressed in terms of the tensor components.

6. Anisotropic crystals. The formulae derived in Sect. 5 are based on the form of the permittivity tensor in a cubic crystal, Eq. (2.4), and in the weak field or non-resonant case this leads to some effects being isotropic. Theoretical treatments for anisotropic crystals have been given by AUSTIN (1959)[1] for Faraday effect in a uniaxial crystal, by GUREVICH and IPATOVA (1959)[2] who derive refractive indices for specific directions in hexagonal and cubic crystals and by DONOVAN and WEBSTER (1962/63)[3] in a series of papers on Faraday and Voigt effect. In uniaxial crystals Faraday effect is best measured with propagation along the optic axes when the effect is a maximum. In other directions the crystal birefringence tends to quench the effect.

7. Dispersion relations. The foregoing macroscopic discussion allows us to use any available information about the permittivity or conductivity tensors to calculate the various magneto-optical effects. This information may be simple classical theory, as in Sect. 1. Boltzmann transport theory results (Sect. 13) or full quantum mechanical calculation (Sects. 8, 9, 10) according to the circumstances in which these theories are useful. In the latter case in particular absorption

[1] I. G. AUSTIN: J. Electronics and Control **6**, 271 (1959).

[2] L. E. GUREVICH, and I. P. IPATOVA: Zhur. Eksptl. i Teoret. Fiz. **37**, 1324 (1959); Translation into English 1960, Soviet Phys. JETP **10**, 943.

[3] B. DONOVAN, and J. WEBSTER: Proc. Phys. Soc. (London) **79**, 107, 1081 (1962); **81**, 90 (1963).

effects are understood in terms of transitions between allowed energy states of the system concerned. No such simple picture exists for dispersion and it is necessary to use the relation between absorption and dispersion to treat the dispersive effects. The appropriate formulae are variously known as the Kramers-Kronig or Kramers-Heisenberg dispersion relations. The real and imaginary parts of the complex permittivity (i.e., dielectric constant) or conductivity tensors are related merely as a consequence of causality and the properties of a complex variable. The causality condition is that if $\boldsymbol{E}(t)=0$ for all $t<0$ then $\boldsymbol{D}(t)\,(=\underline{\varepsilon}\,\boldsymbol{E})$ is also zero for $t<0$. The resultant dispersion relations are quite general and may be applied to any system, classical or quantum mechanical. The dispersion relations may be written, after BOSWARVA, HOWARD and LIDIARD (1962)[1] and BENNETT and STERN[2]

$$\varepsilon_{ij}^{R}(\omega)-\delta_{ij}=-\frac{1}{\pi}P\int_{-\infty}^{+\infty}\frac{\varepsilon_{ij}^{I}(\omega')\,d\omega'}{\omega'-\omega} \tag{7.1}$$

and

$$\varepsilon_{ij}^{I}(\omega)=\frac{1}{\pi}P\int_{-\infty}^{\infty}\frac{(\varepsilon_{ij}^{R}(\omega')-\delta_{ij})\,d\omega'}{\omega'-\omega} \tag{7.2}$$

where P denotes the principal value of the integral. These forms do not hold between the real and imaginary parts of ε_+ and ε_- [defined in Eq. (3.2)] and this has led to error in certain theoretical treatments (see Sect. 23).

In terms of conductivity we have

$$\sigma_{ij}^{R}\omega=-\frac{2}{\pi}P\int_{0}^{\infty}\frac{\omega'\sigma_{ij}^{I}(\omega')\,d\omega'}{\omega'^{2}-\omega^{2}}, \tag{7.3}$$

$$\sigma_{ij}^{I}(\omega)=\frac{2}{\pi}P\int_{0}^{\infty}\frac{\omega\,\sigma_{ij}^{R}(\omega')\,d\omega'}{\omega'^{2}-\omega^{2}}. \tag{7.4}$$

The dispersion relations can be transformed into other forms, e.g., the relation between Faraday rotation and magnetoabsorption or Faraday rotation and ellipticity. For the former we have

$$\Theta=-\frac{\omega^{2}l}{2\pi n}\int_{0}^{\infty}\frac{(n_-\alpha_--n_+\alpha_+)}{\omega'(\omega'^{2}-\omega^{2})}\,d\omega'. \tag{7.5}$$

III. Quantum mechanical theory.

8. Introduction. From a theoretical point of view, the problem of the motion of electrons in crystals in the presence of a magnetic field has not been solved in a complete and closed form, although a form analogous to the Bloch function has been given by HARPER[3]. One or two problems have been solved exactly and others by semi-classical approximations. For most purposes, including magneto-optical effects, such theoretical calculations are however sufficiently accurate.

The necessity to treat the problem quantum mechanically can be appreciated by noting that the same effects responsible for the free electron diamagnetic

[1] I. M. BOSWARVA, R. E. HOWARD, and A. B. LIDIARD: Proc. Roy. Soc. (London) A **269**, 125 (1962).
[2] H. S. BENNETT, and E. A. STERN: Phys. Rev. **137**, A 448 (1965).
[3] P. G. HARPER: Proc. Phys. Soc. (London) A **68**, 879 (1955).

susceptibility are those causing the magneto-optical effects which are the subject of this review and that classical theory predicts a zero magnetic susceptibility[1] in this case. Some of the theory originally developed to explain magnetic susceptibility of free electrons can be used only slightly modified to treat magneto-optical effects in semiconductors.

The success of the one-electron theories for the motion of electrons in metals and semiconductors is well known and we commence by considering the Hamiltonian ($\mathscr{H}$) for a single electron in a magnetic field. This is

$$\mathscr{H} = \frac{1}{2m}(\boldsymbol{p} + e\boldsymbol{A})^2 + V(r) + \frac{e\hbar}{2m}\boldsymbol{\sigma}\cdot\boldsymbol{B} \tag{8.1}$$

where $\boldsymbol{p}$ is the momentum operator, $\boldsymbol{A}$ the magnetic vector potential related to $\boldsymbol{B}$ by $\boldsymbol{B} = \nabla_x \boldsymbol{A}$. $V(r)$ is the crystal potential function and the last term represents the interaction between the electron spin of the field in which the spin moment $\boldsymbol{S} = \frac{1}{2}\hbar\,\boldsymbol{\sigma}$ and the components of the Pauli spin matrix $\boldsymbol{\sigma}$ are

$$\sigma_x = \begin{pmatrix} 0 & 1 \\ 1 & 0 \end{pmatrix}, \quad \sigma_y = \begin{pmatrix} 0 & -i \\ i & 0 \end{pmatrix}, \quad \sigma_z = \begin{pmatrix} 1 & 0 \\ 0 & -1 \end{pmatrix}.$$

Electrons in solids may be separated into two groups: those that remain in atomic or molecular orbitals — as in ionic or molecular crystals and those whose orbitals extend throughout the crystal and can be treated by band theory — as in metals and semiconductors. Magnetic fields affect these states in different ways[2]. If $\boldsymbol{B}$ does not vary with position a possible choice for $\boldsymbol{A}$ is $\boldsymbol{A} = \frac{1}{2}\boldsymbol{B}\times\boldsymbol{r}$.

We can then write the magnetic terms in the Hamiltonian as

$$-\frac{l\hbar}{2m}\boldsymbol{B}(\boldsymbol{l} + 2\boldsymbol{s}) + \frac{l^2}{8m}\{\boldsymbol{B}^2\boldsymbol{r}^2 - \boldsymbol{B}r^2\}$$

where $\boldsymbol{l}$ and $\boldsymbol{s}$ are orbital and spin angular momenta. The first term, linear in $\boldsymbol{B}$ is the paramagnetic term, the second the diamagnetic term. For atomic or molecular orbitals $\boldsymbol{r}$ will be $\sim 10^{-8}$ cm and the second term will be small. It is of the order $10^{-8}\,H^2$ (eV) if H is in oersted while the linear term is $\sim 10^{-8}\,H$ (eV) and so the paramagnetic term dominates for $H < 10^6$ oe.

When $\boldsymbol{r}$ is ~ 1 mm and extends throughout the crystal, the second term dominates and it is with this type of effect that we are concerned. The total magnetic susceptibility can be expressed

$$\chi_{\text{total}} = \chi_{\text{spin}}^{\text{band}} + \chi_{\text{diamagnetic}}^{\text{band}} + \chi_{\text{core}}.$$

We should not discuss χ_{core} as giving negligible effect in this context. $\chi_{\text{diamagnetic}}^{\text{band}}$ is the "orbital" effect and $\chi_{\text{spin}}^{\text{band}}$ is the paramagnetic effect arising from electron spin. The wave-function may be expressed as the product of a space function and a spin function and the energy as a sum of orbital and spin terms. The latter are just $\pm e\hbar B/2m$ and we omit them in the first instance, and consider the orbital terms alone. The spin terms may be readily reintroduced at a later stage. The starting point is then LANDAU's solution of the Schrödinger equation for a free electron in a magnetic field.

9. The Landau levels of free electrons. Taking $V(r) = 0$ and neglecting the spin term we have

$$\mathscr{H} = \frac{1}{2m}(\boldsymbol{p} + e\boldsymbol{A})^2 \tag{9.1}$$

---

[1] J. H. VAN VLECK: Electric and Magnetic Susceptibilities, p. 102. London: Oxford University Press 1959.

[2] R. J. ELLIOTT: Brit. J. Appl. Phys. **14**, 770 (1963).

and we may take, for a magnetic field along the z-direction

$$A_x = -B y, \quad A_y = A_z = 0. \tag{9.2}$$

Substituting for $\boldsymbol{A}$, and writing $p_x = \frac{\hbar}{i}\frac{\partial}{\partial x}$, etc, the Schrödinger equation becomes

$$-\frac{\hbar^2}{2m}\left[\left(\frac{\partial}{\partial x} - \frac{i e B y}{\hbar}\right)^2 + \frac{\partial^2}{\partial y^2} + \frac{\partial^2}{\partial z^2}\right]\psi(r) = E\psi(r) \tag{9.3}$$

in Cartesian coordinates. The equation is separable if we put

$$\psi(r) = e^{i(k_x x + k_z z)}\, g(y) \tag{9.4}$$

which on substitution into (9.3) yields

$$\frac{d^2 g}{d y^2} + \frac{2m}{\hbar^2}\left[E - \frac{\hbar^2 k_z^2}{2m} - \frac{1}{2m}(\hbar k_x - e B y)^2\right] g = 0 \tag{9.5}$$

If we now introduce the classical cyclotron resonance frequency $\omega_c = \frac{eB}{m}$ and define a quantity $y_0 = \frac{\hbar k_x}{eB}$ we may re-write (9.5) as:

$$\frac{d^2 g}{d y^2} + \frac{2m}{\hbar^2}\left[\left(E - \frac{\hbar^2 k_z^2}{2m}\right) - \frac{1}{2} m\omega_c^2 (y - y_0)^2\right] g(y) = 0.$$

This is the equation of a simple harmonic oscillator of frequency ω_c, centred at y_0 which in quantum mechanics has eigen values of $\left(E - \frac{\hbar^2 k_z^2}{2m}\right)$ given by

$$\left(E - \frac{\hbar^2 k_z^2}{2m}\right) = \left(n + \frac{1}{2}\right)\hbar\omega_c$$

where n is any positive integer. Thus the energy levels in the field B are

$$E_{n,k_z} = \frac{\hbar^2 k_z^2}{2m} + \left(n + \frac{1}{2}\right)\hbar\omega_c \tag{9.6}$$

and are known as LANDAU[1] levels (Fig. 1b)[2].

For the motion of electrons and holes near band extrema in semiconductors it is, of course, well known that the free electron result is valid with the substitution of an appropriate effective mass. Eq. (9.6) therefore predicts that the three dimensional parabolic band structure is split up into a series of lines [the oscillator levels $(n + \frac{1}{2})\,h\omega_c$] which may be associated with classical cyclotron resonance motion of the carriers, i.e., circles in a plane *perpendicular* to the field together with a one-dimensional motion in a direction *parallel* to the field unchanged from the zero field case. The energy of the lowest state is however raised to $\frac{1}{2}\hbar\omega_c$.

The Landau levels are highly degenerate with respect to the choice of "orbit centre", y_0 and the number of states with a particular quantum number, n, increases with B. We can estimate this degeneracy by considering the system to be contained in a box of sides L_x, L_y, L_z. The number of possible values of k_x in a small interval Δk_x is

$$L_x \frac{\Delta k_x}{2\pi} \tag{9.7}$$

and similarly for k_y and k_z. The maximum value of k_x is determined from $y_0 = \frac{\hbar k_x}{e B}$ by the requirement that y_0, the orbit centre lies inside the box, i.e.,

$$-L_y/2 \leqq y_0 \leqq L_y/2.$$

[1] L. LANDAU: Z. Physik **64**, 629 (1930).

[2] See, for example, J. CALLAWAY: Energy Band Theory, p. 252. New York: Acad. Press 1964.

So, substituting for y_0

$$\frac{-e\,B\,L_y}{2\hbar} \lesssim k_x \lesssim \frac{e\,B\,L_y}{2\hbar}.$$

The number of states in a single Landau level is then, using (9.7)

$$L_x \frac{\Delta k_x}{2\pi} = \frac{e\,B\,L_x\,L_y}{2\,\pi\,\hbar}. \tag{9.8}$$

With $L_z \frac{\Delta k_z}{2\pi}$ states in an interval Δk_z, the number of Landau levels associated with n and Δk_z is then

$$\frac{e\,B\,V}{4\,\pi^2\,\hbar} \Delta k_z \tag{9.9}$$

where $V = L_x\,L_y\,L_z$. The degeneracy is directly proportional to the applied field B.

These results provide all that is required to understand the magneto-optic experiments in a *qualitative* manner since it emerges that we may apply the results to semiconductor energy bands with relatively small modifications. Suppose we assume that the above results hold if we introduce the effective mass m^* in place of the free electron mass m. The average energy of two neighbouring Landau states, n and $n+1$ is then, if $k_z = 0$,

$$\overline{E} = \frac{E_n + E_{n+1}}{2} = (n+1)\,\hbar\,\omega_c$$

but we can write this kinetic energy, semi-classically[1] as

$$\overline{E} = \tfrac{1}{2}\,m^*\,v^2$$

whence

$$v = \sqrt{\frac{2(n+1)\hbar\omega_c}{m^*}}.$$

The radius of the orbit is then

$$R = \frac{v}{\omega_c} = \sqrt{\frac{2(n+1)\hbar}{m^*\,\omega_c}} \tag{9.10}$$

with $\omega_c = \frac{e\,B}{m^*}$. For B equal to 1 Weber/sq. metre (10,000 gauss) r is ~ 200 Å, when m^*/m is ~ 1, i.e., many times greater than the lattice constants. These are therefore circumstances in which the use of an effective mass is valid. A further useful formula is that

$$\hbar\,\omega_c = \hbar\left(\frac{e\,B}{m^*}\right) = 1.1577 \times 10^{-4}\,\frac{B}{(m^*/m)}\ \text{eV} \tag{9.11}$$

where B is again in Webers/sq. metres, m is the mass of the free electron and one works with the effective mass ratio.

It is therefore permissible to use effective masses for both electrons m_c^* and holes m_v^* and so obtain the relations (1.6) and (1.7) to describe orbital effects in conduction and valence bands in a simple model. Electron spin may now be readily re-introduced by simply including an extra term

$$g\left(\frac{e\hbar}{2m}\right) B\,M_J$$

where g is an effective g-factor, which may vary considerably from 2.00 the free electron value, $\frac{\hbar e}{2m}$ is β, the gyromagnetic ratio or Bohr magneton, and M_J is

---

[1] B. Lax: Fermi School XXII.

the spin quantum number which may take values $\pm\frac{1}{2}$. The energy bands corresponding to 1.6 and 1.7 then become (Fig. 2 and Fig. 6.)

$$E_c = E_{c0} + \frac{\hbar^2 k_z^2}{2m_c^*} + \left(n + \frac{1}{2}\right)\hbar\,\omega_{cc} + g_c \beta B M_J, \tag{9.12}$$

$$E_v = E_{v0} - \frac{\hbar^2 k_z^2}{2m_v^*} - \left(n + \frac{1}{2}\right)\hbar\,\omega_{cv} + g_v \beta B M_J. \tag{9.13}$$

These relations describe satisfactorily the behaviour in a magnetic field of simple, non-degenerate bands with extrema at $\boldsymbol{k}=0$ in the region in which they are

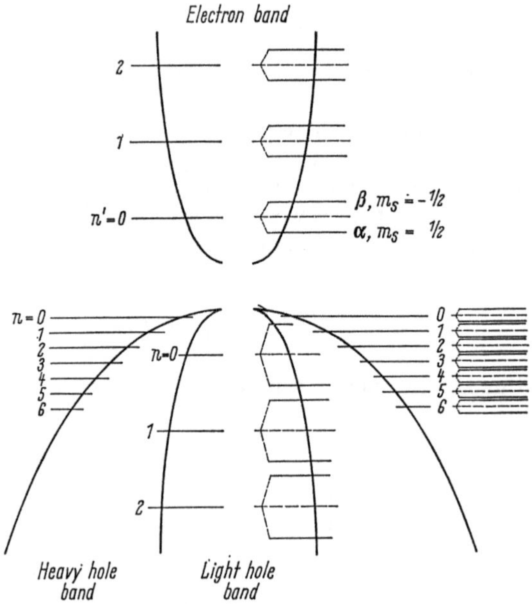

Fig. 2. Schematic diagram of light and heavy hole "ladders". Landau levels are shown in the left half and their spin splitting in the right half of the figure. The diagram applies only to the classical limit of high quantum numbers (after ZWERDLING et al., Ref. 1, p. 291).

parabolic, i.e., $E \propto k^2$ and m^* is a constant. Since the effective g factors are related to the effective mass of the carrier this also implies that g_c and g_v are constant.

The properties of bands of this nature were extensively discussed by DINGLE (1952)[1] in a series of papers on the magnetic properties of free electrons in metals. In the first of these papers Dingle solves the Schrödinger equation in cylindrical polar coordinates instead of Cartesians and obtains wave functions $\psi_\perp$ for the transverse motion given by

$$\psi_\perp = e^{il\varphi}\, y^{\frac{1}{2}l}\, e^{-\frac{1}{2}y}\, L_{n+l}^{l}(y) \tag{9.14}$$

where, using DINGLE's notation, $y = \frac{eB}{2\hbar} r^2$,

$$L_{n+l}^{l}(y) = (-1)^l \frac{\{(n+1)!\}^2}{n!\,l!} M(-n, l+1, y)$$

and

$$M(-n, l+1, y) = 1 - \frac{n}{l+1} y + \frac{n(n-1)}{(l+1)(l+2)\,2!} y^2 - \text{(etc.)}$$

[1] R.B. DINGLE: Proc. Roy. Soc. (London) A **211**, 500, 517 (1952); A **212**, 38, 47 (1952).

With $l < l_{\max} = \frac{eB}{\hbar} A$ whence A is the cross sectional area $(L_x\, L_z)$ of the system and thus l describes the "orbit centre" degeneracy.

We now have

$$E_\perp = \frac{e\hbar B}{m}\left(n + \frac{1}{2}\right) \tag{9.15}$$

with $n = 0, 1, 2$ while the component of angular momentum can take values 0, 1, 2, up to $l_{\max}$.

With these wave functions DINGLE[1] discussed what he called "diamagnetic resonance absorption" by calculating the matrix elements between different Landau levels, i.e., states with various values of n. We now call this effect "cyclotron resonance". It was discussed simultaneously and independently by DORFMAN[2] in 1951.

The probability of transition between states (n, l) and (n', l') is proportional to the square of the interaction energy. For the interaction of an electric field $\boldsymbol{E}$ of the radiation and the electric dipole moments of the electrons orbiting around B, the transition probability $(n, l) \rightleftharpoons (n'\, l')$ is proportional to

$$\left[e \iint \psi_{nl} \psi^*_{n'l'} \cdot \begin{Bmatrix} E_x \cdot r \cdot e^{\pm i\varphi} \\ E_y \cdot r \cdot e^{\pm i\varphi} \\ E_z \cdot z \end{Bmatrix} \cdot r\, dr\, d\varphi \right]^2 . \tag{9.16}$$

Dingle then finds on substitution of (9.14) that the only transition leading to absorption of energy gives a dipole moment $D(n, l; n+1, l-1)$

$$D(n, l; n+1, l-1) = \left\{ \frac{eh}{\pi B} (n+l) \right\}^{\frac{1}{2}} . \tag{9.17}$$

This result may also be obtained using the "orbit radius" from LAX's semiclassical form [Eq. (9.10)]. For 4kgauss $\hbar\omega_c = 10^{-4}$ eV and $m^* = 0.01\, m_0$ where $r \sim 2 \times 10^{-5}$ cm. D is found to be $\sim 10^{-4}$ esu and represents a strong absorption implying that $\sim 10^6$ carriers/cc can be detected.

The *magnetic* dipole moment is also non-zero for transitions between the spin split levels of the same Landau state, but is some 10^6 times smaller. The electric dipole transition between Landau levels is the fundamental transition responsible for all the *free carrier* magneto-optical effects.

10. Bloch functions — the treatment of LUTTINGER and KOHN. In semiconductors in the absence of a magnetic field the electron states are expressed in terms of Bloch functions

$$\psi(\boldsymbol{k}) = u_k\, e^{i\boldsymbol{k}\cdot\boldsymbol{r}} \tag{10.1}$$

and the eigenvalues $E(\boldsymbol{k})$ are such that the $E - \boldsymbol{k}$ relation varies with direction in the reciprocal lattice which defines $\boldsymbol{k}$ and, for example in Gp IV elements and Gp III—V compounds, can have its minima or maxima away from $\boldsymbol{k} = 0$. In addition for valence states a point of degeneracy occurs at $\boldsymbol{k} = 0$ which is usually a maximum in the materials mentioned, there being two bands degenerate at this point with the third split off by spin orbit coupling (Fig. 3). All three of these bands are double degenerate.

More recently it has been shown[3] that in the presence of spin-orbit coupling the orbital and spin resonances can combine, when, for example, an electric dipole

[1] R. B. DINGLE: Proc. Int. Conf. on Very Low Temperatures, Oxford 1951.
[2] J. DORFMAN: Doklady Acad. Sci. (U.S.S.R.) **81**, 765 (1951).
[3] Y. YAFET: Solid State Physics **14**, 1 (1963).

can excite the spin transition. These are, in principle, then four possible effects since the magnetic dipole could also couple to the orbital resonance.

The theory of the effect of a magnetic field on such systems is due to LUTTINGER and KOHN (1955)[1]. The method is sometimes known as an "effective mass" theory and proceeds by an expansion in terms of the Bloch functions at $\boldsymbol{k}=0$, or of the functions at the band edge if this is not at $\boldsymbol{k}=0$.

$$\psi(\boldsymbol{r})=\sum_{\alpha}\int d\boldsymbol{k}\, A_{\alpha}(\boldsymbol{k})\, u_{\alpha,0}(\boldsymbol{r})\, e^{i\boldsymbol{k}\cdot\boldsymbol{r}} \tag{10.2}$$

where $u_{\alpha,0}(\boldsymbol{r})$ are the band edge functions.

The $A_{\alpha}(\boldsymbol{k})$ are then found by substitution of (10.2) in the Hamiltonian (8.1) and require the evaluation of all the matrix elements of the Hamiltonian between

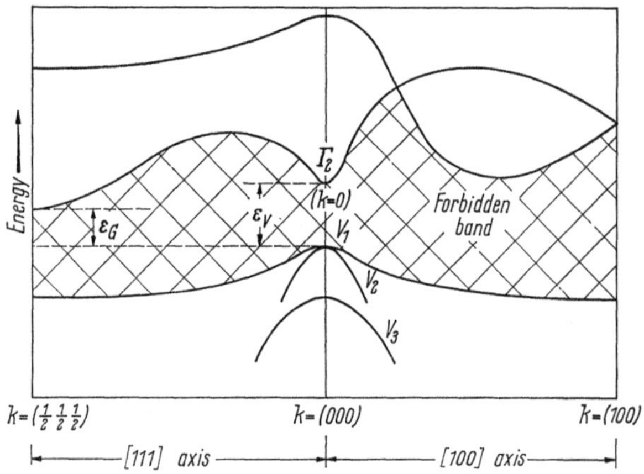

Fig. 3. Band structure in Germanium showing, at $\boldsymbol{k}=0$, the Γ_2 conduction band minimum, the four-fold degenerate valence bands (V_1, V_2) and the split off two-fold degenerate band (V_3)[2].

pairs of states $(\alpha, \boldsymbol{k})$. The resulting transformation is chosen so as to make inter-band (i.e. off-diagonal band coupling) elements vanish to 1st order in (a/R) where a is the lattice parameter and R the orbit radius $[R\sim 1/\sqrt{B}$, Eq. (9.10)]. The entire error of this expansion is of order $(a/R)^2$.

The required transformation is e^s giving a transformed Hamiltonian

$$\mathscr{H}=e^{-s}\,\mathscr{H}\,e^{s}$$

where the matrix elements of S are

$$\langle\alpha, \boldsymbol{k}|\; S\,|\gamma, \boldsymbol{k}'\rangle=\frac{\hbar}{i m\,(\varepsilon_{\alpha}(0)-\varepsilon_{\gamma}(0))}\left\{\boldsymbol{k}\cdot\boldsymbol{p}_{\alpha\gamma}(0)-\frac{eB}{2\hbar}\,(\boldsymbol{p}_{\alpha\gamma}(0)\times\nabla_k)_z\right\}\delta(\boldsymbol{k}'-\boldsymbol{k})$$

when

$$\varepsilon_{\alpha}(0)\neq\varepsilon_{\gamma}(0)$$

and the matrix elements of S vanish when

$$\varepsilon_{\alpha}(0)=\varepsilon_{\gamma}(0)\,.$$

$\varepsilon_{\alpha}(\boldsymbol{k})$ is the energy of an electron in band α and wave vector $\boldsymbol{k}$ and $p_{\alpha\gamma}(\boldsymbol{k})$ is the momentum matrix element

$$p_{\alpha\gamma}(\boldsymbol{k})=-i\hbar\,[(2\pi)^3/\Omega]\int_{\substack{\text{unit}\\\text{cell}}} d\boldsymbol{r}\; u^{*}_{\alpha k}(\boldsymbol{r})\,\nabla\, u_{\gamma k}(\boldsymbol{r}) \tag{10.3}$$

[1] J. M. LUTTINGER, and W. KOHN: Phys. Rev. 97, 869 (1955).
[2] F. HERMAN: Proc. Inst. Radio Engrs. N.Y., 43, 1703 (1955).

and Ω is the volume of the unit cell. The wave functions are then

$$\psi(\boldsymbol{r}) = \sum_{\alpha} \int d\boldsymbol{k}\, e^{i\boldsymbol{k}\cdot\boldsymbol{r}}\, u_{\alpha' 0}(\boldsymbol{r}) \left[B_\alpha(\boldsymbol{k}) - \sum_{\gamma} \left\{\frac{\hbar \boldsymbol{k}\cdot\boldsymbol{p}_{\alpha\gamma}}{m\omega_{\alpha\gamma}} + \frac{i e B}{2m} (\boldsymbol{p}_\alpha \times \nabla_k)_z\right\} B_\gamma(\boldsymbol{k})\right]$$

where

$$\hbar\,\omega_{\alpha\gamma} = \varepsilon_\alpha(0) - \varepsilon_\gamma(0)$$

and $B_\alpha(\boldsymbol{k})$ is the appropriate amplitude to the transformed Hamiltonian. $A_\alpha(\boldsymbol{k})$ is related to $B_\alpha(\boldsymbol{k})$ by

$$A_\alpha(\boldsymbol{k}) = \sum_{\gamma} \int d\boldsymbol{k}' \langle\alpha, \boldsymbol{k}|\, e^{s}\, |\gamma, \boldsymbol{k}'\rangle\, B_\gamma(\boldsymbol{k}')$$

to 1st order in (a/R).

Substitution into Schrödinger equation

$$\left[\frac{1}{2m} (\boldsymbol{p} + e\boldsymbol{A})^2 + V(\boldsymbol{r})\right]\psi = E\,\psi$$

then gives a differential equation for $B_\alpha(\boldsymbol{k})$ in momentum space which can be transferred into coordinate space by the relation

$$F_\alpha(\boldsymbol{r}) = \int d\boldsymbol{k}\, B_\alpha(\boldsymbol{k})\, e^{i\boldsymbol{k}\cdot\boldsymbol{r}}.$$

For the case of simple spherical energy surfaces, with only spin degeneracy, i.e., at zero B expressed by

$$\varepsilon_\alpha(\boldsymbol{k}) = \varepsilon_\alpha(0) + \frac{\hbar^2 k^2}{2m_\alpha}$$

the energy levels in the magnetic field are then solutions of

$$\varepsilon_\alpha\left(1/_i \nabla + \frac{e\boldsymbol{A}}{\hbar}\right) F_\alpha(\boldsymbol{r}) = \varepsilon F_\alpha(\boldsymbol{r}) \tag{10.4}$$

and are given in Eqs. (9.12) and (9.13) including the spin term.

The effective mass m^* can then be written to extend the theory to ellipsoidal energy surfaces as

$$\frac{1}{m^*} = \frac{1}{\sqrt{m_1 m_2 m_3}} \left(\sqrt{\alpha_1^2 m_1 + \alpha_2^2 m_2 + \alpha_3^2 m_3}\right)$$

where α_i are direction cosines of the magnetic field direction to the principal axes of the ellipsoid and m_i are the components of the mass tensor in the ellipsoidal system.

For the spherical case the wave functions are

$$\psi(r) = \sum_{\alpha} \left[F_\alpha(r) - \sum_{\beta \neq \alpha} \left\{\frac{p_{\alpha\beta} \nabla r\, F_\beta(r)}{i m \omega_{\alpha\beta}} + \frac{e B}{2\hbar} \frac{(p_{\alpha\beta}^{x} y - p_{\alpha\beta}^{y} x)}{m\omega_{\alpha\beta}} F_\alpha(r)\right\}\right] x\, u_{\alpha 0}(\boldsymbol{r}) \tag{10.5}$$

where $F_\alpha(r)$ are the free electron functions but containing effective mass parameters derived by DINGLE and given in Eq. (9.14). We may also express this result in the form

$$\psi(r) = F_\alpha(r)\, \varphi_{M_J}^{(J)}(r) \tag{10.6}$$

where $\varphi_{M_J}^{(J)}$ are Bloch functions at $\boldsymbol{k} = 0$ having symmetry properties characterised by the angular momentum quantum numbers J and M_J of the *atomic* states to which they reduce in the tight binding limit.

The non-degenerate case is illustrative but of more practical interest is the case of degenerate valence bands as in the diamond and zinc-blende structure semiconductors. The theory has been developed by LUTTINGER and KOHN (1955)[1], LUTTINGER (1956)[2], ELLIOTT, MCLEAN and MACFARLANE (1958)[3] and WALLIS and BOWLDEN (1960)[4].

For degenerate bands with Bloch functions $u_{\alpha 0}$, $u_{\beta 0}$ etc. the wave functions at zero magnetic field are (SHOCKLEY, 1950)[5]

$$\psi = \sum_{\alpha} a_\alpha \, u_{\alpha 0}(r) \, e^{i\boldsymbol{k}\cdot\boldsymbol{r}} \tag{10.7}$$

where the a_α are constants determined by the matrix equation

$$\sum_{\beta} (D_{\alpha\beta}(\boldsymbol{k}) - E\,\delta_{\alpha\beta})\, a = 0 \tag{10.8}$$

the sums in (10.7) and (10.8) being over the degenerate states. $D_{\alpha\beta}(\boldsymbol{k})$ is a quadratic function of $\boldsymbol{k}$

$$D_{\alpha\beta}(\boldsymbol{k}) = \frac{\hbar^2 k^2}{2m}\,\delta_{\alpha\beta} + \frac{\hbar^2}{m^2} \sum_{\gamma} \frac{\boldsymbol{k}\cdot\boldsymbol{p}_{\alpha\gamma}(0)\,\boldsymbol{k}\cdot\boldsymbol{p}_{\gamma\beta}(0)}{\varepsilon_\alpha(0) - \varepsilon_\gamma(0)} \tag{10.9}$$

in which the sum is over all bands excluding the degenerate set $u_\alpha\, u_\beta$. For a diamond-type lattice, symmetry requirements give $D_{\alpha\beta}$ the form:

$$D = \begin{Bmatrix} A k_x^2 + B(k_y^2 + k_z^2) & C k_x k_y & C k_x k_z \\ C k_x k_y & A k_y^2 + B(k_x^2 + k_z^2) & C k_y k_z \\ C k_x k_z & C k_y k_z & A k_y^2 + B(k_x^2 + k_z^2) \end{Bmatrix} \tag{10.10}$$

where A, B, C are real constants, determinable from experiment. The introduction of spin orbit splitting lifts the degeneracy partially to form a four-fold (V_1, V_2) degenerate band at $\boldsymbol{k}=0$ and a two-fold degenerate band (V_3) split away by an energy Δ (Fig. 3).

In the presence of the magnetic field the wave functions are again written as in Eq. (10.2). The zero order functions are

$$\sum_{\alpha} F_\alpha(r)\, u_{\alpha 0}(r) \tag{10.11}$$

where F_α are solutions of

$$\sum_{\beta} \left[D_{\alpha\beta}\left(-i\nabla + \frac{e\boldsymbol{A}}{\hbar}\right) - E\,\delta_{\alpha\beta} \right] F_\beta(r) = 0. \tag{10.12}$$

If the spin orbit splitting is large compared to the energy level separations of the solution of (10.12) the theory can be applied to each multiplet or set of degenerate levels.

Eq. (10.12) cannot be explicitly solved except in special cases. LUTTINGER (1956) has given the solution for the case of two spherical energy surfaces, degenerate at $\boldsymbol{k}=0$ and taking $k_z=0$. This approximation is reasonable for the

[1] J. M. LUTTINGER, and W. KOHN: See footnote 1, p. 252.
[2] J. M. LUTTINGER: Phys. Rev. **102**, 1030 (1956).
[3] R. J. ELLIOTT, T. P. MCLEAN, and G. G. MACFARLANE: Proc. Phys. Soc. (London) **72**, 553 (1958).
[4] R. F. WALLIS, and H. BOWLDEN: Phys. Rev. **118**, 456 (1960).
[5] W. SHOCKLEY: Phys. Rev. **78**, 173 (1950).

valence bands of germanium. The effects of finite k_z and warped energy surfaces can be treated by perturbation theory but most of the features of the magneto-optical effects are contained in the simplified theory. The matrix D can then be written as

$$D = D_0 + D_1 + D_2$$

where $D_1 = D_2 = 0$ when $k_z = 0$ and

$$D_0 = \frac{1}{2m}\begin{pmatrix} (\gamma_1+\bar\gamma)(p_x^2+p_y^2) + 3\hbar^2\kappa, & -\sqrt{3}\,\bar\gamma\,(p_x - i p_y)^2 & 0 & 0 \\ -\sqrt{3}\,\bar\gamma\,(p_x + i p_y)^2, & (\gamma_1-\bar\gamma)(p_x^2+p_y^2) - \hbar^2 k & 0 & 0 \\ 0 & 0 & (\gamma_1-\bar\gamma)\,p_x^2 + p_y^2 + \hbar^2\kappa, & -\sqrt{3}\,\gamma\,(p_x - i p_y)^2 \\ 0 & 0 & -\sqrt{3}\,\bar\gamma\,(p_x + i p_y)^2, & (\gamma_1+\bar\gamma)(p_x^2+p_y^2) - 3\hbar^2\kappa \end{pmatrix}.$$

The parameters γ_1 and $\bar\gamma$ are related to the A, B, C of (10.10) by

$$\frac{\gamma_1}{2m} = -\frac{A+2B}{3}, \qquad \frac{\gamma_2}{2m} = -\frac{(A-B)}{6}, \qquad \frac{\gamma_3}{2m} = -\frac{C}{6}. \tag{10.13}$$

The assumption of spherical bands makes $\gamma_2 = \gamma_3 = \bar\gamma$. DRESSELHAUS, KIP and KITTEL[1] derived the relation

$$\kappa = \gamma_3 + \frac{2\gamma_2}{3} - \frac{\gamma_1}{3} - \frac{2}{3} \tag{10.14}$$

for the remaining parameter κ. The wave functions and energy levels can then be found from Eq. (10.12). To zero order in $\boldsymbol{k}$ and B

$$\left.\begin{aligned} \psi = {} & a_1 F_{n-2,\,l+2,\,k_z}\, u_{\frac{3}{2}} + a_2 F_{n,\,l,\,k_z}\, u_{-\frac{1}{2}} + \\ & + b_1 F_{n-2,\,l+2,\,l\,k_z}\, u_{\frac{1}{2}} + b_2 F_{n,\,l,\,k_z}\, u_{-\frac{3}{2}}. \end{aligned}\right\} \tag{10.15}$$

The $F_{n,\,l,\,k_z}$ are the free-electron in magnetic field type functions of Eq. (9.14), while the u's are the four degenerate Bloch functions at $\boldsymbol{k}=0$ which may be expressed as

$$\left.\begin{aligned} u_{\frac{3}{2}} &= \frac{1}{\sqrt{2}}\,[(X+iY)\,\alpha], \qquad u_{\frac{1}{2}} = \frac{1}{\sqrt{6}}\,[(X+iY)\,\beta - 2Z\,\alpha], \\ u_{-\frac{1}{2}} &= \frac{1}{\sqrt{6}}\,[(X-iY)\,\alpha + 2Z\beta], \qquad u_{-\frac{3}{2}} = \frac{1}{\sqrt{2}}\,[(X-iY)\,\beta]. \end{aligned}\right\} \tag{10.16}$$

X, Y, Z have the symmetry of atomic p_x, p_y, p_z wave functions and α, β are spin components. The coefficients $a(n)$ are given by LUTTINGER from

$$\left.\begin{aligned} [(\gamma_1+\bar\gamma)(n-\tfrac{3}{2}) + \tfrac{3}{2}\kappa]\, a_1^{\pm}(n) - \bar\gamma\,[3n(n-1)]^{\frac{1}{2}}\, a_2^{\pm}(n) &= \varepsilon_1^{\pm}(n)\, a_1^{\pm}(n), \\ -\bar\gamma\,[3n(n-1)]^{\frac{1}{2}}\, a_1^{\pm}(n) + [(\gamma_1-\bar\gamma)(n+\tfrac{1}{2}) - \tfrac{1}{2}\kappa]\, a_2^{\pm}(n) &= \varepsilon_1^{\pm}(n)\, a_2^{\pm}(n). \end{aligned}\right\} \tag{10.17}$$

The coefficients b are obtained also from (10.17) by writing $-\bar\gamma$ instead of γ, b for a, and $-b$ for a_2.

Finally the energies are

$$\varepsilon_1^{\pm}(n) = \gamma_1 n - (\tfrac{1}{2}\gamma_1 + \bar\gamma - \tfrac{1}{2}\kappa) \pm \{[\bar\gamma n - (\gamma_1 + \tfrac{1}{2}\bar\gamma - \kappa)]^2 + 3\bar\gamma^2 n(n-1)\}^{\frac{1}{2}} \tag{10.18}$$

[1] G. DRESSELHAUS, A. F. KIP, and C. KITTEL: Phys. Rev. 98, 368 (1955).

and a second set

$$\varepsilon_2^{\pm}(n) = \gamma_1 n - \left(\tfrac{1}{2}\gamma_1 - \bar{\gamma} + \tfrac{1}{2}\kappa\right) \pm \left\{\left[\bar{\gamma} n + \left(\gamma_1 - \kappa - \tfrac{1}{2}\bar{\gamma}\right)\right]^2 + 3\bar{\gamma}^2 n(n-1)\right\}^{\frac{1}{2}}. \quad (10.19)$$

These equations define four sets of energy levels, often called "ladders" and will be referred to as 1^+, 2^+, 1^- and 2^- (Fig. 4).

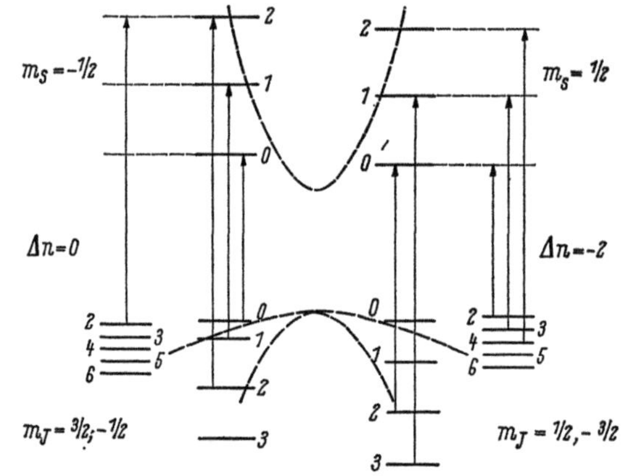

Fig. 4. Illustrating the spin splitting of the magnetic levels in a $\boldsymbol{k}=0$ conduction band and degenerate valence bands. The transitions shown are these allowed for $E \| B$ (after Roth et al., Ref. 3, p. 257).

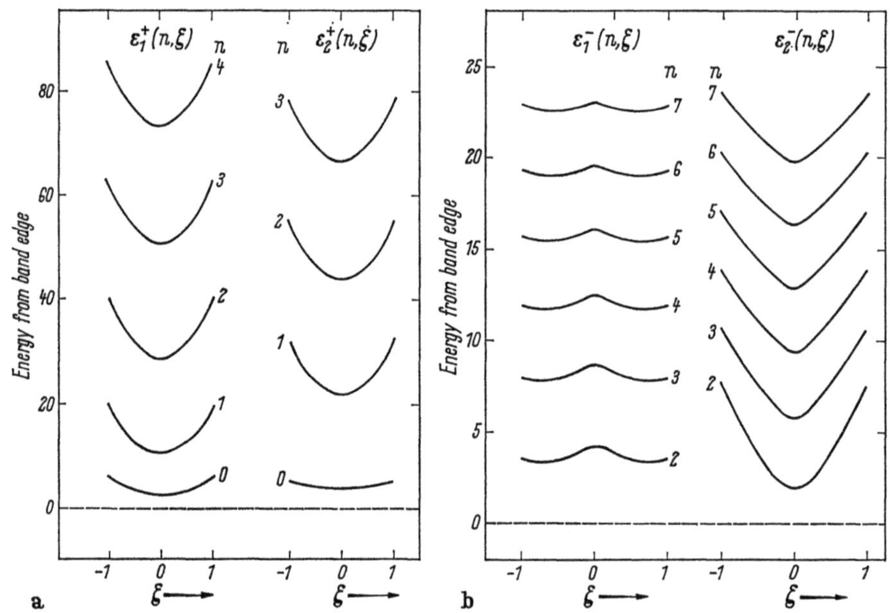

Fig. 5a and b. Details of (a) light hole, (b) heavy hole magnetic sub bands as calculated by Wallis and Bowlden. Energies are measured from the band edge in limits of $\hbar e B/m$ (Ref. 1, p. 257).

Unlike the Landau levels of equation 9.13 the *spacing of the levels is irregular for small values of n* — this is known as the "quantum effect" arising as the theory shows from the mixing of the states. For $n=0$, only ε_1^+ and ε_2^+ are defined.

At high values of n, the levels become *equally spaced* and are given, like the Landau levels by equations like (9.13), with appropriate effective mass and

g factors inserted. In this limit the 1^+ and 2^+ ladders are derived from "light hole" valence states and the 1^- and 2^- set from the heavy hole states. The sets are thus named "light-hole ladders" and "heavy-hole ladders" respectively (Fig. 4). In using Eq. (9.13)

$$\hbar^2 \frac{m}{m_v^*} = (\gamma_1 \pm 2\bar{\gamma}) \tag{10.20}$$

and the g-factor will be anomalous. We have assumed that $k_z=0$ in the above; the dependence of energy on k_z will be quadratic near $k_z=0$ and may therefore be described by an effective mass parameter m_z. This will depend on n and an expression in terms of n, γ_1 and $\bar{\gamma}$ has been given by LUTTINGER, and KOHN from second order perturbation theory. Difficulties are found for 1^- and 2^- levels that become very close together as n increases and a more recent method of WALLIS and BOWLDEN (1960)[1] overcomes these problems and gives detailed results (Fig. 5) which show anomalous shapes for the 1^- ladder in germanium.

11. Transition probabilities in a magnetic field. The selection rules of transition probabilities for various possible *interband* processes have been discussed by ELLIOTT, MCLEAN and MACFARLANE (1958)[2], ROTH, LAX and ZWERDLING (1959)[3] and BURSTEIN, PICUS, WALLIS and BLATT (1959)[4], using the Luttinger-Kohn results as described previously for the wave functions and energy levels.

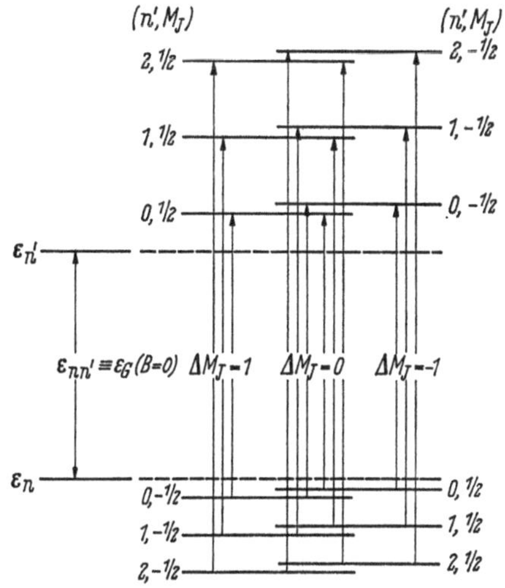

Fig. 6. Allowed transitions for simple bands (after BURSTEIN et al. [4]).

For direct transitions (i.e., $\Delta k=0$) which give the first order interband magneto-optic effects the probability of transition between states i and f is proportional to the square of the matrix element

$$\langle f | \boldsymbol{\alpha}\cdot\boldsymbol{p} | i\rangle = -i\hbar \int \psi_f^*(r)\, \alpha\cdot\nabla\psi(r)\, dr \tag{11.1}$$

where α is a polarisation vector.

The occasions when the matrix element is non zero are determined by the state of polarisation and direction of propagation of the radiation in relation to the direction of the applied magnetic field and the character of the valence and conduction bands. The simplest effect can be illustrated using simple, non-degenerate spherical bands described by (9.12) and (9.13) and the allowed transitions are illustrated in Fig. 6 taken from BURSTEIN et al. (1959)[4].

The selection rules are $\Delta n=0$, $\Delta M=0$, ± 1. The $\Delta M=\pm 1$ rules occur in the Faraday ($E \perp B$) orientation, $+1$ and -1 occurring for left- and right-circularly polarised modes respectively, while $\Delta M=0$ occurs in the Voigt orientation for plane polarised radiation with $E \parallel B$, and $\Delta M=\pm 1$ for $E \perp B$. Further details of selection rules with complex bands are discussed in the section on interband effects.

[1] R. F. WALLIS, and H. J. BOWLDEN: Phys. Rev. **118**, 456 (1960).
[2] See footnote 3, p. 254.
[3] L. M. ROTH, B. LAX, and S. ZWERDLING: Phys. Rev. **114**, 90 (1959).
[4] E. BURSTEIN, G. S. PICUS, R. F. WALLIS, and F. BLATT: Phys. Rev. **113**, 15 (1959).

These selection rules may be contrasted with those for free carrier cyclotron resonance, i.e.,

$$\Delta n = \pm 1, \qquad \Delta M = 0.$$

The circularly polarised modes enable us to define velocity matrix elements

$$V_{\pm} = V_x \pm i V_y$$

which are non-zero for $\Delta M = \pm 1$. From the theory of ELLIOTT, MCLEAN and MACFARLANE, BOSWARVA and LIDIARD (1964)[1] have developed expressions for the velocity matrix elements. The velocity operator can be written

$$m\boldsymbol{v} = \boldsymbol{P} + \frac{e}{2}\boldsymbol{B}\times\boldsymbol{r} \tag{11.2}$$

where $\boldsymbol{P}$ is the momentum operator $-i\hbar\nabla$ plus a spin orbit term and the vector potential has been chosen as $\boldsymbol{A} = \frac{1}{2}\boldsymbol{B}\times\boldsymbol{r}$. The matrix elements of the velocity operator between a valence and conduction band state becomes[2]

$$m v_{cv}(\pm) = \left(1 \pm \frac{eB}{m\omega_g}\right) \sum_i \langle u_c | P(\pm) | u_{vi}\rangle \int F_c^* F_{vi}\, d\boldsymbol{r} \tag{11.3}$$

where $P(\pm) = P_x \pm i P_y$ and the matrix elements $\langle u_c | P(\pm) | u_{vi}\rangle$ are evaluated between the conduction state at $\boldsymbol{k}=0$, and the i-th valence state with amplitude with which it appears in Eq. (10.15). To obtain the field term use is made of the relation

$$\langle u_c | \boldsymbol{r} | u_{vi}\rangle = \langle u_c | \boldsymbol{P} | u_{vi}\rangle / i m\,\omega_g .$$

In treating the second term in (11.2) ELLIOTT et al. evaluate $\boldsymbol{P}$ (not $m\boldsymbol{v}$) and obtain a field dependent term $\frac{2m\,\omega_g}{eB}$, the remaining half in (11.3) comes from the $\boldsymbol{B}\times\boldsymbol{r}$ term in (11.2).

The mass occurring in these expressions is the free mass, m, so that $\Delta M = +1$ and $\Delta M = -1$ transitions differ in strength by a field dependent term which is usually small. It can be significant however, in some magneto-optical effects, notably Faraday rotation.

For transitions between valence states $1^{\pm}$, $2^{\pm}$ and conduction states $C^{\pm}$ only certain matrix elements

$$\langle u_c | \boldsymbol{P} | u_v\rangle$$

are non-zero. This determines the selection rules summarised below:

$E \parallel B$	$1^{\pm}(n) \to C^{-}(n)$
	$2^{\pm}(n) \to C^{+}(n-2)$
$E \perp B$	$1^{\pm}(n) \to C^{+}(n)$
	$1^{\pm}(n) \to C^{+}(n-2)$
	$2^{\pm}(n) \to C^{-}(n)$
	$2^{\pm}(n) \to C^{-}(n-2)$.

The restrictions on Δn, Δk_z are due to the orthogonality of the F functions for different values of n, l, k_z. The allowed values of $\Delta n = 0$ or -2 (and $\Delta l = 0$ and $+2$) arise due to the mixing of two F functions in Eq. (10.15).

[1] I. M. BOSWARVA, and A. B. LIDIARD: Proc. Roy. Soc. (London) A **278**, 588 (1964).
[2] See also Sect. 23 for fuller discussion.

The above transitions account for the most prominent interband phenomena excepting those arising from the third split off valence band. Free carrier phenomena arise from cyclotron resonance transitions; there remains spin resonance of free carriers — $\Delta n=0$, $\Delta M=\pm 1$.

12. Band models and g-factors. The Luttinger and Kohn treatment has been described in the approximation of spherical energy surfaces and quadratic dependence of energy upon wave vector, i.e., parabolic bands. For semiconductors with small energy gaps, non-parabolic effects are significant for any experiment that depends upon the properties of electrons or holes away from the band extrema where the effective mass will depend strongly on energy.

The conduction band non-parabolicity in III—V compounds has been very successfully treated by the $\boldsymbol{k}\cdot\boldsymbol{p}$ perturbation calculation (KANE, 1957)[1] for zero magnetic field. A semiclassical theory for conduction electron cyclotron resonance in a non-parabolic band was proposed by WALLIS (1958)[2] using an expansion for the energy up to the fourth power of $\boldsymbol{k}$ in a simplified form of the effective mass equations of KJELDAAS and KOHN. There are some doubts as to whether the method is sufficiently accurate in practical cases.

The $\boldsymbol{k}\cdot\boldsymbol{p}$ result has been generalised to include a magnetic field by BOWERS and YAFET (1959)[3] (see also Sect. 14β) and a less general calculation from the Luttinger and Kohn formulation made by LAX et al.[4].

A simpler method of treating magnetic states of electrons in crystals has been given recently by HARPER[5] which leads easily to the Kohn Luttinger equations by using an extension of the Bloch theorem. HARPER has shown quite generally[6] that crystalline magnetic states have the form

$$\psi(\boldsymbol{r})=e^{i\boldsymbol{\kappa}\cdot\boldsymbol{r}}U(\boldsymbol{\kappa},\boldsymbol{r})$$

where $\boldsymbol{\kappa}=\boldsymbol{k}-\frac{e}{\hbar}\boldsymbol{A}(\boldsymbol{r})$ and $\boldsymbol{B}=\nabla_x\boldsymbol{A}(\boldsymbol{r})$.

The effect of the dependence of $\boldsymbol{\kappa}$ on $\boldsymbol{r}$, through the operation of $\boldsymbol{p}\left(=\frac{h}{i}\nabla_r\right)$ is to introduce the terms $\left(\frac{1}{m}\boldsymbol{\pi}\cdot\boldsymbol{p}+\frac{1}{2m}\boldsymbol{\pi}^2\right)$ into the Hamiltonian where

$$\boldsymbol{\pi}=\hbar\boldsymbol{\kappa}+e\boldsymbol{A}(-i\nabla_k).$$

The Schrödinger equation now becomes

$$\left\{\frac{1}{2m}\boldsymbol{p}^2+V(\boldsymbol{r})+\frac{1}{m}\boldsymbol{\pi}\cdot\boldsymbol{p}+\frac{1}{2m}\boldsymbol{\pi}^2-E\right\}U(\boldsymbol{\kappa},\boldsymbol{r})=0$$

which is six-dimensional. The fourth term taken alone, gives the Landau levels of *free* electrons, the first two terms are the usual crystal Hamiltonian in the absence of a field. The *third* term, *directly analogous to* $\boldsymbol{k}\cdot\boldsymbol{p}$, at zero field, couples the magnetic and crystal terms together and so expresses interaction between bands in the presence of a magnetic field. The $\boldsymbol{\pi}\cdot\boldsymbol{p}$ term is, in fact, just the interaction treated in the conventional Luttinger-Kohn method. The $\boldsymbol{\pi}\cdot\boldsymbol{p}$ interaction can be removed to 1st order for distant bands and groups of adjacent bands treated exactly as in the Luttinger-Kohn case. Such calculations for III—V compounds, treating conduction bands together with the degenerate valence band

[1] E. O. KANE: J. Phys. Chem. Solids **1**, 249 (1957).
[2] R. F. WALLIS: J. Phys. Chem. Solids **4**, 101 (1958).
[3] R. BOWERS, and Y. YAFET: Phys. Rev. **115**, 1165 (1959).
[4] B. LAX, J. G. MAVROIDES, H. J. ZEIGER, and R. J. KEYES: Phys. Rev. **122**, 31 (1961).
[5] P. G. HARPER: To be published.
[6] P. G. HARPER: Proc. Phys. Soc. (London) A **68**, 879 (1955).

set exactly and higher bands to order k^2 have been made by PIDGEON and BROWN[1] and the $\boldsymbol{k}\cdot\boldsymbol{p}$ method extended to the energy levels of the lead salts in a magnetic field by MITCHELL and WALLIS[2].

For the simple band picture described by Eqs. (9.12) and (9.13) the spin splitting of the electron and hole states was described by an effective g-factor. The splitting in practical cases is anomalous with respect to a free electron g-factor of 2.00.

To explain this effect we consider a conduction band edge at $\boldsymbol{k}=0$, having only time reversal (spin) degeneracy. In the absence of a magnetic field, the energy can be written[3] in terms of coefficients $D_{\alpha\beta}$ [see also Eqs. (10.8) and (10.9)] as

$$E(k)=\sum_{\alpha\beta} D_{\alpha\beta}\, k_\alpha\, k_\beta \quad (\alpha,\beta=x,y,z)$$

and this includes symmetric terms of the form

$$(D_{\alpha\beta}+D_{\beta\alpha})\, k_\alpha\, k_\beta .$$

In the presence of a field we can replace $\boldsymbol{k}$ by the operator $(\boldsymbol{p}-e\boldsymbol{A})$ which we write as $\bar{\boldsymbol{k}}$[4]. The energy then becomes

$$E(\bar{k})=\sum_{\alpha\beta} D_{\alpha\beta}\, \bar{k}_\alpha\, \bar{k}_\beta . \tag{12.1}$$

$D_{\alpha\beta}$ can now be written in the form

$$\left.\begin{aligned} D_{\alpha\beta}=\tfrac{1}{2}(D_{\alpha\beta}+D_{\beta\alpha})\,[\bar{k}_\alpha\bar{k}_\beta+\bar{k}_\alpha\bar{k}_\beta]+{}\\ +\tfrac{1}{2}(D_{\alpha\beta}-D_{\beta\alpha})\,[\bar{k}_\alpha\bar{k}_\beta-\bar{k}_\alpha\bar{k}_\beta].\end{aligned}\right\} \tag{12.2}$$

When $\boldsymbol{B}=0$, the commutator $[\bar{k}_\alpha\bar{k}_\beta-\bar{k}_\beta\bar{k}_\alpha]\equiv[\bar{k}_\alpha,\bar{k}_\beta]=0$, but for $\boldsymbol{B}\neq 0$, some components are non zero e.g.

$$\text{if}\quad \boldsymbol{A}=\boldsymbol{B}(0,x,0),\ [k_x,k_z]=0 \quad\text{and}\quad [k_y,k_z]=0$$
$$\text{but}\quad [k_x,k_y]=ieB_z .$$

Using this result we can express Eq. (12.1) as the Hamiltonian

$$\mathscr{H}=\sum_{\alpha\beta} D^s_{\alpha\beta}[\bar{k}_\alpha\bar{k}_\beta+\bar{k}_\beta\bar{k}_\alpha]+ie\,D^A_{xy}\,B_z+\mu_\beta\,\boldsymbol{\sigma}\cdot\boldsymbol{B} \tag{12.3}$$

in which the tensor $D_{\alpha\beta}$ has been separated into symmetric ($D^s_{\alpha\beta}$) and anti-symmetric ($D^A_{\alpha\beta}$) parts according to Eq. (12.2) and a spin orbit term has been added ($\mu_\beta\,\boldsymbol{\sigma}\cdot\boldsymbol{B}$). The form of the anti symmetric part is

$$D^A_{\alpha\beta}=\begin{bmatrix} 0 & D_{xy} & -D_{xz}\\ -D_{xy} & 0 & D_{yz}\\ D_{xz} & -D_{yz} & 0\end{bmatrix}$$

and represents an axial vector.

Thus the *antisymmetric* part of the mass tensor $D_{\alpha\beta}$ together with the spin orbit term in Eq. (12.2) constitute the coefficient of B_z; by analogy with the form familiar in atomic theory, $\boldsymbol{L}\cdot\boldsymbol{B}+\boldsymbol{S}\cdot\boldsymbol{B}$, they give orbital ($\boldsymbol{L}$) and spin ($\boldsymbol{S}$) contributions to angular momentum and between them give the effective g-factor.

---

[1] C. R. PIDGEON, and R. N. BROWN: Phys. Rev. **146**, 515 (1966).

[2] D. L. MITCHELL, and R. F. WALLIS: To be published.

[3] See, for example, C. KITTEL: Quantum Theory of Solids, p. 279. et seq. New York: John Wiley & Sons 1963.

[4] $\bar{\boldsymbol{k}}$ corresponds to $\boldsymbol{\pi}$ of the previous discussion.

The *symmetric* part of the tensor $(D^s_{\alpha\beta})$ gives the Landau levels in terms of the effective mass. The coefficients $D_{\alpha\beta}$ are obtained from $\boldsymbol{k}\cdot\boldsymbol{p}$ perturbation theory. Combining the orbital and spin terms we have

$$E(k)=\sum_{\alpha\beta} D^s_{\alpha\beta}[\bar{k}_\alpha \bar{k}_\beta+\bar{k}_\beta \bar{k}_\alpha]-\mu^* \boldsymbol{\sigma}\cdot\boldsymbol{B}$$

where the anomalous magnetic moment μ^* is defined by

$$\frac{\mu^*}{\mu_B}=(\langle\gamma|L_z|\gamma\rangle+1)$$

where

$$i D^A_{xy}=\frac{1}{2m}\langle\gamma|L_z|\gamma\rangle$$

and the effective g-factor is

$$g=\mu^*/\mu_B.$$

Using the results of $\boldsymbol{k}\cdot\boldsymbol{p}$ perturbation theory

$$\frac{\mu^*}{\mu_\beta}=1+\frac{1}{2im}J\left(\sum_\delta \frac{\langle\gamma|p_x|\delta\rangle\langle\delta|p_y|\gamma\rangle}{\varepsilon_\gamma-\varepsilon_\delta}\right)$$

where J represents "imaginary part of".

Using this treatment, Roth[1] has given an expression for g_c in terms of the effective mass m^*, the energy gap E_G and the spin orbit splitting of the valence states, Δ, as follows:

$$g_c=2\left[1-\left(\frac{m}{m^*}-1\right)\frac{\Delta}{3E_G+2\Delta}\right]. \tag{12.4}$$

This expression shows that $g_c\to 2$ as Δ, the spin orbit splitting $\to 0$, so that the presence of spin orbit interaction is essential for anomalous g-factors. Eq. (12.3) applies to cases where we have p-like valence states and s-like conduction states as in Ge, Si and InSb.

The effective g-factor may be either positive or negative and as large as -50 (e.g. InSb) or as small as $+0.2$ (e.g., GaAs) according to the effective mass, energy gap and spin orbit splitting. Thus the spin splitting can be quite significant and occasionally comparable with the orbital splittings between Landau levels. In a non-parabolic band the g-factor will, like the effective mass, vary with energy.

In the degenerate valence bands the arrangement of the $1^\pm$, $2^\pm$ ladders does not lend itself to the definition of effective g-factors as the first few levels are irregularly spaced and the levels arising by the mixing of wavefunctions considered as labelled by their quantum numbers as in Eqs. (10.18) and (10.19).

A detailed discussion of g-factors is given by Yafet (1963)[2].

IV. Free carrier magneto-optical effects.

a) General theory.

13. Boltzmann theory — low field and high field limits. The distribution of electrons over the energy states in a semiconducting crystal is governed by the Fermi-Dirac distribution function

$$f=\frac{1}{1+e^{(E-E_F)/kT}} \tag{13.1}$$

---

[1] L. M. Roth, B. Lax, and S. Zwerdling: Phys. Rev. **114**, 90 (1959).

[2] Y. Yafet: Solid State Physics **14**, 1 (1963).

where E_F is the Fermi level at which the probability of occupation f, is $\frac{1}{2}$. The position of the Fermi level, normally in the forbidden gap for a pure crystal, may be adjusted by doping the crystal so that E_F may be anywhere from deep inside the conduction band for high donor concentrations to deep inside the valence bands with high acceptor concentrations. At high doping levels when E_F is within the conduction or valence bands the concentration of the states in the bands ensures that f changes rapidly (in energy) from one to zero, that is E_F is well defined: the situation is said to be "degenerate". Electrons below the Fermi level are then, by the Pauli exclusion principle, unable to contribute to conduction in the filled part of the band and the situation is similar to that in a metal, i.e., only electrons at the Fermi surface (where f is changing) take part in conduction.

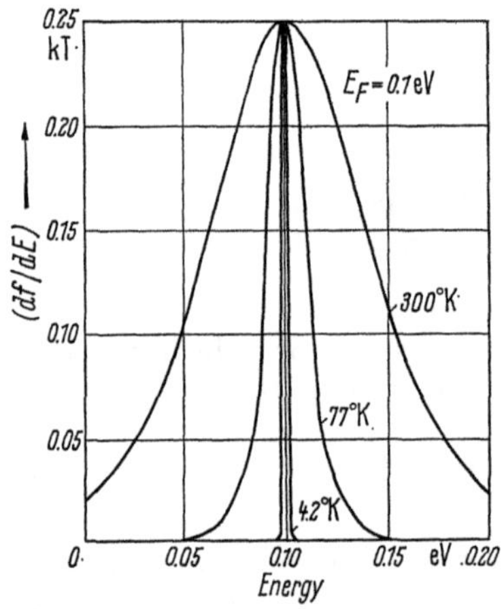

Fig. 7. Energy derivative of the Fermi function at three temperatures.

At the other extreme in pure or weakly doped material the Fermi level may be far from the band edge and the distribution is spread throughout all the populated states — then the situation may be described by Maxwell-Boltzmann statistics, i.e.,

$$f = e^{-(E-E_F)/kT}$$

and all the states in the "tail" of the distribution take part in conduction. Since from the macroscopic formulae developed in Sect. II, the magneto-optical effects can be expressed in terms of a high (i.e., optical or infra-red) frequency conductivity these considerations are relevant to magneto-optical effects.

In the quantum mechanics described in Sect. III no account was taken of the occupancy of the quantized levels. For free carrier effects we are concerned with "cyclotron resonance transitions" between Landau levels (or the corresponding states in the light and heavy hole ladders) in which

$$\hbar\omega_c = \frac{eB\hbar}{m^*}$$

and $\Delta n = \pm 1$, $\Delta M_J = 0$.

Two interesting conditions arise according to whether $\hbar\omega_c$ is much greater or smaller than the energy range over which df/dE is a significant quantity (see Fig. 7). In the former (high field) case only a few Landau levels take part in any free carrier processes. Effects are dominated, in fact, by one transition between levels n and $n+1$ when the state n is largely occupied and $n+1$ largely unoccupied. The full quantum mechanical treatment is appropriate in these circumstances which applies to most cases of observation of cyclotron resonance at infra-red frequencies (Sect. 14β).

At the other extreme, if the field is so low that many Landau levels occur over the range of df/dE finite it is not necessary to consider the magnetic splitting of the levels explicitly and the problem can be satisfactorily treated from the Boltzmann transport equation modified for the presence of the magnetic field. This is a wave packet approach and introduces the band structure parameters through the group velocity of the electron waves, i.e.,

$$\frac{d\omega}{dk} = \frac{1}{\hbar}\frac{dE}{dk}. \tag{13.2}$$

This treatment applies to microwave cyclotron resonance (Sect. 14α), non-resonant Faraday effect (Sect. 15) and Voigt effect (Sect. 17) etc. and has important advantages in that in some cases no prior knowledge of the relation between E and $\boldsymbol{k}$ is required, and explicit information about the band structure may therefore be deduced from experiment. In all such cases there is no necessity for a theory of the behaviour of the levels in a magnetic field. We can conveniently write the two conditions as $\hbar\omega_c \gg kT$ or $\hbar\omega_c \ll kT$.

Low field Boltzmann theory. The Boltzmann expression for the distribution function f in the presence of external fields is

$$\left(\frac{\partial f}{\partial t}\right)_{\text{applied field}} - \left(\frac{\partial f}{\partial t}\right)_{\text{collisions}} = \left(\frac{\partial f}{\partial t}\right)_{\text{total}} = -\boldsymbol{v}\,\mathrm{grad}_r\, f. \tag{13.3}$$

Now the force on an electron described by a wave vector $\boldsymbol{k}$ is $\hbar\frac{d\boldsymbol{k}}{dt}$, thus in an electric field $\boldsymbol{E}$ and magnetic field $\boldsymbol{B}$ we have

$$\frac{d\boldsymbol{k}}{dt} = \frac{e}{\hbar}[\boldsymbol{E}+\boldsymbol{v}\times\boldsymbol{B}].$$

Thus

$$\begin{aligned}\left(\frac{\partial f}{\partial t}\right)_{\text{applied field}} &= -\mathrm{grad}_k\, f\cdot\frac{d\boldsymbol{k}}{dt}\\ &= -\mathrm{grad}_k\, f\cdot\frac{e}{\hbar}[\boldsymbol{E}+\boldsymbol{v}\times\boldsymbol{B}].\end{aligned}$$

With the assumption that a collision time τ may be defined so that $\left(\frac{\partial f}{\partial t}\right)_{\text{collisions}} = -\frac{f-f_0}{\tau}$ where f_0 is the equilibrium value of the distribution function, the Boltzmann expression becomes

$$-\frac{\hbar}{e}[\boldsymbol{E}+\boldsymbol{v}\times\boldsymbol{B}]\cdot\mathrm{grad}_k\, f+\boldsymbol{v}\cdot\mathrm{grad}_r\, f = -\frac{f-f_0}{\tau}. \tag{13.4}$$

In the presence of a magnetic field along the z-axis and writing

$$f = f_0 - \varphi\frac{\partial f_0}{\partial E}, \tag{13.5}$$

an equation for φ has been given by WILSON (1936)[1] as

$$\frac{\varphi}{\tau} + \frac{e}{\hbar}\boldsymbol{E}\,\mathrm{grad}_k\,\boldsymbol{E} + \frac{eB}{\hbar^2}\nabla_k E\,\nabla_k\varphi = 0 \tag{13.6}$$

and the solution for φ can be written in ascending powers of B. (For certain cases of anisotropic semiconductors such an expansion is inadequate but DONOVAN and WEBSTER[2] have shown that a closed form may be found for ellipsoidal energy surfaces.)

The current density is given by

$$\boldsymbol{j} = \frac{2}{(2\pi)^3}\int e\,\boldsymbol{v}\,f(\boldsymbol{k})\,d\boldsymbol{k}.$$

Substituting for $\boldsymbol{v} = \frac{1}{\hbar}\frac{\partial E}{\partial\boldsymbol{k}}$ and for $f(\boldsymbol{k})$ from (13.5), we obtain

$$\boldsymbol{j} = \frac{-e}{4\pi^3\hbar}\int \mathrm{grad}_k\, E\,\varphi\,\frac{df_0}{dE}\,d\boldsymbol{k} \tag{13.7}$$

noting that f_0 makes no contribution to the integral. By substituting $\tau/(1+i\omega)$ for τ, the high frequency current caused by a sinusoidal applied field $\boldsymbol{E}$ can be

[1] A. H. WILSON: Theory of Metals. London: Cambridge 1953.

[2] B. DONOVAN, and J. WEBSTER: Proc. Phys. Soc. (London) **81**, 90 (1963).

calculated and hence the high frequency conductivity tensor from

$$\boldsymbol{j} = \underline{\sigma}\,\boldsymbol{E}. \tag{13.8}$$

From the relations given in Part II the magneto-optical effects can then be obtained in terms of Boltzmann theory. This is a convenient formulation as we can make use of the tensor description of the current density in the presence of a magnetic field given by ABELÉS and MEIBOOM[1].

b) The experimental phenomena.

14. Cyclotron resonance.

α) *Microwave cyclotron resonance.* The first observation of cyclotron resonance of free carriers in semiconductors were made at microwave frequencies by DRESSELHAUS, KIP and KITTEL (1953 and 1955)[2] and then by LAX, ZEIGER, DEXTER and ROSENBLUM (1954)[3], following the theory and suggestions of DORFMAN, DINGLE and SHOCKLEY mentioned earlier (Parts I and III). These experiments which gave very explicit information about the energy surfaces near the extrema of the valence and conduction bands in very pure silicon and germanium have been extensively reviewed by LAX and MAVROIDES (1960)[4]. In this article, therefore, we shall discuss these observations only in so far as they relate to infrared cyclotron resonance and other free carrier effects.

Microwave observations typically imply a field of ~1000 gauss and the splitting of the Landau levels is relatively small. Even at liquid helium temperatures this can be in the classical region that can be treated by Boltzmann theory. The experiments were carried out at low temperatures to satisfy the condition

$$\omega_c \tau \gg 1$$

required for the observation of resonance in which τ is the collision time for the carriers. In pure materials τ can be made long by cooling and suppressing the lattice vibrations. This criterion also restricted the early measurements to pure Ge and Si, but it has more recently been possible to work at higher temperatures and with compounds, e.g., InSb — see for example BAGGULEY, STRADLING and WHITING (1961)[5].

Since the range of the band explored by CR experiments is restricted to near the band extrema, the dependence of energy upon $\boldsymbol{k}$ is accurately quadratic, i.e., the bands are parabolic. Under these circumstances, the Boltzmann expression containing terms of the form $\frac{d^2E}{dk^2}$ which are constant, i.e., the effective mass $\left(\text{given by } \frac{1}{m^*} = \frac{1}{\hbar^2}\frac{\partial^2 E}{\partial k^2}\right)$ is a constant with energy. The Boltzmann expression then reduces to a single particle analysis, independent of the statistics and CR can be described by Eq. (1.1), with m^* in place of m and $\omega_0 = 0$. The anisotropy of the mass tensor is readily measured, since for ellipsoidal surfaces as for electrons in germanium, ω_c varies with direction and is given by

$$\omega_c = \frac{eB}{m^*} = eB\sqrt{\frac{m_1\alpha^2 + m_2\beta^2 + m_3\gamma^2}{m_1 m_2 m_3}} \tag{14.1}$$

where the α's are direction cosines to axes 1, 2 and 3.

---

[1] B. ABELÉS, and S. MEIBOOM: Phys. Rev. **95**, 31 (1954).

[2] G. DRESSELHAUS, A. F. KIP, and C. KITTEL: Phys. Rev. **93**, 827 (1953); **98**, 368 (1955).

[3] B. LAX, H. J. ZEIGER, R. N. DEXTER, and E. S. ROSENBLUM: Phys. Rev. **96**, 222 (1954).

[4] B. LAX, and J. G. MAVROIDES: Solid State Physics, **11**, 261 (1960).

[5] D. M. S. BAGGULEY, R. A. STRADLING, and J. S. S. WHITING: Proc. Roy. Soc. (London) A **262**, 342 (1961); — Phys. Letters **6**, 143 (1963).

From the results of resonance experiments at various orientations it is found experimentally that the conduction band in germanium can be expressed as

$$E=\frac{\hbar^2}{2}\left(\frac{k_1^2}{m_t}+\frac{k_2^2}{m_t}+\frac{k_3^2}{m_l}\right) \tag{14.2}$$

where

$$m_t=(0.0819\pm0.0003)\,m_0,$$
$$m_l=(1.64\pm0.03)\,m_0.$$

The line shape of the resonance absorption can be obtained by writing an expression for the current density

$$\boldsymbol{j}=Ne\frac{d\boldsymbol{r}}{dt}=Ne\boldsymbol{v}$$

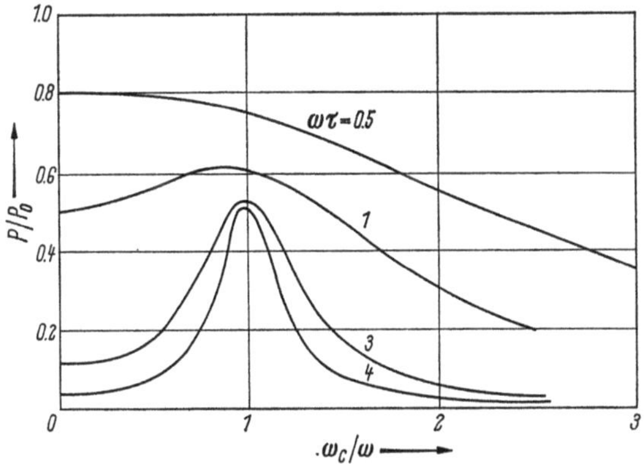

Fig. 8. Line width of cyclotron resonance absorption as a function of relaxation time [after LAX et al., Ref. Physica 20, 818 (1954)].

where $\boldsymbol{r}$ is obtained from Eq. (1.1). Using the relation

$$\boldsymbol{j}=\underline{\sigma}\boldsymbol{E}=Ne\boldsymbol{v}$$

and writing out the components of $\boldsymbol{v}$ we can obtain the elements of the conductivity tensor and hence the conductivity for circularly polarised modes $\sigma_\pm$ [Eq. (3.3)]. Using Eq. (5.2), the absorption coefficient is obtained. In terms of fractional power absorbed, this gives

$$\frac{P}{P_0}=\frac{1+(\omega^2+\omega_c^2)\,\tau^2}{1+(\omega^2-\omega_c^2)^2+4\omega^2\tau^2} \tag{14.3}$$

for the line shape. Graphs of this function for various values of $\omega_c\,\tau$ are shown in Fig. 8. At 25 kM cps if $\omega_c\,\tau$ is to be greater than 4, τ must be $>3\times10^{-11}$ secs.

Using p-type material, the effective masses and energy surfaces for holes may be obtained. From $\boldsymbol{k}\cdot\boldsymbol{p}$ theory, the valence bands in diamond, silicon and germanium are given, near $\boldsymbol{k}=0$, by

$$E=-\frac{\hbar^2}{2m}\left\{A\,k^2\pm\left[B^2\,k^2+C^2\,(k_x^2\,k_y^2+k_y^2\,k_z^2+k_x^2\,k_z^2)\right]^{\frac{1}{2}}\right\} \tag{14.4}$$

for the light (—) and heavy (+) holes and

$$E=-\varDelta-\frac{\hbar^2}{2m}A\,k^2 \tag{14.5}$$

for the split off band. Some recent values of A, B and C (from STICKLER, ZEIGER and HELLER, 1962)[1] are:

	A	B	C
Ge	-13.12	8.2	13.3
Si	-4.22	1.0	4.34

If $\hbar\omega_c$ is increased and/or the sample temperature reduced the quantum situation can be reached and the quantum effects predicted from the Luttinger-Kohn theory (Part III) observed. The resonance spectrum becomes complex and presents interpretational problems since theory predicts many more lines than have been observed. Quantum effects were first noticed by FLETCHER, YAGER and MERRITT[2] in germanium and a recent extension was made by HENSEL[3] working at 4 mm (50—60 GC/S).

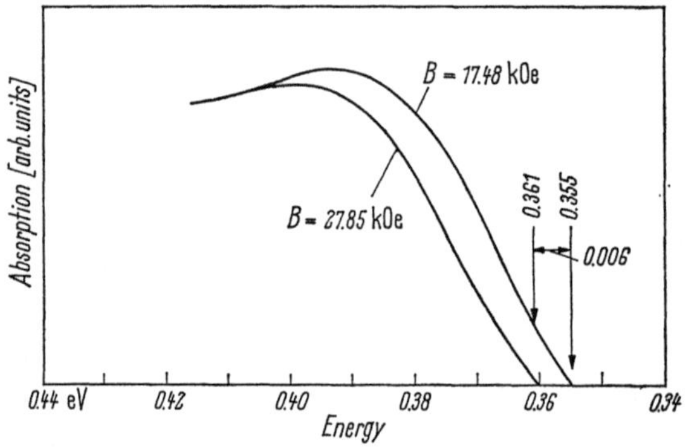

Fig. 9. The dependence of cyclotron resonance absorption strength in semi-conducting diamond as a function of the photon energy of the radiation used to create free holes (after RAUCH[4]).

A further important experiment on CR at 4 mm has extended CR technique by using *monochromatic* infra radiation to excite the carriers. This experiment, carried out by RAUCH[4] on Type IIb (p-type) diamond, has determined the masses of the light hole, heavy hole and split off bands, giving m^*/m 0.76, 2.18 and 1.06 respectively and also determined the spin-orbit splitting Δ, as 0.006 eV. This is achieved by setting the magnetic field resonance in one of the degenerate bands and varying the photon energy of the radiation used to excite the holes into the valence band from the acceptor centres. A plot of CR absorption strength against photon energy corresponding to difference between the acceptor level and the valence band is thus obtained. The resonance is then tuned to that appropriate to the split off band and the experiment repeated Fig. 9. Now, no CR absorption is seen until a photon energy corresponding to the energy difference from the *split off band to the acceptor* is reached. From the energy difference of these "edges" the spin-orbit splitting is determined. The spectrum of CR absorption versus exciting radiation energy showed structure; this is due to preferential trapping of the carriers by optical phonon processes and is also observed in photoconductivity. There was some considerable experimental difficulty in obtaining a diamond with a sufficiently long relaxation time τ even at 1.2 °K and the field required for resonance of heavy holes at 4 m was 54 kilogauss. Values for A, B and C obtained were

$$A = 0.94, \quad B^2 \simeq 0.2, \quad \text{and} \quad C^2 < 0.16.$$

[1] J. J. STICKLER, H. J. ZEIGER, and G. S. HELLER: Phys. Rev. **127**, 1077 (1962).
[2] R. C. FLETCHER, W. A. YAGER, and F. R. MERRITT: Phys. Rev. **100**, 747 (1955).
[3] J. C. HENSEL: Proc. Int. Conf. on Physics of Semiconductors, Exeter, 1962, p. 281.
[4] C. J. RAUCH: Proc. Int. Conf. on Physics of Semiconductors, Exeter, 1962, p. 276.

β) Infra-red cyclotron resonance. At infra-red frequencies the conditions for the observation of quantum effects, $\hbar\omega_c \gg kT$, and for observation of CR, $\omega_c \tau > 1$ can be readily met at low temperatures. However high magnetic fields are required to raise $\omega_c = eB/m^*$ into this frequency range so that observations have been restricted to compound semiconductors in which, at least in the conduction band, effective masses are relatively low.

The earliest infra-red experiments were carried out at Naval Research Laboratory (NRL), Washington, by BURSTEIN, PICUS and GEBBIE[1] on n-type InSb at room temperature. Using fields ~ 40 kilogauss they obtained resonance at 41.1 μ giving an effective mass of 0.015 m. This value was later considered to be rather inaccurate. In this experiment $\hbar\omega_c \sim 2\,kT$ so that several Landau levels would be involved. Immediately following this experiment, resonance was observed in the same material by the Lincoln Laboratory group at MIT[2] and also in bismuth and InAs. Wavelengths in the range 10—22 μ were used with pulsed fields up to 320 kilogauss, working at room temperature. For n-InSb masses in the range (0.019—0.03) m were obtained depending on the magnetic field. A microwave CR experiment (DRESSELHAUS et al.)[3] had yielded an effective mass of 0.013 m and a later infra-red measurement (BOYLE and BRAILSFORD)[4] gave 0.015 m. These differing values could be qualitatively understood in terms of the non-parabolic conduction band of InSb but were by no means consistent with each other.

A more recent determination of effective mass in InSb has been made by PALIK, PICUS, TEITLER and WALLIS[5], by which time KANE's $\boldsymbol{k}\cdot\boldsymbol{p}$ theory and detailed measurements of band shape by Faraday rotation (see Sect. 15iii) were available for comparison. Measurements were made from 155 to 29 μ so determining the cyclotron frequency over a range of 10—70 kilogauss. The material was n-type with $n \sim 10^{15}\,\mathrm{cm}^{-3}$. Temperature dependence at 25, 80 and 300° K was studied and also wavelength dependence in the range 25—40 μ with the higher fields (20—60 kgauss). The observed position, ω, of resonance is influenced by depolarisation effects depending on simple geometry and is related to ω_c by

$$\omega_c = \omega\left[1 - (\omega_L/\omega)^2\right] \tag{14.6}$$

where $\omega_L = \left(\frac{L}{4\pi} N e^2/m^*\right)^{\frac{1}{2}}$, and L is a depolarisation factor depending on geometry. As sample thickness increases $L \to 4\pi/\varepsilon_R$ and

$$\omega_L \to (N e^2/m^* \varepsilon_R)^{\frac{1}{2}} = \omega_p \tag{14.7}$$

where ω_p is the plasma frequency and ε_R is a residual permittivity. In this case $\omega_c \gg \omega_p$ and the effect is shown to be negligible. PALIK et al. analyse their results in terms of an extension of the Wallis treatment referred to in Part III, but obtain better results using the method of BOWERS and YAFET to give the energies of the Landau levels. The effective mass in the magnetic field can be expressed as

$$\frac{1}{m^*(B)} = \frac{1}{eB\hbar}\left[E(1, 0, +) - E(0, 0, +)\right]. \tag{14.8}$$

BOWERS and YAFET[6] obtain for the Landau levels,

$$E(n, 0, \pm) = \tfrac{1}{2}E_G\{1 + [1 + 2(2n+1 \mp \tfrac{1}{2})(\hbar\omega_c/E_G)]^{\frac{1}{2}}\}. \tag{14.9}$$

[1] E. BURSTEIN, G. PICUS, and H. A. GEBBIE: Phys. Rev. **103**, 825 (1956).
[2] R. J. KEYES, S. ZWERDLING, S. FONER, H. H. KOHN, and B. LAX: Phys. Rev. **104**, 1804 (1956).
[3] G. DRESSELHAUS, A. F. KIP, C. KITTEL, and G. WAGONER: Phys. Rev. **98**, 556 (1955).
[4] W. BOYLE, and A. D. BRAILSFORD: Phys. Rev. **107**, 903 (1957).
[5] E. D. PALIK, G. PICUS, S. TEITLER, and WALLIS: Phys. Rev. **122**, 475 (1961).
[6] R. BOWERS, and Y. YAFET: Phys. Rev. **115**, 1165 (1959).

The magnetic field dependence of m^* obtained from Eq. (14.8) and (14.9) is plotted in Fig. 10, values obtained ranging from 0.015 to 0.019. By obtaining a further expression for m^* as a function of energy, PALIK et al. compared their 77° K results to the Faraday rotation determination of band shape by SMITH, MOSS and TAYLOR[1] (see Sect. 15iii). As can be seen from Fig. 11 quite reasonable

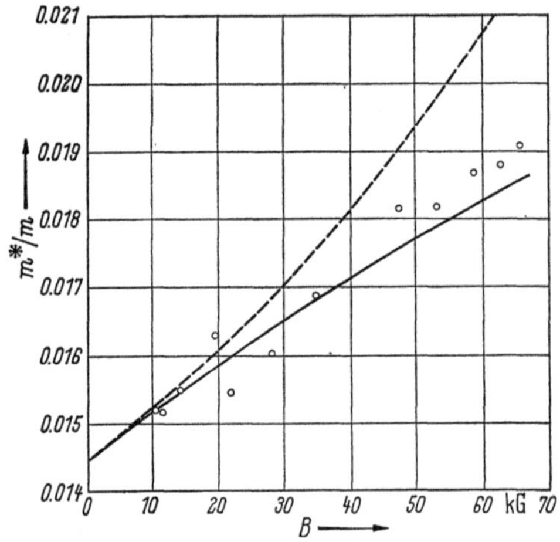

Fig. 10. Variation of effective mass with magnetic field for InSb. The solid line is the theoretical result from the Bowers and Yafet theory and the points are experimental results (after PALIK et al., Ref. 5, p. 267).

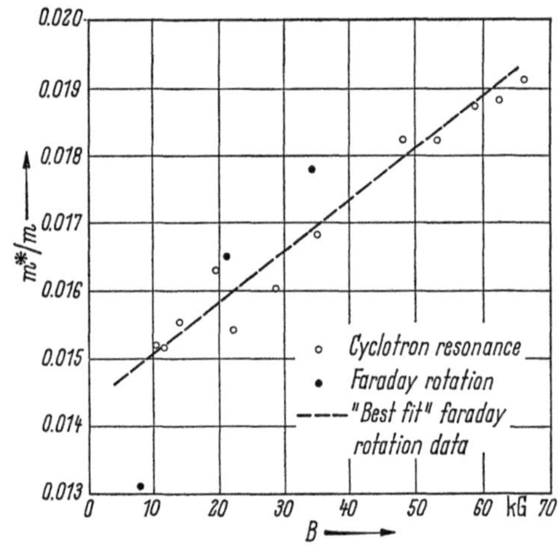

Fig. 11. Comparison of effective mass in n-InSb measured by Cyclotron Resonance and Faraday effect (after PALIK et al., Ref. 5, p. 267).

agreement was found between the two experiments which have the important difference that for CR $\hbar\omega_c \gg [E(k)-E_c]_F$ and $\hbar\,\omega_c \ll [E(k)-E_c]_F$ for Faraday effect. A simple physical picture of the relation between these methods can be obtained by considering the CR as arising primarily from transitions from $\frac{1}{2}\,\hbar\,\omega_c$ to $\frac{3}{2}\,\hbar\,\omega_c$ with a mean energy of $\hbar\,\omega_c$ above the zero field minimum; the masses obtained from the two experiments then coincide if the Fermi level is at $\hbar\,\omega_c$ above E_{c0} in the Faraday case. (Inclusion of a large and energy dependent g-factor complicates the picture).

[1] S. D. SMITH, T. S. MOSS, and K. W. TAYLOR: J. Phys. Chem. Solids **11**, 131 (1959).

LAX[1] reported a very similar analysis of the pulsed field CR data from the Lincoln Laboratory, covering the range 0.02—0.03 m. This data indicated a zero $-\boldsymbol{k}$ mass of 0.0115 m at 300° K.

Thus cyclotron resonance with varying fields can be used to probe band shapes away from the extrema. Interpretation of the data requires a theory of the band shape in the presence of the magnetic field. This is a reasonably satisfactory procedure if the $\boldsymbol{k}\cdot\boldsymbol{p}$ theory is generalised for the presence of the field. We shall compare this technique with the low field Faraday effect method in Sect. 15 iii and with interband methods in Sect. 21.

Owing to the experimental difficulties, notably the high magnetic field required, rather few infra-red CR experiments have been carried out. Resonance in n-type InAs was observed in the pulsed field work of KEYES et al. and gave a mass of 0.03 m; later work indicated a zero-k mass of 0.015 m. The latter is in strong disagreement with the data of PALIK and WALLIS[2] who find a magnetic field dependence in the range 0.020—0.026 in reasonable agreement with Faraday rotation results. The NRL group also report CR in InP with an electron mass of 0.077 ± 0.005 at $\boldsymbol{k}=0$, and in GaAs[3] giving a mass of 0.071 ± 0.005 for electrons. Both the latter cases show only a few per cent charge with field indicating a more parabolic band for InP and GaAs than InSb and InAs (Fig. 12). PALIK et al.[4] also report CR in p-InSb with $m^* = 0.016$ m. Other materials in which CR has been observed include Bi (KEYES et al.), HgCdTe alloy (DICKEY, unpublished) — Lax quoted a total of 13 materials by 1962.

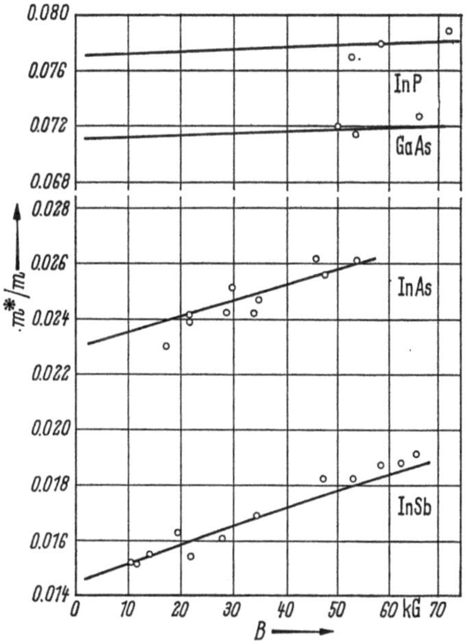

Fig. 12. Variation of effective mass with magnetic field for several n-type III—V compounds (after PALIK, TEITLER and WALLIS[4]).

A further type of cyclotron resonance in spatially inhomogeneous electromagnetic fields, when the ratio of cyclotron radius r to skin depth δ is around unity, has been observed in a semiconductor by STILES et al. (1962)[5]. This is known as Azbel'-Kaner resonance and was observed in p-type PbTe at microwave frequencies. Useful band structure information can be obtained by this method.

15. Free carrier Faraday effect. This effect is caused by the dispersion associated with the cyclotron resonance absorption of free carriers as can be readily seen from the form of the Kramers-Kronig relation given in Eq. (7.5). The rotation is given in terms of conductivity by Eq. (5.6) as

$$\Theta = \frac{\omega l}{2c}(n_- - n_+) = \frac{\sigma^R_{xy}}{2nc}. \tag{15.1}$$

[1] B. LAX: Proc. Int. Conf. on Semiconductor Physics, Prague 1960, p. 322.
[2] E. D. PALIK, and R. F. WALLIS: Phys. Rev. **123**, 131 (1961).
[3] E. D. PALIK, J. R. STEVENSON, and R. F. WALLIS: Phys. Rev. **124**, 701 (1961).
[4] E. D. PALIK, S. TEITLER, and R. F. WALLIS: J. Appl. Phys. **32**, 2132 (1961).
[5] P. J. STILES, E. BURSTEIN, and D. N. LANGENBERG: Proc. Int. Conf. on Physics Semiconductors, Exeter 1962, p. 577.

It is therefore possible to obtain a relation between Θ and the band structure parameters by using the Boltzmann expression [Eq. (13.6)] to find this component of the conductivity tensor in the low field case.

For the case of spherical energy surfaces and a parabolic band, the Boltzmann expression reduces to single particle classical dynamics as in the case of CR. The indices n_- and n_+ may then be deduced directly from Eq. (1.2) with $\omega_c=0$ and with the same approximation as Eq. (5.6) yields

$$\Theta=\frac{B N e^3 l}{2c\,\varepsilon_0\, m^{*2}\omega^2} \tag{15.2}$$

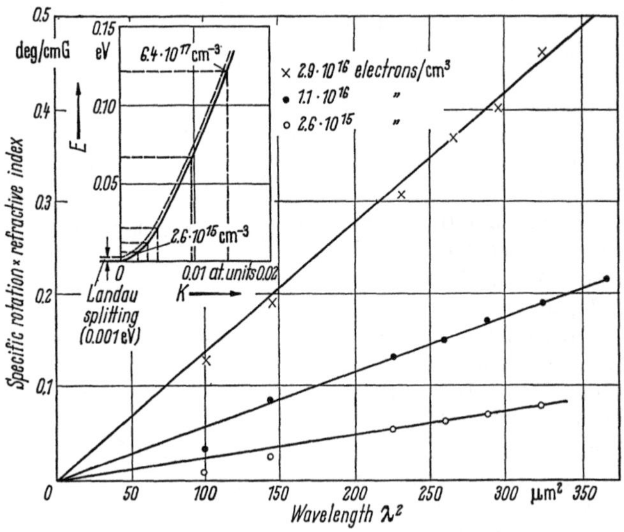

Fig. 13. Free carrier Faraday rotation in n-type InSb. Inset Fermi levels in variously doped specimens (after SMITH et al., Ref. 1, p. 268).

which is the expression obtained by MITCHELL (1955)[1]. The dependence of Θ upon the inverse square of the mass implies that Faraday rotation is an accurate method for determination of effective mass, and as noted by MITCHELL, the relaxation time τ does not appear in the expression. This is a consequence of choosing the meaning frequency $\omega \gg \omega_c$ the cyclotron frequency. The *infra-red* free carrier effect for which this applies must be distinguished from the *microwave* effect which depends upon the square of the relaxation time and gives information about the mobility ($e\tau/m^*$) rather than the mass of the carriers[2].

We may obtain further insight into the problem by noting that we may obtain Eq. (15.2) by inserting values of $\alpha_\pm$ in Eq. (7.5) where $\alpha_\pm$ are obtained from Eq. (14.3) and the corresponding equation for the non-resonant component $(\omega+\omega_c)$. Eq. (14.3) gives a Lorentzian line shape with half-width $1/\tau$. In the approximation quoted, the *integral* of the absorption over this line is independent of τ. Thus τ can be chosen to be so short that the line width is large and the CR would be unobservable, i.e., $\omega_c\,\tau>1$ does not have to be satisfied. The Faraday rotation will still be observable and given by Eq. (15.2) under these conditions — a valuable property as high temperatures and large impurity content (factors which make τ small) can be tolerated.

[1] E. W. J. MITCHELL: Proc. Phys. Soc. (London) B **68**, 973 (1955).

[2] See, for example, R. RAU, and M. E. CASPARI, Phys. Rev. **100**, 632 (1955), and J. K. FURDYNA, and S. BROERSMA, Phys. Rev. **120**, 1995 (1960).

The earliest infra-red free carrier measurements, by SMITH and MOSS in 1958[1], showed that the rotation was accurately proportional to B and to λ^2 as predicted by Eq. (15.2) and that the rotation could be easily measured (see Fig. 13). The simplicity and flexibility of the Faraday effect experiment has subsequently resulted in the performance of many more such experiments than the more difficult cyclotron resonance — some 20 materials having been examined by 1965.

Given that the rotation is accurately proportional to magnetic field and to the square of the wavelength, the effective mass deduced can be interpreted more generally in the case of non-spherical energy surfaces and/or non-parabolic bands by comparison between (15.2) and the appropriate Boltzmann theory for the rotation. The success of the early experiments stimulated STEPHEN and LIDIARD[2] to calculate expressions for the rotation due to free carriers in this from and to give general results for cubic crystals. Use is made of the tensor notation of ABELÉS and MEIBOOM (1954)[3] in which the current density is written

$$\left.\begin{aligned} \boldsymbol{j} &= \varepsilon_{ij} E_j + \varepsilon_{ijk} E_j H_k \\ &= \sigma\, \delta_{ij} E_j + \sigma\, \varepsilon_{ijk} E_j H_k \end{aligned}\right\} \tag{15.3}$$

where δ_{ij} is the Kronecker delta and

$$\left.\begin{aligned} \varepsilon_{123} = \varepsilon_{321} = \varepsilon_{312} = +1 \\ \varepsilon_{213} = \varepsilon_{132} = \varepsilon_{321} = -1 \end{aligned}\right\} \text{ all others zero.}$$

The operator $\operatorname{grad}_k E \times \operatorname{grad}_k$ in Eq. (13.6), now becomes

$$\nabla_k E\, \nabla_k = \varepsilon_{ikl} \frac{\partial E}{\partial k_k} \cdot \frac{\partial}{\partial k_l} \tag{15.4}$$

so that φ [Eq. (13.6)] can be expanded to second order in B as

$$\begin{aligned} \varphi = -\frac{e}{\hbar}\Big[\tau \boldsymbol{E}\cdot \operatorname{grad}_k E - \frac{e}{\hbar^2 c}\Big\{\tau \boldsymbol{B}\cdot \varepsilon_{ikl} \frac{\partial E}{\partial k_k} \cdot \frac{\partial}{\partial k_l} (\tau \boldsymbol{E}\cdot \operatorname{grad}_k E)\Big\} + \\ + \frac{e^2}{\hbar^4 c^2}\, \tau \boldsymbol{B}\cdot \varepsilon_{ikl} \frac{\partial E}{\partial k_k} \cdot \frac{\partial}{\partial k_l}\Big\{\tau \boldsymbol{B}\cdot \varepsilon_{ikl} \frac{\partial E}{\partial k_k}\, \frac{\partial}{\partial k_l} (\tau \boldsymbol{E}\cdot \operatorname{grad}_k E)\Big\} + \cdots\Big]. \end{aligned}$$

Thus substituting in Eq. (13.6) and comparing with (15.3) σ and σ' may be found. Since

$$\Theta = \frac{\sigma_{xy}^R\, l}{2nc} = \frac{\sigma'^R\, l}{2n\, \varepsilon_0\, c}$$

relating to Eq. (5.6), we obtain the general expression for Faraday rotation in a cubic crystal:

$$\frac{n\Theta}{Bl} = \frac{e^3}{8\pi^3 \hbar^4 \varepsilon_0 c\, \omega^2} \int \frac{\partial f_0}{\partial E}\, \frac{\partial E}{\partial k_x} \left[\frac{\partial E}{\partial k_y}\, \frac{\partial^2 E}{\partial k_x \partial k_y} - \frac{\partial E}{\partial k_x} \cdot \frac{\partial^2 E}{\partial k_y^2}\right] d^3 \boldsymbol{k}. \tag{15.5}$$

The carrier density, N, can be included in (15.5) from the relation

$$N = \frac{2}{(2\pi)^3} \int_0^\infty f_0\, d^3 \boldsymbol{k} \tag{15.6}$$

---

[1] S. D. SMITH, and T. S. MOSS: Brussels Solid State Conference, Vol. II, p. 671. New York: John Wiley & Sons 1958.

[2] M. J. STEPHEN, and A. B. LIDIARD: J. Phys. Chem. Solids 9, 43 (1958).

[3] B. ABELÉS, and S. MEIBOOM: Phys. Rev. 95, 31 (1954).

so that Eqs. (15.5) and (15.2) can be compared. This comparison gives an interpretation of the effective mass measured in the Faraday rotation experiments. We note that so far no assumption has been made about the form of the $E-\boldsymbol{k}$ relation; this relation is required to perform the integral in Eq. (15.5) and so evaluate the Faraday rotation. There are four particular cases of interest:

(i) When E has a quadratic dependence on $\boldsymbol{k}$ and the energy surfaces are spherical. Comparison between (10.9) and (11.2) and including (11.3) then gives

$$\frac{1}{m^*} = \frac{1}{\hbar^2} \frac{d^2 E}{d k^2} \tag{15.7}$$

and the result is independent of the distribution function f, i.e., electrons in all the populated states behave with the same effective mass since d^2E/dk^2 is a constant in a parabolic band.

(ii) With a quadratic dependence of E upon k, but with (say) ellipsoidal energy surfaces such as for electrons in germanium. A combination of transverse and longitudinal masses is then measured in the Faraday effect given by

$$\frac{1}{m^*} = \left[\frac{K(K+2)}{3}\right]^{\frac{1}{2}} \frac{1}{m_l} \tag{15.8}$$

where $K = m_l/m_t$.

(iii) In a semiconductor with spherical energy surfaces in which the carrier concentration and temperature are such that $d f_0/dE$ is finite only at the Fermi level, E_F, and negligible elsewhere and that the Fermi level lies within one of the bands. Under these circumstances (15.5) yields

$$\frac{1}{m^*} = \frac{1}{\hbar^2 k_F} \left(\frac{\partial E}{\partial k}\right)_{k_F}. \tag{15.9}$$

The carrier distribution in such cases is said to be "degenerate". This result is very important since with the degenerate distribution the wave vector at the Fermi surface, k_F, can be found from the carrier density, N, by the relation

$$N = \frac{2}{(2\pi)^3} \cdot \frac{4\pi k_F^3}{3} \tag{15.10}$$

and so can be found from a Hall effect experiment. Thus values of k_F and $(\partial E/\partial k)_{k_F}$, the slope of the band at k_F can be found from the Faraday rotation. Hence the $E-\boldsymbol{k}$ relation can be determined directly over a range of k by varying the doping and progressively filling up the band since this result holds when the band is non-parabolic. In the parabolic case the result reduces to that in (i), Eq. (15.7).

(iv) If the band is non-parabolic and the distribution non-degenerate, df/dE is finite over a significant range of the $E-\boldsymbol{k}$ curve. Thus electrons at different energies have different effective masses and an average is measured. Evaluation of the integrals in (15.5) is now needed and requires the insertion of a known relation between E and $\boldsymbol{k}$.

Interpretation of Faraday rotation under conditions (i) is straightforward; probably the most useful case is (iii) as the band shape can be directly determined (by numerical integration of the experimentally determined quantity dE/dk) *without the use of a theory of the band shape*. The results can then be used as a direct test of such theory. Such a comparison is also important for comparison and checking with high field experiments, such as CR, in non-parabolic bands, the interpretation of which requires both use of a band shape theory and its extension to the case of high magnetic fields.

The most favourable material for such investigation is InSb since CR results have shown that the energy surfaces for electrons are spherical[1] and the small effective mass ensures a low density of states. Such an experiment has been carried out by SMITH, MOSS and TAYLOR (1959)[2] and further refined by SMITH, PIDGEON and PROSSER[3]. The range of carrier and Fermi levels is shown in Fig. 13, and the experimental values of dE/dk against k in Fig. 14. An interesting feature of this plot is that the slope of the dE/dk curve at $\boldsymbol{k}=0$, which is of course d^2E/dk^2, gives the mass at the band minimum, where in the limit the band will be parabolic. This gives a useful technique for finding the zero-k mass.

The results are compared with the $\boldsymbol{k}\cdot\boldsymbol{p}$ perturbation calculation of KANE[4] which yields the result

$$E'(E'-E_G)(E'+\Delta)-k^2 P^2(E'+2\Delta/3)=0 \tag{15.11}$$

when the conduction-valence band interaction is treated exactly and the interactions with higher band are neglected, and

$$E'=E-\frac{\hbar^2 k^2}{2m}$$

where m is the electron mass. The parameters P, E_G, Δ are determined from experiment. The energy gap, E_G, may be found from absorption edge analysis or, better, from IMO measurements (Sect. 21) at low temperatures where the $\boldsymbol{k}\cdot\boldsymbol{p}$ theory is appropriate. At higher temperatures, as will be seen later, some corrections to the optically measured energy gap are required. The expression is insensitive to Δ, the spin-orbit splitting and values derived from atomic data may be used without significant error. The parameter P is the momentum matrix element between the valence and conductor states and is related to the previously defined quantity p_{cv} [Eq. (10.3)]

$$p_{cv}=-\sqrt{\tfrac{2}{3}}\,P. \tag{15.12}$$

Determination of the band shape is therefore essentially a determination of the matrix element P. P may be found directly from the zero-k mass since for small k (parabolic region) Eq. (15.11) may be expanded in powers of k. Retaining only the term in k^2

$$E=\left\{\frac{\hbar^2}{m_0^{*2}}-\frac{\hbar^2}{m^2}\right\}k^2=\frac{2P^2}{3}\left\{\frac{2}{E_G}+\frac{1}{E_G+\Delta}\right\}k^2 \tag{15.13}$$

whence neglecting $1/m^2$ in comparison with $1/m_0^{*2}$ where m_0^* is the effective mass at $\boldsymbol{k}=0$:

$$\frac{\hbar^2}{m_0^{*2}}=\frac{2P^2}{3}\left(\frac{2}{E_G}+\frac{1}{E_G+\Delta}\right). \tag{15.14}$$

Alternatively P may be found from a best fit over a range of the $E-k$ curve. Such a fit is shown in Fig. 14 from an analysis by PIDGEON[5]. A very good fit between experiment and theory is obtained yielding a value of $P^2=0.395$ atomic units and corresponding to a zero-k mass 0.0145 m. This differs from the values originally taken by KANE (0.44 a.u.) based on the zero-k microwave CR value of 0.013 which is probably in error. The experiment therefore confirms with some precision the accuracy of the $\boldsymbol{k}\cdot\boldsymbol{p}$ formula and gives an accurate value for P^2.

[1] An anisotropy of early $2\frac{1}{2}\%$ has recently been reported at electron densities of 3×10^{18} cm^{-3} [(G. A. ANTCHIFFE and R. A. STRADLING, Physics Letters **20**, 119 (1966)].

[2] S. D. SMITH, T. S. MOSS, and K. W. TAYLOR: J. Phys. Chem. Solids **11**, 131 (1959).

[3] S. D. SMITH, C. R. PIDGEON, and V. PROSSER: Proc. Int. Conf. on Physics of Semiconductors, Exeter, 1962, p. 301.

[4] E. O. KANE: J. Phys. Chem. Solids **1**, 249 (1957).

[5] C. R. PIDGEON: Ph. D. Thesis, University of Reading. 1962.

An important comparison has been made by the NRL group between high field CR and Faraday rotation determinations of conduction band shape in InSb[1, 2], over the region in which both experiments overlap (the Faraday measurement extends considerably further into the band). They find the measured mass to be the same when

$$[E-E_F]=[1+\tfrac{1}{2}(k_0/k_2)]\hbar\,\omega_c$$

where $k_0/k_2=0.41$ for InSb. Since this depends upon an expansion for E of the form

$$E=A\,k^2+B\,k^4$$

this result only applies for rather low values of field (up to ~30 kgauss) since such a form can be shown to be inadequate higher in the band. Nevertheless satisfactory agreement between the two sets of results is obtained within experimental error, particularly with a "best fit" Faraday rotation curve (Fig. 11). The comparison shows that the high field extension of the $\boldsymbol{k}\cdot\boldsymbol{p}$ theory of BOWERS and YAFET provides a satisfactory interpretation of high field CR results in agreement with the low field Faraday results which give the zero-field band shape directly. The two sets of experiments imply a zero-k mass in range (0.0145 ± 0.0002) m for conduction electrons in InSb at 77° K.

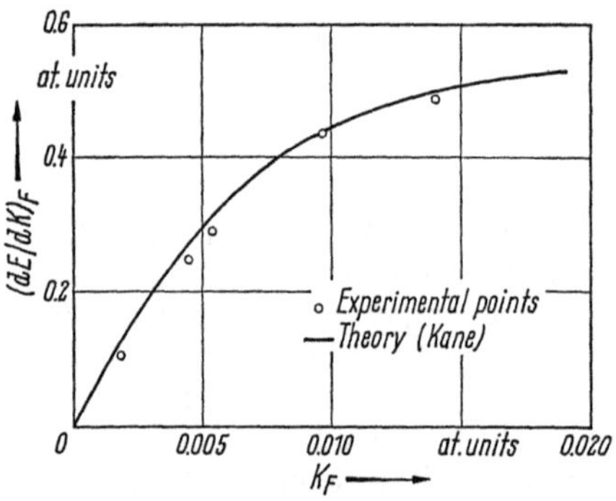

Fig. 14. Curvature of conduction band in InSb. The solid line represents the best fit to KANE's $k\cdot p$ theory giving a value of $P^2=0.395$ a. u. (after PIDGEON, Ph. D. thesis University of Reading 1962).

At levels deep in the band, where the mass has risen to values ~0.03 the Faraday results agree well with masses deduced from free carrier dispersion by SPITZER and FAN[3]. In the degenerate case the same quantity $\left(\frac{1}{k}\frac{dE}{dk}\right)_{k_F}$ is determined, the free carrier dispersion is, however, only separable at high doping levels where it is large compared with other contributions to dispersion.

Faraday rotation due to conduction electrons in GaAs and InP has been measured by MOSS and WALTON[4] (1959) and in GaAs and InAs by CARDONA (1961)[5]. For GaAs the former obtain, at room temperature,

$$m^*/m=0.072\,(+0.008-0.005)$$

but found inconsistent behaviour with doping in the range $3-7\times10^{16}\,\text{cm}^{-3}$ although Cardona obtains an increase of mass with doping. For InP, a mass ratio of 0.075 ± 0.008 was obtained at $N=1.07\times10^{16}\,\text{cm}^3$. Both these results are consistent with subsequent CR experiments of PALIK, TEITLER and WALLIS (1961) shown in Fig. 12. No Boltzmann-type analysis was performed on these experiments so that the exact part of the band examined is not determined. However the donor concentrations are sufficiently low to ensure that the regions are much nearer the band minima than the high field CR and Fig. 12. shows that over this

---

[1] E. D. PALIK, G. S. PICUS, S. TEITLER, and R. F. WALLIS: Phys. Rev. **122**, 475 (1961).
[2] E. D. PALIK, S. TEITLER, and R. F. WALLIS: J. Appl. Phys., Suppl. to **32**, 2132 (1961).
[3] W. G. SPITZER, and H. Y. FAN: Phys. Rev. **106**, 882 (1957).
[4] T. S. MOSS, and A. K. WALTON: Proc. Phys. Soc. **74**, 131 (1959); — Physica **25**, 1142 (1959).
[5] M. CARDONA: Phys. Rev. **121**, 756 (1961).

range the bands are not markedly nonparabolic in GaAs and InP. This is in agreement with the prediction of $\boldsymbol{k}\cdot\boldsymbol{p}$ theory due to the larger values of E_G in these materials.

Temperature dependence of effective mass. The flexibility of the Faraday rotation experiment due to its independence from the scattering time τ has been exploited to determine the temperature dependence of effective mass. Experiments and analysis have been performed by CARDONA (1961)[1] on n-type GaAs and InAs in the range 77—300° K and by SMITH, PIDGEON and PROSSER[2] on n-type InSb in the range 5—300° K. At the higher temperatures the condition is as in (iv), at least for low donor concentrations. It is then necessary to insert the function $E(\boldsymbol{k})$ in (15.5) to evaluate the integral.

There are two origins for the temperature dependence of mass: (a) from Eq. (15.13), the zero-k mass changes as the energy gap, E_G, changes with temperature. Since, for most materials, E_G decreases with increasing temperature, so then does m^*. (b) At higher temperatures the electrons are distributed over a greater range of energies in the band; thus in a non-parabolic band more electrons are at higher levels where the mass is larger. The two effects therefore compete.

Cardona develops expansions from KANE's $\boldsymbol{k}\cdot\boldsymbol{p}$ expression (15.11) appropriate for GaAs and InAs in the form

$$E = -E_G/2 + \left((E_{G/2})^2 + k^2 P^2\right)^{\frac{1}{2}}, \qquad \text{(GaAs)}$$

$$E = -E_G/2 + \left((E_{G/2})^2 + k^2 \gamma^2\right)^{\frac{1}{2}} \qquad \text{(InAs)}$$

where

$$\gamma^2 = P^2 \left\{ \frac{E_G + \frac{2}{3}\Delta}{E_G + \Delta} \right\}$$

and hence analytically evaluates the problem obtaining, for GaAs

$$\frac{1}{m^*(T)} = \frac{1}{m_0^*} \left[1 - \frac{10}{3} \frac{kT}{E_G} \frac{\int_0^\infty x^{\frac{3}{2}} f(x-\eta)\, dx}{\int_0^\infty x^{\frac{1}{2}} f(x-\eta)\, dx} \right]$$

where f is the Fermi function and $\eta = E_F/kT$. The net increase of m^* in GaAs is thus estimated at $\sim 3\%$ between 100 and 300° K.

Experimentally, he finds a $(2\pm 2)\%$ increase in m^* for GaAs with $N = 1.48\times 10^{17}$ in good agreement with theory. For InAs he finds an experimental 6% increase between 77 and 300 °K. CARDONA's results were taken over a rather limited wavelength range near the absorption edge and recent more detailed measurements by SUMMERS (1965)[3] have shown that the change is in fact 12% over the same temperature range (Fig. 15). Theory predicts too large a change (16%) if the optically determined energy gaps at 77 and 300° K are used for the InAs case.

For InSb, the expansion $E = A k^2 + B k^4$ is quite inadequate and SMITH, PIDGEON and PROSSER use a numerical method with a more general band shape[4]. This method has general application. In the limit $\Delta \gg k^2 P^2$ and E_G

$$E = \alpha k^2 - \tfrac{1}{2} E_G + \tfrac{1}{2} (E_G^2 + \beta k^2)^{\frac{1}{2}}$$

[1] See footnote 5, p. 274.

[2] See footnote 3, p. 273.

[3] C. J. SUMMERS: Ph. D. Thesis, University of Reading 1965.

[4] Independently, J. KOLODZIEJCZAK: Acta Phys. Polon. **21**, 637 (1962), has given a theoretical discussion of this problem. The application of generalised Fermi-Dirac integrals is described by W. ZAWADZKI, R. KOWALCZYK, and J. KOLODZIEJCZAK, Phys. Staat. Sol. **10**, 513 (1965), and in earlier papers referred to.

with $\alpha = \hbar^2/2m$, $\beta = 8P^2/3$. Thus obtaining partial differentials (15.5) gives

$$\frac{n\Theta}{Bl} = \frac{4e^3}{3\pi\hbar^4 c\,\varepsilon^2 kT} I_1$$

where

$$I_1 = \frac{1}{2}\int_0^\infty \frac{k^4[\alpha + \frac{1}{4}\beta(E_G^2 + \beta k^2)^{-\frac{1}{2}}]^3}{1 + \cosh[(E - E_F)/kT]}\,dk.$$

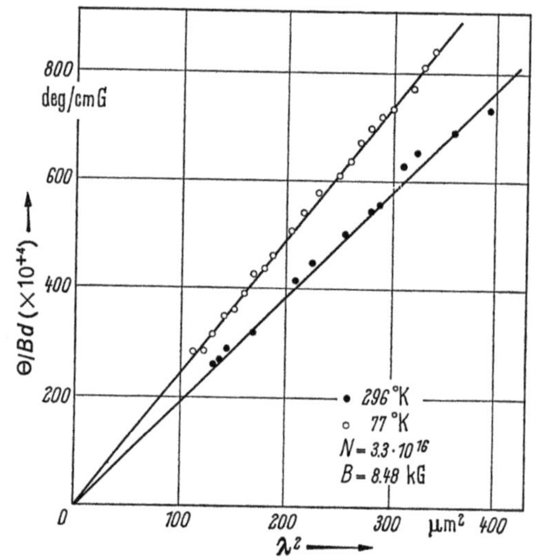

Fig. 15. Temperature dependance of free carrier Faraday rotation in InAs (after SUMMERS).

By including the carrier density N from Eq. (15.6) and comparing with (15.2), an expression for the effective mass, as measured is obtained:

$$\frac{1}{m^{*2}} = \left(\frac{8}{3}\hbar^4 kT\right)\frac{I_1}{I_2} \tag{15.15}$$

where

$$I_2 = \pi^2 N = \int_0^\infty \frac{k^2\,dk}{1 + \exp[(E - E_F)/kT]}.$$

Both I_1 and I_2 depend on the band shape and are evaluated numerically, varying P^2 and E_G for best fit. The process is sensitive to P^2 and yields this quantity to the same accuracy as the mass determination. At low doping levels the fitting process is independently sensitive to P^2 and E_G but at high levels insensitive to E_G. The low doping level results (Fig. 16) indicate a better fit if an "effective" energy gap E_G is used in KANE's formula. This is consistent with the dilational contribution to the energy gap alone being responsible for change in band curvature and is required by both sets of temperature dependent data described for both InSb and InAs.

The high pulsed field data of LAX et al.[1] yield a room temperature zero-k mass of 0.0116 m for InSb and are in apparent disagreement with the Faraday results which give 0.0135 m. The former value however is the result of a long extrapolation from data deep within the band, using an analysis similar to the Bowers and Yafet treatment. The low value of m_0^* arises from use of the optical gap $E_G = 0.18$ eV (instead of $E_G^* = 0.21$ eV) in Eq. (15.14). Since the fit at deep

[1] B. LAX, J. MAVROIDES, H. J. ZEIGER, and R. J. KEYES: Phys. Rev. **122**, 31 (1961).

levels is *insensitive* to E_G, this experiment could be fitted with various values of E_G and so cannot determine m_0^* accurately by (15.14). This does not therefore invalidate the above conclusions. Temperature dependent CR results from NRL also support the conclusions of the Faraday experiments.

Temperature dependence of effective mass in n-type InSb, InAs, GaAs, Ge and Si has also been studied at temperatures above room temperature using the Faraday rotation method. UKHANOV and MAL'TSEV[1] have made measurements in the range 117 to 600° K and using a simple analysis according to Eq. (15.2) find an increase of mass for all materials except InAs in the range 10—20%. For

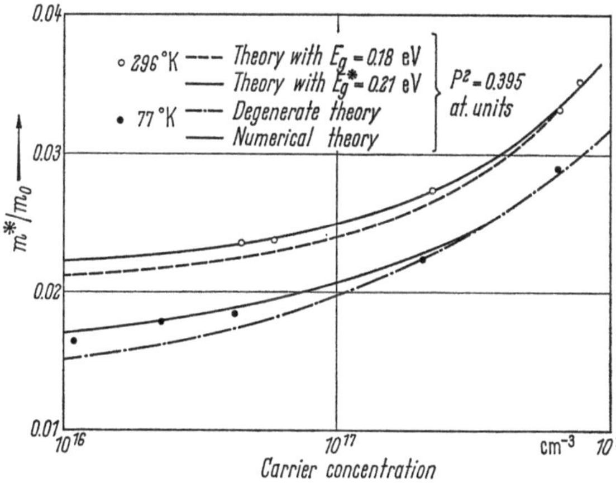

Fig. 16. Temperature and impurity dependance of m^* in n-type InSb (after SMITH, PIDGEON and PROSSER, Ref. 3, p. 273).

InAs the mass increases up to 370° K and then decreases upon further heating. No Boltzmann-type analysis is attempted.

All the rotation experiments so far discussed except the last, have been performed on conduction electrons in bands assumed to have spherically symmetric energy surfaces. In some cases this has been checked by CR experiments; any departure from such symmetry is not detectable by low field Faraday effect.

Non-spherical energy surfaces. Rotation in n-type germanium has been measured by WALTON and MOSS (1959)[2] and by HARTMAN and KLEMAN (1960)[3]. The energy surfaces are ellipsoids, as determined from CR (Sect. 3.2), and in these circumstances the low-field Faraday effect is isotropic in such a cubic crystal and a combination of transverse and longitudinal masses given by (15.8) is measured[4]. The experiment therefore gives no further band structure information beyond that obtained from CR but the mass obtained $m^* = (0.135 \pm 0.004)$ m is consistent with that calculated from the CR masses which give 0.134 m so checking the theory. The experiment is probably most usefully interpreted as giving information about the form of the Hall constant R, from which the carrier density is determined. When degenerate, the Hall constant is given by

$$R = \frac{1}{Ne}$$

[1] YU. I. UKHANOV, and YU. V. MAL'TSEV: Soviet Phys. — Solid State 5, 2144 (1964) and earlier papers quoted.

[2] A. K. WALTON, and T. S. MOSS: J. Appl. Phys. **30**, 351 (1959).

[3] B. HARTMAN, and B. KLEMAN: Arkiv Physik **18**, 75 (1960).

[4] A. K. WALTON and C. R. EVERETT have however recently shown that information on anisotropy *can* be obtained on applying uniaxial stress [Solid State Comm. **4**, 222 (1966)].

but under other conditions is modified by a scattering factor, depending upon mobility, which can vary in the range 1.3—1.0. Thus m^*, depending upon $N^{\frac{1}{3}}$, could be in error by as much as 15% if the factor is left undetermined. In most experiments the uncertainty from the carrier density is less than 5% and is probably the biggest source of error in Faraday effect determinations of mass.

Free carrier rotation in GaSb is reported by PILLER (1964)[1]. In this case two minima of the conduction band can be populated, one at $\boldsymbol{k}=0$, [0, 0, 0] and one, very similar to germanium, in ⟨111⟩ direction. The total rotation is then a sum:

$$\Theta = \frac{B e^3 l}{2 n c \varepsilon_0 \omega^2}\left[\frac{N_0}{m_0^{*2}} + \frac{N_1}{m_1^{*2}}\right]$$

where N_0 N_1 are the relative carrier densities and m_0^* the mass in the [000] band. m_1^*, the mass in the second minimum is given by Eq. (15.8). Piller analyses data at two temperatures, which vary the relative concentrations and obtaining consistency with theory with $m_0^* = (0.049 \pm 0.004)$ m, $m_t = 0.14$ m and $m_l = 1.2$ m. The carrier density was $\sim 3.8 \times 10^{17}\ \mathrm{cm}^{-3}$. Other III—V compounds studied include GaP and AlSb both giving rather high effective masses[2].

Faraday rotation measurements have also been reported in II—VI compounds: in CdS, by BALKANSKI and HOPFIELD[3] (1962) and in heavily doped CdTe and ZnSe by MARPLE (1963, 1964)[4]. For CdS, which has a spherical conduction band with maximum at $\boldsymbol{k}=0$ a mass of 0.20 ± 0.01 was found. The paper contains an interesting discussion of polaron effects. MARPLE's data on CdTe and ZnSe are both obtained by treating combinations of Faraday rotation and reflectivity data. The latter determines the free carrier dispersion and Eq. (5) is then used to determine m^*. This technique had been previously demonstrated on InSb in the original work of SMITH and MOSS (1958) referred to earlier; it has particular application to high doping levels. In MARPLE's case these are in the range $0.5-15 \times 10^{19}\ \mathrm{cm}^{-3}$. Masses obtained, assuming a band minimum at $\boldsymbol{k}=0$, are

$$n\text{-type CdTe} \qquad m^* = (0.11 \pm 0.01)\ \mathrm{m},$$

$$n\text{-type ZnSe} \qquad m^* = (0.17 \pm 0.025)\ \mathrm{m}$$

with very little non-parabolicity.

Studies of the lead salts have been made by WALTON, MOSS and ELLIS (1962)[5] at 300 °K and by PALIK, TEITLER, HENVIS and WALLIS[6] at 300 and 77° K in n-type material. Masses are obtained as follows:

Table 2.

	N (cm^{-3})	T. °K	m^*/m	Observer
PbS . .	$1-2 \times 10^{19}$	300	(Av.) 0.176 ± 0.012	WME
	$1-17 \times 10^{17}$	300	0.16 ± 0.02	PTHW
	$1-17 \times 10^{17}$	77	0.12 ± 0.01	PTHW
PbSe . .	$1-2 \times 10^{17}$	300	(Av.) 0.114 ± 0.007	WME
PbTe .	3×10^{17}	300	0.086 ± 0.004	WME
	2×10^{18}	300	0.13 ± 0.008	WME

[1] H. PILLER: Proc. Int. Conf. on Physics of Semiconductors, Paris, 1964, p. 297. Paris: Dunod 1964.

[2] T. S. MOSS, A. K. WALTON, and B. ELLIS: Proc. Int. Conf. on Physics of Semiconductors, Exeter 1962.

[3] M. BALKANSKI, and J. J. HOPFIELD: Phys. Stat. Solidi 2, 623 (1962).

[4] D. T. F. MARPLE: Phys. Rev. **129**, 2466 (1963); J. App. Phys. **35**, 1879 (1964).

[5] A. K. WALTON, T. S. MOSS, and B. ELLIS: Proc. Phys. Soc. (London) **79**, 1065 (1962).

[6] E. D. PALIK, S. TEITLER, B. HENVIS, and R. F. WALLIS: Proc. Int. Conf. on Physics of Semiconductors, Exeter 1962, p. 288.

These results, which indicate that only PbTe shows effects of doping on apparent effective mass are deduced on the basis of Eq. (15.2) which implies spherical energy surfaces. However, more recent magneto-absorption and other studies (Sect. 21) have shown that the band minima occur at the zone boundary in these materials and give ellipsoidal energy surfaces. The above data should therefore be re-interpreted according to Eq. (15.8).

Bismuth telluride was the first anisotropic (uniaxial) material in which free carrier Faraday effect has been studied, observations on both *n*- and *p*-type material and theory being given by AUSTIN (1959, 1960)[1], considering particularly propagation along the optic axis and more recent examples are hexagonal SiC[2] and *n*-type tellurium[3]. Nearly all the measurements reported have been on *n*-type material. Effects of free holes have been observed in Ge[4], InSb[5] and diamond[6]. Rotations are usually very small and interpretation is difficult since both light and heavy holes (and in the case of diamond the split off band) contribute to the total rotation. In the latter case a rotation of only $\sim 1/100^\circ$ was found for a *p*-type specimen at 600° K with a hole concentration $\sim 1\times 10^{15}$, roughly consistent with the hole masses found by RAUCH (Sect. 4.2).

We may summarise this report on free carrier Faraday effect by noting that since the first observations in 1958 at least 20 materials have been studied and, more pertinently, most of them over a wide range of impurity content. The technique has therefore become established as perhaps the most flexible and widely used method of obtaining band structure parameters in semiconductors. These remarks apply to the low field non-resonant type of measurement characterised by independence from τ, being isotropic in cubic crystals and derived from the integral of all free carrier magneto-absorption (e.g., CR of carriers of various masses). None of these conditions apply to *resonant free carrier Faraday effect*. This has been observed by PALIK[7] and provides another method of observing the CR frequency ω_c. In this case the resonant frequency rather than the magnitude of the effect is the important quantity. Such an experiment also implies that the quantum conditions are required in treating the problem theoretically; this has been discussed by GUREVICH, IPATOVA and URICKIJ (1960)[8].

16. Faraday ellipticity. Faraday rotation is always, in principle, accompanied by the presence of some ellipticity, caused by differential attenuation of the two circularly polarised components making up the plane polarised incident wave. This is given, for small absorption, by Eq. (5.7) as

$$\Delta = \frac{\omega l}{2c}\,(\kappa_+ - \kappa_-)\,.$$

For the case of parabolic bands and spherical energy surfaces the single particle classical equation gives

$$\Delta = \frac{N e^3 B l (1/\tau)}{n c \varepsilon_0 m^{*2} \omega^3}\,. \tag{16.1}$$

[1] I. G. AUSTIN: J. Electronics **6**, 271 (1959); — Proc. Phys. Soc. (London) **76**, 169 (1960).
[2] T. S. MOSS, and B. ELLIS: Proc. Int. Conf. on Physics of Semiconductors, Paris 1964, p. 311.
[3] J. L. CALLIS, and C. RIGAUX: Proc. Int. Conf. on Physics of Semiconductors, Paris 1964, p. 305.
[4] A. K. WALTON, and T. S. MOSS: Proc. Phys. Soc. (London) **78**, 1393 (1961).
[5] S. D. SMITH, C. R. PIDGEON, and V. PROSSER: Proc. Int. Conf. on Physics of Semiconductors Exeter 1962,, p. 301.
[6] V. PROSSER, and S. D. SMITH: To be published.
[7] E. D. PALIK: Appl. Optics **2**, 527 (1963).
[8] L. E. GUREVICH, I. P. IPATOVA, and Z. I. URICHIJ: Proc. Int. Conf. on Semiconductor Physics, Prague 1960, p. 328.

This equation was implicitly given by AUSTIN (1959)[1], and gives the *amplitude* ratio of the axes of the ellipse (mistakenly quoted as the intensity ratio by AUSTIN).

The effect was observed by SMITH and PIDGEON (1960)[2] in *n*-type InSb, both free carrier and interband effects being noted. Combining Eqs. (16.1) and (15.2) we obtain the useful relation

$$\Theta/\Delta = \omega\,\tau/2 \tag{16.2}$$

showing that measurements of Θ and Δ yield the relaxation time τ, independently of the mass parameters. This is illustrated in plots of Θ vs. Δ taken from PIDGEON and SMITH (1963)[3] (Fig. 17) the differing slopes showing the temperature dependence of the relaxation time. In more complicated cases the Boltzmann treatment must be applied and complete separation of m^* and τ will not always be possible [see for example DONOVAN and WEBSTER (1961)[4]].

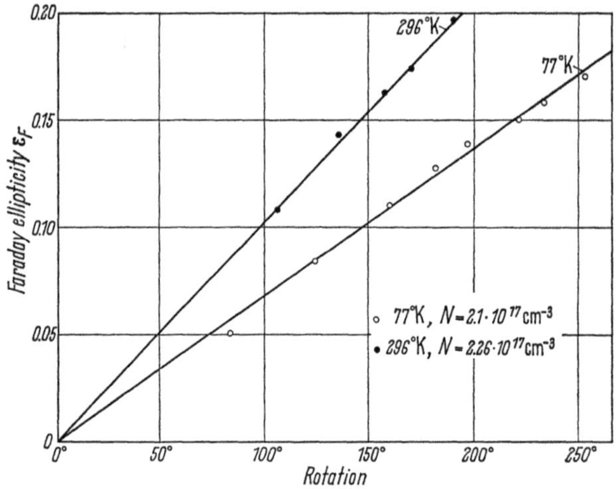

Fig. 17. Free carrier ellipticity and rotation in *n*-type InSb (after PIDGEON and SMITH [3]).

17. Voigt effect. The Voigt effect is magnetic birefringence and might also have been called the Cotton-Mouton effect. It arises from the difference between $n_\parallel$ and $n_\perp$ as defined in Sect. 5 δ and Eqs. (5.10), (5.11) and (5.12).

On applying classical theory in the usual approximations of parabolic bands, spherical energy surfaces and $\omega \gg \omega_c$, $\omega\,\tau \gg 1$ and also $\omega \gg \omega_p$ (where $\omega_p = n\,e^2/m^*\,\varepsilon_0$, the plasma frequency), we obtain for the Voigt phase shift

$$\delta = \frac{\omega l}{c}\,(n_\parallel - n_\perp) = \frac{N e^4\, l\, B^2\, \lambda^3}{16\pi^3 c^4\, n\, m^{*3}\, \varepsilon_0}\,. \tag{17.1}$$

This effect differs from the other non-resonant magneto-dispersive effect, Faraday rotation, in several ways: the direction of the magnetic field relative to the radiation, its dependence on magnetic field, effective mass and frequency.

The effect was first observed and Eq. (17.1) given by TEITLER and PALIK (1960[5] and 1961[5]). Further theoretical discussion has been given by CARDONA[6],

[1] I. G. AUSTIN: J. Electronics **6**, 271 (1959).
[2] S. D. SMITH, and C. R. PIDGEON: Proc. Int. Conf. on Semiconductor Physics, Prague 1960, p. 342.
[3] C. R. PIDGEON, and S. D. SMITH: Infra-red Physics **4**, 13 (1963).
[4] B. DONOVAN, and J. WEBSTER: Proc. Phys. Soc. (London) **78**, 170 (1961).
[5] S. TEITLER, and E. D. PALIK: Phys. Rev. Letters **5**, 546 (1960). — S. TEITLER, E. D. PALIK, and R. F. WALLIS: Phys. Rev. **123**, 1631 (1961).
[6] M. CARDONA: Helv. Phys. Acta **34**, 796 (1961).

WEBSTER and DONOVAN[1] and TEITLER[2]. Experimental results for InSb showing the dependence of δ on B^2 are shown in Fig. 18.

Two interesting possibilities arise from the difference between free carrier Faraday and Voigt effects. Firstly, combining Eqs. (17.1) and (15.2) we have

$$\delta/\Theta = \omega_c/\omega \tag{17.2}$$

with $\omega_c = \frac{e B}{m^*}$, so that isotropic masses could be determined by measurements of the two effects without the necessity to determine N, rather similarly to the combination of Θ and $dn/d\omega$ referred to earlier. The approximation given in (17.1) is rather restrictive and PALIK, TEITLER and WALLIS expand in terms of ω_c and ω_p to give

$$\delta/\Theta = \omega\, \omega_c/(\omega^2 - \omega_p^2)$$

a more general result, and then use an iterative procedure to deduce masses for n-InSb and n-InAs. Eq. (17.2), however, shows that δ is much smaller than Θ for small magnetic fields and is consequently difficult to measure in the region in which low field theory is valid. The technique is therefore rather restricted in application.

The second possibility lies in the differing tensor components responsible for the Voigt effect, which, related to the squared dependence on magnetic field, implies that anisotropic effects must be present for cubic semiconductors with ellipsoidal energy surfaces. CARDONA, for example, noted that, in consequence, by measuring Voigt effect as a function of sample orientation more than one component of the mass tensor can be determined by this low field non-resonant technique. TEITLER[2] however in a detailed discussion concludes that the application will be rather limited due to the small size of the effect. One experimental determination of energy surfaces in germanium using this technique has been reported by PALIK[3].

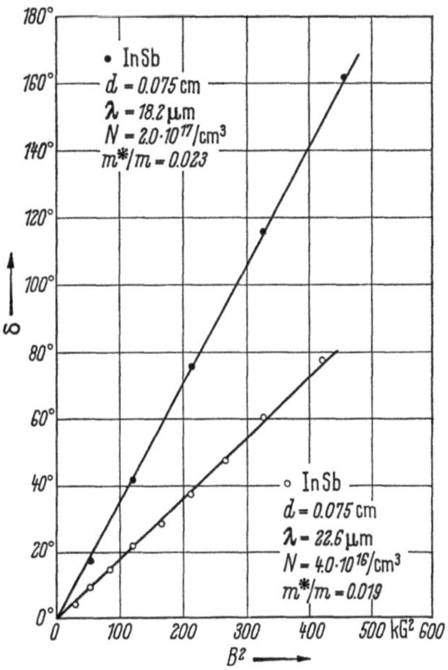

Fig. 18. Voigt shift in n-type InSb at liquid nitrogen temperature (after TEITLER and PALIK, Ref. 5, p. 280).

18. Magneto-reflection phenomena.

α) *Magneto-plasma reflection.* We obtain expressions for these effects by substituting in Eqs. (5.13) and (5.14) for $n_\pm$ and $n_\parallel$, $n_\perp$ assuming that $k^2 \ll n^2$ and can be neglected.

In zero field, with the usual classical isotropic mass theory, we may write for the refractive index

$$n^2 = \varepsilon_R \left(1 - \frac{N e^2}{m^* \varepsilon_0 \omega^2}\right) \tag{18.1}$$

[1] J. WEBSTER, and B. DONOVAN: Phys. Letters 2, 330 (1962).
[2] S. TEITLER: J. Phys. Chem. Solids 24, 1487 (1963).
[3] E. D. PALIK: J. Phys. Chem. Solids 25, 167 (1964).

where ε_R is the dielectric constant contributed by non-free carrier effects — e.g., interband and lattice vibrational transitions. Substituting

$$\omega_p^2 = \frac{N e^2}{m^* \varepsilon_0}$$

where ω_p is called the plasma frequency (18.1) becomes

$$n^2 = \varepsilon_R \left(1 - \frac{\omega_p^2}{\omega^2}\right). \tag{18.2}$$

As the frequency is decreased from large ω, a point is reached where $n=1$, at which frequency

$$R = \left(\frac{n-1}{n+1}\right)^2 \to 0.$$

From (18.2) this reflectivity minimum occurs at a frequency $\omega_{\min}$ given by

$$\omega_{\min} = \frac{\omega_p}{(1 - 1/\varepsilon_R)^{\frac{1}{2}}} \simeq \omega_p \left(1 + \frac{1}{2\varepsilon_R}\right). \tag{18.3}$$

A small further decrease in ω (3% in InSb doped at $N=10^{17}$ cm^{-3}) causes the reflectivity to rise steeply to unity as $\omega \to \omega_p$ and $n \to 0$. Experimental determination of $\omega_{\min}$ and hence ω_p from (18.3) then leads to the effective mass from $\omega_p^2 = N e^2/m^* \varepsilon_0$ requiring auxiliary experiments to determine N, the carrier density and ε_R the residual dielectric constant. SPITZER and FAN[1] have used this zero field technique on highly doped semiconductors and, as discussed in Sect. 4.3, it is effective when combined with Faraday rotation measurements.

The elegant extension of the technique, using magnetic fields, is due to LAX and WRIGHT (1960, 1961)[2,3]. For magnetic field normal to the reflecting surface we have "Faraday" configuration and the indices of refraction become

$$n_\pm^2 = \varepsilon_R \left[1 - \frac{\omega_p^2}{\omega(\omega \mp \omega_c)}\right]. \tag{18.4}$$

From this result we now have two frequencies for which $n_\pm = 0$ given by the positive solutions of

$$\omega^2 \mp \omega\,\omega_c - \omega_p^2 = 0$$

and the same result is obtained for the Voigt configuration.

The experimentally useful quantities are the zero reflection ($n=1$) frequencies which are

$$\omega_\pm = \omega_p \left\{1 + \frac{1}{2\varepsilon_R} \pm \frac{\omega_c}{2} + \frac{1}{8}\,\frac{\omega_c^2}{\omega_p}\left(1 - \frac{1}{2\varepsilon_R}\right)\right\}.$$

There are now *two* minima, separated by a frequency ω_c, apart from a small correction term and from this separation the effective mass may be found ($\omega_c = e B/m^*$). The experiment has the advantage of being independent of any auxiliary electrical measurements but it is usually necessary to use high doping levels to bring ω_p into a conveniently accessible spectral region. Results for n-InSb and n-HgSe are shown in Fig. 19 and 20. Effective masses for the former agree with other methods and for the latter a value of

$$m^*/m = 0.045 \pm 0.003$$

is obtained when $N = 2.45 \times 10^{18}$ cm^{-3}.

[1] W. G. SPITZER, and H. Y. FAN: Phys. Rev. **106**, 882 (1957).
[2] B. LAX, and G. B. WRIGHT: Phys. Rev. Letters **4**, 16 (1960).
[3] G. B. WRIGHT, and B. LAX: J. Appl. Phys. Suppl. **32**, 2113 (1961).

β) Rotation and ellipticity effects with polarised radiation. The relations giving these effects were given in Sect. 5, in Eqs. (5.12) to (5.15) on substituting for $n_\pm$ $\kappa_\pm$ from a more general form of (18.1), viz

$$(n - i\kappa)^2_\pm = \varepsilon_R \left[1 - \frac{\omega_p^2}{\omega(\omega \mp \omega_c - i/\tau)}\right]. \tag{18.5}$$

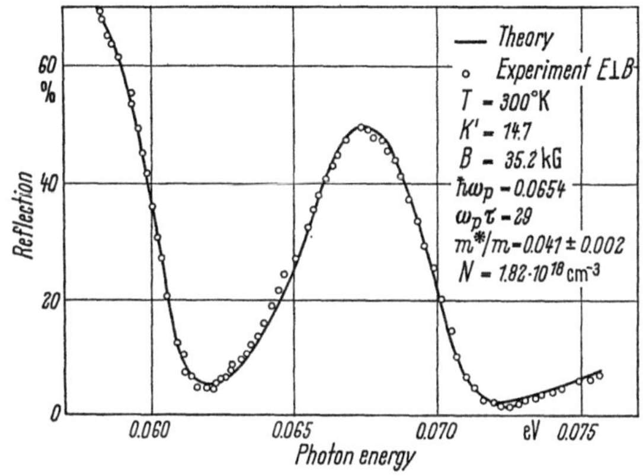

Fig. 19. Transverse magnets plasma reflection in n-InSb (after WRIGHT and LAX Ref. 3, p. 282).

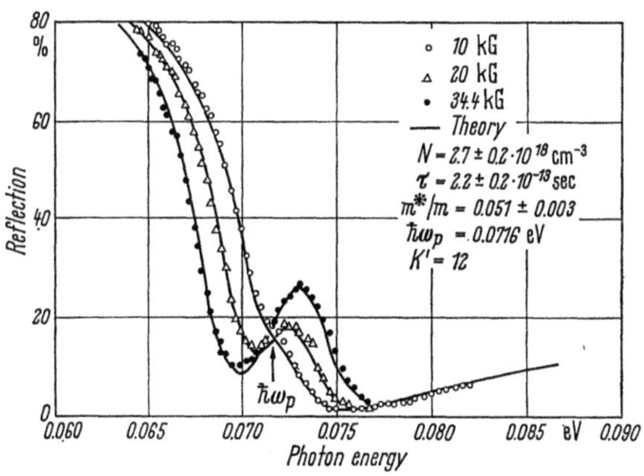

Fig. 20. Transverse magneto plasma reflection in HgSe (after WRIGHT and LAX, Ref. 3, p. 282).

The effects which arise from the differential reflectivities and phase shift were observed by PALIK, TEITLER, HENVIS and WALLIS (1962)[1] in both longitudinal (Fig. 21) and transverse orientations. *Rotation* is due to phase shifts, arising from $\kappa_\pm$ and therefore τ, in the longitudinal case. Ellipticity originates from differential reflectivity, correspondingly $n_\pm$, and therefore N/m^*. The opposite occurs, but with incomplete separation, in the transverse configuration. The results are fitted satisfactorily by appropriate values of N, m^* and τ.

All the reflection phenomena discussed here have been analysed by the classical theory; it should be pointed out that a Boltzmann analysis as described for the Faraday effect in Sect. 15 must be used if band parameters are to be properly deduced from such measurements. Magneto-reflection techniques,

[1] E. D. PALIK, S. TEITLER, B. W. HENVIS, and R. F. WALLIS: Proc. Int. Conf. on Physics of Semiconductors, Exeter 1962, p. 288.

which tend to be used with rather highly doped specimens, have some points of experimental convenience but so far are rather less used than, for example, low field Faraday effect.

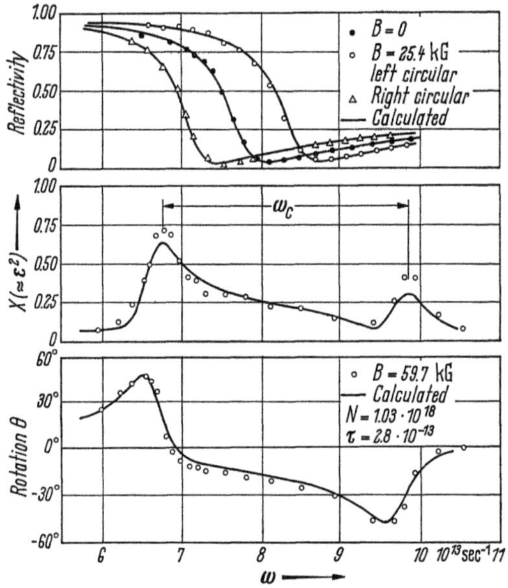

Fig. 21. Longitudinal plasma reflectivity and longitudinal rotation and ellipticity in reflection from n-type InSb at room temperature (after PALIK, TEITLER, HENVIS and WALLIS, Ref. 1, p. 283).

19. Free carrier absorption and dispersion effects in the high field limit. Free carrier effects, other than cyclotron resonance, have been relatively unexplored in the high field limit. Some theoretical discussion has been given by GUREVICH and URITSKII (1960)[1] and GUREVICH, IPATOVA and URITSKII (1960)[2] and observations and discussion by PALIK and WALLIS (1963)[3]. There appear to be a number of possible effects. If the Fermi level is deep inside a band and well defined, effects might be expected both in free carrier[4] and interband absorption, analogous to the de Haas-van Alphen and Shubnikov-de Haas effects when

$$E_F = \hbar\,\omega_c (n + \tfrac{1}{2}).$$

However a sharp Fermi level is not a necessity for observation of oscillatory effects and PALIK and WALLIS have observed such oscillations in n-type InSb at room temperature (Fig. 22). They ascribe these oscillations to a breakdown in the selection rule for CR, that

$$\Delta n = \pm 1$$

which could occur by several processes, among them being interaction with low energy acoustic phonons. Absorption maxima could then occur at frequencies given by

$$\hbar\,\omega = n\,\hbar\,\omega_c$$

[1] L. E. GUREVICH, and Z. I. URITSKII: Soviet Phys. — Solid State **1**, 1188 (1960).
[2] L. E. GUREVICH, I. P. IPATOVA, and Z. I. URITSKII: Proc. Int. Conf. on Semiconductor Physics, Prague 1960, p. 328.
[3] E. D. PALIK, and R. F. WALLIS: Phys. Rev. **130**, 41 (1963).
[4] E. D. PALIK, and D. L. MITCHELL: Bull. Am. Phys. Soc. **10**, 368 (1965) report oscillations in Faraday rotation in PbS, periodic in $(1/B)$.

where n is an integer. This equation can be re-written

$$n = \frac{\omega_c}{\omega} = \left(\frac{\omega m^*}{e}\right)\left(\frac{1}{B}\right)$$

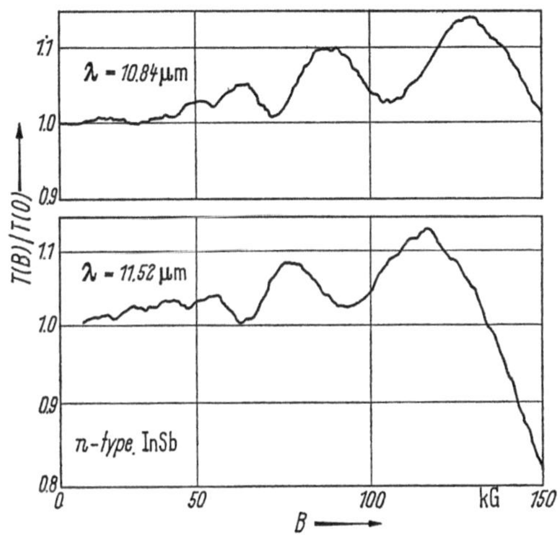

Fig. 22. Oscillations in the free carrier absorption of InSb at room temperature (after PALIK and WALLIS, Ref. 3, p. 284).

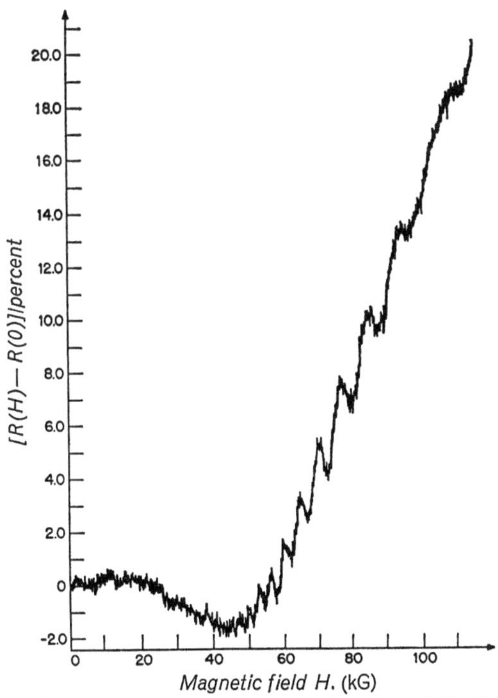

Fig. 23. Optical de Haas-Shubnikov oscillations at the plasma edge in antimony (after DRESSELHAUS and MAVROIDES, Ref. 1, p. 286).

and the absorption plotted as a function of $(1/B)$ will be periodic. PALIK and WALLIS deduce masses in reasonable agreement with other measurements and the method is capable of yielding effective mass, carrier density and mass anisotropy.

The optical analog of the Shubnikov-de Haas effect has been observed in the semi-metal antimony by DRESSELHAUS and MAVROIDES[1] and the results, observed in reflection, are shown in Fig. 23. Anisotropy and mass parameters were obtained.

Another possibility, distinct from the above, is the use of high field experiments in which the cyclotron resonance frequency is swept through the frequency of the optical modes of lattice vibrations. In the case of polar crystals interaction with optical modes can cause "polaron" effects in the effective mass and also enhanced scattering influencing the relaxation time τ. Faraday rotation and ellipticity may be used as "remote" measuring techniques for m^* and τ to observe the effect of tuning the cyclotron and lattice resonances by observing at higher frequencies. Such effects have bein observed at Reading in PbTe[2].

All the free carrier effects have been discussed as appropriate to infra-red rather than microwave frequencies. This implies that measuring frequencies $\omega \gg 1/\tau$ the relaxation time. Quite different information is obtained in the low frequency region where this condition no longer holds and all effects are strongly dependent upon τ and hence require a full treatment of scattering mechanism. In this review, therefore, we have restricted the discussion to the cases where band structure information can be inferred, i.e., where the dependence upon τ is weak, without detailed theory of scattering process. It is the possibility of such separation that gives the infra-red methods their particular importance.

V. Interband effects.

20. Introduction. The transitions between valence states and conduction states which are responsible for interband effects were discussed in Part III and the energy for non-degenerate parabolic bands with spherical energy surfaces are given in Eqs. (9.12) and (9.13). Transitions of this type are responsible for a variety of resonant or oscillatory phenomena such as interband magnetoabsorption (IMO) which may be considered the primary phenomenon, and derived effects such as resonant Faraday and Voigt effects and interband magneto-reflection. These effects occur at photon energies greater than the minimum energy gap between conduction and valence bands. At lower photon energies there are non-resonant effects, particularly Faraday and Voigt effects caused by the summed effects of all the allowed interband transitions. As mentioned in the introduction the resonant effects have been observed only since 1956 whereas interband Faraday effect, for example, has a long history. It will however be convenient to treat the resonant phenomena first as the origin of the nonresonant effects then becomes apparent on application of the Kramers-Kronig relations given in Part II, Sect. 13 to the interband transitions.

Progress in this field up to 1959 has been reviewed by LAX and ZWERDLING (1960)[3].

21. Direct transitions — interband magneto-absorption. In the absence of a magnetic field the dominant interband absorption process is the direct, or vertical, transition in which the initial and final states have the same $\boldsymbol{k}$ vector. The absorption coefficient is then obtained by summing the squares of the matrix elements over all states arising from valence and conduction bands and is given by

$$\alpha(\omega) = A \int d^3k \, |\boldsymbol{\alpha} \cdot \boldsymbol{p}_{cv}(\boldsymbol{k})|^2 \, \delta[E_c(k) - E_v(k) - \hbar\omega] \tag{21.1}$$

[1] See B. LAX: Proc. Int. Conf. on Physics of Semiconductors, Paris, 1964, p. 262.

[2] See Sect. 21 for similar effects in IMO.

[3] B. LAX, and S. ZWERDLING: Progr. in Semiconductors 5, 221 (1960).

where

$$A = \frac{\pi e^2}{n m^2 c \omega \varepsilon_0} \frac{2}{(2\pi)^3}$$

$\boldsymbol{\alpha}$ is a polarisation vector and $\boldsymbol{p}_{cv}$ was defined in Eq. (10.3) substituting c, v for α, γ. The integration over the states defined in the δ-function introduces the density of combined conduction and valence states. For spherical surfaces and parabolic bands this varies as $(\hbar\omega - E_G)^{\frac{1}{2}}$ and for *allowed* transitions where $\boldsymbol{p}_{cv}$ is non zero at $\boldsymbol{k}=0$ the absorption coefficient is

$$\alpha(\omega) = A'(\hbar\omega - E_G)^{\frac{1}{2}} \tag{21.2}$$

where

$$A' = \frac{e^2}{2\pi n m^2 c \omega c \varepsilon_0} \left(\frac{2m^*}{\hbar^2}\right)^{\frac{3}{2}} |\boldsymbol{\alpha} \cdot \boldsymbol{p}_{cv}(\boldsymbol{k})|^2$$

and

$$m^* = m_c m_v / (m_c + m_v) \tag{21.3}$$

is the *reduced* effective mass.

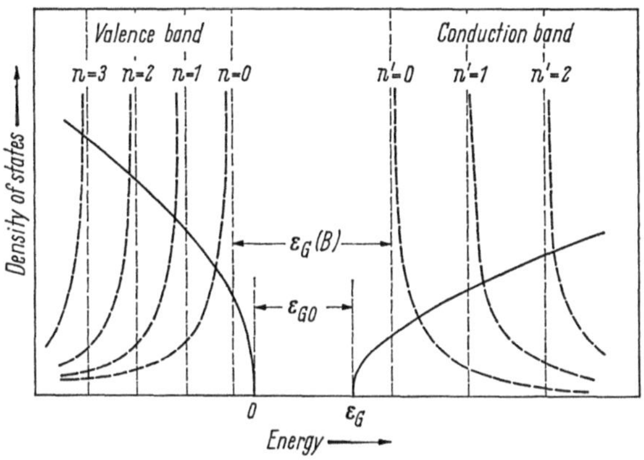

Fig. 24. Density of states at Landau levels (dotted) and the absence of a magnetic field (full line) for simple bands (after BURSTEIN et al., Ref. 4, p. 257).

In the corresponding magnetic case, the density of states changes drastically due to the high degeneracy of the Landau levels as discussed in Part III, Eqs. (9.9), (9.10). For bands as described by Eqs. (1.6) and (1.7), i.e., neglecting spin, the densities of states with and without magnetic fields are illustrated in Fig. 24, and it is these forms which primarily determine the nature of the interband-magneto (IMO) absorption spectrum for direct transitions. The absorption coefficient is formed from

$$\alpha(\omega, B) = A \sum_{nn'} \sum_{ll'} \int dk_z\, dk_z\, |\langle c|\boldsymbol{\alpha}\cdot\boldsymbol{p}|v\rangle|^2\, \delta[E_{cn'}(k_z) - E_{vn}(k_z) - \hbar\omega] \tag{21.4}$$

and gives

$$\alpha(\omega, B) = A' \left(\frac{\hbar e B}{2m^*}\right) \sum_n (\hbar\omega - \varepsilon_n)^{-\frac{1}{2}} \tag{21.5}$$

for the allowed direct transition, where $\varepsilon_n = E_G + (n+\frac{1}{2})\hbar\omega_c + (g_c M_c - g_v M_v)\beta B$ and $\hbar\omega_c = \frac{eB}{m^*}$, m^* is the *reduced* mass as above. The spins have been re-introduced as in (9.13) and (9.14) and $\beta = \frac{e\hbar}{2m}$.

The selection rules for direct interband transitions were discussed in Sect. 11. For simple bands

$$\Delta n = 0 \quad \text{and} \quad \Delta M = 0, \ \pm 1$$

and the particular values of ΔM are dependent upon the state of polarisation and relative orientation of the magnetic field as discussed in Sect. 11. The shape of the absorption spectrum will be modified by broadening of the energy levels. Each electronic transition can be empirically treated by including Lorentzian broadening. The IMO lines are also broadened by transitions away from $k_z=0$ which convert the IMO lines of $k_z=0$ into extended absorption bands. The tails of these bands results from the quasi-continuous nature of the energy bands in the z direction and the different curvature (in sign as well as magnitude) of the valence and conduction bands. Typical calculated curves of IMO absorption for a simple semiconductor are shown in Fig. 25.

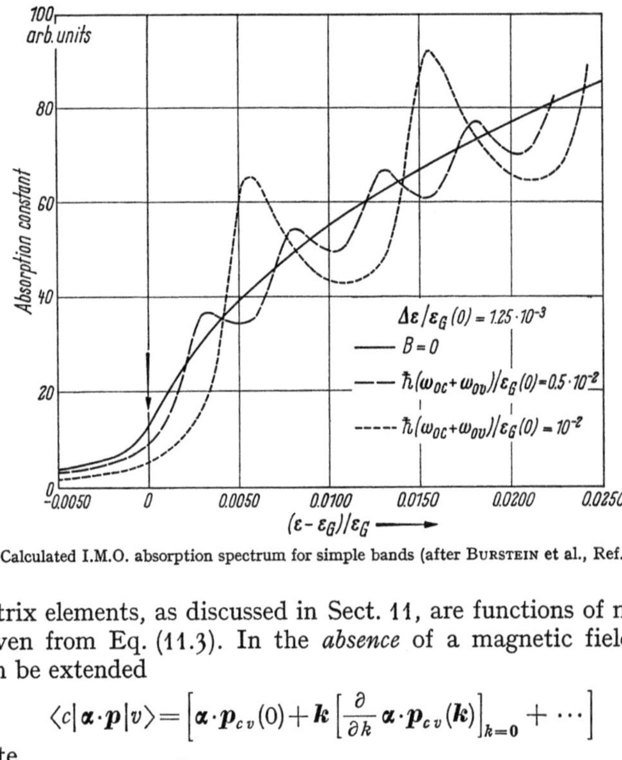

Fig. 25. Calculated I.M.O. absorption spectrum for simple bands (after BURSTEIN et al., Ref. 4, p. 257).

The matrix elements, as discussed in Sect. 11, are functions of magnetic field and are given from Eq. (11.3). In the *absence* of a magnetic field the matrix element can be extended

$$\langle c|\boldsymbol{\alpha}\cdot\boldsymbol{p}|v\rangle=\left[\boldsymbol{\alpha}\cdot\boldsymbol{p}_{cv}(0)+\boldsymbol{k}\left[\frac{\partial}{\partial k}\,\boldsymbol{\alpha}\cdot\boldsymbol{p}_{cv}(\boldsymbol{k})\right]_{k=0}+\cdots\right] \tag{21.6}$$

and we write

$$\frac{\partial}{\partial k}\,[\boldsymbol{\alpha}\cdot\boldsymbol{p}_{cv}(\boldsymbol{k})]_{k=0}=\mathrm{M}_{cv}(0)\,.$$

If $\boldsymbol{p}_{cv}(0)\neq 0$ we have the case of *allowed transitions* and all terms except $\boldsymbol{\alpha}\cdot\boldsymbol{p}_{cv}(0)$ will be negligible. In the presence of a magnetic field the matrix elements for $v_\pm$ from (11.3) imply a difference from the term $\left(1+\frac{eB}{m\,\omega_g}\right)$ in which m is the free mass. For a field of 100 kgauss this term differs from 1 by 10^{-2}—10^{-3} for most semiconductors and hence is negligible in absorption processes. It can however be significant in processes depending upon differences between v_+ and v_-, such as Faraday effect.

When $p_{cv}(0)=0$, the second term in the expansion, $\mathrm{M}_{cv}(0)$, becomes important and we have the case of *forbidden transitions*. There are two cases:

(a) $E\,||\,B$

$$\alpha(\omega, B)=\frac{3\hbar\,\omega_c A''}{2}\sum_n\,[\hbar\,\omega_c-\varepsilon_n]^{\frac{1}{2}}$$

where

$$A'' = \frac{e^2 \hbar^2}{6\pi n m^2 c \omega \varepsilon_0} |\mathrm{M}_{cv}(0)|^2 \left(\frac{2m^*}{\hbar^2}\right)^{\frac{3}{2}}$$

and

$$\varepsilon_n = E_G + (n+\tfrac{1}{2})\hbar\,\omega_c + (M_c\, g_c - M_v\, g_v)\beta B$$

with selection rules $\Delta n=0$, $\Delta M=0$.

(b) $E \perp B$

$$\alpha(\omega, B) = \frac{3}{2}\left(\frac{\hbar\omega_c}{2}\right)^2 A'' \sum_n (n+1)\,[(\hbar\,\omega - \varepsilon_{n_1})^{-\frac{1}{2}} + (\hbar\,\omega - \varepsilon_{n_2})^{-\frac{1}{2}}]$$

where

$$\varepsilon_{n_{1,2}} = E_G + (n+\tfrac{1}{2})\hbar\,\omega_c + (M_c\, g_c - M_v\, g_v)\beta B + \hbar\,\omega_{c_{1,2}}$$

with selection rules $\Delta n = \pm 1$ and $\Delta M = \pm 1$.

This predicts two series of massing (neglecting spin) with intensity increasing with n. Forbidden transitions apply, for example, to the case of cuprous oxide[1].

For Ge, InSb etc. the allowed transition case is appropriate but must be modified to include the degenerate valence bands, and also non-parabolic bands when necessary. The Luttinger-Kohn results for the magnetic levels in the valence states given by (10.18) and (10.19) are then combined with conduction states given by

$$E_c = E_{c0} + \frac{\hbar^2 k_z^2}{2m_c} + \hbar\,\omega_{cc} + g_c\,\beta B \tag{21.7}$$

which parabolic form is relevant to the Γ_2 conduction band minimum at $\boldsymbol{k}=0$ in Ge (i.e., not the lowest minimum occurring in the ⟨111⟩ direction which gives rise to indirect processes). From (10.18), (10.19) and (21.7) the energies of the IMO transitions can be predicted, and taking into account the polarisation of radiation and relative direction of the magnetic field the appropriate transition probabilities have been calculated using the wave functions given by (10.15), (10.16) and (10.17). The resulting selection rules were summarised in Sect. 11. Analysis of experimental results has taken the form of a comparison of the position and strengths of the IMO lines with the observed spectrum rather than a calculation of the detailed shape.

The earliest measurements and the detailed analysis of results are contained in a series of papers from the N.R.L. and M.I.T. Laboratories published almost simultaneously and containing nearly identical material. The first experiments were carried out on InSb[2] and InSb and InAs[3] in which the specimens were thick so that only an edge shift was observed. With thinner specimens the resonant oscillatory effects were then observed in InSb[4] and for the direct transition in Ge[5]. Detailed results and analysis for the germanium case have been given by both groups[6–8], the high resolution low temperature data from the Lincoln Laboratory[7] being the most complete experimental results.

Results for plane polarised radiation with $E \parallel B$ and $E \perp B$ when B was along the [100] axis are shown in Fig. 26, together with predicted positions and strengths of the lines from the Lincoln results[8]. The theoretical line positions were predicted

[1] E. F. Gross: J. tech. Phys. (Moscow) **27**, 2177 (1957).
[2] E. Burstein, G. S. Picus, H. A. Gebbie, and F. Blatt: Phys. Rev. **103**, 826 (1956).
[3] S. Zwerdling, R. F. Keyes, S. Foner, H. H. Kohn, and B. Lax: Phys. Rev. **104**, 1805 (1956).
[4] E. Burstein, and G. Picus: Phys. Rev. **105**, 1123 (1957).
[5] S. Zwerdling, B. Lax, and L. Roth: Phys. Rev. **108**, 1402 (1957).
[6] E. Burstein, G. S. Picus, R. F. Wallis, and F. Blatt: Phys. Rev. **113**, 15 (1959).
[7] S. Zwerdling, B. Lax, L. M. Roth, and K. J. Button: Phys. Rev. **114**, 80 (1959).
[8] L. M. Roth, B. Lax, and S. Zwerdling: Phys. Rev. **114**, 90 (1959).

using the method outlined making use of Eq. (10.18), (10.19) and (21.7). The valence band parameters γ_1, γ_2, γ_3 and κ were obtained from DEXTER, ZEIGER and LAX[1] and the spectra had to be "stretched" by 1.11% to obtain the fit illustrated, taking the band edge at 0.9870 eV. The prominent features (minima in I_B/I_0) then fit quite reasonably except lines 1, 1' and 3, 3'. Quantitative results are best deduced from such data by plotting the positions of successive minima as a function of magnetic field. The lines 1, 1' and 3, 3' are found to have a quadratic

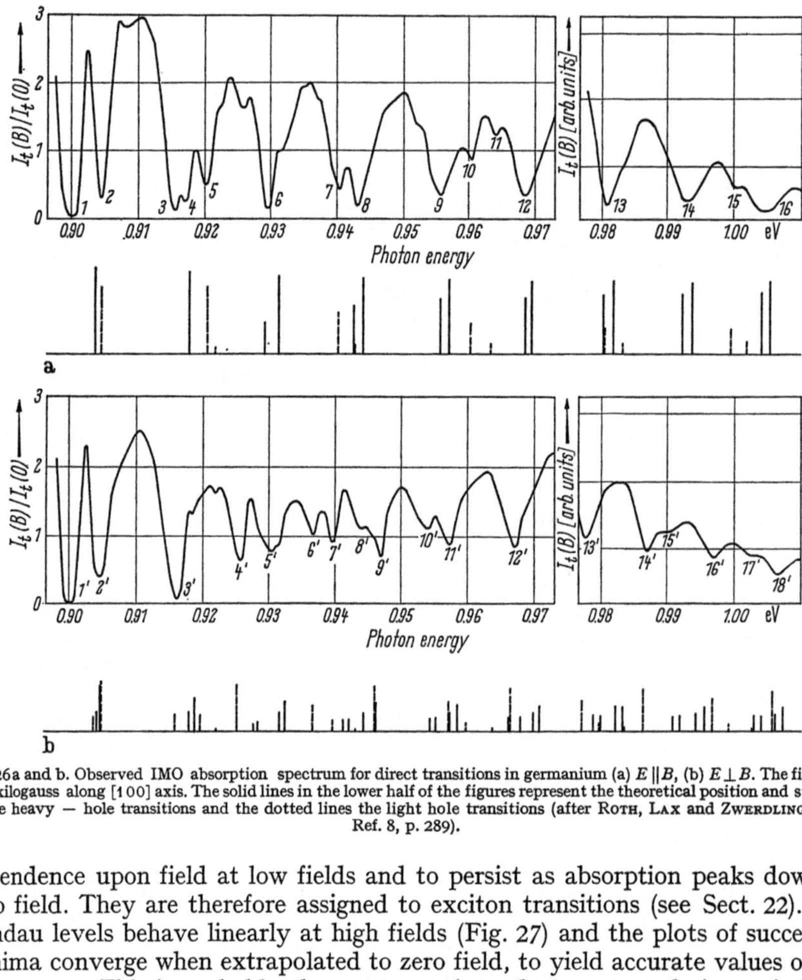

Fig. 26a and b. Observed IMO absorption spectrum for direct transitions in germanium (a) $E \parallel B$, (b) $E \perp B$. The field was 38.9 kilogauss along [100] axis. The solid lines in the lower half of the figures represent the theoretical position and strength of the heavy — hole transitions and the dotted lines the light hole transitions (after ROTH, LAX and ZWERDLING et al., Ref. 8, p. 289).

dependence upon field at low fields and to persist as absorption peaks down to zero field. They are therefore assigned to exciton transitions (see Sect. 22). The Landau levels behave linearly at high fields (Fig. 27) and the plots of successive minima converge when extrapolated to zero field, to yield accurate values of the energy gap. This is probably the most certain and accurate technique of all for obtaining energy gaps in semiconductors.

The selection rules for the $E \parallel B$ and $E \perp B$ spectra of Fig. 26a and b differ with the consequence that the prominent minima in the $E \parallel B$ case correspond to transitions from *heavy hole* levels while for $E \perp B$ (Fig. 26b) the transitions from *light hole* levels appear most strongly. Thus an important separation can be made (see Fig. 28).

The experiment was also carried out with B along [110] and [111] axes giving a measure of the anisotropy of the energy surfaces.

[1] R. N. DEXTER, H. J. ZEIGER, and B. LAX: Phys. Rev. **104**, 637 (1956).

The *slope* of the plot of position of IMO peaks versus magnetic field can be directly interpreted in terms of the cyclotron frequency corresponding to the *reduced* mass m^*. Considering then the heavy hole transitions, which have the smaller splitting, and including the heavy hole mass from cyclotron resonance results ($m^{111}=0.54$, $m^{100}=0.46$) the *conduction band mass* can be deduced. In the case of direct transitions this is the mass associated with the Γ_2 minimum at $\boldsymbol{k}=0$ and the Lincoln Laboratory results give a value of $(0.037\pm0.001)\,m_2$. This experimental determination is an example of the usefulness of IMO techniques as it would not have been possible by CR methods to reach such a higher conduction band minimum. However some independent knowledge of the valence band is required.

When the IMO lines have been identified it is also possible to obtain values for the conduction band spin splitting. The transitions 7 and 7′ come from the same

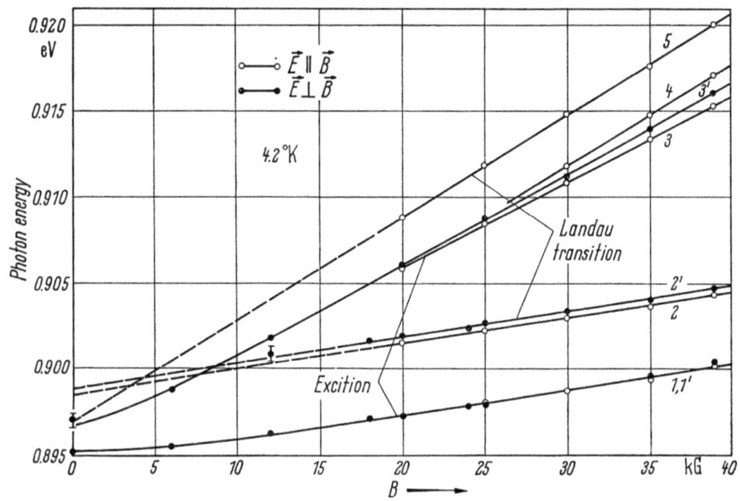

Fig. 27. Magnetic field dependence of position of I.M.O. peaks in germanium (after ZWERDLING et al., Ref. 7, p. 289).

valence state to the upper and lower states spin split from the same conduction band Landau level. The separation yields a g-factor of -3.1. ROTH et al. find that the energy levels can be fitted by the expression

$$\varepsilon_n=(27.1\pm0.6)\,(n+\tfrac{1}{2})\,\hbar\,\omega_0-(0.33\pm0.1)\,(n+\tfrac{1}{2})^2(\hbar\,\omega_0)^2\mp 1.2\,\beta B \qquad (21.8)$$

where ω_0 is the free mass CR frequency. The second term is due to the non-parabolicity of the conduction band which is a fairly small effect over the range of the experiments described. The expression (21.8) yields a g-factor of -2.5.

The effects of anomalous g-factor, and of non-parabolic conduction bands are much greater in InSb than in Ge. The early measurements on InSb have been referred to previously, while in Fig. 29 is shown the more recent results of ZWERDLING, KLEINER and THERIAULT[1,2], in which polarised radiation, low temperatures and high resolution grating spectroscopy were used. Detailed spectra extending deep into the band were obtained in the range from 0.2300 eV, the 5° K energy

[1] S. ZWERDLING, W. H. KLEINER, and J. P. THERIAULT: J. Appl. Phys., Suppl. **32**, 2118 (1961).

[2] S. ZWERDLING, W. H. KLEINER, and J. P. THERIAULT: Proc. Int. Conf. on Physics of Semiconductors, Exeter, 1962, p. 455 (1962).

gap, to 0.4530 eV. When plotted against magnetic field the positions of the minima lay on a curve, indicating the non-parabolicity of the conduction band. Transitions involving heavy hole quantum numbers up to $n=27$ were observed. An earlier

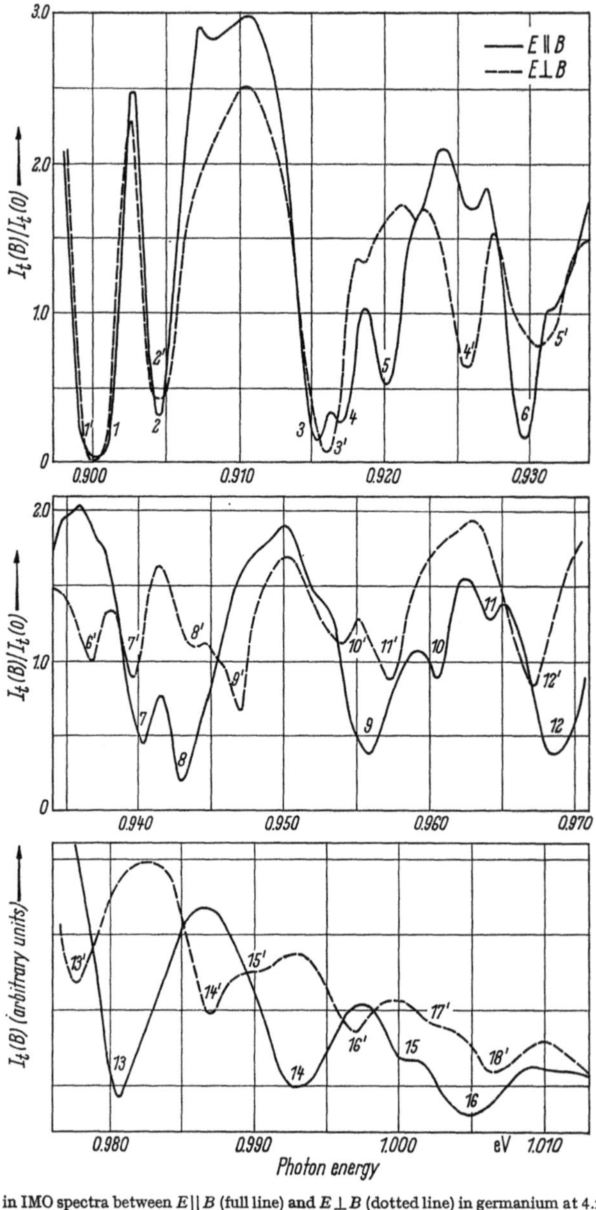

Fig. 28. The change in IMO spectra between $E \parallel B$ (full line) and $E \perp B$ (dotted line) in germanium at 4.2° K and 38.9 kgauss (after ZWERDLING et al., Ref. 7, p. 289).

convergence plot, at room temperature is shown in Fig. 30, giving a zero field energy gap of 0.180 ± 0.002 eV at room temperature. This graph also shows the anomalous spin splitting of the 1st Landau level. The g-factor determined from the separation is approximately -54. This value is in agreement with the prediction of ROTH [Eq. (12.1)], but care must be taken to use an appropriate value for m_c^*. Since m_c^* varies in the non-parabolic band, so too does the g-factor.

Using polarised radiation with $E \| B$ to pick out the heavy hole transitions, the anisotropy of the valence band was investigated (assuming the conduction band

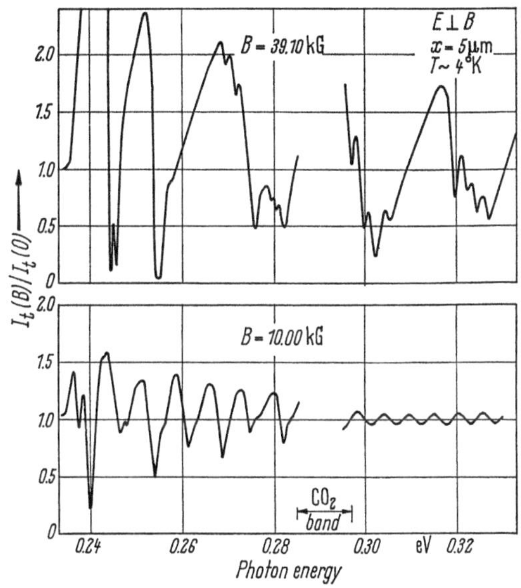

Fig. 29. High resolution IMO spectra for InSb at liquid helium temperature with $E \perp B$ (after ZWERDLING et al., Ref. 1, p. 291).

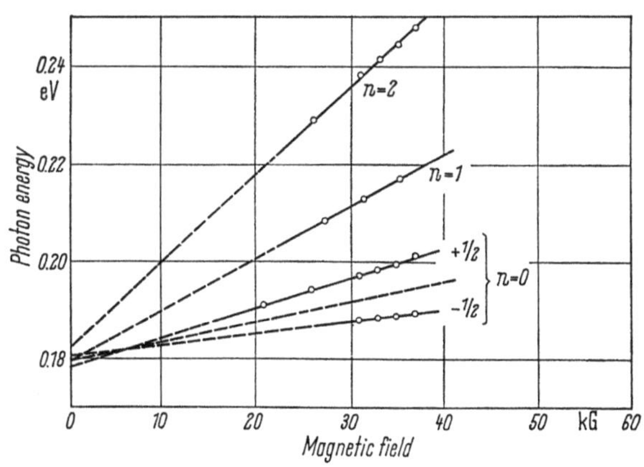

Fig. 30. Convergence plot for IMO spectra of InSb at room temperature. The first level ($n=0$) is split by spin ($+\frac{1}{2}$, $-\frac{1}{2}$). The separations giving a g value for the conduction electron of -54 (after ZWERDLING et al., Ref. 3, p. 286).

to have spherical energy surfaces). The results for three principal crystallographic directions are given in Fig. 31. They can be expressed in terms of the parameters of the Luttinger-Kohn theory discussed earlier. The valence band anisotropy is found to be somewhat smaller than in Si and Ge. If the valence bands are approximated to spheres, a result we require in Sect. 23, the results can be expressed in terms of the three parameters γ, $\bar{\gamma}$ and κ [Eq. (9.11)] as follows:

γ	$\bar{\gamma}$	κ	
25.0	11.0	10.13	CR
36.0	15.35	13.87	IMO

in which the second set is derived from the IMO measurements. The first set is from recent CR measurements of BAGGULEY and STRADLING[1] and differs significantly from the IMO results. In Sect. 23 we shall see that the CR values give a better fit to resonant interband Faraday rotation. A heavy hole mass $\sim 0.02\, m_0$ is implied by the IMO results[2,3].

Other materials studied by IMO methods include InAs[4] and GaSb[4], resembling InSb and germanium respectively, CdS[5] and the lead salts PbS, PbSe, PbTe[6]. The two latter cases have somewhat different band structure than Ge, InSb etc., with the CdS case showing strong exciton effects which will be further discussed in Sect. 22.

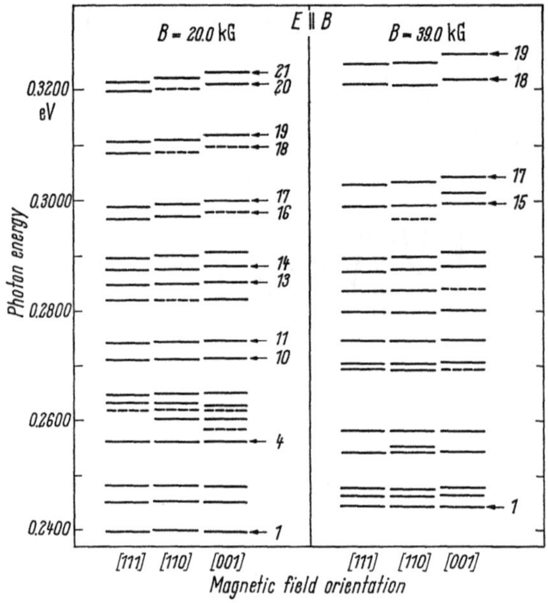

Fig. 31. Magnetic energy levels for $E \| B$ as a function of crystallographic direction in InSb, illustrating the anisotropy of the valence band (after ZWERDLING et al., Ref. 2, p. 291).

The lead salt studies were made with epitaxial evaporated films 2—4 microns thick both with samples freed from the substrate and others still attached. In the latter case there was strain present. The band structure of PbS, PbTe and PbSe has attracted a great deal of attention for some years. Only recently has the band edge structure become clear due to experiments on de Haas-van Alphen oscillatory magnetic susceptibility[7], Shubnikov-de Haas oscillatory magneto resistance[8],

[1] D. M. S. BAGGULEY, and R. A. STRADLING: Phys. Letters 6, 143 (1963).

[2] More complete IMO and Resonant Faraday rotation data have recently been given by C. R. PIDGEON, and R. N. BROWN, Phys. Rev. 146, 515 (1966). The results are analysed by a much more adequate theory in which conduction and valence band coupling is treated exactly and higher bands to order k^2. Band parameters at 20° K are quoted as $m_c = 0.0145$ m, $m_{lh} = 0.0160$ m, $m_{hh}[100] = 0.32$ m, $m_{hh}[110] = 0.42$ m, $m_{hh}[111] = 0.44$ m.

[3] E. J. JOHNSON, and D. M. LARSEN: Phys. Rev. Letters 16, 655 (1966), have observed non-linear shifts of IMO peaks when $\hbar\omega_{cc} \simeq \omega_{LO}$; they assign the peaks to a polaron doublet.

[4] See B. LAX, and S. ZWERDLING: Progr. in Semiconductors 5, 231 (1960).

[5] A. MISA, K. AOYAGI, and G. KUWABARA: Proc. Int. Conf. on Physics of Semiconductors, Paris, 1964, p. 317.

[6] D. L. MITCHELL, E. D. PALIK, and J. N. ZEMEL: Proc. Int. Conf. on Physics of Semiconductors, Paris, 1964, p. 325.

[7] See W. W. SCANLON: Solid State Physics 9, 83 (1959) for a review.

[8] K. F. CUFF, M. R. ELLETT, and C. D. KUGLIN: Proc. Int. Conf. on Physics of Semiconductors, Exeter, 1962, p. 316.

tion of ROTH [Eq. (12.1)], but care must be taken to use an appropriate value for m_c^*. Since m_c^* varies in the non-parabolic band, so too does the g-factor.

Using polarised radiation with $E \| B$ to pick out the heavy hole transitions, the anisotropy of the valence band was investigated (assuming the conduction band

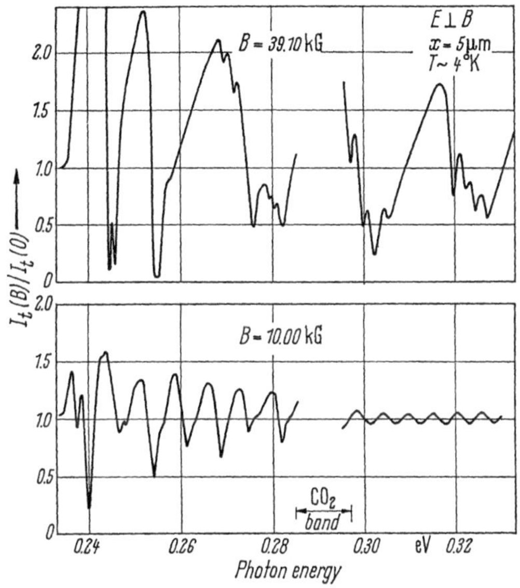

Fig. 29. High resolution IMO spectra for InSb at liquid helium temperature with $E \perp B$ (after ZWERDLING et al., Ref. 1, p. 291).

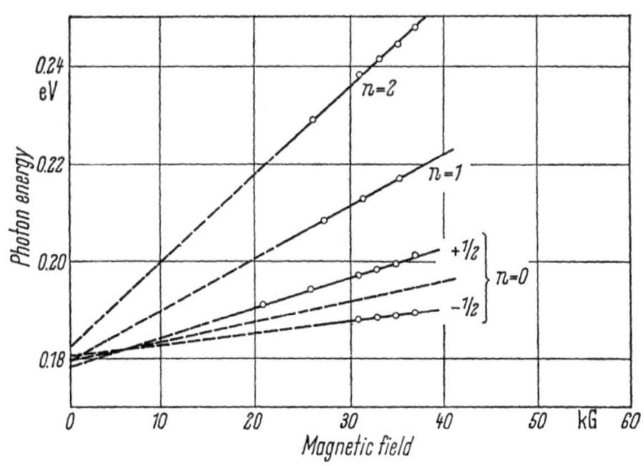

Fig. 30. Convergence plot for IMO spectra of InSb at room temperature. The first level ($n=0$) is split by spin ($+\frac{1}{2}$, $-\frac{1}{2}$). The separations giving a g value for the conduction electron of -54 (after ZWERDLING et al., Ref. 3, p. 286).

to have spherical energy surfaces). The results for three principal crystallographic directions are given in Fig. 31. They can be expressed in terms of the parameters of the Luttinger-Kohn theory discussed earlier. The valence band anisotropy is found to be somewhat smaller than in Si and Ge. If the valence bands are approximated to spheres, a result we require in Sect. 23, the results can be expressed in terms of the three parameters γ, $\bar{\gamma}$ and κ [Eq. (9.11)] as follows:

γ	$\bar{\gamma}$	κ	
25.0	11.0	10.13	CR
36.0	15.35	13.87	IMO

It is apparent that the IMO technique is very rich in the interpretable quantitative information that can be extracted from measurements of direct transition processes. We have, so far, excluded the effects of *indirect transitions* and *excitons* which, in certain materials, notably Ge and CdS, complicate the spectra in certain regions.

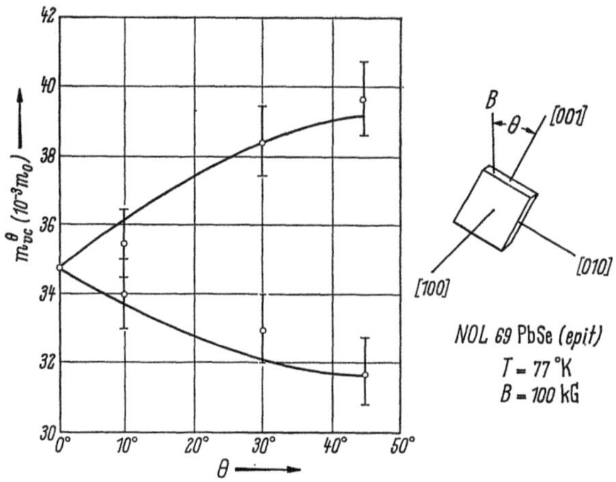

Fig. 33. Anisotropy of reduced effective mass (after MITCHELL et al., Ref. 6, p. 294).

22. Indirect magneto-absorption and excitons. In the description of the primary interband magnetoabsorption processes due to direct transitions we have, so far, neglected effects concerned with the Coulomb interaction between the electron and hole. These interactions give rise to exciton effects, particularly strong in ionic materials of large energy gap, such as Cu_2O and CdS, but present to some extent in all crystals.

Further complication occurs when the minimum energy gap occurs between band extrema located at different points in k-space, and excitons can be associated with both direct and indirect energy gaps.

α) *Indirect transitions and magneto-absorption.* The indirect gap gives rise to absorption in the zero field case which is weak compared to the direct transition effect. This arises as it is a second order process made possible by the interaction of phonons. It is therefore only observable when the indirect gap is less than the direct gap, and it shows a characteristic temperature dependence associated with the phonon populations. The absorption coefficient is of the form

$$\alpha(0) = C_{\pm}\,[\hbar\,\omega - E_G \mp \hbar\,\omega_q]^2$$

where $\hbar\,\omega_q$ is the phonon energy and C involves the phonon population and fundamental constants. Such a situation applies to germanium with the conduction band minima occurring in the $\langle 111\rangle$ direction at the zero boundary. In the presence of a field the absorption is given by

$$\alpha(B) = 2C_{\pm}\,(\hbar^2\,\omega_{c_1}\,\omega_{c_2}) \sum_{n_1 n_2} S\,(h - \varepsilon_{n_1\,n_2})$$

where

$$\varepsilon_{n_1 n_2} = E_G + (n_1 + \tfrac{1}{2})\,\hbar\,\omega_{c_1} + (n_2 + \tfrac{1}{2})\,\hbar\,\omega_{c_2} \pm \hbar\,\omega_q + (M_c\,g_c - M_v\,g_v)\,\beta\,B \qquad (22.1)$$

and $S\,(\hbar\,\omega - \varepsilon_{n_1 n_2})$ is a step function.

The presence of phonon interaction relaxes the selection rules on the Landau orbital quantum numbers n_1 and n_2 but the rules for spin quantum number

$M(\Delta M=0\ E\parallel B,\ \Delta M=\pm 1\ E\perp B)$ still hold. Eq. (22.1) predicts a spectrum consisting of a series of "steps". Between each step the absorption is roughly constant with frequency and is proportional to B^2. Convergence plots may be made from the step positions (found to be linear with field) which yield accurate values of indirect energy gaps[1]. A further "step" in the spectrum is interpreted as an indirect exciton, and shows a quadratic dependence of its position upon magnetic field[2].

β) *Zeeman effect of excitons.* An exciton consists of an electron and hole bound together electrostatically; when the pair has energy less than that of an energy gap they orbit around each other. If the orbits are large compared with the lattice constant they can be treated approximately as two point charges having effective masses and bound together by a Coulomb potential to give rise to a hydrogen-like series of energy levels and hence a line spectrum at energies lying in the forbidden gap. In the presence of magnetic fields excitons give rise to Zeeman effects analogous to those in atomic spectra. Fine structure can occur due to motions other than the simple orbiting of electron and hole: the carriers can have intrinsic motion, motion around an atom, spin motion and motion of the complete exciton through the lattice. Some of these motions may be coupled together. The importance of exciton effects depends upon the binding energy given by

$$E_{\text{ex}} = -(\mu/m)\,\varepsilon_0^2/n^2\,\varepsilon^2 \times 13.5 \text{ eV}$$

where μ is the reduced hole-electron mass and ε the permittivity.

This energy tends to be small in materials of small energy gap (e.g., 10^{-4} eV in InSb) and increases with energy gap. Experimental work has tended to concentrate upon favourable systems, in particular cuprous oxide (Cu_2O) and cadmium sulphide (CdS). The Zeeman effect in Cu_2O was observed by GROSS and ZAHARČENJA[3] in 1957. It was found that the lines in the exciton series did not display a paramagnetic Zeeman effect proportional to the magnetic field but a diamagnetic shift proportional to the *square* of the field. The experiments were in agreement with the theory of the quadratic Zeeman effect, the shift being proportional to the square of the electron orbit radius and so to the fourth power of the quantum number. Quite large (400 Å) exciton orbits were implied. The interband transitions in Cu_2O are an example of a forbidden transition (Sect. 21) and the excition spectrum has been interpreted by ELLIOTT[4,5] in terms of indirect processes and quadrupole transitions. Further extension of the experiments[6,7] using also Stark effects and strain have enabled some band parameters to be deduced.

The general problem requires the solution of the Schrödinger equation

$$\left[-\frac{\hbar^2}{2\mu}\nabla^2 + \frac{i e\hbar}{2}\frac{m_e m_h}{(m_e - m_h)}\boldsymbol{B}\cdot\boldsymbol{r}\cdot\times\nabla + \frac{e^2}{8\mu}(\boldsymbol{B}\times\boldsymbol{r})^2 - \frac{e^2}{\varepsilon r}\right] u(r) = E\,u(r)$$

[1] See, for example, B. LAX, and S. ZWERDLING: Progr. in Semiconductors 5, 221 (1960). — R. J. ELLIOTT, T. P. MCLEAN, and G. G. MACFARLANE: Proc. Phys. Soc. (London) 72, 553 (1958).

[2] New and extended results for the indirect transition in germanium have been given recently by J. HALPERN and B. LAX, J. Phys. Chem. Solids **26**, 911 (1965).

[3] E. F. GROSS, and B. P. ZAHARČENJA: J. Phys. Radium **1**, 68 (1957). — E. F. GROSS: J. Phys. Chem. Solids **8**, 172 (1959).

[4] R. J. ELLIOTT: Proc. Int. Conf. on Semiconductor Physics, Prague 1960, p. 408.

[5] R. J. ELLIOTT: Phys. Rev. **124**, 340 (1961).

[6] E. F. GROSS, B. P. ZAHARČENJA, and A. A. KAPLANSKY: Proc. Int. Conf. on Physics of Semiconductors, Exeter 1962, p. 409.

[7] S. NIKITINE et al.: Proc. Int. Conf. on Physics of Semiconductors, Exeter 1962, p. 409.

where m_e is the effective electron mass, m_h is the effective hole mass, $\mu = \frac{m_e + m_h}{m_e m_h}$, the reduced mass, and ε is the dielectric constant of the crystal. The term in B^2 gives the quadratic Zeeman splitting which becomes dominant as B increases.

A detailed study and analysis for the case of CdS has been made by HOPFIED and THOMAS[1]. In this material the valence band is split by spin orbit and crystal field effects into three nearly degenerate bands at $\boldsymbol{k}=0$. Excitons formed from the top valence band and the conduction band are considered and the spectrum was analysed quantitatively. An electron mass (0.20 m), almost isotropic was found, with the assumption of a lowest minimum at $\boldsymbol{k}=0$ and hole masses of $m_{h\perp}=0.7$ m and $m_{h\parallel}=5$ m where $\perp$ and $\parallel$ refer to the hexagonal c-axis of the crystal. Electron and hole g-factors were also deduced. CdSe has been similarly studied by DIMMOCK and WHEELER[2].

Exciton effects in germanium were mentioned in Sect. 21. The characteristic quadratic dependence of the position of the exciton line and field is shown in Fig. 27. Excitons in germanium can arise from both the indirect and direct energy gaps. These studies have been extended by EDWARDS, LAZAZZERA and PETERS[3] who consider the effects of strain and high magnetic field upon the direct exciton in germanium. Using unstrained samples they make precise observations and note that the interband Landau levels are affected by discrete exciton effects as predicted by ELLIOTT and LONDON[4] and also HOWARD and HASEGAWA[5]. These effects are dominant when $\hbar\omega_c \gg \hbar\omega_{ex}$, and $\hbar\omega_{ex}$ is the exciton binding energy. At zero field $\hbar\omega_{ex}=0.0017$ eV for Ge but increases quadratically with field.

Exciton effects can also influence interband Faraday and Voigt effects as well as magneto-absorption (Sect. 23α).

23. Interband Faraday rotation and Voigt effects. The non-resonant Faraday effect arises from the dispersion associated with the interband magneto-absorption which was described in Sect. 21. Each IMO absorption peak has a corresponding dispersion resonance and the summation of the tails of these dispersion curves which entend to photon energies lying in the band gap gives rise to the non-resonant interband effects in both Faraday rotation and Voigt effects. It is therefore instructive to consider the resonant effects first although they have only recently been observed. The resonant effect is simpler theoretically in that the effect is dominated by the nearest magneto-optical transition whereas off-resonance the effect of a large number of transitions has to be estimated.

α) *Resonant interband Faraday effect.* Faraday rotation spanning large oscillations at the energies of the Landau levels was observed simultaneously and independently by NISHINA, KOLODZIEJCZAK and LAX[6] in Ge and by SMITH, PIDGEON and PROSSER in InSb[7] both experiments being reported at the Exeter Semiconductor Conference in 1962. A further observation on Ge including also ellipticity measurements was made by MITCHELL and WALLIS[8]. The germanium case is somewhat complicated by exciton effects whereas for InSb with an exciton

[1] J. J. HOPFIELD, and D. G. THOMAS: Phys. Rev. **122**, 35 (1961).
[2] J. O. DIMMOCK, and R. G. WHEELER: J. Appl. Phys. **32**, 227 (1961).
[3] D. F. EDWARDS, V. J. LAZAZZERA, and C. W. PETERS: Proc. Int. Conf. on Semiconductor Physics, Prague 1960, p. 335.
[4] R. J. ELLIOTT, and R. LOUDON: J. Phys. Chem. Solids **15**, 196 (1960).
[5] H. HASEGAWA, and R. E. HOWARD: J. Phys. Chem. Solids **21**, 179 (1961).
[6] Y. NISHINA, J. KOLODZIEJCZAK, and B. LAX: Phys. Rev. Letters **9**, 55 (1962).
[7] S. D. SMITH, C. R. PIDGEON, and V. PROSSER: Proc. Int. Conf. on Physics of Semiconductors, Exeter 1962, p. 301.
[8] D. L. MITCHELL, and R. F. WALLIS: Phys. Rev. **131**, 1965 (1963).

binding energy $\sim 10^{-4}$ eV such effects are negligible. The oscillating rotation in InSb is illustrated in Fig. 34 together with the corresponding IMO spectrum. The energies and strengths of the allowed transitions for right and left circularly polarised radiation are given in Fig. 35. As computed by BOSWARVA[1] using the Luttinger-Kohn model and the valence band parameters discussed in Sects. 14 and 21, together with $m_c^* = 0.0145$ m, $g_c = -47.2$ to describe the conduction band levels and $E_G = 0.228$ eV at 77° K. The shape of the Faraday rotation curve at resonances can be understood by taking a simple case. Since the Faraday rotation is given by

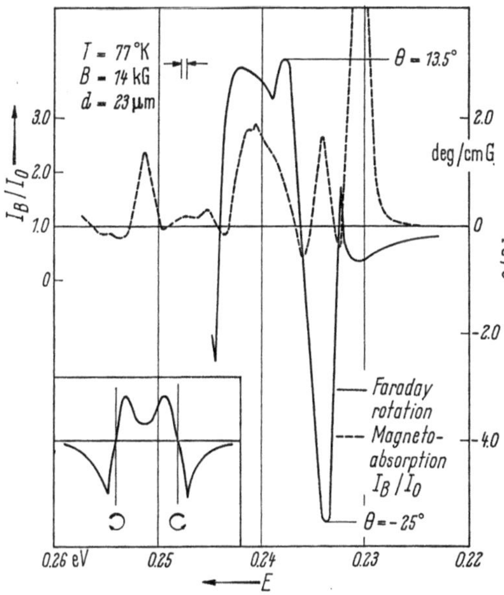

Fig. 34.

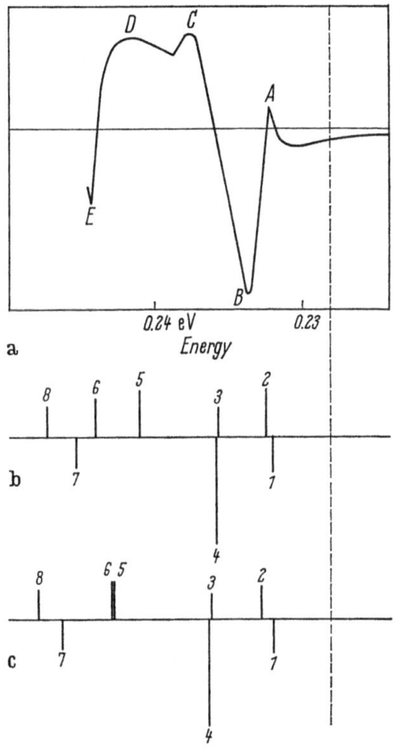

Fig. 35.

Fig. 34. Resonant interband Faraday rotation in InSb (full line) and for comparison the IMO spectrum (dotted line). Inset: resonant Faraday rotation at a simple pair of transitions (after SMITH et al., Ref. 7, p. 298).

Fig. 35. (a) Faraday rotation through the absorption edge region in InSb at 77° K using $B = 14$ kgauss. (b) Energies and strengths of allowed transitions, at $K = 0$, computed from the Kohn-Luttinger model using the band parameters of BAGGULEY and STADDLING. (c) As (b) using band parameters of ZWERDLING et al. (after BOSWARVA[1]).

$$\Theta = \frac{\omega l}{2c}(n_+ - n_-)$$

n_+ and n_- will respectively show a characteristic dispersion resonance at the energy appropriate to a transition for r.c.p. and l.c.p. radiation respectively and since the *difference* is measured one dispersion resonance will be inverted. Thus the rotation will behave as in Fig. 34 (inset) for single energy levels. In real cases the transitions are between magnetic sub-bands, involving nearby states from heavy hole and light hole ladders. These may sometimes be of opposite sign. This sensitivity to sign enables the oscillatory Faraday effect to distinguish between different types of transition unlike magneto-absorption in which effects are, of course, only additive. For InSb, BOSWARVA also calculated detailed shape, including Lorentzian broadening and, as shown in Fig. 35, obtains good agreement

[1] I. M. BOSWARVA: Proc. Phys. Soc. (London) **84**, 389 (1964).

between the positions of the magnetic levels and the observed resonances. The structure in the region A is due to the effects of a positively contributing pair of bands with a negative pair preceding them by 0.001 eV and the analysis is consistent with the first four transitions being respectively from light hole, heavy hole, heavy hole, light hole states in agreement with magneto absorption results. Better agreement is obtained using the valence band parameters of BAGGULEY and STRADLING than those of ZWERDLING et al. (Sect. 21) in the approximation that the valence bands are considered to be spherical.

Experimental results for oscillatory Faraday rotation and ellipticity in Ge, measured by MITCHELL and WALLIS, are shown in Fig. 36 for a field of 86 kgauss.

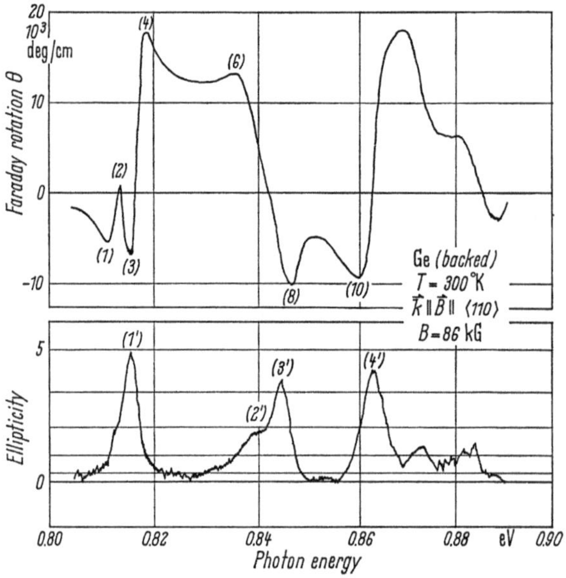

Fig. 36. Resonant Faraday rotation and ellipticity for germanium (after MITCHELL and WALLIS, Ref. 8, p. 298).

Since this field is in the strong field range ($\hbar\,\omega_c \gg \hbar\,\omega_{ex}$) for exciton effects, these authors examine the line shape of the peak labelled l' in ellipticity and find it consistent with the exciton line shape given by ELLIOTT and LOUDON (Sect. 5.2). The Faraday rotation is also influenced by such exciton effects through the dispersion relation

$$\Theta = -\frac{\omega^2 l}{2\pi n}\int_0^\infty \frac{(n_+\alpha_+ - n_-\alpha_-)}{\omega'(\omega'^2-\omega^2)}\,d\omega' \tag{23.1}$$

and such relation is experimentally illustrated in Fig. 36. The frequency dependence of Faraday rotation also shows reasonable agreement with that expected from discrete exciton transitions. A further interesting observation is that of strong field saturation which is found when rotation is measured as a function of field at the position of the zero-field band edge. This occurs for fields greater than 30 kgauss and is also consistent with a model based on discrete exciton transitions. Effects of strain were also observed. Further measurements at low temperatures of oscillatory Faraday effect in the region of the direct exciton in germanium are given by NISHINA, KOLODZIEJCSAK and LAX[1] and detailed ex-

[1] Y. NISHINA, J. KOLODZIEJCZAK, and B. LAX: Proc. Int. Conf. on Physics of Semiconductors, Paris 1964, p. 867.

pressions for the resonant line shapes as a function of relaxation time τ are given by HALPERN, LAX and NISHINA[1]. For each *exciton* absorption peak the Faraday rotation Θ is given as

$$\Theta_k = A\left\{\frac{X_k+Y_k}{(X_k+Y_k)^2+1} - \frac{X_k-Y_k}{(X_k-Y_k)^2+1}\right\} \tag{23.2}$$

where $X_k=(\omega_k-\omega)\tau_k$, $Y_k=\gamma_k B\,\tau_k$, A is a constant, ω is the photon frequency, ω_k the frequency of an exciton in state k and γ_k is an effective gyromagnetic constant such that $2\hbar\gamma_k=\mu_B\,g_k$ with g_k the effective g-factor. τ_k is a phenomenological relaxation time; the actual shape observed will depend strongly on the relation between $1/\tau_k$ and the size of the magnetic splitting.

The corresponding expression for the resonant line shape at *Landau* levels is more complicated:

$$\Theta_n = \frac{\omega_c}{4\pi\sqrt{\tau}}\left(\frac{\mu}{\hbar}\right)^{\frac{3}{2}} A\left[\left\{\frac{\sqrt{(X_n+Y_n)^2+1}+X_n+Y_n}{(X_n+Y_n)^2+1}\right\}^{\frac{1}{2}} - \left\{\frac{\sqrt{(X_n-Y_n)^2+1}+X_n-Y_n}{(X_n-Y_n)^2+1}\right\}^{\frac{1}{2}}\right]. \tag{23.3}$$

For germanium, NISHINA et al. fit both Eqs. (23.2) and (23.3) to their Faraday rotation measurements after subtracting background contribution from other levels. They find the better fit for the exciton case, in agreement with the conclusions of MITCHELL and WALLIS.

NISHINA et al. also report measurements of resonant Voigt shift and give corresponding expressions for the line shapes. Studies of resonant line shapes can therefore estimate the extent of exciton and Landau level contributions to the observed effects and, at high fields, combination of rotation $\Theta_{\max}$ and Voigt shift $\delta_{\max}$ at resonance yields

$$\delta_{\max}/\Theta_{\max} = \text{constant} \times \gamma B \tau$$

where $\gamma=(g_c+g_v)\mu_B/2\hbar$ and simple bands are considered. Since γ can be found from appropriate splitting, this gives a method of investigating τ. Apart from its sign sensitivity and hence ability to separate sets of magnetic levels, resonant Faraday rotation also shows greater sensitivity to structure which appears as minor slope changes in magneto-absorption.

β) *Non-resonant low field interband Faraday effect.* Owing to its experimental simplicity this effect has attracted much experimental attention in all types of crystals with transparent regions since the earliest days of magneto-optics. More recently, with the increasing knowledge of the band structure of certain materials the effect has become recognisable as due to the dispersive effects of interband transitions responsible for IMO absorption, but measured in the transparent region below the energy gap. In semiconductors in particular the band structure is sufficiently well known for quantitative theoretical interpretation to be attempted. Consequently the effect has also attracted considerable theoretical interest, possibly more than the effect justifies since it is not a very specific method of probing band structure. Much of the early theory has subsequently proved to be either incorrect or totally inadequate. The subject has been enlivened by a certain amount of controversy, notably between LIDIARD at Reading and LAX at MIT, and despite rapid progress agreement between theory and experiment is still qualitative rather than quantitative.

We will, therefore, in this article begin by reviewing the experimental situation in semiconducting materials with known band structure. Earlier measurements

[1] J. HALPERN, B. LAX, and Y. NISHINA: Phys. Rev. **134**, A 140 (1964).

on insulating solids were found to conform quite well to BECQUEREL's formula, Eq. (1.5), but showed variation in the sign of the effect (see RAMASHESHAN and SIVARAMAKRISHNAN[1] for a review).

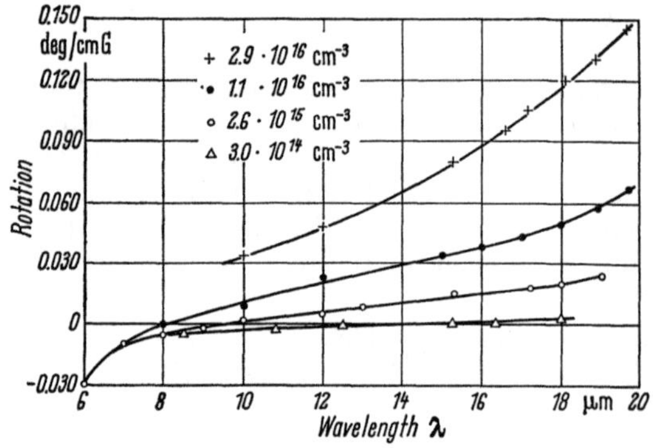

Fig. 37. Interband (negative) and free carrier Faraday rotation in InSb (after SMITH et al.[3]).

The earliest measurements in semiconductors were reported by KIMMEL[2] in 1957 in GaP, Si and InP but were very inadequate, being made with interference filters at a few spot points so that no interpretation was attempted.

SMITH, MOSS and TAYLOR (1959) observed the interband effect in InSb[3] (Fig. 37), finding that the interband rotation was of opposite sense to the free electron case. (It is usual to refer to this as negative rotation, most workers taking the free electron effect as positive). They also showed that pure specimens and low temperatures were re-

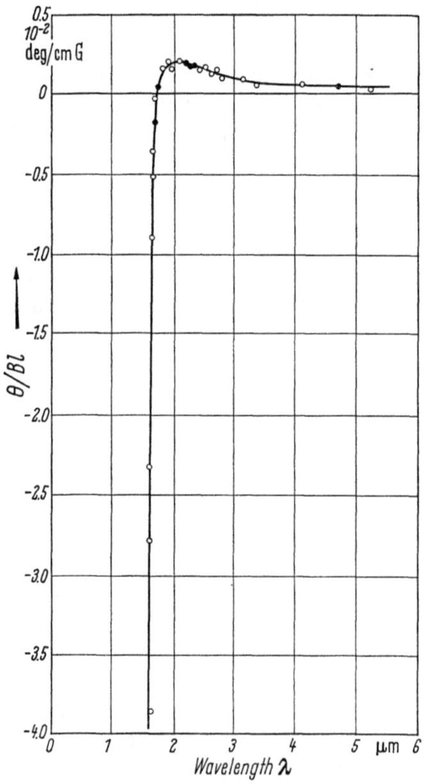

Fig. 38. Interband Faraday rotation in pure germanium (after HARTMANN and KLEMAN, Ref. 2, p. 303).

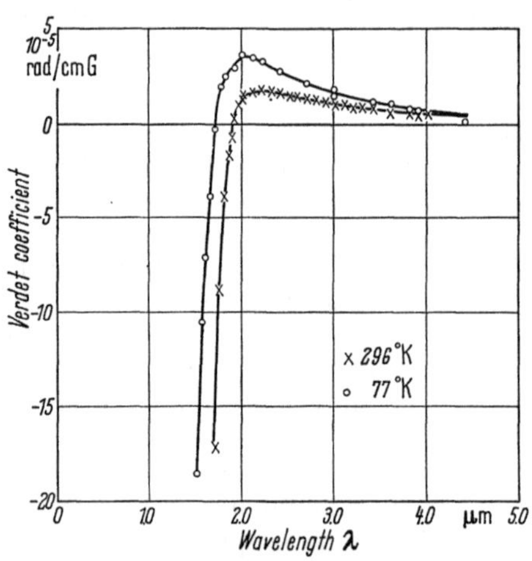

Fig. 39. Interband Faraday rotation in GaSb (after PILLER and PATTON, Ref. 7, p. 303).

[1] S. RAMASHESHAN, and V. SIVARAMAKRISHNAN: Progr. in Crystal Physics 1, 168 (1958).
[2] H. KIMMEL: Z. Naturforsch. 12, 1016 (1957).
[3] S. D. SMITH, T. S. MOSS, and K. W. TAYLOR: J. Phys. Chem. Solids 11, 131 (1959).

quired to separate the interband effect from the relatively large free carrier effect (Fig. 37). This result was independently confirmed by LAX and NISHINA[1] (1959/61) although their published measurements (1960) made with a high free electron concentration did not actually extend to negative rotations, but imply such by their divergence from the λ^2 dependence of the free carrier effect.

The next, and a most interesting observation, was the interband effect in intrinsic germanium (Fig. 38) by HARTMAN and KLEMAN (1960)[2] which shows that although the interband effect is large and negative near the band edge it rises to *positive* values before falling towards zero at long wavelengths. CARDONA (1961)[3] then showed that InAs behaved like InSb, with negative rotation but that GaAs, by contrast gave a positive and apparently monotonically increasing rotation as the band edge was approached.

These results established qualitatively the experimental situation requiring explanation — the existence of positive or negative rotation and, in some materials, of both together with a theory of their frequency dependence. Such effects cannot be obtained from a classical model of a bound electron oscillator and consequently early attempts at interpretation in classical terms are of little value. The early data is also, in most cases, inadequate and more refined measurements have since been made on the materials mentioned above and further materials examined.

For the materials showing negative rotation detailed measurements on pure InSb at 5 and 77° K are given by SMITH, PIDGEON and PROSSER[4] and for InAs 77 and 300 °K by SUMMERS[5]. Further measurements on Ge at 300 and 77° K are given by WALTON and MOSS[6] and PILLER and PATTON[7] and the latter also find that GaSb behaves almost identically to Ge (Fig. 39), giving positive rotation at long waves, changing to negative near the absorption edge.

The materials giving apparently monotonically increasing positive rotation include Si (PILLER and POTTER)[8], InP and GaP[7] but recent measurements on GaAs[9] show that the positive rotation in fact turns over quite near the absorption edge and becomes negative (Fig. 40). In addition to this, GaSb[7] and PbS[9] show a reversal of the sign of interband rotation as a function of the free carrier density while alloys of GaAs and InAs in varying proportion show effects intermediate between the positive and negative effects of the constituent compounds[5].

We now consider the interpretation and theory of the interband effect: this may be approached in a low field limit and at energies far enough below the energy gap to ignore exciton effects. All theories essentially calculate the differential dispersive effect of the transitions responsible for interband magneto-absorption associated with right- and left-circularly polarised radiation. The low field limit may imply that the magnetic splitting is small compared with the line broadening and so the magneto-absorption would be unobservable under these

---

[1] R. N. BROWN, and B. LAX: Bull. Am. Phys. Soc. **4**, 133 (1959). — B. LAX, and Y. NISHINA: J. Appl. Phys. Suppl. **32**, 2128 (1961); — Proc. Int. Conf. on Semiconductor Physics, Prague 1960, p. 321.

[2] B. HARTMAN, and B. KLEMAN: Arkiv Fysik **18**, 75 (1960).

[3] M. CARDONA: Phys. Rev. **211**, 756 (1961).

[4] S. D. SMITH, C. R. PIDGEON, and V. PROSSER: Proc. Int. Conf. on Physics of Semiconductors, Exeter 1962, p. 301.

[5] C. J. SUMMERS: To be published.

[6] A. K. WALTON, and T. S. MOSS: Proc. Phys. Soc. (London) **78**, 1393 (1961).

[7] H. PILLER, and V. A. PATTON: Phys. Rev. **129**, 1169 (1963).

[8] H. PILLER, and R. F. POTTER: Phys. Rev. Letters **9**, 203 (1962).

[9] H. PILLER: Proc. Int. Conf. on Physics of Semiconductors, Paris 1964, p. 301.

conditions. As with the free carrier effect, however, the dispersive effect away from resonance is relatively unaffected by line broadening and is observable.

It was noted quite early that the positive and negative rotations could arise from simple Landau levels, including spin [Eqs. (9.13) and (9.14)] if the sum of the effective g-factors, (g_c and g_v) was positive or negative respectively [SMITH and PIDGEON (1960)[1]; CARDONA[2]] the sign of the combined g-factor simply determining the order in energy in which the resonances for r.c.p. or l.c.p. occur.

The earliest attempts to predict frequency dependence were made by LAX and NISHINA (1960, 1961)[3] and SUFFCZYNSKI (1960, 1961)[4]. These authors used a

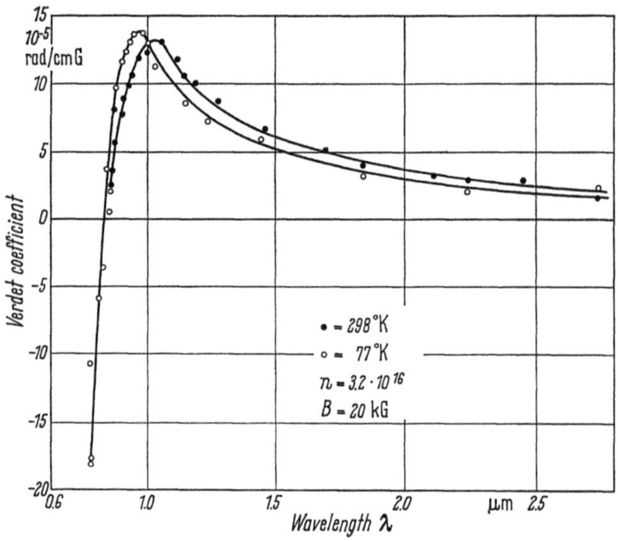

Fig. 40. Interband Faraday rotation in GaAs (after PILLER, Ref. 9, p. 301).

dispersion relation between the real and imaginary parts of the dielectric constant defining the propagation and absorption of circularly polarised modes, i.e., $\varepsilon_\pm$, as defined in Eq. (3.2), i.e.,

$$\varepsilon_\pm = \varepsilon_{xx} \mp i\,\varepsilon_{xy}.$$

The dispersion was then found by substituting $\varepsilon_{xx}^I \mp \varepsilon_{xy}^R$ which characterises the absorption in a single Kramers-Kronig dispersion relation, making analogy with a classical oscillator and hence the Faraday rotation calculated on assumption of simple bands and direct transitions. It was found by LAX and NISHINA to have a leading term of the form

$$\Theta = A\mu^{\frac{3}{2}} |p_{cv}^x|^2 \gamma [\omega^{-1}(\omega_g - \omega)^{-\frac{1}{2}}] \tag{23.4}$$

where A is a constant, μ is the reduced mass and γ contains the effective g-factor $(g_c + g_v)$. This leading term in fact describes the rotation near the edge in, for example, InSb quite well but diverges at lower energies. The important constant γ was left to be determined empirically. LAX and NISHINA also considered indirect

[1] S. D. SMITH, and C. R. PIDGEON: Proc. Int. Conf. on Semiconductor Physics, Prague 1960.
[2] M. CARDONA: Phys. Rev. **121**, 156 (1961).
[3] B. LAX, and Y. NISHINA: Proc. Int. Conf. on Physics of Semiconductors, Prague 1960; — J. Appl. Phys. **32** (Suppl.) 2128 (1961).
[4] M. SUFFCZYNSKI: Proc. Phys. Soc. (London) **78**, 1393 (1961).

transitions and attributed the positive rotation region for Ge to the effects of these transitions.

The problem has been rigorously formulated in macroscopic and quantum mechanical formulae by BOSWARVA, HOWARD and LIDIARD (1962)[1] who showed that there was an error in the use of the Kramers-Kronig relations by both LAX and NISHINA, and, SUFFCZYNSKI. This arises as the real and imaginary parts of the quantity $\varepsilon_\pm$ are asymetric in dispersion relations and separate consideration has to be given to the tensor components ε_{xx} and ε_{xy} through Eqs. (7.1) and (7.2). This error is recognised in later publications of the M.I.T. authors[2]; it gives rise to an incorrect low frequency behaviour but leaves the leading term, as quoted in Eq. (23.4), unchanged.

BOSWARVA et al. develop quantum mechanical formulae for the dielectric constant tensor using time dependent perturbation theory. In the Bloch approximation the Hamiltonian is, including spin-orbit interaction and magnetic field effects, of the form

$$\mathscr{H}=\frac{1}{2m}(\boldsymbol{p}+e\boldsymbol{A})^2+V(\boldsymbol{r})+\frac{1}{2m^2}(\boldsymbol{S}\times\nabla V)\cdot(\boldsymbol{p}+e\boldsymbol{A})+\frac{e}{m}(\boldsymbol{S}\cdot B)$$

where $V(\boldsymbol{r})$ is the one-electron potential, $\boldsymbol{S}$ the spin operator and $\boldsymbol{B}$ the external magnetic induction of vector potential $\boldsymbol{A}$. m is the free electron mass. BHL define an electron velocity operator

$$\boldsymbol{v}=\frac{1}{m}(\boldsymbol{p}+e\boldsymbol{A})+\frac{1}{2m^2}(\boldsymbol{S}\times\nabla V).$$

The interaction with the radiation field is then of the form

$$e\sum_\nu v^{(\nu)}\cdot\boldsymbol{A}'$$

where $\boldsymbol{A}'$ is the vector potential of the radiation and

$$\boldsymbol{A}'=\left(\frac{1}{\omega}\right)\boldsymbol{E}_0\exp i(\omega t-\boldsymbol{k}\cdot\boldsymbol{r})+\text{conj.}$$

in terms of the electric field.

HALPERN, LAX and NISHINA (1964)[3] (HLN) also give a quantum mechanical theory in which they define a momentum operator $\boldsymbol{P}$ from

$$\boldsymbol{P}=\boldsymbol{p}+e\boldsymbol{A}+\frac{1}{2m'}(\boldsymbol{S}\times\nabla V). \tag{23.5}$$

The usual methods of time-dependent perturbation theory then lead BHL to

$$\left.\begin{aligned}\varepsilon_{ij}=\delta_{ij}&-\frac{e^2}{\hbar\omega\varepsilon_0}\sum_k^0\sum_{k'}^u\frac{1}{\omega_{k'k}}\left(\frac{(v_i)_{kk'}(v_j)_{k'k}}{\omega+\omega_{k'k}}-\frac{(V_j)_{kk'}(V_i)_{k'k}}{\omega_{k'k}-\omega}\right)-\\&-\frac{ie^2}{\hbar\omega\varepsilon_0}\sum_k^0\sum_{k'}^u\frac{1}{\omega_{k'k}}\big((V_i)_{kk'}(v_j)_{k'k}\,\delta(\omega+\omega_{k'k})+\\&+(v_j)_{kk'}(v_i)_{k'k}\,\delta(\omega_{kk'}-\omega)\big).\end{aligned}\right\}\tag{23.6}$$

The summation k is over all occupied states and k' over all unoccupied states. The last term in (23.6) determines the energy absorption; the velocity operator

[1] I. M. BOSWARVA, R. E. HOWARD, and A. B. LIDIARD: Proc. Roy. Soc. (London) A **269**, 125 (1962).
[2] J. KOLODZIEJCZAK, B. LAX, and Y. NISHINA: Phys. Rev. **128**, 2655 (1962).
[3] J. HALPERN, B. LAX, and Y. NISHINA: Phys. Rev. **134**, A 140 (1964).

is related to the position operator by

$$(\boldsymbol{v})_{k'k} = i\,\omega_{k'k}(\boldsymbol{r})_{k'k}$$

and on substituting this, the second term gives the Kramers-Heisenberg dispersion formula. In solids, with periodic wave functions however it is more convenient to formulate in velocity matrix elements. Calculating the component ε_{xy} from (23.6) and noting that in a cubic crystal $\varepsilon_{xy} = -\varepsilon_{yx}$ BHL find the Faraday rotation from Eq. (5.5) as

$$\Theta = -\frac{i e^2}{2n\hbar\varepsilon_0 c}\sum_k\sum_{k'}\frac{(V_x)_{kk'}(v_y)_{k'k}-(v_y)_{kk'}(v_x)_{k'k}}{\omega^2_{k'k'}-\omega^2}$$

in m.k.s. units. The δ-function terms in (23.6) have been neglected for Θ at frequencies well away from the allowed transitions. This result can be generally applied to various band models and can be written in terms of the velocity matrix elements for right and left circularly polarised radiation as

$$\Theta = \frac{-e^2}{4n\hbar\varepsilon_0 c}\sum_k^0\sum_{k'}^u\left\{\frac{\omega^2}{\omega^2_{k'k}}\left[\frac{|v_{k'k}(+)|^2-|v_{k'k}(-)|^2}{\omega^2_{k'k}-\omega^2}\right]\right\} \tag{23.7}$$

in which use is made of a sum rule on the matrix elements v_x and v_y. This form predicts that as $\omega\to 0$, $\Theta\propto\omega^2$ which fact was known by Verdet 100 years ago. The denominators containing the term $(\omega^2_{k'k}-\omega^2)$ ensure that the allowed transitions close to the absorption edge dominate the non-resonant rotation, particularly near the absorption edge. BHL calculate Θ for simple bands [Eqs. (9.13) and (9.14)], making the assumption that $v(+)\doteq v(-)$, obtaining

$$\Theta = \frac{\sqrt{2}\,e^2\mu^{\frac{3}{2}}|p^x_{cv}|^2\gamma B}{4\pi m^2\hbar^{\frac{5}{2}} n c\varepsilon_0}\left[\omega^{-1}(\omega_g-\omega)^{-\frac{1}{2}}-\omega^{-1}(\omega_g+\omega)^{\frac{1}{2}}-\omega_g^{-\frac{3}{2}}\right]. \tag{23.8}$$

This expression agrees in its leading term with the incorrect one of LAX and NISHINA. LAX, however, has criticised the assumption that $v(+)=v(-)$ was an adequate approximation for the necessary summations, since $v(+)$ and $v(-)$ are field dependent quantities. It should be noted however that the frequency dependence of interband rotation in InSb in the range $\frac{3}{4}\omega_g$ to $\frac{1}{4}\omega_g$ was found experimentally by SMITH, PIDGEON and PROSSER (Fig. 41) to agree very well with Eq. (23.8). The BHL model does not however provide a basis for predicting the relative signs and magnitude of rotation in different materials.

In a second paper, BOSWARVA and LIDIARD[1] (BL) take into account the field dependence of the velocity matrix elements from Eq. (11.3) and adopt a more realistic band model by using the Luttinger-Kohn expressions for the degenerate valence states. Machine computations of the sums indicated in Eq. (23.7) are made for the cases of GaAs, GaSb, Ge, InAs and InSb, using experimental values for the energy gaps and conduction and valence band parameters.

The main physical conclusions can be easily stated: inclusion of the degenerate valence band gives rise to transitions arising from light hole heavy hole levels which give contributions to the Faraday rotation of opposite sign. This competition allows the possibility of both signs and also that the rotation can pass through a positive maximum and then become negative near the edge as occurs so prominently in Ge and GaSb. Thus the model, invoking only band edge wave functions and direct transitions gives the first qualitatively satisfactory explanation of the observation on interband Faraday effect. The calculation is too complicated to quote formulae and the results are therefore presented in Figs. 42a

[1] I. M. BOSWARVA, and A. B. LIDIARD: Proc. Roy. Soc. (London) A **278**, 588 (1964).

to c. It may be noticed that quantitative agreement is still not very satisfactory. This is not surprising considering that two relatively large terms of opposite sign must be calculated and small errors will be magnified on taking the difference. Damping is not included; recent calculations and measurements on InAs by BOSWARVA and SUMMERS[1] including damping have given the best quantitative agreement to date (Fig. 43).

It would seem, therefore, that the main effects can be accounted for by direct transitions near the zone centre. However, deeper lying transitions give rise to

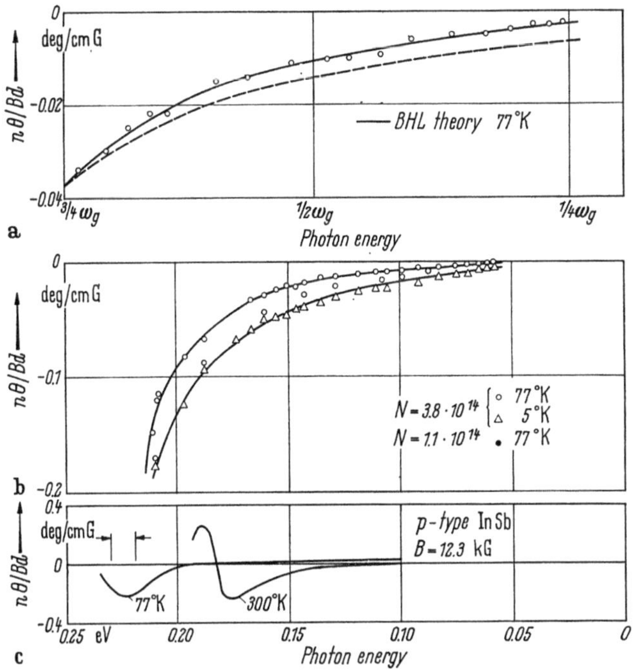

Fig. 41. (a) Comparison with theory of interband Faraday rotation for InSb ($N = 3.8 \times 10^{14}$ cm^{-3}, $T = 77°$ K). (b) Interband rotation as function of temperature and impurity for *n*-type material. (c) As (b) for *p*-type material (after SMITH et al., Ref. 4, p. 303).

very intense absorption due to the larger number of states available and may give significant contributions to the Faraday rotation as may be estimated from the Kramers-Kronig relations in the form of Eq. (7.5). The BL model of parabolic bands does not sensibly apply to deep levels. These workers estimate that, with reasonable parameters for the gyromagnetic splitting, transitions near the L-point (~ 2.0 eV) make some contribution in large gap materials such as Ge, GaAs, and GaSb but suggest that it is qualitatively unimportant. An earlier paper considering these transitions is in error both qualitatively and quantitatively[2]. Experimental results tend to support the conclusions that the zone centre transitions dominate since effects of filling the conduction and valence bands and changing the composition of alloys of GaAs and InAs can produce profound effects on the interband rotation. Effects of strain are rather inconclusive[3].

[1] C. J. SUMMERS, and I. M. BOSWARVA: To be published.

[2] I. M. BOSWARVA, and A. B. LIDIARD: Proc. Int. Conf. on Physics of Semiconductors, Exeter 1962, p. 308.

[3] C. R. PIDGEON, C. J. SUMMERS, T. ARAI, and S. D. SMITH: Proc. Int. Conf. on Physics of Semiconductors, Paris, 1964, p. 289.

A systematic trend is predicted in that small band gap materials (e.g., InAs, InSb) will have small valence band masses and hence a favoured light hole (negative) contribution to Θ, in agreement with experiment. The overall agreement and use of Eq. (7.5) seem to rule out the theory put forward by LAX and

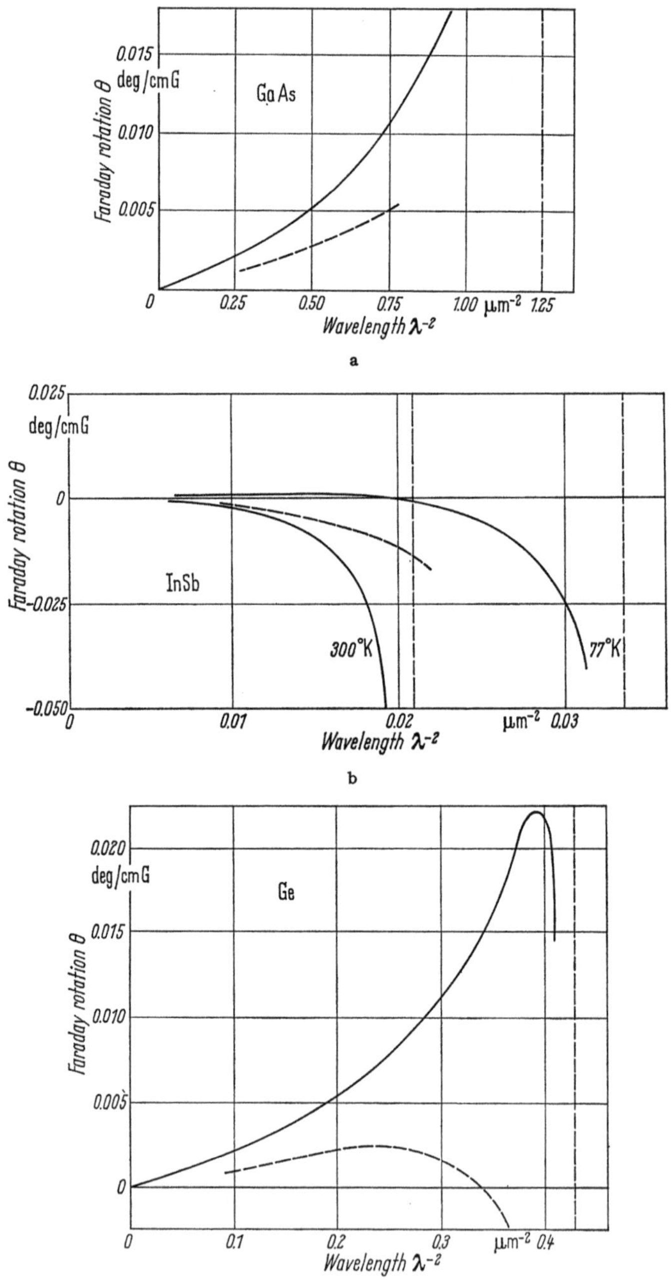

Fig. 42a–c. Comparison of theory (full line) with experiment (dashed) for (a) GaAs; (b) InSb; (c) Ge (after BOSWARVA and LIDIARD, Ref. 1, p. 306). The vertical dashed line is drawn at the position of the energy gap.

NISHINA that indirect transitions are responsible for the positive effects and used by several experimenters in interpretating measurements on Ge and GaSb.

The same problem has also been treated by ROTH[1] and by BENNETT and STERN[2]. ROTH uses a modified Bloch representation and a different formalism. She applies the results to similar band models as BL but fails to predict positive rotation for GaAs although her theory suggests that large gaps favour positive rotation.

There exist, in the published literature, discrepancies over the velocity matrix elements (but now changed from that stated by LAX[3]). HALPERN, LAX

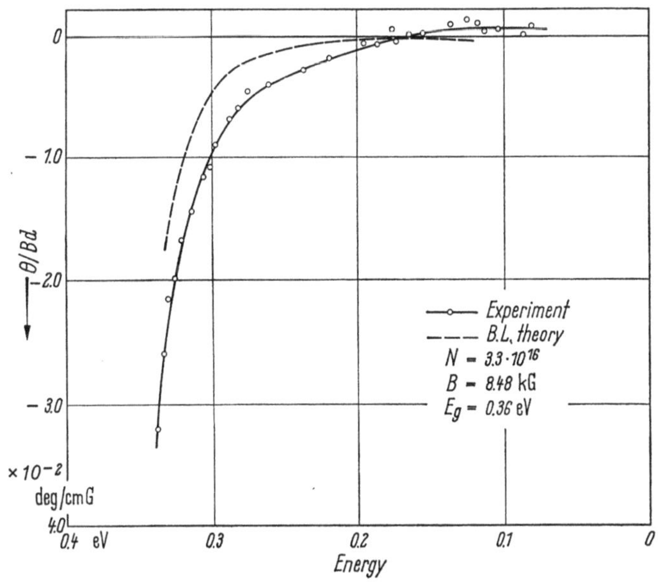

Fig. 43. Interband Faraday rotation in InAs compared with theory (after SUMMERS and BOSWARVA, Ref. 1, p. 307).

and NISHINA express the Faraday rotation in terms of the momentum matrix element $P_{kk'}$ as defined in Eq. (23.5) as follows:

$$\Theta = A \sum_k \sum_{k'} \frac{|P^+_{kk'}|^2}{(\omega^+_{kk'})^2 - \omega^2} - \frac{|P^-_{kk'}|^2}{(\omega^-_{kk'})^2 - \omega^2}$$

where A is a constant, giving a different denominator to each of the sets of transitions. Since $\Theta \to 0$ as $\omega \to 0$ it follows that

$$\sum_k \sum_{k'} \frac{|P^+_{kk'}|^2}{(\omega^+_{kk'})^2} = \sum_k \sum_{k'} \frac{|P^-_{kk'}|^2}{(\omega^-_{kk'})^2}$$

and they argue further that this equality holds for *each* pair of transitions kk', i.e.,

$$\frac{P^+_{kk'}}{\omega^+_{kk'}} = \frac{P^-_{kk'}}{\omega^-_{kk'}} \qquad (23.9)$$

on the basis that the levels are discrete, and equivalent to a classical oscillator so that the correspondence principle applies. This is at variance with results of BL who state that the matrix elements are given by Eq. (11.3). The result (23.9) can be obtained from (11.3) as follows: the velocity matrix elements contain the

[1] L. M. ROTH: Phys. Rev. **133**, A 542 (1964).
[2] H. S. BENNETT, and E. A. STERN: Phys. Rev. **137**, A 448 (1965).
[3] B. LAX: Proc. Int. Conf. on Physics of Semiconductors, Paris 1964, p. 253.

factor $(1 \pm \omega_c/\omega_{k k'})$ where $\omega_c = \frac{eB}{m}$ and m is the free mass of the electron. It is then necessary to assume that the transition frequencies have a gyromagnetic splitting of $\pm\omega_c$ for the HLN equality to hold. This splitting is, of course $\pm\gamma B$ where $\gamma = (g_c + g_v)\beta$ and the g-factors are anomalous so the HLN result is contrary to the magnetic field dependence of the matrix elements as given by ELLIOTT, McLEAN and MACFARLANE.

BENNET and STERN (1965) give a discussion of the elements of the complex conductivity tensor responsible for the Faraday rotation (i.e. $\sigma_{xy}^{(R)}$) which in an earlier form provided some of the basis for the BHL treatment, but they discuss the effect of the field dependence of the matrix elements more completely. They obtain the following results [from (23.6)]:

$$\sigma_{xy}^{(R)} = \text{Constant} \times \sum_k \sum_{k'} \left[\frac{2|\langle k|\pi^-|k'\rangle|^2 \omega_{kk'}^{-2}}{(\omega_{kk'}^- - \omega^2)^2} \frac{\partial \omega_{kk'}^-}{\partial B} - \frac{1}{(\omega_{kk'}^- - \omega^2)} \frac{\partial}{\partial B} |\langle k|\pi^-|k'\rangle|^2 \right\} \tag{23.10}$$

and corresponding terms with R.C.P. Selection rules. The operator $\boldsymbol{\pi}$ includes a spin term, i.e.

$$\boldsymbol{\pi} = (\boldsymbol{p} + e\boldsymbol{A}) + \frac{\hbar}{4m} (\boldsymbol{\sigma} \times \nabla V(r)). \tag{23.11}$$

The expression (23.10) shows clearly that the field dependence of the matrix elements influence the Faraday rotation in addition to level splitting. Using the BHL expression for the matrix elements from Eq. (11.3), we note that V_+ and V_- differ by about 1 part in 10^4 for fields around 10,000 gauss. The matrix element change will therefore only become important at frequencies removed from the transition frequency such that $\omega_{kk'} - \omega$ is around 10^4 times $(\omega_{kk'}^+ - \omega_{kk'}^-)$, the magnetic splitting frequency, when considering the same value of applied field. In practice for semiconductors this applies only to frequencies very close to zero and so only affects the zero field limit for interband effects. BENNETT and STERN note that the expansion leading to Eq. (11.3) is made only to 1st order in (a/R), where a is a lattice constant and R the orbit radius [Eq. (9.10)], and, when including the spin orbit term [Eq. (23.11)] and expanding further they find that terms in $(a/R)^2$ make contributions of equal magnitude. When applied to a non-degenerate band model their calculation gives a correct zero frequency limit and in this respect is more satisfactory than the treatment of BOSWARVA and LIDIARD. The Roth treatment also illustrates the dual contribution of level splitting matrix elements and considers the effects of degenerate bands. Neither treatment supports the HLN equality for each pair of levels.

It should be noted however that in the experimentally significant region for non-resonant Faraday effect — say from ω_g to $\frac{1}{4}\omega_g$ the matrix element term has little effect and the main conclusions of BOSWARVA and LIDIARD concerning the competition of light and heavy hole states appears to stand. Similarly, the main results of HLN for resonant Faraday and Voigt effects [Eqs. (23.2), (23.3) etc.] are essentially unaffected. Contributions from the matrix elements of very deep lying levels may have a small effect in both cases.

Relevant to the low frequency limit is a recent result of MITCHELL, PALIK and WALLIS[1], who find that the effect of populating the conduction band of PbS with electrons is to change the *inter-band* rotation in such a way that a non-zero low frequency limit is implied.

---

[1] D. L. MITCHELL, E. D. PALIK, and R. F. WALLIS: Phys. Rev. Letters **14**, 827 (1965).

Very recently new results on interband Faraday rotation have been reported for ZnO, ZnS, ZnSe, ZnTe, CdS and CdTe at room temperature and 107° K[1]. All give positive effects.

24. Interband magneto-reflection. Resonant interband effects can be observed in reflection as well as transmission. Since the reflectivity depends upon both n and κ from Eqs. (15.13) and (15.14) complete separation of absorptive and dispersive effects is not always possible in this type of measurement. However, even for high absorption, n is usually greater than κ and thus magneto-reflection at resonance has much in common with resonant interband Faraday effect in its dependence upon optical constants. We can write

$$R = 1 - \frac{4n}{(n+1)^2 + \kappa^2},$$

$$\Delta R = \frac{4(n^2 - \kappa^2 - 1)\,\Delta n + 8n\kappa\,\Delta\kappa}{[(n+1)^2 + \kappa^2]^2}$$

where changes Δn and $\Delta\kappa$ are introduced by the magnetic field. If $\kappa \ll n$, $\Delta R \sim \Delta n$, but typical values may be $n \sim 4$, $\kappa \sim 2$. As is common, the experiment was first pioneered using InSb, by WRIGHT and LAX (1961)[2]. A maximum charge in R of around 10% was obtained on applying the field. The positions of the Landau levels were readily distinguished and the results enabled the usual convergence plots to be made from which energy gaps, effective masses and g-factors could be deduced.

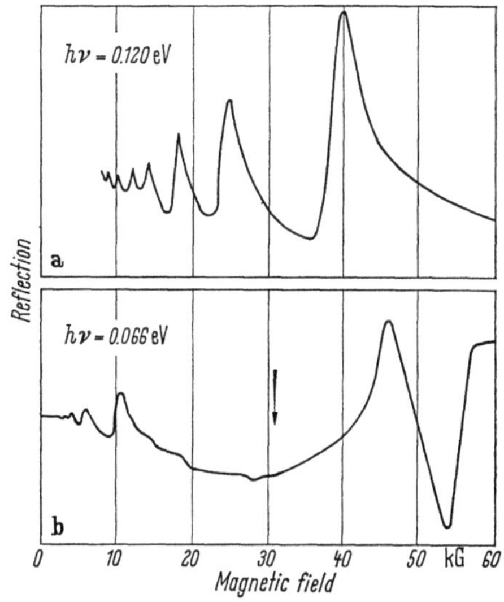

Fig. 44a and b. Magneto reflection from Bismuth at 4.2° K as a function of magnetic field. (a) is at a photon energy of 0.120 eV, and (b) is at 0.066 eV (after BROWN et al., Ref. 3, p. 311).

The particular application of interband reflection techniques (IMR) is to materials having no region of transparency, such as bismuth and graphite. It is not necessary to construct very thin specimens, thus avoiding experimental difficulties and strain effects when using IMR techniques although care must be taken to ensure that the surface is typical of the bulk crystal in structure.

Interesting results have been obtained for bismuth by BROWN, MAVROIDES and LAX[3] (Fig. 44). The peaks (taken at 0.12 eV) can be fitted to a non-quadratic two-band model given by

$$\varepsilon_n\left(1 + \frac{\varepsilon_n}{\varepsilon_g}\right) = \left(n + \frac{1}{2}\right)\hbar\,\omega \pm g_c\,\mu_B\,B$$

with $m_1^* = 0.0021$ and $m_2^* = 0.0046$ for values of an anisotropic conduction band mass and $\varepsilon_g = 0.015$ eV.

For graphite[4] two series of reflection peaks are obtained. These are associated with two different interband transitions at different points of the Brillouin zone. Quantitative interpretation in terms of non-parabolic bands is again possible.

[1] A. EBINA, T. KODA, and S. SHIONOYA: J. Phys. Chem. Solids **26**, 1497 (1965) and M. BALKANSKI, E. AMZALLAG, and D. LANGER: J. Phys. Chem. Solids **27**, 299 (1966).
[2] G. B. WRIGHT, and B. LAX: J. Appl. Phys., Suppl. **32**, 2113 (1961).
[3] R. N. BROWN, J. G. MAVROIDES, and B. LAX: Phys. Rev. **129**, 2055 (1963).
[4] M. S. DRESSELHAUS, and J. G. MAVROIDES: See B. LAX, Proc. Int. Conf. on Semiconductor Physics, 1964, p. 253.

Interband magneto-polarimetric effects in rotation have been observed by LAX and NISHINA[1] on InSb and by GOBRECHT, TAUSEND and HERTEL[2] with selenium.

25. Cross-field magneto absorption. Interband resonant effects in the simultaneous presence of magnetic and electric fields have been investigated theoretically by ARONOV[3] and experimentally by VREHEN and LAX[4]. Aronov originally suggested that fields around 10^4 V/cm and 100 kgauss would be required. In practice effects are observable using oscillating electric fields in the range 200 to 500 v/cm and magnetic fields ~25 kgauss. The basic cross field effect consists

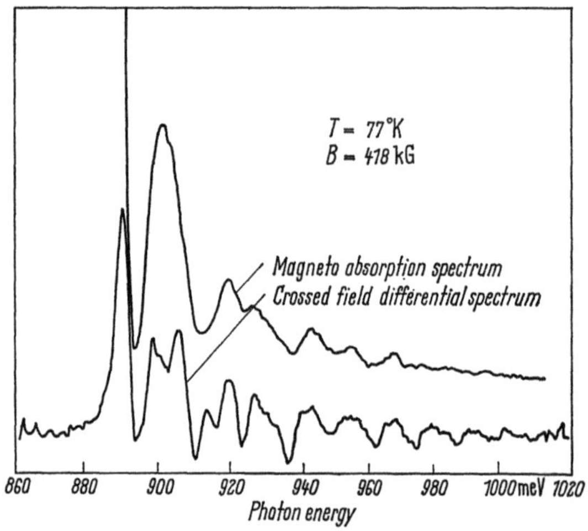

Fig. 45. (a) IMO spectrum of germanium at $B=41.8$ kgauss and zero electric field. (b) Differential cross-field absorption at second harmonic (i.e., 1400 cps) at same value of B, and $E=500$ volts per cm. "Positive" and "Negative" lines occur (after VREHEN and LAX, Ref. 4, p. 312).

of two parts: (i) a modification of the energies of Landau levels so that the energy separation between valence and conduction bands is given by ARONOV's analysis is

$$\Delta\varepsilon_{nn'}=E_G+\left(n+\frac{1}{2}\right)\hbar\omega_{cc}+\left(n'+\frac{1}{2}\right)\hbar\,\omega_{cv}-\frac{E^2}{2B^2}(m_c+m_v)$$

although the expression does not give a realistic low field limit, and (ii) modification to the matrix elements so that the previously forbidden transitions $\Delta n=\pm 1$, ± 2, become allowed. This enables the effective masses of electrons and holes to be determined separately, rather than the reduced mass. A comparison between the magneto absorption and the differential cross field absorption as observed in germanium is shown in Fig. 45. Both positive (increased absorption) and negative (decreased absorption) lines were found, the latter corresponding to coincide with normally allowed transitions.

26. Emission from semiconductor diodes in a magnetic field. Semiconductor diodes as radiation sources have become of great interest since the discovery of laser action in the GaAs diode. The effect of a magnetic field upon an InSb diode, emitting non-coherently, was reported by BENOIT À LA GUILLAUME and LAVAL-

[1] B. LAX, and Y. NISHINA: J. Appl. Phys. (Suppl.) **32**, 2128 (1961).
[2] H. GOBRECHT, A. TAUSEND, and J. HERTEL: Z. Physik **178**, 19 (1964).
[3] A. G. ARONOV: Soviet Physics — Solid State **5**, 402 (1963).
[4] Q. H. F. VREHEN, and B. LAX: Phys. Rev. Letters **12**, 471 (1964).

LARD[1], line splitting and shifting associated with the Landau levels being observed. This observation has been extended to stimulated emission from both InSb[2] and GaAs[3] diodes and provides a method of timing the frequency of the emission (Fig. 46). A further important result is that the threshold currenct for laser action

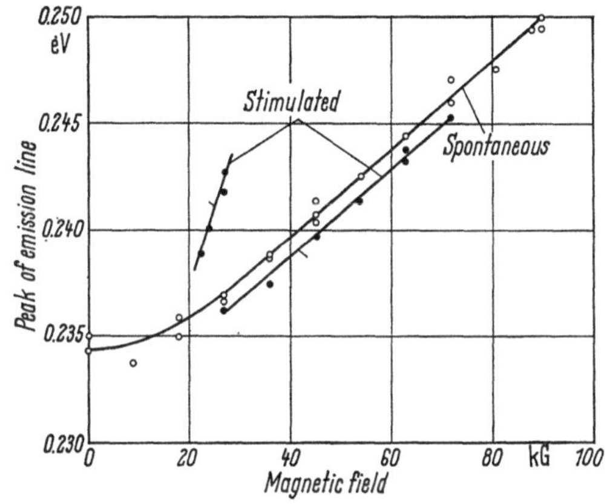

Fig. 46. Emission of InSb diode and laser as a function of magnetic field (after REDIKER and PHELAN, Ref. 2, p. 313).

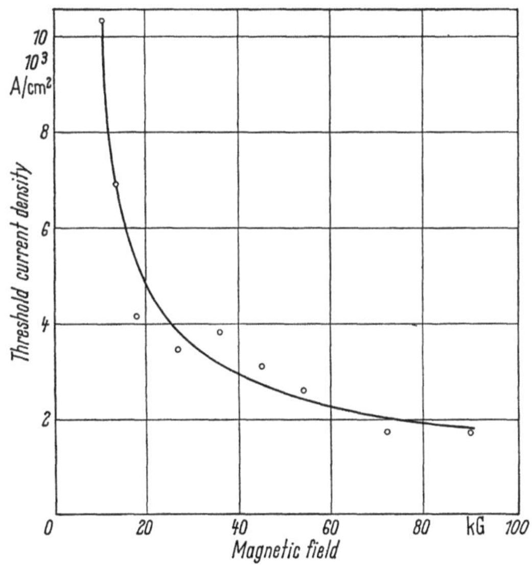

Fig. 47. Threshold current density for laser action as a function of magnetic field for InSb laser at 1.7° K (after REDIKER and PHELAN, Ref. 2, p. 313).

is greatly reduced in a magnetic field (Fig. 47). Magneto-dispersive effects on the refractive index also have effect upon the interferometric modes of the crystal étalon which shift in frequency. In the GaAs case[3], energy shifts of emission lines are found to vary quadratically with field, suggesting that the transitions proceed through the ground states of donors (see Sect. 27).

[1] C. BENOIT À LA GUILLAUME, and P. LAVALLARD: Proc. Int. Conf. on Physics of Semiconductors, Exeter, 1962, p. 875.
[2] R. H. REDIKER, and R. J. PHELAN: Proc. I.E.E.E. **52**, 91 (1964).
[3] F. L. GALEENER, et al.: Phys. Rev. Letters **10**, 472 (1964).

Various schemes for the generation of far infra-red radiation by exciting stimulated emission from the cyclotron-resonance transitions have been proposed. Optical pumping between interband Landau levels in InSb using He-Ne maser at 3.1 μ or InAs at 3.5 μ has been suggested which together with a field of 100 kgauss is said to have the possibility of inverting the population of the $n=2$ and $n=1$ levels and so giving emission at about 30 μ. Alternatively, JAVEN has suggested optical pumping at two frequencies differing by $\hbar\omega_c$[1].

VI. Impurities and magnetic materials.

27. Impurity magneto-optic effects. The simplest model for the energy states associated with impurities is that of the hydrogen atom with an electron which has an isotropic effective mass m^* and with the nuclear charge reduced by e/ε where ε is the high frequency dielectric constant of the crystal. This leads to a hydrogen-like series of energy levels leading up to a photo-ionisation continuum commencing at an energy such that the electron (or hole) is excited into the conduction (or valence) band.

Application of a magnetic field gives rise to two optical effects — (i) a Zeeman effect for the transitions between ground and excited states directly analogous to that in the free atom and (ii) an impurity photo-ionisation magneto-optic effect.

The low-field theory of the Zeeman effect is well known — linear and quadratic effects are obtained; the energy shift for the quadratic effect may be written

$$\Delta\varepsilon = \left\{\frac{\hbar^2 e^2/m^2}{13.6}\right\} \frac{\varepsilon^2}{(m^*/m)^3} \frac{B^2}{8}$$

in the hydrogenic-effective mass approximation. Due to the dependence upon m^* and ε, this splitting can readily reach a very high field limit in semiconductors where the usual theory does not apply. Such a situation has been studied by YAFET, KEYES and ADAMS[2]. An important consequence is that the ionisation energy of the impurity increases as the magnetic field increases.

The theory of the photo-ionisation effect has been considered by WALLIS and BOWLDEN[3]. Since impurity ionisation energies can be very small two cases can be distinguished: when the ionisation energy (E_I) is greater or less than $\frac{1}{2}\hbar\omega_c$ — the change in energy gap. In the low case ($E_I > \frac{1}{2}\hbar\omega_c$) the spectrum consists of a series of oscillations, assuming that the ground state of the impurity is unaffected by the magnetic field and that the band Landau states are unaffected by the impurity potential. The oscillations are just those of the band Landau levels in such a case, with separation $\hbar\omega$ and commencing when $\hbar\omega = E_I$.

The observation of Zeeman splitting of impurity levels in germanium has been reported in 1958 by FAN and FISHER[4] and BOYLE[5] for arsenic and phosphorus impurities. In germanium, the energy surfaces for electrons are, of course, ellipsoids and thus the situation is more complicated than the isotropic hydrogenic case. However, both investigators used magnetic fields in (100) directions which is symmetric with respect to ⟨111⟩ oriented ellipsoids leaving the Landau levels in the four valleys degenerate and the simplest situation for the band state splitting. Results are shown in Fig. 48 — a linear shift being observed. Values for the transverse mass were obtained, in agreement with other experiments. BOYLE's

[1] See B. LAX, Proc. Int. Conf. on Physics of Semiconductors, Paris, 1964, p. 258.
[2] Y. YAFET, R. W. KEYES, and E. N. ADAMS: J. Phys. Chem. Solids 1, 137 (1956).
[3] R. F. WALLIS, and H. J. BOWLDEN: J. Phys. Chem. Solids 7, 78 (1958); 9, 318 (1959).
[4] H. Y. FAN, and P. FISHER: J. Phys. Chem. Solids 8, 270 (1959).
[5] W. BOYLE: J. Phys. Chem. Solids 8, 321 (1959).

results also showed the photo-ionisation oscillatory effect. Zeeman spectra of both *n*- and *p*-type silicon have been studied by ZWERDLING, BUTTON and LAX[1] and the linear and quadratic effects separated.

The change of ionisation energy in a magnetic field has been demonstrated in InSb, for example in the photoconductive detector described by PUTLEY[2]. Zeeman effects from impurity levels associated with the "split-off" valence band in silicon, called internal impurity levels, have been observed in boron doped silicon and the spin orbit splitting (0.0442 eV) deduced[3].

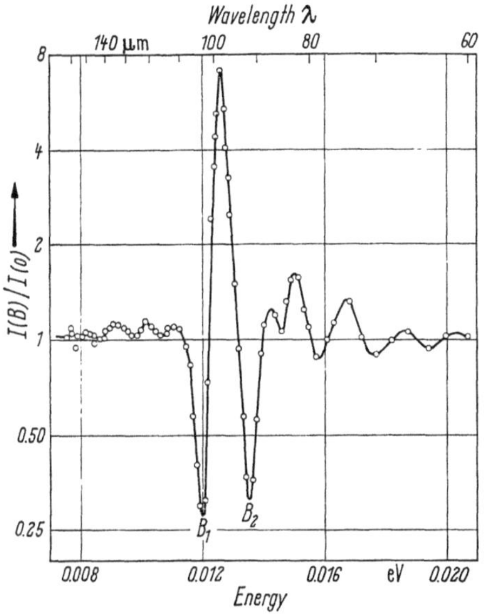

Fig. 48.

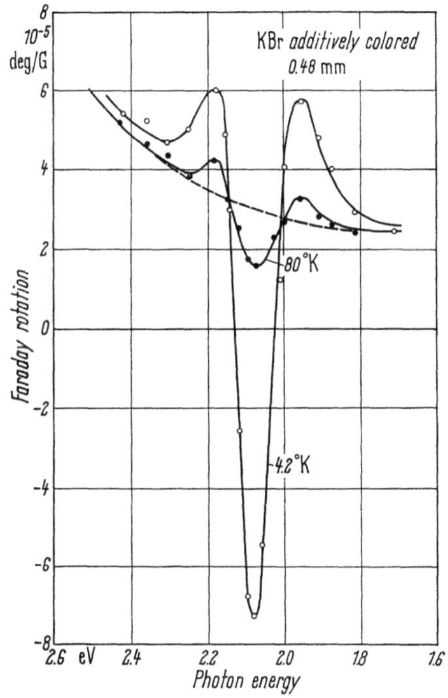

Fig. 49.

Fig. 48. Zeeman splitting of the impurity line of boron in Silicon (after FAN and FISHER, Ref. 4, p. 314).

Fig. 49. Faraday rotation through the F band in KBr (7.8×10^{16} F centres per cc and with a field of 50k gauss giving a maximum rotation of 5° at Helium temperatures; after MORT et al., Ref. 4, p. 315).

Faraday rotation has been observed at frequencies passing through the absorption resonance of the *F*-centre (i.e. an electron trapped at a negative ion vacancy) in alkali halide crystals[4]. This is a beautiful example of the greater sensitivity of the Faraday effect since the Zeeman splitting is $\sim 10^{-4}$ eV and with the *F*-band width ~ 0.2 eV the primary absorption effect would be very difficult to observe. Nevertheless well defined Faraday rotations were observed (Fig. 49) in KCl, NaCl, KBr, KI, RbBr and CsBr.

The results are interpreted in this of the electronic structure of the *F*-centre including spin orbit splitting.

28. Magnetic materials. We have concentrated in this review upon systems, which due to their simplicity, have enabled considerable progress to be made in the understanding of magneto-optical effects during the past ten years. For ferro-

[1] S. ZWERDLING, K. BUTTON, and B. LAX: Phys. Rev. **118**, 975 (1960).
[2] E. PUTLEY: J. Phys. Chem. Solids **22**, 241 (1961).
[3] S. ZWERDLING, et al.: Phys. Rev. Letters **4**, 173 (1960).
[4] J. MORT, F. LÜTY, and F. C. BROWN: Phys. Rev. **137**, A 566 (1965).

magnetic materials, by contrast, the effects on the state of polarisation of both transmitted (Faraday effect) and reflected (Kerr magneto-optic effect) radiation have been known since the last century. Experiments show that the effects are proportional to the net magnetisation of the sample and not to the external field. In terms of *external* applied fields the magnitude of the effects are also vastly greater in ferromagnetic material (e.g. 2° per cm in 10 Kilogauss in quartz compared with 380,000° per cm in iron under the same conditions). Such magnitudes require the presence of internal fields of around 10^6—10^7 gauss, comparable with the postulates of the Weiss field explanation of ferromagnetic materials. A theoretical account of Faraday and Kerr effects in 'metallic' ferromagnetics has been given by ARGYRES[1] using band theory concepts. He points out that the Weiss field cannot affect the motion of electrons as an equivalent field and that spin orbit interaction, introduced to the problem by HULME, is required to explain the optical effects. The theory is developed and shows why this interaction is operative only in ferromagnets and is found to be in reasonable agreement with experiment. Measurements suffer from the difficulties associated with highly absorbing materials and are almost all of the non-resonant polarimetric type, often on thin evaporated films. Recently, insulating ferromagnets such as CrI_3, $CrBr_3$ and $CrCl_3$ have been investigated[2]. When magnetised they show very large rotations of the order of 2.4×10^5 degrees per cm. CrI_3 has an absorption edge near 1 μ and it would be very interesting to investigate resonant phenomena such as IMO on thin layers of such material.

Antiferromagnetic resonance is a far infra-red or microwave phenomenon in which, with two equal anti-parallel sublattices of spins present, the spins of single ions can turn over against the exchange field and cause a resonance. The frequency of resonance can be used as a method to determine the size of the anisotropy field of a material[3].

VII. Experimental techniques.

29. Standard experimental techniques of generating high magnetic fields, high resolution spectroscopy (mainly in the infra-red) cryogenics and sample preparation are required in magneto-optical studies and it would be inappropriate to review these in detail. Some particular parts concerning each merit some discussion however.

Magnetic fields up to 35 kgauss are usually obtained from iron-core magnets and crystals and specimens are often miniaturised to ensure the smallest possible gap. Fields between 50—250 kgauss usually imply a Bitter-type solenoid with massive subsidiary power and cooling installations and are therefore only available in large laboratories, notably the National Magnet Laboratory at MIT[4] and the Naval Research Laboratory in Washington. Pulsed field techniques have not found general favour. It is worthy of note that the vast majority of experiments have been performed with fields less than (say) 65 kgauss and recent developments in superconducting solenoids have made fields of this size available at a relatively modest cost of a few thousand pounds. An example of such a solenoid specifically designed for magneto-optical work at liquid helium, liquid nitrogen and room temperatures is shown in Fig. 50. This solenoid which gives 66 kgauss using

[1] P. N. ARGYRES: Phys. Rev. **97**, 334 (1955).
[2] J. F. DILLON, and C. E. OLSON: J. App. Phys. **36**, 3, 2, 1259 (1965).
[3] R. C. OHLMANN, and M. TINKHAM: Phys. Rev. **123**, 425 (1961).
[4] F. BITTER: Brit. J. Appl. Phys. **14**, 759 (1963).

niobium-tin wire was developed by the Oxford Instrument Company in conjunction with C. J. SUMMERS at the University of Reading. Improvements in superconducting wire may well raise the practical limit to 100 kgauss within a few years.

Sample preparation development enables single crystals down to a few microns to be prepared; problems of strain in such samples at low temperature cause problems. Magneto-polarimetric experiments have stimulated some developments. Polarisers are still a problem beyond 2 μ up to which HR polaroid may be used. Brewster angle stacks of mica polythene and other plastics are commonly used and also reflection from Ge or Si plates but all are unsatisfactory in some ways. Metal wire grids, useful beyond 2 μ are a recent development[1]. Measuring techniques can however be quite powerful even in the infra-red — Faraday rotation can be measured to $\sim \frac{1}{200}^{\circ}$ [1,2], and ingenious methods of measuring Voigt effects and ellipticity have been devised[2-4].

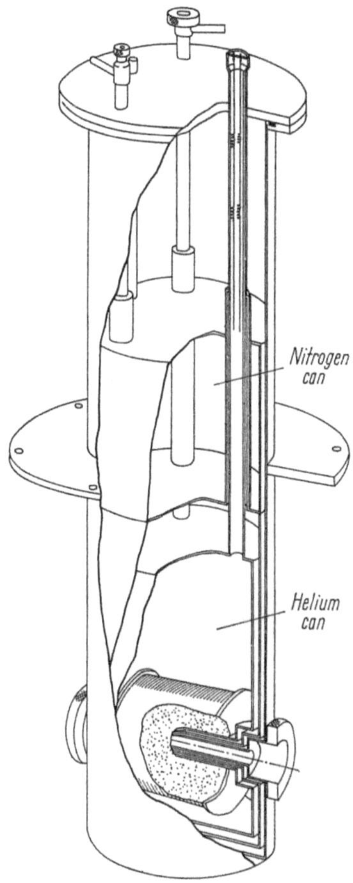

Fig. 50. Superconducting solenoid for magneto-optical experiments, constructed by Oxford Instrument Company for the University of Reading.

VIII. Summary.

30. The enormous expansion of this subject since the crucial resonance experiments in 1956 must be apparent from the number of topics covered in this review. The spread of crystal magneto-optics to semiconductors has enabled quantitative interpretation to be made of many effects in terms of band structure and we have concentrated on such uses. With most of the possible effects now observed we can now reverse the latter statement and say that the purpose of magneto-optical experiments is to explore quantitatively the electronic band structure of semiconductors (and insulators). Thus the subject has a role similar to inelastic neutron scattering and phonon dispersion curves and to oscillatory magnetic susceptibility and energy surfaces in metals. However, the variety of magneto-optic experiments is much greater and each technique is, as yet, rather limited in application as compared with the other two examples cited. We conclude therefore with a list of parameters which it is possible, or desirable, to measure together with what appear at present the most suitable techniques or combination of techniques.

---

[1] G. R. BIRD, and M. PERRISH: J. Opt. Soc. Am. **50**, 886 (1960).
[2] C. R. PIDGEON, and S. D. SMITH: Infra-red Physics **4**, 13 (1964).
[3] D. L. MITCHELL, and R. F. WALLIS: Phys. Rev. **131**, 1965 (1963).
[4] E. D. PALIK: Appl. Optics **2**, 527 (1963).

Table 3.

Determination	Experiment	Comment
Effective mass at band extrema including anisotropy	Cyclotron resonance (microwave)	Extends only $\sim 10^{-3}$ eV from extrema
Effective mass and position of band extrema lying above or below the minimum gap	CR (microwave) with selective optical excitation of carriers from *n*- or *p*-type impurities	Extends only $\sim 10^{-3}$ eV from extrema
Effective mass in materials of short relaxation time and/or heavily doped	High field infra-red CR. Free carrier Faraday effect (with reflectivity minimum)	No anisotropy observable
Temperature dependence of effective mass	Free carrier Faraday effect	No anisotropy observable
Exploration of band shape (non-parabolicity) away from extrema	Free carrier Faraday effect with heavy doping	Only low field theory. No anisotropy
	CR (infra-red with high fields)	Need high field theory
	Interband magneto absorption (IMO)	Involves both valence and conduction bands. Thin samples
Energy gaps	IMO	Convergence plot against B
Effective mass at direct gap minima when indirect gap is smaller	IMO	Involves valence band
Conduction band g-factors	IMO	
	Resonant interband Faraday effect	
Study of light hole and heavy hole valence states	IMO	Using various orientations of B
	Resonant interband Faraday effect	Opposite sign of rotation
Very deep levels, or where absorption is high	Interband magneto reflection	Particularly for semimetals
Relaxation times:		
I. Free carriers	Free carrier ellipticity and Faraday rotation combined	
II. Interband processes	Combination of resonant interband Faraday effect and Voigt effect	

Acknowledgements.

The author wishes to acknowledge discussion with many colleagues in the field particularly at M.I.T. and N.R.L.

Dr. D. L. MITCHELL and Dr. P. G. HARPER gave specific help during the preparation of the manuscript.

The manuscript was completed in August 1965. Important new developments are reported in Proc. Int. Conf. on Physics of Semiconductors, Kyoto 1966: J. Phys. Soc. Japan **21**, Supplement 1966, pp. 165—212, 244—253, 148—150, 443—447, 713—760. New effects of interest include non-linear effects, plasmon-phonon coupling, magneto-Raman effect, magneto-electro reflectance.

Sachverzeichnis.

(Deutsch-Englisch.)

Bei gleicher Schreibweise in beiden Sprachen sind die Stichwörtpr nur einmal aufgeführt.

Subject Index.

(English-German.)

Where English and German spelling of a word is identical the German version is omitted.

FSC
www.fsc.org
MIX
Papier aus verantwortungsvollen Quellen
Paper from responsible sources
FSC® C105338

If you have any concerns about our products,
you can contact us on
ProductSafety@springernature.com

In case Publisher is established outside the EU,
the EU authorized representative is:
Springer Nature Customer Service Center GmbH
Europaplatz 3, 69115 Heidelberg, Germany

Printed by Libri Plureos GmbH
in Hamburg, Germany